Kyoji Sassa
Hiroshi Fukuoka
Fawu Wang
Gonghui Wang
(Editors)

Landslides

Risk Analysis and Sustainable Disaster Management

Kyoji Sassa
Hiroshi Fukuoka
Fawu Wang
Gonghui Wang
(Editors)

Landslides

Risk Analysis and Sustainable Disaster Management

Proceedings of the First General Assembly of the International Consortium on Landslides

With 417 Images

Editors

Sassa, Kyoji

President of the International Consortium on Landslides
Research Centre on Landslides, Disaster Prevention Research Institute,
Kyoto University, Uji, Kyoto 611-0011, Japan
Tel: +81-774-38-4110, Fax: +81-774-32-5597, E-mail: sassa@SCL.kyoto-u.ac.jp

Fukuoka, Hiroshi

Research Centre on Landslides, Disaster Prevention Research Institute,
Kyoto University, Uji, Kyoto 611-0011, Japan
Tel: +81-774-38-4111, Fax: +81-774-38-4300, E-mail: fukuoka@SCL.kyoto-u.ac.jp

Wang, Fawu

Research Centre on Landslides, Disaster Prevention Research Institute,
Kyoto University, Uji, Kyoto 611-0011, Japan
Tel: +81-774-38-4114, Fax: +81-774-38-4300, E-mail: wangfw@landslide.dpri.kyoto-u.ac.jp

Wang, Gonghui

Research Centre on Landslides, Disaster Prevention Research Institute,
Kyoto University, Uji, Kyoto 611-0011, Japan
Tel: +81-774-38-4114, Fax: +81-774-38-4300, E-mail: wanggh@landslide.dpri.kyoto-u.ac.jp

ISBN-13 978-3-642-06682-5 e-ISBN-13 978-3-540-28680-6

Springer is a part of Springer Science+Business Media
springeronline.com

Softcover reprint of the hardcover 1st edition 2005

Cover design: Erich Kirchner, Heidelberg

Printed on acid-free paper 32/3141/as – 5 4 3 2 1 0

Preface

Population growth, increasing urbanization, and mountain and coastal development are magnifying the risk of various kinds of disasters. The imperatives of Earth system risk reduction will be even more pressing for sustainable development and environmental protection in the coming decades.

Landslides are various types of gravitational mass movements of the Earth's surface that pose the Earth-system risk; they are triggered by earthquakes, rainfall, volcanic eruptions and human activities. Landslides cause many deaths and injuries and great economic loss to society by destroying buildings, roads, life lines and other infrastructures; they also pose irrecoverable damage to our cultural and natural heritage. Landslides are multiple hazards, involving typhoons/hurricanes, earthquakes, and volcanic eruptions, and sometimes causing tsunamis. Landslide disaster reduction requires cooperation of a wide variety of natural, social, and cultural sciences.

The International Consortium on Landslides (ICL) was founded at the Kyoto Symposium in January 2002, through the adoption of the Statutes of the ICL and the 2002 Kyoto Declaration "Establishment of an International Consortium on Landslides", with participation from the Divisions of Earth Sciences, Water Sciences, and Cultural Heritage of the United Nations Educational, Scientific and Cultural Organization (UNESCO), the World Meteorological Organization (WMO), the United Nations Secretariat for International Strategy for Disaster Reduction (UN/ISDR), the Ministry of Education, Culture, Sports, Science and Technology (MEXT) and the Ministry of Foreign Affairs (MOFA) of the Government of Japan, and worldwide universities and institutes involved in landslide studies. The ICL has convened its Board of Representatives and Steering Committee meetings at UNESCO in Paris 2002, at the Food and Agriculture Organization of the United Nations (FAO) in Rome 2003, at the United Nations Building "Palace de Nation" in Geneva 2004, and at universities in Vancouver, Bratislava, and Cairo.

On 19 January 2005 in Kobe, Japan, the ICL organized the theme session "New International Initiatives for Research and Risk Mitigation of Floods (IFI) and Landslides (IPL)" at the United Nations World Conference on Disaster Reduction together with UNESCO, WMO, FAO, UNU (United Nations University), MEXT, Kyoto University, and others. In this session, the ICL proposed a "Letter of Intent" to United Nations organizations, as well as to the International Council for Science (ICSU) and the World Federation of Engineering Organizations (WFEO), in order to strengthen learning and research on "Earth System Risk Analysis and Sustainable Disaster Management" within the framework of the United Nations International Strategy for Disaster Risk Reduction (ISDR). These functions must be effective in order to create a sound basis of international cooperation in the field of Earth-system risk reduction, including landslides. This Letter of Intent was approved and signed by seven global stakeholders of UNESCO, WMO, FAO, UN/ISDR, UNU, ICSU, and WFEO by 30 June 2005. The electronically combined Letter of Intent is attached below.

Study and learning of landslide risk analysis and sustainable disaster management, including close cooperation with experts and organizations specializing in other types of disasters, must be our task in the coming decades. For the occasion of the ICL First General Assembly, to be held at the Keck Center of the National Academy of Sciences,

LETTER OF INTENT

"United Nations World Conference on Disaster Reduction (WCDR)", Kobe, Japan, 18-22 January 2005

This 'Letter of intent' aims to provide a platform for a holistic approach in research and learning on 'Integrated Earth system risk analysis and sustainable disaster management'.

Rationale

- Understanding that any discussion about global sustainable development without addressing the issue of Disaster Risk Reduction is incomplete;
- Acknowledging that risk-prevention policies including warning systems related to Natural Hazards must be improved or established;
- Underlining that disasters affect poor people and developing countries disproportionately;
- Stressing that after years of under-investment in preventive scientific, technical and communicational infrastructure activities it is time to change course and develop all activities needed to better understand natural hazards and to reduce the vulnerability notably of developing countries to natural hazards, and
- Acknowledging that a harmful deficiency in coordination and communication measurements related to Disaster Risk Reduction exists.

Proposal

Representatives of United Nations Organisations, as well as the Scientific (ICSU) and Engineering (WFEO) Communities propose to promote further joint global activities in disaster reduction and risk prevention through

Strengthening research and learning on 'Earth System Risk Analysis and Sustainable Disaster Management' within the framework of the 'United Nations International Strategy for Disaster Risk Reduction' (ISDR).

More specifically it is proposed,

based on the existing structural framework of the ISDR and plan of action of the UN-WCDR, as well as other relevant networks and institutional and international expertise,

to establish specific, goal-oriented 'Memoranda of Understanding' (MoUs) between international stakeholders targeting Disaster Risk Reduction, for example focusing on landslide risk reduction, and other natural hazards.

Invitation

Global, regional and national competent institutions are invited to support this initiative by joining any of the specific MoUs following this letter through participation in clearly defined projects related to the issues and objectives of any of the MoUs.

Signatories:

Mr. Koïchiro Matsuura
Director-General
United Nations Educational, Scientific and Cultural Organization
4 MAR 2005
Date

Mr. Michel Jarraud
Secretary-General
World Meteorological Organization
22. 3. 2005
Date

Mr. Jacques Diouf
Director-General
Food and Agriculture Organization of the United Nations
Date

Mr. Sálvano Briceño
Director
UN International Strategy for Disaster Risk Reduction
19.01.05
Date

Mr. Hans van Ginkel
Rector
United Nations University
19.01.05
Date

Ms. Jane Lubchenco
President
International Council for Science
21. 04. 05
Date

Ms Françoise Come
Executive Director
World Federation of Engineering Organizations
24/ 2/ 2005
Date

The International Consortium on Landslides (ICL) proposed the "Letter of Intent" at the thematic session 3.8 "**New International Initiatives for Research and Risk Mitigation of Floods (IFI) and Landslides (IPL)**" of the United Nations World Conference on Disaster Reduction held on 19 January 2005 in Kobe, Japan. This is the Letter of Intent, which was electronically combined based on the original Letters of Intent, formally approved and signed by all parties. All of the original Letters of Intent with signatures are deposited in the secretariat of the International Consortium on Landslides which is located in the Research Centre on Landslides of the Disaster Prevention Research Institute, Kyoto University.

International Consortium on Landslides

Secretariat : Research Centre on Landslides, Disaster Prevention Research Institute, Kyoto University, Kyoto, Japan
Web: http://ICL.dpri.kyoto-u.ac.jp, E-mail: jimu@landslide.dpri.kyoto-u.ac.jp, Tel: +81-774-38-4110, Fax: +81-774-32-5597

Washington, D.C., we decided to organize a panel discussion on "Earth-system risk analysis and sustainable disaster management, especially in regard to landslides". This volume, which includes the proceedings of papers submitted to the First General Assembly, is titled "Landslides – Risk Analysis and Sustainable Disaster Management –" to symbolize our target in the coming decades.

It is hoped that this volume will visualize the objectives and activities of the International Consortium on Landslides and result in intensified international cooperation in learning and research for landslide disaster reduction within global and regional entities involving in landslides. We request cooperation and support from scientists and engineers working on other disasters, and particularly from those organizations and entities that are willing to contribute to Earth-system risk reduction, including that of landslides.

Acknowledgments

I express my gratitude for the cooperation of the staff of the National Cooperative Geologic Mapping Program and the Landslide Hazard Program of the U.S. Geological Survey and the Research Centre on Landslides, Disaster Prevention Research Institute, Kyoto University, for organization of the First General Assembly and for edition of this volume. Thanks also go to UNESCO, WMO, FAO, UN/ISDR, UNU, IUGS, and to the governments of Japan, U.S.A., Italy, Canada and Norway for their continued support of ICL activities. It is acknowledged that the organization of the First General Assembly of ICL and this publication are financially supported by the UNESCO fund supporting IPL/ICL activities, the Presidential leadership fund of Kyoto University, and the twenty-first century COE (Centre of Excellence) fund from the Ministry of Education, Culture, Sports, Science and Technology of the Government of Japan allocated to the Disaster Prevention Research Institute, Kyoto University.

Kyoto University, UNESCO and ICL launched the UNITWIN Cooperation Programme "Landslide Risk Mitigation for Society and the Environment" on 18 March 2003, then jointly constructed the UNITWIN Headquarter building as the activity-base on the Uji campus of Kyoto University. It is acknowledged that worldwide cooperation through the UNITWIN network and the facilities of the UNTIWN Headquarter were very helpful for the preparation of the General Assembly and editing this volume.

Kyoji Sassa

President, International Consortium on Landslides

Director of the Research Centre on Landslides,
Disaster Prevention Research Institute,
Kyoto University, Japan

Welcome Address

Today, many parts of the world, including the United States, are at significant risk from natural disasters. Escalating population and increased development on the coast, fault zones, mountainous areas, and flood plains mean that increasing numbers of people are at risk from hazards. Each year the importance of assessing, preparing for and mitigating the potential effects of natural hazards, including landslides, increases. For this reason, the U.S. Geological Survey (USGS) and the National Research Council (NRC) of the National Academy of Sciences are pleased to host this important General Assembly of the International Consortium on Landslides (ICL).

It is an honor that the first General Assembly of the ICL is meeting in the Washington, DC area – the center of the U.S. government with numerous Federal agency headquarters, including the USGS, NRC, Federal Emergency Management Agency, Federal Transportation Administration, and the National Oceanic Atmospheric Administration (NOAA). The city is also home to numerous international organizations such as the World Bank, the Organization of American States, the Inter American Bank, and others. We hope that you will have an opportunity to visit this very beautiful city and some of the organizations that would benefit from the work of ICL.

Landslides threaten lives and property in every state in the U.S. Fall and winter of 2004–2005, were especially active landslides seasons with numerous landslides caused by hurricanes in the east coast and heavy rainfall in the west coast, throughout the intermountain states of Utah and Colorado and the east coast states of Ohio, Pennsylvania and New York. Landslides in 2004–2005 caused many deaths and extensive property damage. Communities are still cleaning up after some of the most damaging events. Those members of ICL who will visit southern California will be able to see, first hand, two very heavily impacted areas.

The USGS currently has two important efforts to lessen the impact of natural hazards. The first is the "Initiative to Protect Communities and Resources from Natural Disasters." The USGS is working toward implementing this initiative by 2007, which will focus on delivering USGS science to public officials and private industry to help them reduce the vulnerability of communities and the environment to hazards including: earthquakes, droughts, floods, landslides, and wildfires. Landslide hazards and debris-flow hazards following wildfires are important elements of this initiative. The second effort is the NOAA-USGS Debris Flow Warning System, which is a demonstration project which combines the expertise of the two agencies in precipitation forecasting, debris-flow prediction, and debris-flow hazard assessment in order to establish a debris-flow warning system for recently burned areas of southern California. This demonstration project, if successful and if funded, will be expanded to reach other parts of the U.S. which face similar hazards. More detailed explanations about these two important efforts by the USGS will be given during the General Assembly.

International meetings such as this one provide a unique opportunity for managers and researchers to share new research findings, knowledge, and experiences that can lead to better understanding of how to mitigate the devastating effects of natural hazards. I look forward to learning from the presentations and discussions during the next few days and reading the proceedings of this First General Assembly of the International Consortium of Landslides.

Good luck and best wishes for a successful assembly,

P. Patrick Leahy

Honorary Chairman of the First General Assembly of the International Consortium on Landslides

Acting Director of the U.S. Geological Survey

Contents

List of Contributors

Abdrakhmatov, Kanatbek E. · (Chapter 12)

Institute of Seismology, National Academy of Science, Asanbay 52/1, Bishkek, Kyrgyzstan

Abidin, Roslan Zainal · (Chapter 24)

Director, National Soil Erosion Research Centre, Universiti Teknologi MARA, Shah Alam, Malaysia
Corresponding author of Chapter 24:
Tel: +603-5544-2779, Fax: +603-5544-2783, E-mail: roslanza@salam.uitm.edu.my

Abu Hassan, Zulkifli · (Chapter 24)

Postgraduate Student, Faculty of Civil Engineering, Universiti Teknologi MARA, Shah Alam, Malaysia

Antonello, Giuseppe · (Chapter 43)

European Commission, Joint Research Center, Via E. Fermi 1, Ispra (VA), 21020, Italy

Araiba, Kiminori · (Chapter 40)

National Research Institute of Fire and Disaster/14-1, Nakahara 3 chome, Mitaka, Tokyo 181-8633, Japan

Astete, Fernándo V. · (Chapters 2, 4)

Instituto Nacional de Cultura (INC), Calle San Bernardo s/n., Cusco, Peru

Ayalew, Lulseged · (Chapter 22)

Addis Ababa University, P.O. Box 29970, Addis Ababa, Ethiopia

Baldi, Paolo · (Chapter 32)

Dept. of Physics, University of Bologna, Italy

Benavente, Edwin · (Chapter 2)

Instituto Nacional de Cultura (INC), Calle San Bernardo s/n., Cusco, Peru

Best, Mel · (Chapter 5)

BEMEX Consulting, 5288 Cordova Bay Rd., British Columbia, Victoria, Canada

Bhandary, Netra P. · (Chapter 27)

Department of Civil and Environmental Engineering, Ehime University, 3 Bunkyo-cho, Matsuyama, Japan
Corresponding author of Chapter 27:
Tel: +81-89-927-8566, Fax: +81-89-927-8566, E-mail: netra@dpc.ehime-u.ac.jp

Bobrowsky, Peter · (Chapter 5)

NRCan, Geological Survey of Canada, 601 Booth Street, Ontario, Ottawa, Canada
Corresponding author of Chapter 5:
Tel: +1-613-947-0333, Fax: +1-613-992-0190, E-mail: pbobrows@nrcan.gc.ca

Boldini, Daniela · (Chapter 21)

Department of Structural and Geotechnical Engineering, University of Rome "La Sapienza",
Via Monte d'Oro 28, 00186 Rome, Italy

Caillaux, Victor Carlotto · (Chapter 5)

INGEMMET, Instituto Geologico Minero y Metalurgico, Av. Canada, No 1470, San Borja, Lima, Peru

Canuti, Paolo · (Chapters 3, 23)

Department of Earth Sciences, University of Firenze, Via Giorgio La Pira 4, 50121 Firenze, Italy

Cao, Binglan · (Chapter 10)

Environmental Geological Disaster Research Institute, Jilin University, 6 Ximinzhu Street,
Changchun 130026, China
Corresponding author of Chapter 10:
Tel: +86-431-854-0912, Fax: +86-431-556-7570, E-mail: caobl@jlu.edu.cn

Carreño, Raúl · (Chapters 17, 34)

GRUDEC AYAR, Apartado Postal 638, Cusco, Peru
Corresponding author of Chapters 17, 34:
Tel: +51-84-974-1455, Fax: +51-84-26-2590, E-mail: raulcarreno@ayar.org.pe

Casagli, Nicola · (Chapters 3, 6, 23, 43)

Department of Earth Sciences, University of Firenze, Via Giorgio La Pira 4, 50121 Firenze, Italy

Cheibany, Ould Elemine · (Chapter 20)

Research Institute for Hazards in Snowy Areas, Niigata University, 2-8050 Ikarashi, Niigata, 950-2181, Japan

Chiocci, Francesco Latino · (Chapter 32)

Dept of Earth Sciences, University of Rome "La Sapienza", Italy

Christaras, Basile · (Chapter 30)

School of Geology, Aristotle University of Thessaloniki, 54124 Thessaloniki, Greece
Corresponding author of Chapter 30:
Tel: +30-2310-99-8506, Fax: +30-2310-99-8506, E-mail: christar@geo.auth.gr

Colombini, Vittorio · (Chapter 8)

Via Merulana 272, 00185 Rome, Italy

Coltelli, Mauro · (Chapter 32)

INGV, Catania, Italy

Coman, Mihai · (Chapter 45)

ISPIF SA, 35-37, Oltenitei St., Bucharest, Romania

Crippa, Carlo · (Chapter 8)

Trevi S.p.A, Divisione Rodio, Via Pandina 5, 26831 Casalmaiocco, Lodi, Italy

Delmonaco, Giuseppe · (Chapters 3, 13, 26)

ENEA CR Casaccia, Via Anguillarese 301, 00060 S. Maria di Galeria, Rome, Italy
Corresponding author of Chapter 13:
Tel: +39-06-3048-4502, Fax: +39-06-3048-4029, E-mail: delmonaco@casaccia.enea.it

Devoli, Graziella · (Chapter 29)

International Centre for Geohazards, c/o Norwegian Geotechnical Institute, P.O. Box 3930, Ullevaal Stadion, 0806 Oslo, Norway
and Ph.D. Candidate at University of Oslo, Department of Geosciences, Norway
Corresponding author of Chapter 29:
Tel: +47-2202-3045, Fax: +47-2223-0448, E-mail: gde@geohazards.no

Dimitriou, Anastasios · (Chapter 30)

School of Geology, Aristotle University of Thessaloniki, 54124 Thessaloniki, Greece

Dimopoulos, George · (Chapter 30)

School of Geology, Aristotle University of Thessaloniki, 54124 Thessaloniki, Greece

Douma, Marten · (Chapter 5)

NRCan, Geological Survey of Canada, 601 Booth Street, Ontario, Ottawa, Canada

ElShayeb, Yasser · (Chapter 28)

Faculty of Engineering, Cairo University, Giza 12613, Egypt
Corresponding author of Chapter 28:
Tel: +20-10-604-4698, Fax: +20-2-571-0035, E-mail: yasser.elshayeb@tempus-egypt.com

Emamjomeh, Reza · (Chapter 42)

Soil Conservation and Watershed Management Research Institute, P.O. Box 13445-1136, Tehran, Iran

Falconi, Luca · (Chapter 13)

ENEA CR Casaccia, Via Anguillarese 301, 00060 Rome, Italy

Fanti, Riccardo · (Chapters 6, 23)

Department of Earth Sciences, University of Firenze, Via Giorgio La Pira 4, 50121 Firenze, Italy
Corresponding author of Chapters 6, 23:
Tel: +39-055-275-7523, Fax: +39-055-275-6296, E-mail: rfanti@steno.geo.unifi.it

Farina, Paolo · (Chapter 43)

Department of Earth Sciences, University of Firenze, Via Giorgio La Pira 4, 50121 Firenze, Italy
Corresponding author of Chapter 43:
Tel: +39-055-275-6221, Fax: +39-055-275-6296, E-mail: paolo.farina@geo.unifi.it

Ferretti, A. · (Chapter 3)

Tele-Rilevamento Europa, T.R.E., Via Vittoria Colonna 7, 20149 Milano, Italy

Fortuny-Guasch, Joaquim · (Chapter 43)

European Commission, Joint Research Center, Via E. Fermi 1, Ispra (VA), 21020, Italy

Frangov, Georgi · (Chapter 33)

Geological Institute, BAS, St. Acad. G. Bonchev Block 24, Sofia 1113, Bulgaria

Fukuoka, Hiroshi · (Chapters 2, 9, 11, 18, 19)

Research Centre on Landslides, Disaster Prevention Research Institute, Kyoto University, Uji, Kyoto 611-0011, Japan
Corresponding author of Chapters 9, 18:
Tel: +81-774-38-4111, Fax: +81-774-38-4300, E-mail: fukuoka@SCL.kyoto-u.ac.jp

Furuya, Gen · (Chapter 20)

Research Centre on Landslides, Disaster Prevention Research Institute, Kyoto University, Uji, Kyoto 611-0011, Japan

Gallage, Chaminda · (Chapter 16)

Department of Civil Engineering, University of Tokyo, 7-3-1, Hongo, Bunkyo-Ku, Tokyo 113-8656, Japan

Gao, Shihang · (Chapter 39)

Department of Civil Engineering, Xi'an Jiaotong University, Xi'an 710049, China

Gratchev, Ivan B. · (Chapters 15, 46)

Graduate School of Science, Kyoto University, Japan
and Research Centre on Landslides, Disaster Prevention Research Institute, Kyoto University, Uji, Kyoto 611-0011, Japan
Corresponding author of Chapters 15, 46:
Tel: +81-77-438-4107, Fax: +81-77-438-4300, E-mail: gratchev@landslide.dpri.kyoto-u.ac.jp

Guerri, Letizia · (Chapter 43)

Department of Earth Sciences, University of Firenze, Via Giorgio La Pira 4, 50121 Firenze, Italy

Hasegawa, Manabu · (Chapter 35)

Geographic Department, Geographical Survey Institute, Ministry of Land, Infrastructure and Transport, 1 Kitasato, Tukuba, Ibaraki, Japan

Havenith, Hans-Balder · (Chapter 12)

SED-ETH, Swiss Seismological Service, Institute of Geophysics, Hönggerberg, 8093 Zurich, Switzerland

Jezný, Michal · (Chapter 7)

Department of Engineering Geology, Faculty of Natural Sciences, Comenius University Bratislava, 84215 Bratislava, Mlynska dolina, Slovak Republic

Kalafatovich, Susana · (Chapter 34)

GRUDEC AYAR, Apartado Postal 638, Cusco, Peru

Kato, Koji · (Chapter 22)

Shin Engineering Consultant Co. Ltd., Sakaedori 2-8-30 Shiroishiku, Sapporo 003-0021, Japan

Kjekstad, Oddvar · (Chapter 41)

Norwegian Geotechnical Institute, NGI, Sognsveien 72, P.O. Box 3930, Ullevaal Stadion, 0801, Norway
Corresponding author of Chapter 41:
Tel: +47-2202-3002, Fax: +47-2223-0448, E-mail: oddvar.kjekstad@ngi.no

Klimeš, Jan · (Chapter 4)

Institute of Rock Structure and Mechanics, Academy of Sciences, V Holešovièkách 41, 18000 Praha 8, Czech Republic

Kobayashi, Hideji · (Chapter 44)

Shin Engineering Consultants Co. Ltd., Sapporo 062-0931, Japan

Korup, Oliver · (Chapter 12)

Swiss Federal Research Institutes WSL/SLF, 7260 Davos, Switzerland

Koukis, G. · (Chapter 37)

Department of Geology, Section of Applied Geology and Geophysics, University of Patras, 26500 Patras, Greece

Leoni, Gabriele · (Chapter 13)

Consorzio Civita, Via del Corso 300, 00168 Rome, Italy

Leva, Davide · (Chapter 43)

LiSALab s.r.l., V. XX Settembre 34, Legnano (MI), 20025, Italy

Liao, Hongjian · (Chapter 39)

Department of Civil Engineering, Xi'an Jiaotong University, Xi'an 710049, China
Corresponding author of Chapter 39:
Tel: +86-29-8266-3228, Fax: +86-29-8323-7910, E-mail: hjliao@mail.xjtu.edu.cn

Lollino, G. · (Chapter 3)

CNR-IRPI Torino, Strada delle Cacce 73, 10135 Torino, Italy

Loupasakis, C. · (Chapter 37)

Institute of Geology and Mineral Exploration, Engineering Geology Department, Messogion Avenue 70, 11527 Athens, Greece

Mamaev, Yuri A. · (Chapter 46)

Institute of Environmental Geoscience of Russian Academy Science,
Ulansky Pereulok 13, Building 2, P.O. Box 145,
101000 Moscow, Russia

Mamani, Romulo Mucho · (Chapter 5)

INGEMMET, Instituto Geologico Minero y Metalurgico, Av. Canada, No 1470, San Borja, Lima, Peru

Margottini, Claudio · (Chapters 3, 8, 13, 26)

ENEA CR Casaccia, Via Anguillarese 301, 00060 S. Maria di Galeria, Rome, Italy
Corresponding author of Chapters 3, 8, 26:
Tel: +39-06-3048-4688, Fax: +39-06-3048-4029, E-mail: margottini@casaccia.enea.it

Marinos, Paul · (Chapters 30, 31)

National Technical University of Athens, School of Civil Engineering, Geotechnical Department, 9 Heroon Polytechniou Str., 15780 Zografou, Athens, Greece
Corresponding author of Chapter 31:
Tel: +30-210-772-3430, Fax: +30-210-772-3770, E-mail: marinos@central.ntua.gr

Marsella, Maria · (Chapter 32)

Dept of Hydraulics, Transports and Roads, University of Rome "La Sapienza", Italy

Marui, Hideaki · (Chapter 20)

Research Institute for Hazards in Snowy Areas, Niigata University, 2-8050 Ikarashi, Niigata, 950-2181, Japan

Marunteanu, Cristian · (Chapter 45)

Faculty of Geology and Geophysics, University of Bucharest, 6, Traian Vuia St., 020956 Bucharest, Romania
Corresponding author of Chapter 45:
Tel: +40-1-3125-003/34, Fax: +40-1318-1557, E-mail: cristian@gg.unibuc.ro

Matova, Margarita · (Chapter 33)

Geological Institute, BAS, St. Acad. G. Bonchev Block 24, Sofia 1113, Bulgaria
Corresponding author of Chapter 33:
Tel: +359-2979-2212, Fax: +359-272-4638, E-mail: m_matova@geology.bas.bg

Mucho, R. · (Chapter 3)

INGEMMET, Istituto Geologico Minero y Metallurgico, Av. Canadá 1470, San Borja, Lima 41, Peru

Nadim, Farrokh · (Chapter 41)

International Centre for Geohazards, ICG, c/o NGI, Sognsveien 72, P.O. Box 3930, Ullevaal Stadion, 0801, Norway

Nichol, Susan · (Chapter 25)

Geological Survey of Canada, Natural Resources Canada, 601 Booth Street, Ottawa, Ontario, K1A 0E8, Canada

Nikolaou, N. · (Chapter 37)

Institute of Geology and Mineral Exploration, Engineering Geology Department, Messogion Avenue 70, 11527, Athens, Greece
Corresponding author of Chapter 37:
Tel: +30-210-779-6351, Fax: +30-210-778-2209, E-mail: nikolaou@igme.gr

Nocentini, Massimiliano · (Chapter 6)

Department of Earth Sciences, University of Firenze, Via Giorgio La Pira 4, 50121 Firenze, Italy

Ochiai, Hirotaka · (Chapter 11)

Forestry and Forest Products Research Institute, Tsukuba, Japan

Osipov, Victor I. · (Chapter 15)

Institute of Environmental Geoscience, Russian Academy of Sciences, Ulansky Pereulok 13, Building 2, P.O. Box 145, Moscow 101000, Russia

Oviedo, Martin Jhonathan · (Chapter 5)

INGEMMET, Instituto Geologico Minero y Metalurgico, Av. Canada, No 1470, San Borja, Lima, Peru

Pagáčová, Zuzana · (Chapter 7)

Department of Engineering Geology, Comenius University Bratislava, Faculty of Natural Sciences, 84215 Bratislava, Mlynska dolina, Slovak Republic

Paluš, Milan · (Chapter 14)

Institute of Computer Science, Academy of Sciences of the Czech Republic, Pod vodárenskou vě•í 2, 18207 Prague 8, Czech Republic

Pavlides, Spyros · (Chapter 30)

School of Geology, Aristotle University of Thessaloniki, 54124 Thessaloniki, Greece

Peng, Xuanming · (Chapter 40)

Yichang Institute of Geology and Mineral Resources, China Geological Survey, 37 Gangyao Road, Yichang 443003, China

Pinto, Walter Pari · (Chapter 5)

INGEMMET, Instituto Geologico Minero y Metalurgico, Av. Canada, No 1470, San Borja, Lima, Peru

Pompilio, Massimo · (Chapter 32)
INGV, Catania, Italy

Postoyev, G. P. · (Chapter 38)
Institute of Environmental Geosciences RAS, Ulansky Pereulok 13, Moscow 101000, Russia

Puglisi, Claudio · (Chapters 3, 13)
Consorzio Civita, Via del Corso 300, 00186 Rome, Italy

Righini, Gaia · (Chapter 6)
Department of Earth Sciences, University of Firenze, Via Giorgio La Pira 4, 50121 Firenze, Italy

Romagnoli, Claudia · (Chapter 32)
University of Bologna, Dept. of Earth and Environmental Sciences, Italy

Rondoyanni, Theodora · (Chapter 31)
School of Mining and Metallurgical Engineering, Dept. of Geological Sciences,
National Technical University of Athens, 9 Heroon Polytechniou Str., 15780 Zografou,
Athens, Greece

Sabatakakis, N. · (Chapter 37)
Department of Geology, Section of Applied Geology and Geophysics, University of Patras,
26500 Patras, Greece

Sagara, Wataru · (Chapter 20)
SABO Technical Center, 4-8-21 Kudan-minami, Chiyoda-ku, Tokyo, 102-0074, Japan

Sasaki, Ryo · (Chapter 18)
Former Master Course graduate student, Graduate School of Science, Kyoto University, Japan

Sassa, Kyoji · (Chapters 1, 2, 9, 11, 15, 18, 19, 21, 40)
President of the International Consortium on Landslides
Research Centre on Landslides, Disaster Prevention Research Institute, Kyoto University, Uji,
Kyoto 611-0011, Japan
Corresponding author of Chapters 1, 2, 11:
Tel: +81-774-38-4110, Fax: +81-774-32-5597, E-mail: sassa@SCL.kyoto-u.ac.jp

Sato, Hiroshi P. · (Chapter 35)
Geography and Crustal Dynamics Research Center, Geographical Survey Institute,
Ministry of Land, Infrastructure and Transport,
1 Kitasato, Tukuba, Ibaraki, Japan

Shan, Wei · (Chapter 36)
College of Civil Engineering of Northeast Forestry University, Harbin 150040, China

Sheng, Qian · (Chapter 39)
Institute of Rock and Soil Mechanics, Chinese Academy of Sciences, Wuhan 430071, China

Shoaei, Gholamreza · (Chapter 42)
Research Centre on Landslides, Disaster Prevention Research Institute, Kyoto University, Uji, Kyoto
611-0011, Japan
and Natural Disaster Research Institute of Iran (NDRII)

Shoaei, Zieaoddin · (Chapter 42)

Soil Conservation and Watershed Management Research Institute, P.O. Box 13445-1136, Tehran, Iran
Corresponding author of Chapter 42:
Tel: +98-21-4490-1415, Fax: +98-21-4490-5876, E-mail: shoaei58@yahoo.com

Spizzichino, Daniele · (Chapters 13, 26)

Consorzio Civita, Via del Corso 300, 00168 Rome, Italy

Strom, Alexander L. · (Chapter 12)

Institute of Geospheres Dynamics, Russian Academy of Sciences, Leninskiy Av., 38-1, Moscow, Russia
Corresponding author of Chapter 12:
Tel: +7-095-939-7980, Fax: +7-095-137-6511, E-mail: a.strom@g23.relcom.ru

Su, Xueqing · (Chapter 25)

Geological Survey of Canada, Natural Resources Canada, 601 Booth Street, Ottawa, Ontario, K1A 0E8, Canada

Svalova, V. B. · (Chapter 38)

Institute of Environmental Geosciences RAS, Ulansky Pereulok 13, Moscow 101000, Russia
Corresponding author of Chapter 38:
Tel: +7-095-207-4726, Fax: +7-095-823-1886, E-mail: inter@geoenv.ru

Takata, Shuzo · (Chapter 27)

Fukken Civil Engineering Consultants, 2-10-11 Hikarimachi, Higashi-ku, Hiroshima, Japan

Takeuchi, Atsuo · (Chapter 40)

Research Centre on Landslides, Disaster Prevention Research Institute, Kyoto University, Uji,
Kyoto 611-0011, Japan

Tarchi, Dario · (Chapters 3, 43)

European Commission, Joint Research Centre, SERAC Unit, Via E. Fermi 1, TP 723, 21020 Ispra (VA), Italy

Tian, Yongjin · (Chapter 9)

Lishan Landslide Prevention and Control Office, Xian, China

Tommasi, Paolo · (Chapters 21, 32)

Institute for Geo-Engineering and Environmental Geology, National Research Council, c/o Faculty of
Engineering, Via Eudossiana 18, 00184 Rome, Italy
Corresponding author of Chapter 32:
Tel: +39-064-458-5005, Fax: +39-064-458-5016, E-mail: paolo.tommasi@uniroma1.it

Tonoli, Gedeone · (Chapter 8)

Trevi S.p.A, Divisione Rodio, Via Pandina 5, 26831 Casalmaiocco, Lodi, Italy

Towhata, Ikuo · (Chapter 16)

Department of Civil Engineering, University of Tokyo, 7-3-1, Hongo, Bunkyo-Ku, Tokyo 113-8656, Japan
Corresponding author of Chapter 16:
Tel: +81-35-841-6121, Fax: +81-35-841-8504, E-mail: towhata@geot.t.u-tokyo.ac.jp

Tsunesumi, Haruo · (Chapter 35)

Geographic Department, Geographical Survey Institute, Ministry of Land, Infrastructure
and Transport, 1 Kitasato, Tukuba, Ibaraki, Japan
Corresponding author of Chapter 35:
Tel: +81-29-864-6920, Fax: +81-29-864-1804, E-mail: tsune@gsi.go.jp

Uchimura, Taro · (Chapter 16)

Department of Civil Engineering, University of Tokyo, 7-3-1, Hongo, Bunkyo-Ku, Tokyo 113-8656, Japan

Ugarte, David · (Chapter 2)

Instituto Nacional de Cultura (INC), Calle San Bernardo s/n., Cusco, Peru

Vankov, Dmitri A. · (Chapter 46)

Former doctoral student at Graduate School of Science, Kyoto University, Japan
and Research Centre on Landslides, Disaster Prevention Research Institute, Kyoto University, Uji, Kyoto 611-0011, Japan

Vařilová, Zuzana · (Chapter 14)

Bohemian Switzerland National Park Administration, Pra•ská 5, 40746 Krásná Lípa, Czech Republic

Verdel, Thierry · (Chapter 28)

LAEGO, École des Mines de Nancy, France

Vilímek, Vít · (Chapter 4)

Department of Physical Geography and Geoecology, Faculty of Science, Charles University, Albertov 6, 12843 Prague 2, Czech Republic
Corresponding author of Chapter 4:
Tel: +420-22-195-1361, Fax: +420-22-195-1367, E-mail: vilimek@natur.cuni.cz

Vlčko, Ján · (Chapters 7, 4)

Department of Engineering Geology, Faculty of Natural Sciences, Comenius University Bratislava, 84215 Bratislava, Mlynska dolina, Slovak Republic
Corresponding author of Chapter 7:
Tel: +421-26-029-6596, Fax: +421-26-029-6702, E-mail: vlcko@fns.uniba.sk

Wang, Baolin · (Chapter 25)

Geological Survey of Canada, Natural Resources Canada, 601 Booth Street, Ottawa, Ontario, K1A 0E8, Canada
Corresponding author of Chapter 25:
Tel: +1-613-992-8323, Fax: +1-613-992-0190, E-mail: bwang@nrcan.gc.ca

Wang, Fawu · (Chapters 2, 9, 11, 21, 40)

Research Centre on Landslides, Disaster Prevention Research Institute, Kyoto University, Uji, Kyoto 611-0011, Japan
Corresponding author of Chapters 21, 40:
Tel: +81-774-38-4114, Fax: +81-774-38-4300, E-mail: wangfw@landslide.dpri.kyoto-u.ac.jp

Wang, Gonghui · (Chapters 2, 9, 11, 18, 19, 40)

Research Centre on Landslides, Disaster Prevention Research Institute, Kyoto University, Uji, Kyoto 611-0011, Japan
Corresponding author of Chapter 19:
Tel: +81-774-38-4114, Fax: +81-774-38-4300, E-mail: wanggh@landslide.dpri.kyoto-u.ac.jp

Wang, Hui · (Chapter 10)

Environmental Geological Disaster Research Institute, Jilin University, 6 Ximinzhu Street, Changchun 130026, China

Wang, Yong · (Chapter 9)

Lishan Landslide Prevention and Control Office, Xian, China

Watanabe, Naoki · (Chapter 20)

Research Institute for Hazards in Snowy Areas, Niigata University, 2-8050 Ikarashi, Niigata, 950-2181, Japan
Corresponding author of Chapter 20:
Tel: +81-25-262-7058, Fax: +81-25-262-7050, E-mail: jibanken@cc.niigata-u.ac.jp

Yamagishi, Hiromitsu · (Chapters 22, 44)

Department of Environmental Science, Faculty of Science, Niigata University, Igarashi 2-no-cho 8050, Japan
Corresponding author of Chapter 22:
Tel: +81-25-262-6957, Fax: +81-25-262-6957, E-mail: hiroy@env.sc.niigata-u.ac.jp

Yasuda, Tadashi · (Chapter 44)

Public Consultants Co, Ltd., Sapporo 060-0005, Japan
Corresponding author of Chapter 44:
Tel: +81-11-222-2985, Fax: +81-11-222-2579, E-mail: tad_yasuda@public-con.co.jp

Yatabe, Ryuichi · (Chapter 27)

Department of Civil and Environmental Engineering, Ehime University, 3 Bunkyo-cho, Matsuyama, Japan

Ying, Jie · (Chapter 39)

Department of Civil Engineering, Xi'an Jiaotong University, Xi'an 710049, China

Yonekura, Naoshi · (Chapter 20)

Department of Public Works, Niigata Prefectural Government, 4-1 Shinko-cho, Niigata, 950-8570, Japan

Zhang, Lijun · (Chapter 36)

Northeast Institute of Geography and Agricultural Ecology, Chinese Academy of Sciences, Changchun, 130012, China
and Information Center of Ministry of Land and Resources, Beijing, 100812, China
Corresponding author of Chapter 36:
Tel: +86-10-6655-8719, Fax: +86-10-6655-8613, E-mail: ljzhang@infomail.mlr.gov.cn

Zhang, Yeming · (Chapter 40)

Yichang Institute of Geology and Mineral Resources, China Geological Survey, 37 Gangyao Road, Yichang 443003, China

Zheng, Xiaoyu · (Chapter 10)

Environmental Geological Disaster Research Institute, Jilin University, 6 Ximinzhu Street, Changchun 130026, China

Zvelebil, Jiří · (Chapters 4, 14)

Geo-tools, NGO, U Mlejnku 128, 25066 Zdiby, Czech Geological Survey, Klárov 3, 11821 Prague, Czech Republic
Corresponding author of Chapter 14:
Tel: +420-60-225-9921, Fax: +420-25-753-1376, E-mail: jiri.zvelebil@geo-tools.cz

Part I International Consortium on Landslides

Chapter 1

ICL History and Activities

Kyoji Sassa, President of the International Consortium on Landslides

The International Consortium on Landslides (ICL), created during the Kyoto Symposium in January 2002, is an international non-governmental and non-profit scientific organization, which is supported by the United Nations Educational, Scientific and Cultural Organization (UNESCO), the World Meteorological Organization (WMO), the Food and Agriculture Organization of the United Nations (FAO), the United Nations International Strategy for Disaster Risk Reduction (UN/ISDR), and intergovernmental programs such as the International Hydrological Programme of UNESCO, the International Union of Geological Sciences (IUGS), the Ministry of Education, Culture, Sports, Science and Technology (MEXT) of the Government of Japan, U.S. Geological Survey, and other governmental bodies. ICL was registered as a legal body under the Japanese law for non-profit organizations in August 2002 in the Government of Kyoto Prefecture, Japan.

The objectives of the consortium are to:

1. promote landslide research for the benefit of society and the environment, and capacity building, including education, notably in developing countries;
2. integrate geosciences and technology within the appropriate cultural and social contexts in order to evaluate landslide risk in urban, rural and developing areas including cultural and natural heritage sites, as well as contribute to the protection of the natural environment and sites of high societal value;
3. combine and coordinate international expertise in landslide risk assessment and mitigation studies, thereby resulting in an effective international organization which will act as a partner in various international and national projects; and
4. promote a global, multidisciplinary program on landslides.

The central activity is the International Programme on Landslides (IPL). Other activities planned include international co-ordination, exchange of information and dissemination of research activities and capacity building through various meetings, dispatch of experts, landslide database, and publication of "Landslides": Journal of the International Consortium on Landslides. The necessity of establishment of a new Research Centre on

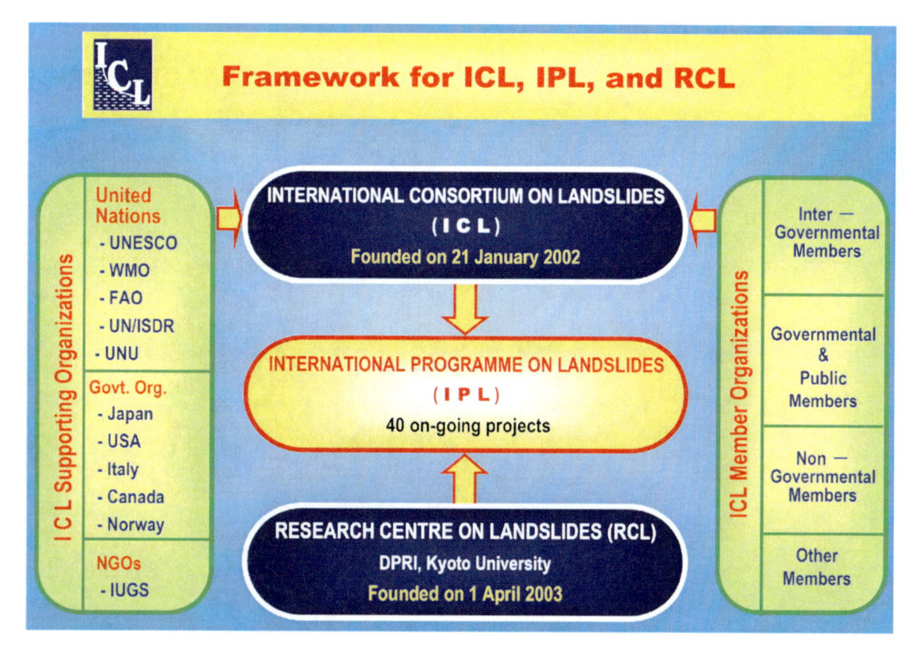

Fig. 1.1.
Relationship of ICL, IPL, and RCL

Landslides to support IPL was proposed in the inaugurated meeting of International Consortium on Landslides on 21 January 2002. Upon request, the new Research Centre on Landslides (RCL) was established on 1 April 2003 in the Disaster Prevention Research Institute, Kyoto University (DPRI/KU). The secretariat of the International Programme on Landslides as well as the International Consortium on Landslides is located in the Research Centre on Landslides.

The relationship of ICL, IPL and RCL is illustrated in Fig. 1.1. The ICL was briefly introduced by Sassa (2004a). ICL has developed for these three years and organizes the First General Assembly in Washington D.C., U.S.A. on 13–14 October 2005. As a reference to have a perspective for reviewing the past history and to plan the future activities at the meeting, I would summarize the background and the history of the International Consortium on Landslides including its major activities with photographs and illustrations in chronological order.

1.1 July 13–17, 1997 International Symposium on Landslide Hazard Assessment, Xian, China

As a part of the Japanese contribution to the IDNDR (International Decade for Natural Disaster Reduction) in the last decade of the twentieth century, the Ministry of Education, Culture, Sports, Science and Technology of the Government of Japan (MEXT) conducted international joint research projects. The projects included a Japan-China Joint Project "Assessment of Landslide Hazards in Lishan (Yang-Que-Fe Palace), Xian, China", which was proposed by Kyoji Sassa, Disaster Prevention Research Institute, Kyoto University. The goal of the project was to investigate the landslide risk to the Lishan Resort Palace of the Tang Dynasty (A.D. 618–907) (Fig. 1.2). The rear slope of Lishan Palace has been stable since the Tang Dynasty. The base rock is Precambrian gneiss (hard rock). At this project, there was spirited discussion by Japanese and Chinese landslide researchers on the possibility of landslides activity. The Palace is one of the most important Cultural Heritage sites attracting more than three million visitors per year. This special project by MEXT continued for eight years in 1991–1999. The project group organized the International Symposium on Landslide Hazard Assessment, Xian, China, in July 1997 (Fig. 1.3) as the Committee for Prediction of Rapid Landslide Motion of the IUGS Working Group on Landslides (WGL/RLM). The symposium received support from the United Nations Educational, Scientific and Cultural Organization (UNESCO), the International Union of Forest Research Organizations (IUFRO), the State Planning Commission of the Government of China, the Embassy of Japan, the Japan Landslide Society and others. The report on the investigation clearly presented evidence of the risk of a large-scale rock slide, based on detailed monitoring and observation of investigation tunnels. Mr. Qiyuan An, the honorary chairman of the symposium and also the Secretary-General of the Communist Party of the Shaanxi Provincial Government, understood the landslide risk at the Lishan Palace (which was also called the Huaqing Palace). Mr. An took a key role in initiation of landslide prevention works, based on landslide risk analysis, by investing about three million U.S.$ with funds from the municipal, regional, and national governments of China.

Fig. 1.2. Front view of the Lishan slope from Lishan Palace

Probably this is the first case in the world of the initiation of extensive landslide remedial measures at a Cultural Heritage site for mitigation of potential landslides at the precursor stage. This investigation of landslides at the precursor stage was evaluated as a contribution of geoscientists to protection of Cultural Heritage.

Products of the Symposium

1. Participants released the 1997 Xian Appeal "'97 Xian Appeal for Protection of the Cultural Heritage (Huaqing Palace) in Xian and Promotion of Worldwide Landslide Hazard Assessment and Risk Mitigation" under the authorship of symposium panelists (Appendix 3.1)
2. Proceedings of the International Symposium on Landslide Hazard Assessment (ISBN4-9900618-0-2 C3051), A4-size, 421 pages (edited by K. Sassa 1997)
3. Special Programme and Video "For the Protection of Huaqing Palace in Xian") produced by Xian Television (1997) (Chinese, English, and Japanese versions, 20 minutes)
4. Invitation to propose a new International Geological Correlation Programme (IGCP) project to promote this research by Prof. Edward Derbyshire (Chairman of the Scientific Board of IGCP)

1.2 February 1998

Based on an invitation by Edward Derbyshire and encouragement by Hideo Noguchi (Division of Cultural Heritage of UNESCO) and other colleagues, Kyoji Sassa proposed an IGCP project. This proposal was approved by the IGCP Board in February 1998. Then, the UNESCO–IUGS joint project, International Geological Correlation Programme (IGCP) no. 425, "Landslide Hazard Assessment and Mitigation for Cultural Heritage Sites and Other Locations of High Societal Value" began. Figure 1.4 presents four components of IGCP-425. The first meeting on this project was organized at the 8th Congress of the International Association of Engineering Geology and the Environment, Vancouver, Canada. The project proposer, Kyoji Sassa, was chosen as the Project Coordinator, and Paolo Canuti (Italy) and Raul Carreño (Peru) became deputy coordinators. Upon a call for participation by the readers of "Landslide News" published by the Japan Landslide Society (JLS), 31 subprojects were proposed to join this project worldwide.

Fig. 1.3. Photos of the Xian Symposium. **a** Symposium at the Grand Castle Hotel, Xian; **b** honorary Chairman of the symposium, Mr. Qiyuan An; **c** participants in the field trip to the Lishan slope viewing posters and panels; **d** release of the '97 Xian Appeals at the end of the symposium

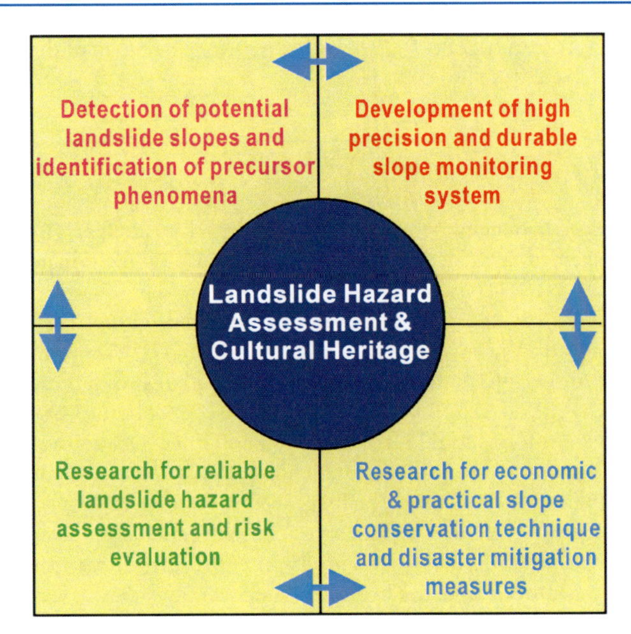

Fig. 1.4. Illustration of IGCP-425 "Landslide Hazard Assessment and Cultural Heritage"

1.3 November 30–December 1, 1998

The UNESCO-IUGS-IGCP Joint Symposium on Natural Hazards and Cultural Heritage was held at the Canadian Embassy in Tokyo from 31 November to 1 December 1998. The organizers were the IGCP National Committee of Japan, the Landslide Research Council of Japan, Kyoto University, the Division of Cultural Heritage of UNESCO, the Canada-Japan S&T Agreement Groups, IGCP-425 "Landslide Hazard Assessment and Cultural Heritage", and others. Participants released "Natural Hazards, Society and Cultural Heritage Approaches for the Next Millennium", the 1999 Tokyo Appeal.

Products of the Symposium

1. 1999 Tokyo Appeal, "Natural Hazards, Society and Cultural Heritage Approaches for the Next Millennium" (Appendix 3.2)
2. Preliminary discussion on an idea to establish "a global entity on landslides" to promote this initiative

1.4 September 20–24, 1999

The International Conference "Cultural Heritage at Risk" was organized by UNESCO and the IGCP-425 group at UNESCO Headquarters, Paris, France (Fig. 1.5). The first half of the conference was the IGCP-425 meeting for "Landslide Hazard Assessment and Cultural Heritage". The activities of 20 subprojects were reported at the meeting. The program budget for IGCP-425 was approximately U.S.$4 000, which was shared by participants from the 20 subprojects as partial travel support. Participants acknowledged that a small grant from an international program is very effective in promoting their subproject research. They wished to found an International Programme on Landslides similar to the IGCP. As its first step, a cooperation agreement was proposed between UNESCO and the institute of the IGCP-425 leader, the Disaster Prevention Research Institute, Kyoto University.

Products of the Conference

1. Draft of a "Memorandum of Understanding for Landslide Risk Mitigation and Protection of the Cultural Heritage" between UNESCO and the Disaster Prevention Research Institute, Kyoto University (The final version of MoU is in Appendix 3.3)
2. Proceedings of Reports and Sub-Project Proposals "Landslide Hazard Assessment and Mitigation for Cultural Heritage Sites and Other Locations of High Soci-

Fig. 1.5. Group photo of the IGCP-425 meeting held at UNESCO Headquarters, Paris

etal Value", B5-size, 156 pages (UNESCO document number: CLT-99/CONF.806/proceedings)

1.5 November 26–December 3, 1999

The Memorandum of Understanding between the United Nations Educational, Scientific and Cultural Organization (UNESCO) and the Disaster Prevention Research Institute (DPRI), Kyoto University, Japan concerning "Cooperation in Research for Landslide Risk Mitigation and Protection of the Cultural and Natural Heritage as a Key Contribution to Environmental Protection and Sustainable Development in the First Quarter of the Twenty-First Century" was signed by Koïchiro Matsuura, Director-General of UNESCO, on 26 November and by Shuichi Ikebuchi, Director of the Disaster Prevention Research Institute, Kyoto University, on 3 December 1999 (Fig. 1.6).

1.6 March 14–20, 2000

Landslides at the Cultural Heritage sites in the vicinity of Cusco, Peru, had been reported on by Raul Carreño in the IGCP-425 meetings in 1998 and 1999. His reports included the Machu Picchu area. His report and his previous research referring other researchers described landslides in the area of Hiram Bingham Road (Carreño and Bonnard 1997), but not mentioned the landslide risk for the Inca's archaeological citadel at Machu Picchu.

K. Sassa, H. Fukuoka, H. Shuzui, and Mutsumi Ishizuka (IGP: Instituto Geofisico del Peru) investigated the slopes of Machu Picchu by a chartered helicopter with the special permission of the INC (Instituto Nacional de Cultura, Peru) (Sassa et al. 2001). The citadel area seemed to be stable. However, at the end of the first day of the investigation, Sassa noted that there was no intact structure or rock located along a line passing through the Plaza area (flat area in the center of the citadel). On the contrary, all existing structures were broken along the line passing through the Plaza (Fig. 1.7a). Thus, the impression was that this site is not an active landslide, but possibly at the precursor stage of landslide activity. Figure 1.7b presents the Hiram Bingham Road area (block no. 1, currently an active landslide) and the Citadel area (block no. 2, a precursor stage of landslide activity). Probably the smooth ground surface was the sliding surface of a previous landslide along a shear band (fault), and another similar shear band is hidden at the bottom of landslide block no. 1 (Sassa et al. 2002; Sassa 2005).

Fig. 1.6. Exchange of the Memorandum of Understanding between Dr. Wolfgang Eder (Director of Division of Earth Sciences of UNESCO) and Prof. Shuichi Ikebuchi, Director of DPRI, on 3 December 1999 at the Director's office of DPRI

Products of this Investigation

1. Understanding of the research needs for landslide risk at the Inca World Heritage of Citadel site
2. The first proposal of the Japanese Fund in Trust to UNESCO in order to investigate the Machu Picchu Citadel site (June 2000)

1.7 January 15–19, 2001

The UNESCO/IGCP Symposium on Landslide Risk Mitigation and Protection of Cultural and Natural Heritage was jointly organized by UNESCO, the IGCP-425 group, the International Union of Geological Sciences (IUGS), the Instituto Nacional de Cultura, Peru (INC), the Instituto Nacional de Recursons Naturales (INRENA), Peru, and the IGCP National Committee of the Science Council of Japan at the headquarters of the Science Council of Japan, Tokyo. The symposium was co-sponsored by the Ministry of Foreign Affairs of the Government of Japan, the Japanese National Commission for UNESCO, the Technical Committee on Landslides (TC-11) of the International Society for Soil Mechanics and Geotechnical Engineering (ISSMGE), two commissions related to landslides and cultural heritage (no. 2 and no. 16) of the International Association of Engineering Geology and the Environment (IAEG), Division 8 "Forest Environment" of the International Union of Forest Research Organizations (IUFRO) and others. Participants, including three representatives (Wolfgang Eder, Hideo Noguchi, and Christian Manhart) from the Divisions of Earth Sciences and Cultural Heritage of UNESCO, the National Director of INC (Luis Enrique Tord), the President of the IUGS (Ed de Mulder), and Paolo Canuti and other IGCP-425 colleagues (Fig. 1.8), agreed on and issued the 2001 Tokyo Declaration "Geoscientists tame landslides" to propose the establishment of the International Consortium on Landslides (Fig. 1.9).

Products of the Symposium

1. Symposium Proceedings, A4-size, 267 pages (edited by K. Sassa 2001)
2. Tokyo Declaration, "Geoscientists tame landslides"

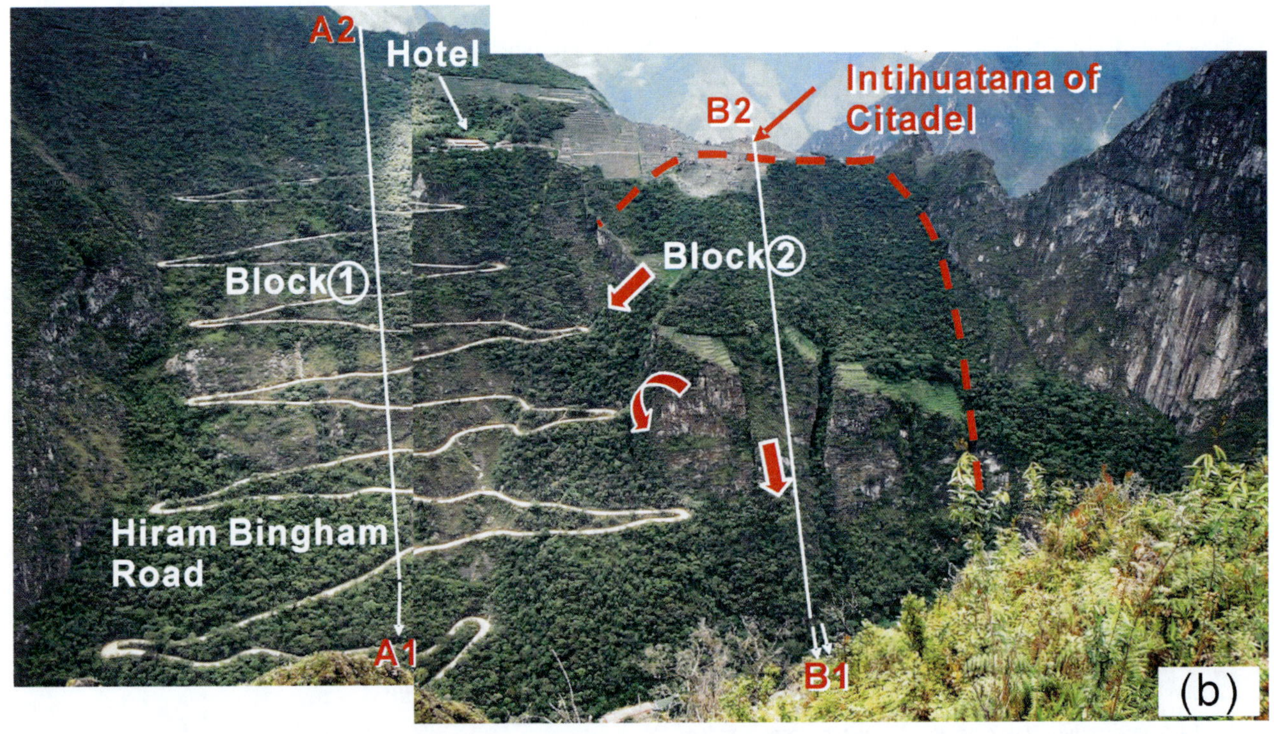

Fig. 1.7. a Machu Picchu Citadel and a line crossing the Plaza as noted taken by Sassa (2000); **b** Machu Picchu slope with interpretation of block no. 1 (Hiram Bingham area) and block no. 2 (Citadel area) (from Sassa 2005, UNESCO brochure)

Fig. 1.8. Group photo of participants in the UNESCO/IGCP Symposium on Landslide Risk Mitigation and Protection of Cultural and Natural Heritage

Fig. 1.9.
2001 Tokyo Declaration signed by symposium participants

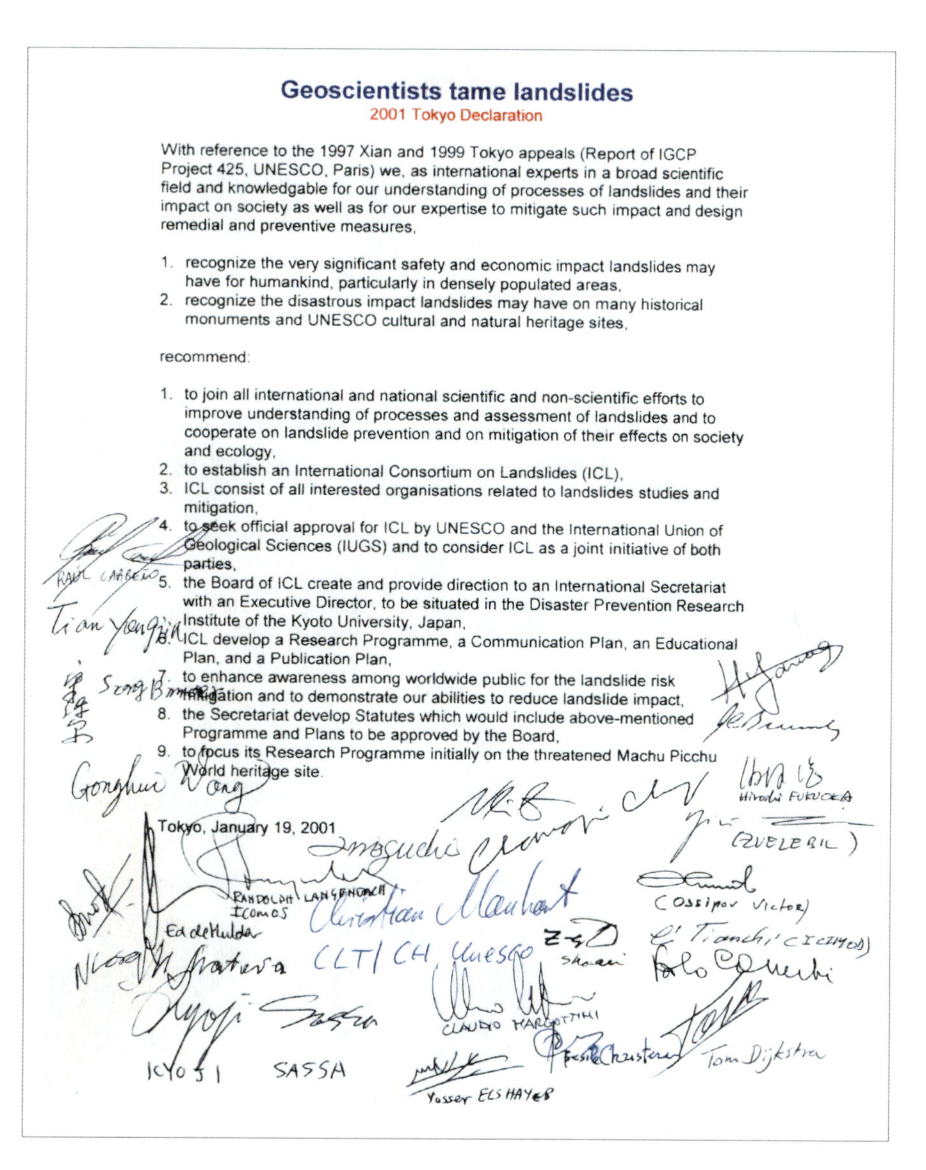

Geoscientists tame landslides
2001 Tokyo Declaration

With reference to the 1997 Xian and 1999 Tokyo appeals (Report of IGCP Project 425, UNESCO, Paris) we, as international experts in a broad scientific field and knowledgable for our understanding of processes of landslides and their impact on society as well as for our expertise to mitigate such impact and design remedial and preventive measures,

1. recognize the very significant safety and economic impact landslides may have for humankind, particularly in densely populated areas,
2. recognize the disastrous impact landslides may have on many historical monuments and UNESCO cultural and natural heritage sites,

recommend:

1. to join all international and national scientific and non-scientific efforts to improve understanding of processes and assessment of landslides and to cooperate on landslide prevention and on mitigation of their effects on society and ecology,
2. to establish an International Consortium on Landslides (ICL),
3. ICL consist of all interested organisations related to landslides studies and mitigation,
4. to seek official approval for ICL by UNESCO and the International Union of Geological Sciences (IUGS) and to consider ICL as a joint initiative of both parties,
5. the Board of ICL create and provide direction to an International Secretariat with an Executive Director, to be situated in the Disaster Prevention Research Institute of the Kyoto University, Japan,
6. ICL develop a Research Programme, a Communication Plan, an Educational Plan, and a Publication Plan,
7. to enhance awareness among worldwide public for the landslide risk mitigation and to demonstrate our abilities to reduce landslide impact,
8. the Secretariat develop Statutes which would include above-mentioned Programme and Plans to be approved by the Board,
9. to focus its Research Programme initially on the threatened Machu Picchu World heritage site.

Tokyo, January 19, 2001

1.8 January 21–25, 2002

The UNESCO-Kyoto University joint symposium "Landslide Risk Mitigation and Protection of Cultural and Natural Heritage" was organized in Kyoto. The symposium was co-sponsored by the Ministry of Foreign Affairs and the National Commission for UNESCO of Japan, IUGS, committees of ISSMGE, IAEG, IUFRO, the Japan Landside Society, and others (Sassa 2002). Eight representatives of UNESCO, the World Meteorological Organization (WMO), and the United Nations Secretariat for the International Strategy for Disaster Risk Reduction (UN/ISDR) participated in the symposium. The founding meeting of the International Consortium on Landslides (ICL) was chaired by Andras Szollosi-Nagy (Director of Water Sciences of UNESCO) on behalf of Walter Erdelen, Assistant Director-General of UNESCO (Fig. 1.10). At the meeting, it was decided to establish the ICL. The Statutes of the ICL were adopted and the first President (Kyoji Sassa) and interim steering committee members were nominated. By releasing the 2002 Kyoto Appeal "Establishment of a New International Consortium on Landslides", the International Consortium on Landslides was inaugurated on 21 January 2002, as an international non-governmental and non-profit scientific organization. Figure 1.11 is a group photo of participants.

Landslides have not usually been treated as central issues by various professional and scientific societies and entities of the world, although they have been dealt with in diverse fields of the natural sciences, engineering, and social sciences. In contrast, landslides are the central issue of the ICL. Therefore, those who regard "landslides" as the most important issue in their professions are the anticipated members of the ICL. Namely, the ICL is an entity of landslide researchers and by landslide researchers for integrated landslide science and technology for disaster reduction. Necessarily, management of the ICL should basically be self-financed by the landslide researchers and their organization. On this basis, the membership fee for the ICL was estimated from the ratio of the level of annual budget needed to maintain the management and activities of the ICL on a global scale divided by the possible number of institutions and entities which are willing to support its objectives and share a not-small membership fee. To estimate this, the experience of 16 years of publication of the international newsletter, "Landslide News", as well as 4 years of activities of IGCP-425, was effective.

Yasushi Taguchi, Director of the Disaster Prevention Research Office of MEXT, encouraged me before the founding meeting of the ICL with a declaration that MEXT would support the initiative of the ICL. During the founding meeting, Andras Szollosi-Nagy (Director of Water Sciences, UNESCO) and Wolfgang Eder (Director of Earth Sciences, UNESCO) led the meeting and played a major role. Badaoui Rouhban (Chief of Engineering Sciences and Technology, UNESCO) supported the management of the meeting. Michel Jarraud (Deputy Secretary-General of WMO), Pedro Basabe (UN/ISDR), three members of the Division of Cultural Heritage of UNESCO (Laurent Levi-Strauss, Galia Saouma-Forero and Christian Manhart), three representatives of the Government of Japan from the Ministry of Foreign Affairs (Multilateral Cultural Co-operation Division) and MEXT (Offices of the Disaster Prevention Research and the Director-General for International Affairs) also participated. Robert Schuster of the U.S. Geological Survey, Paolo Canuti, and other colleagues

Fig. 1.10. Discussion at the founding meeting of ICL

Fig. 1.11. Group photo commemorating the establishment of the International Consortium on Landslides on 23 January 2005 at the Kyoto Campus Plaza

joined this founding meeting. Participants are shown in the group photo of Fig. 1.11, and a list of current ICL members is presented in the Appendix 1.3.

The most important objective of the ICL is to establish a new International Programme on Landslides (IPL), similar to the IGCP. To coordinate and support IPL activities, a secretariat is necessary. Thus, a new Research Centre on Landslides (RCL) was proposed by the participants to be established in Kyoto. Upon this proposal, the Research Centre on Landslides was established on 1 April 2003 as a new center of DPRI/KU, as the result of an agreement among MEXT, Kyoto University, and DPRI. The current relationship between ICL, IPL, and RCL is introduced in Fig. 1.1, which is the same structure proposed in this ICL founding meeting in 2002.

In parallel to this agreement, the outline and principles of cooperation on the landslide investigation at Machu Picchu, including the budget application of the Japanese Fund in Trust to UNESCO, were also agreed upon between the invited group representing the Government of Peru headed by Edwin Benavente (Executive Director for Culture at INC) and the group of DPRI/KU (later expanded to ICL) headed by K. Sassa under the direction of Ms Galia Saouma-Forero in the Division of Cultural Heritage of UNESCO. One of the major projects (C101-1 Landslide Investigation in Machu Picchu) of the IPL was outlined during this meeting.

1.9 November 19–21, 2002

After the founding meeting of the ICL in Kyoto, two other meetings were organized. The first was the Interim Steering Committee Meeting, during the first European Conference on Landslides, held in Prague on 24–26 June 2002. The second was the ICL-IPL Task Force Meeting on 30 October 2002 at UNESCO Headquarters, Paris. During these two meetings, the basic principles of the ICL and IPL management and bylaws were established to ensure the development of this new consortium.

The First Session of the Board of Representatives (BOR) of ICL was organized at UNESCO Headquarters on 19–21 November 2002 (Fig. 1.12). The representatives of the initial 33 ICL members who had paid their membership fees by then were included in the gathering. As a matter of record, all business of the ICL is decided at Board of Representatives meetings. The BOR is composed of the representatives of ICL members with equal rights. At this meeting, the Board decided the initial officers and the initial IPL projects. The first officers of the ICL as decided by the Board were as follows:

- President: Kyoji Sassa (Kyoto University, Japan)
- Vice Presidents: Peter Bobrowsky (Geological Survey of Canada), Paolo Canuti (University of Firenze, Italy), Romulo Mucho (Instituto Geologico Minero y Metalurgico, Peru), John Pallister (U.S. Geological Survey), Executive Director: Kaoru Takara (Kyoto University, Japan), Treasurer: Claudio Margottini (Italian Agency for New Technologies, Energy and Environment/CIVITA Consortium, Italy)
- Assistants to the President: Rafi Ahmad (University of the West Indies, Jamaica), Nicola Casagli (University of Firenze, Italy), Yasser Elshayeb (Cairo University, Egypt), Hiroshi Fukuoka (Kyoto University, Japan), Oddvar Kjekstad (International Centre for Geohazards, Norway), Zieaoddin Shoaei (Soil Conservation and Watershed Management Research Institute, Iran), Alexander Strom (Institute Hydroproject, Russia) and Fawu WANG (Kanazawa University, Japan/China)

The present officers, including coordinators of committees, are listed in the Appendix 1.3.

Fig. 1.12. Group photos of participants in the first session of the BOR at UNESCO Headquarters, Paris

1.10 March 13–18, 2003

In the Round Table Discussion on the International Programme on Landslides organized as a part of the Kyoto Symposium on 24 January 2005, Mr. Christian Manhart, Division of Cultural Heritage of UNESCO, suggested the establishment of a UNESCO's UNITWIN /UNESCO Chairs Programme in the field of landslides.

The UNITWIN Programme was established in 1992 following a relevant decision of the General Conference of UNESCO taken at its 26th session. UNITWIN is the acronym for the UNIVERSITY TWINNING and NETWORKING scheme. It was launched with the aim of developing inter-university cooperation, while emphasizing the transfer of knowledge between universities and the promotion of academic solidarity around the world. The UNITWIN program deals with training and research activities and covers all major fields of learning within UNESCO's competence such as Education, Natural Sciences, Social and Human Sciences, Culture and Communication, and Information. The principal beneficiaries of this program are institutions of higher learning in developing countries and countries in transition.

Upon the proposal at the Kyoto Symposium, Kyoto University, UNESCO, and the ICL examined the possibility of a UNITWIN program on "Landslide Risk Mitigation for Society and the Environment". Agreement was reached by the three parties. The Director-General of UNESCO, Mr. Koïchiro Matsuura, signed the agreement on 10 March, and Mr. Dimitri Beridze of the UNESCO Division of Higher Education visited Kyoto University

Fig. 1.13. Signing of the agreement of the UNITWIN Programme "Landslide Risk Mitigation for Society and the Environment" and a group photo of the President of Kyoto University, the Director of DPRI/KU, two directors of MEXT, two representatives from UNESCO, and the President and three Vice Presidents of the ICL

Fig. 1.14. Round table discussion in the headquarters of FAO

with three sets of documents singed by Mr. Koïchiro Matsuura. At Kyoto University, the President of Kyoto University, Makoto Nagao, and the President of the ICL, Kyoji Sassa, signed the agreement on 18 March 2003 (Fig. 1.13), on which date the agreement came into effect. Prior to this signing ceremony, the ICL organized its secretarial meeting at the DPRI/KU campus in Uji City and on the main campus of Kyoto University in Kyoto City, a field visit to the Bicchu Matsuyama Castle, which is at landslide risk (something like a miniature of the Machu Picchu slope), and an open forum "Cultural Heritage and Landslides" in Takahashi City, Okayama Prefecture.

1.11 May 20–21, 2003

The 2003 First ICL Steering Committee Meeting was organized at the headquarters of the Food and Agriculture Organization of the United Nations (FAO) in Rome, Italy (Fig. 1.14). At this meeting, the followings were planned: (1) the second session of the BOR in Vancouver, (2) the steering committee meeting in 2004, and (3) the new ICL journal, "Landslides". In addition, an International Year of Mountains Medal was conferred by Hosny ElLakany, Assistant Director-General of FAO, to Kyoji Sassa for his contributions to mountain conservation activities.

1.12 Discussion with Springer-Verlag on the Publication of "Landslides"

Discussions with Springer-Verlag on publication of "Landslides" were conducted at their main office in Heidelberg, Germany. The agreement on publication of the full-color quarterly journal of "Landslides" was reached in a visit by K. Sassa in March 2003, and the contract between the ICL and Springer-Verlag was signed on 30 April to 7 May 2003 by post. In the July 2003 meeting, K. Sassa (Editor-in-Chief), P. Bobrowsky, P. Canuti, G. Wieczorek (associate editors), and H. Fukuoka (Secretary General for the Journal) jointly examined edition, publication, distribution, and marketing of "Landslides" with the staff of Springer-Verlag (Fig. 1.15).

Fig. 1.15. Discussion on the journal "Landslides" at Springer-Verlag offices in Heidelberg

K. Sassa discussed "Landslides" preliminarily with the chief of Springer-Verlag at UNESCO, Paris, in 2002, and the company understood the significance of a full-color journal specifically on landslides. Full-color printing is vitally important for publications of landslide research, because color photos of landslides provide important information. In addition, the quality and quantity of information between mono-color and full color are quite different.

The publication of "Landslides" is successful and satisfactory for both of the ICL and Springer-Verlag. For this reason, Springer-Verlag is willing to publish the proceedings of the first General Assembly of ICL as a full-color book with its format being the same as that of "Landslides" journal.

Fig. 1.16.
Photos of BOR meeting in Vancouver: **a** presentation of the first Varnes Medal to Robert Schuster and **b** group photo for the BOR meeting

Fig. 1.17.
Group photo celebrating the establishment of UNITWIN Headquarter at Kyoto University

1.13 October 28–November 1, 2003

The Second Session of the Board of Representatives of ICL was held at Simon Fraser University, Vancouver, Canada. The session included the 2003 Second ICL Steering Committee Meeting, the ICL Special Symposium on Landslides and Natural Resources, an editorial meeting for the ICL Journal "Landslides", and a one-day field trip.

Fig. 1.18. Group photos of participants at the 2004 First ICL Steering Committee Meeting, Palais des Nations Building, United Nations, in Geneva

The first Varnes medal of ICL was presented to Dr. Robert Schuster (U.S.A.) and twenty-six people attended the Board of Representatives meeting (Fig. 1.16), and many people from outside of ICL joined the Special Symposium on Landslides and Natural Resources and the field trip.

1.14 January 21–24, 2004

Three consecutive ICL meetings were held at Kyoto University, Japan, on 21–24 January 2004:

1. International Symposium on Landslide Risk Mitigation and Protection of Cultural and Natural Heritage jointly organized by the IPL C101, and M101 groups, the IGCP-425 group, and the Research Centre on Landslides, Disaster Prevention Research Institute, Kyoto University, on 21–22 January
2. Memorial Meeting for Establishment of the UNESCO-Kyoto University-ICL UNITWIN Cooperation Programme Headquarters on 23 January
3. Secretariat Meeting of the International Consortium on Landslides on 23–24 January

The main meeting was the 23 January meeting celebrating the establishment of the UNESCO-Kyoto University-ICL UNITWIN Cooperation Programme Headquarters. Major ICL officers, Badaoui Rouhban and Dimitri-Beridze of UNESCO, Koichi Nagasaka (Director of the Japanese Meteorological Agency) on behalf of Michel Jarraud (Secretary-General of WMO), Pedro Basabe of UN/ISDR, two people representing MEXT (Disaster Prevention and International Affairs), representatives from the Embassies of Italy and Peru, as well as the President/Vice Presidents of Kyoto University, and the Director of DPRI/KU, were in attendance (Fig. 1.17).

1.15 May 3–4, 2004

The 2004 First ICL Steering Committee Meeting was held in the Palais des Nations Building at the United Nations in Geneva (Fig. 1.18). ICL Board Members and representatives from UNESCO, WMO, and UN/ISDR discussed contributions to the United Nations World Conference on Disaster Reduction (WCDR), which was scheduled to be held in Kobe, Japan, on 18–22 January 2005. K. Sassa and some ICL members participated in the Inter-Agency Task Force Meeting for the United Nations World Conference on Disaster Reduction, which was held immediately after the ICL Steering Committee meeting.

1.16 September 3, 2004

The main building of the UNITWIN Headquarters was constructed jointly by funding from Kyoto University and the ICL at the Uji Campus of Kyoto University (Fig. 1.19). The opening ceremony was organized on 3 September; it began with welcoming address by President Kazuo Oike (Kyoto University) and the Director of the DPRI-KU, Kazuya Inoue, and concluded with a talk on the progress of the UNITWIN programme by the UNITWIN coordinator, Kyoji Sassa. In the ceremony, congratulatory speeches recommending further cooperation were presented by representatives of relevant offices as follows: (1) Wolfgang Eder, Director of the Division of Earth Sciences of UNESCO, on behalf of ADG Walter Erdelen, (2) Winsome Gordon, Section Head of Higher Education in UNESCO, (3) Badaoui Rouhban, Section Head of Disaster Reduction in UNESCO, (4) Kasuo Akiyama, on behalf of the Director-General for International Affairs of MEXT, Masayuki Inoue, (5) Satoru Nishikawa, Director of the International Office for Disaster Management of the Cabinet Office of the Government of Japan, (6) Takayuki Nakamura, Director of Office of Disaster Prevention Research of MEXT, (7) Luis J. Macchiavello, Ambassador of Peru in Tokyo, (8) Hans van Ginkel, Rector of the United Nations University and United Nations Under-Secretary General, (9) Silvio Vita on behalf of Mario Bova (Italian Ambassador in Tokyo), (10) Kenzo Toki, Director of the Research Center for Disaster Mitigation of Urban Cultural Heritage, Ritsumeikan University, Kyoto, and others (Fig. 1.20). Details of the opening ceremony for the UNITWIN Headquarter building were introduced by Sassa (2004b).

Fig. 1.19. Photo of the UNITWIN Headquarter building

1.17 September 12–25, 2004

Project groups of the IPL organized various meetings in 2003 and 2004. A major function was the joint investiga-

Fig. 1.20. Group photo of participants in the opening ceremony of the UNITWIN Headquarter building

Fig. 1.21.
Group photo of participants in the Machu Picchu Stakeholders Meeting, Cusco, Peru, September 2004

tion of Machu Picchu, Cusco, Peru, by the IPL C101-1 Project "Landslide Investigation in Machu Picchu". The groups of this project from Japan, Italy, the Czech Republic, and Slovakia conducted field investigations on 12–25 September 2004, and organized the Machu Picchu Stakeholder's Meeting on 21 September, which was held at the Instituto Nacional de Cultura (INC), Cusco Office, for the presentation of investigation reports and future plans (Fig. 1.21). The meeting was reported by Peruvian newspapers and television, generally welcoming ICL activities for the protection of Machu Picchu with close cooperation from Peruvian counterparts, including technology transfer and capacity building in the field of landslide disaster reduction.

1.18 October 19–22, 2004

The Third Session of the Board of Representatives of the ICL was held at Comenius University, Bratislava, Slovakia. This session included the 2004 Second ICL Steering Committee Meeting, an ICL Symposium, and a Field Trip on Cultural Heritage and Landslides in Slovakia. The second Varnes Medal was conferred to Prof. John Hutchinson at this session (Fig. 1.22).

1.19 January 15–19, 2005

The World Conference on Disaster Reduction (WCDR) was held on 18–22 January in Kobe, Japan. Before the WCDR, the fourth International Symposium on Landslide Risk Mitigation and Protection of Cultural and Natural Heritage and the ICL Secretariat Meeting were organized at Kyoto University. Participants from the ICL, including Hans van Ginkel and Wolfgang Eder, discussed management and proposal of ICL in the thematic conference Session 3.8 on Floods (IFI) and Landslides (IPL).

A session titled "New International Initiatives for Research and Risk Mitigation of Floods (IFI) and Landslides (IPL)" was held on 19 January. Joint organizers were the United Nations Educational, Scientific and Cultural Organization (UNESCO), the World Meteorological Organization (WMO), the Food and Agriculture Organization of the United Nations (FAO), the Ministry of Education, Culture, Sports, Science and Technology of the Government of Japan (MEXT), the United Nations University (UNU), Kyoto University (KU), the Public Works Research Institute (PWRI) of Japan, the International Consortium on Landslides (ICL), and the International Association of Hydrological Sciences (IAHS). Figure 1.23 presents the joint photo before the session *(a)* and addresses by UNESCO Director-General Koïchiro Matsuura *(b)*, by WMO Secretary-General Michel Jarraud *(c)*, presentation by Badaoui Rouhban (Section Chief of UNESCO Natural Disaster Reduction *(d)*, the first signature to the Letter of Intent by Rector Hans van Ginkel (UNU) *(e)* and the reception to major participants for Session 3.8 and key participants to WCDR by Mr. Matsuura in the evening *(f)*.

(1) Opening remarks were presented by Mr. Koïchiro Matsuura (UNESCO), Mr. Michel Jarraud (WMO), Mr. Kazuya Inoue (Kyoto University), and Mr. Tadahiko Sakamoto (PWRI). (2) Presentations on IFI were made by Mr. Slobodan Simonovic (Western Ontario University), Mr. Akira Terakawa (PWRI), and Mr. Kuniyoshi Takeuchi (IAHS). (3) Presentations on IPL were made by Mr. Badaoui Rouhban (UNESCO), Mr. Peter Lyttle (USGS), and Mr. Kyoji Sassa (DPRI/KU and ICL). (4) The discussion was chaired by Hans van Ginkel (UNU) with Andras Szollosi-Nagy (UNESCO) as moderator. (5) Wolfgang Eder (UNESCO Consultant and ICL technical advisor)

Fig. 1.22.
Presentation of the Varnes Medal to Prof. John Hutchinson and group photo of participants at the Third Session of the Board of Representation, October 2004

proposed a Letter of Intent to promote further joint global activities in disaster reduction and risk prevention through "Strengthening research and learning on 'Earth System Risk Analysis and Sustainable Disaster Management' within the framework of the 'United Nations International Strategy for Disaster Risk Reduction' (ISDR)" by global partners: UNESCO, WMO, FAO, UNEP, UNDO, UNU, ICSU, WFEO etc. based on previous discussion at the ICL meeting in Kyoto University.

This Letter of Intent can be an umbrella for all initiatives of Earth-system risk reduction. The ICL presented its model and proposed a Memorandum of Understanding concerning strengthened cooperation in research and education on "Earth System Risk Analysis and Sustainable Disaster Management within the framework of the United Nations International Strategy of Disaster Reduction (ISDR)" in particular regard to landslides. The relationship between the Letter of Intent and each Memorandum of Understanding for a specific disaster field is presented in Fig. 1.24.

1.20 May 15–16, 2005

The 2005 First ICL Steering Committee Meeting was held at Cairo University, Cairo, Egypt. Due to terrorism in Cairo just before the meeting, the number of participants was very small. The organization of the first General Assembly of ICL in Washington, D.C., and the idea of a European Centre of ICL were discussed. Thereafter, participants visited the Valley of the Kings with special attention being paid to the risk of landslides and structural cracks.

1.21 "Letter of Intent"

The Letter of Intent proposed at the WCDR was formally approved and signed by UNESCO, WMO, FAO, UN/ISDR, UNU, ICSU (International Council for Science) and WFEO (World Federation of Engineering Organizations) by

Fig. 1.23.
Photos of thematic session 3.8 of the World Conference on Disaster Reduction: **a** group photo before session; **b** address by K. Matsuura; **c** M. Jarraud; **d** B. Rouhban; **e** the first signature to the Letter of Intent by H. van Ginkel, immediately after the session; **f** a reception was held at the invitation of K. Matsuura in the evening. The photo shows Hans van Ginkel (UNU), Michel Jarraud (WMO), Salvano Briceno (UN/ISDR), Walter Erdelen (UNESCO), Kyoji Sassa (ICL), and others

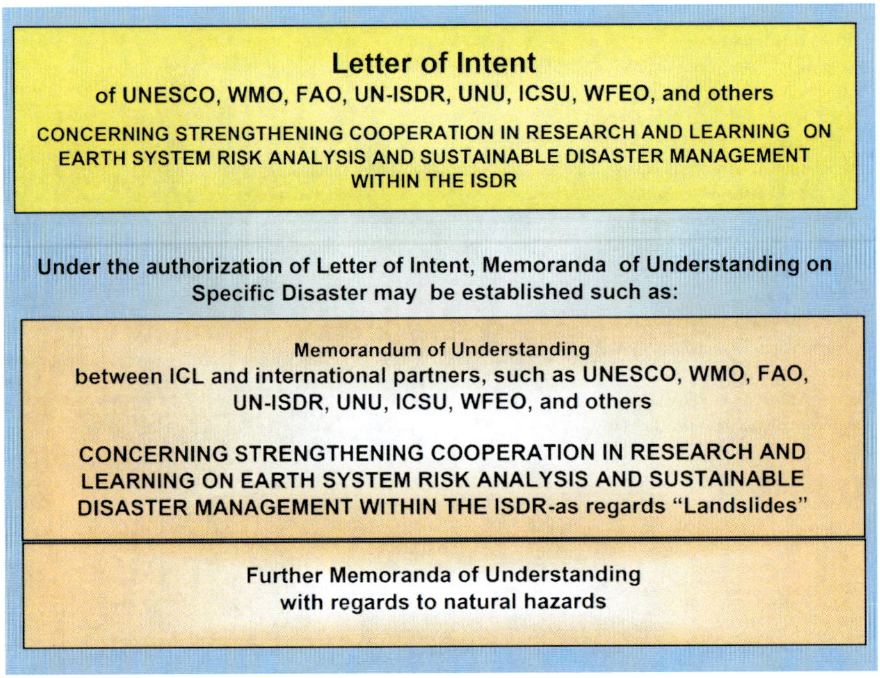

Fig. 1.24.
Structure of Letter of Intent and Memorandum of Understanding

30 June 2005. Then, the Letter was approved by all of proposed global stakeholders of UN organizations and communities of science and technology. Kyoji Sassa and Kaoru Takara (Executive Director of the ICL) visited all six partners of this Letter in Europe with the completed Letter of Intent and discussed further cooperation as stated in the Letter on 14–19 July 2005.

The electronic version of the Letter of Intent can be seen in the *Preface* of this volume. The original Letters of Intent signed by seven partners are deposited in the Research Centre on Landslides of the Disaster Prevention Research Institute, Kyoto University, where the Secretariat of the ICL is located.

This Letter provides a strong platform for the promotion of cooperation within the ISDR in the coming decades. The ICL will have a panel discussion on this cooperation on 13 October 2005 during the First General Assembly of the ICL in Washington, D.C. A Round Table Discussion meeting is planned to jointly examine how to develop the initiative of the Letter of Intent on 18–20 January 2006 with participants from all seven global stakeholders as well as ICL and Kyoto University. Those who are willing to participate in this meeting are welcome.

Acknowledgments

Those organizations and individuals, who created the infra-structural ground and supported the initiatives leading to the present ICL before the Xian Symposium in 1997, are reviewed and acknowledged last, but not least in this article.

The Japan Landslide Society (JLS) was founded in 1965 as the first national society on landslides. There was no other national society on landslides except the Nepal Landslide Society, which was established in 2003. The JLS has a 40-year history and over 2 000 members. The major activities of the society are domestic, but it initiated the first International Symposium on Landslide Control (1st ISL) at the Kyoto International Conference Hall in 1972 and organized the Second International Symposium on Landslides (2nd ISL) in Tokyo in 1977. The society organized the Fourth International Conference and Field Workshop on Landslides (4th ICFL) in Tokyo in 1985 as a development of the U.S.A.-Japan Joint Field Trip and Workshop on Landslides, which began in 1979. Beginning with this series of meetings, the ICFL has met every 3 years.

Based on discussions at the 4th ICFL in Tokyo and some money left from this conference, an international newsletter, "Landslide News", was initiated by the Japan Landslide Society. 4 000–5 000 copies of each issue were printed and distributed within Japan and throughout the world. Around 2 000 copies were distributed outside of Japan free of charge every year. Financial Support for "Landslide News" was provided by the JLS and donations from official Japanese supporters of this newsletter for 16 years (1987–2003). The newsletter was also supported by UNESCO, FAO, UNDRO (present UN/ISDR) from the beginning, and later by IUFRO, TC-11 of ISSMGE, IAEG Commission No. 2, and others. The Landslide Section of the Disaster Prevention Research Institute, Kyoto University, worked from the founding to the end as the Secretariat of "Landslide News". I, myself, worked as chairman of the Publication Committee, including the role of Editor-in-Chief

(Sassa 1999). Robert Schuster contributed as the Chief International Editor for this newsletter and provided support to keep the newsletter at international quality.

The International Consortium on Landslides is supported by various activities of the Japan Landslide Society and also by the 16 years of Landslide News and worldwide supporters of this newsletter. We appreciate all colleagues of the Japan Landslide Society, and the worldwide readers and supporters of Landslide News. Thanks go to colleagues of the Technical Committee on Landslides (TC-11) of ISSMGE with whom I worked from 1985–2001. Finally, all colleagues of IGCP-425 and ICL members are deeply appreciated for their efforts in the founding and development of the International Consortium on Landslides.

References

Carreño R, Bonnard C (1997) Rock slide at Machupicchu, Peru. Landslide News 10:15–17

Japan Landslide Society (1987–2003) Landslide News, no. 1–15

Sassa K (1997) Proc International Symposium on Landslide Hazard Assessment. 421 p

Sassa K (ed) (1999) Landslides of the World. Kyoto University Press, ISBN 4-87698-073-X, C3051, 406 p

Sassa K (ed) (2001) Landslide Risk Mitigation and Protection of Cultural and Natural Heritage Proc. UNESCO/IGCP Symposium, 15–19 January 2001, 268 p

Sassa K (2002) Proceedings of International Symposium on Landslide Risk Mitigation and Protection of Cultural and Natural Heritage. 21–25 January 2002, Kyoto University, 750 p

Sassa K (2004a) The International Consortium on Landslides. Landslides 1(1):91–94

Sassa K (2004b) Opening ceremony of the UNESCO-Kyoto University-ICL UNITWIN Programme Headquarters building. Landslides 1(4):315–323

Sassa K (2005) Precursory stage of landslides in Inca's World Heritage in Machu Picchu, Peru. Independent brochure for UNITWIN Programme, Education Sector of UNESCO, 4 p

Sassa K, Fukuoka H, Kamai T, Shuzui H (2001) Landslide risk at Inca's World Heritage in Machu Picchu, Peru. In: Sassa K (ed) Landslide Risk Mitigation and Protection of Cultural and Natural Heritage, Proc UNESCO/IGCP Symp, 15–19 January 2001, Tokyo, pp 1–14

Sassa K, Fukuoka H, Shuzui H, Hoshino M (2002) Landslide risk evaluation in the Machu Picchu World Heritage, Cusco, Peru. In: Proc of Int Symp on Landslide Risk Mitigation and Protection of Cultural and Natural Heritage, pp 469–488

Xian Television (1997) For the Protection of Huaqing Palace in Xian. Special Programme of Xian TV, China

List of Abbreviated Names (in Alphabetical Order)

DPRI/KU	Disaster Prevention Research Institute, Kyoto University
FAO	Food and Agriculture Organization of the United Nations
IAEG	International Association for Engineering Geology and the Environment IAEG Commission No. 2: Landslides and other mass movement
ICFL	International Conference and Field Workshop on Landslides
ICL	International Consortium on Landslides
ICSU	International Council for Science
IGCP	International Geological Correlation Programme
INC	Instituto Nacional de Cultura, Peru
IPL	Internationsl Programme on Landslides
ISL	International Symposium on Landslides
ISSMGE	International Society for Soil Mechanics and Geotechnical Engineering
IUFRO	International Union of Forest Research Organizations
IUGS	International Union of Geological Sciences
JLS	Japan Landslide Society
MEXT	Ministry of Education, Culture, Sports, Science and Technology of the Government of Japan
RCL	Research Centre on Landslides, DPRI/KU
TC-11	Technical Committee on Landslides of ISSMGE
UNESCO	United Nations Educational, Scientific and Cultural Organization
UN/ISDR	United Nations International Strategy for Disaster Risk Reduction
UNITWIN	University Twinning and Networking Program of UNESCO
UNU	United Nations University
USGS	United States Geological Survey
WCDR	United Nations World Conference on Disaster Reduction
WFEO	World Federation of Engineering Organizations
WMO	World Meteorological Organization

Part II International Programme on Landslides

Chapter 2

Landslide Investigation in Machu Picchu World Heritage, Cusco, Peru (C101-1)

Kyoji Sassa* · Hiroshi Fukuoka · Gonghui Wang · Fawu Wang · Edwin Benavente · David Ugarte · Fernándo V. Astete

Abstract. The Japanese landslide expert team conducted landslide investigation in and around Machu Picchu Citadel since March 2000 in cooperation with the Instituto Nacional de Cultura (INC) and the Instituto Nacional de Recursos Naturales (INRENA). The investigation results and the cooperation scheme between ICL and the Government of Peru are introduced.

In the past, probably a series of retrogressive landslides scraped a part of the mountain ridge of Machu Picchu slope along a shear band almost parallel to the present slope. The flat area was formed by landslides on the mountain ridge. Inca people were likely to have constructed a citadel on this flat part of mountain ridge. Landslide debris provided them weathered debris and soils possible for farming. When undercutting by river erosion reached the level of another shear band, another series of retrogressive landslides have proceeded along the shear band near or a little bit higher than the present river bed. The process was faster in the landslide block (no. 1) including the Hyram Bingham road, and delayed in the landslide block (no. 2) including the Inca's citadel because the river erosion to the slope was stronger for block no. 1 due to sharp curvature of river route. The slope deformation affecting the citadel part is not real landslide at present, but it is a precursor stage of landslides, namely it can become to be a real landslide as the result of retrogressive development of landslides from the Urubamba River and from the block no. 1 side.

The initial slope monitoring using extensometers by the Japanese team of the Disaster Prevention Research Institute of Kyoto University (DPRI/KU) started with cooperation from INC and INRENA from November 2000. One year monitoring in 2001 was presented. After the establishment of the International Consortium on Landslides (ICL) and the International Programme on Landslides (IPL) in 2002, the initial cooperation agreement on Machu Picchu between the Government of Peru and DPRI/KU has developed to the cooperation between the Government of Peru and the ICL. The International Programme on Landslides (IPL) C101-1 'Landslide investigations in Machu Picchu' consists of six groups including Japanese, Italian, Czech-Slovakian, Peruvian-Canadian groups in 2005. The Japanese team installed new four sets of long-span extensometers, three sets of GPS receivers, a Total Station with three prism mirror targets in 2004 and started monitoring of the displacement.

Keywords. Risk evaluation, prediction, monitoring, precursor stage of landslide

2.1 Background and History of Machu Picchu Project

Prediction or identification of precursor phenomena of large scale landslides is not an easy task. Large-scale landslides do not occur in the same place in a short return period compatible to life period of human beings. It is a kind of geological process and the return period of large-scale landslides is usually very long in the order of thousand years or tens of thousand years or even longer.

Before the Hyogoken-Nanbu earthquake took place, most of the Japanese people regarded that earthquakes occurring in the order of thousand years was almost out of scope in present planning of disaster prevention measures. However, such earthquake caused great damages to mega-city area of Kobe. So now it is clearly understood that even infrequent phenomena such as movement of active faults and earthquakes should be seriously considered and people prepared for that. Frequency of active faults and large-scale landslides are rather similar, though the casual forces of phenomena are different; faults by crustal horizontal stress, landslides by gravitational vertical stress.

Protection of mega-city from earthquakes is very important, and we, researchers involved in the field of disaster prevention should focus our study toward a more reliable prediction of site and time. Though the disaster caused by landslides is not so great in the number of death comparing to earthquakes, the Mayuyama landslide in Unzen, Japan, 1792 killed 15 000 people (Sassa 1999) and the most recent Las Colinas landslide killed around 600 people in El Salvador in 2001. So landslides cause not a little disaster.

Entering twenty-first century, we are more and more aware of the value of the environment, especially regarding the invaluable cultural and natural heritage. Those are very fragile treasure for humanity, which cannot be rebuilt once they were destroyed. People worked for economic development in the last century, and the industrial progress and economic development is still very important for the base of society, but at the same time, we have noticed that we should protect and leave our invaluable treasures of humanity to the next generation so long as possible.

Fortunately the progress of geosciences is approaching to a level to identify precursor phenomena of large-

* The corresponding author of each chapter is marked with an asterisk.

scale landslides, and assess the location, size, velocity, and hazard area of these landslides. It was recognized during the International Decade for Natural Disaster Reduction to create less hazardous world in the last decade of twenty-first century. The landslide hazard assessment in Lishan (resort palace of Tang Dynasty), Xi'an, China by DPRI/KU and the Xi'an Construction Committee was successful. The landslide risk assessment based on the joint research has convinced the landslide risk to the Chinese government as well as the Shaanxi Provincial Government and the Xi'an City. Based on the landslide risk assessment research, landslide prevention works were initiated before occurrence of any disaster due to landslides. Then, Prof. Edward Derbyshire of the IGCP scientific committee invited DPRI/KU to propose a project, IGCP-425 "Landslide Hazard Assessment and Cultural Heritage". The project was adopted and now on-going. The Machu Picchu landslide was introduced by Carreño and Bonnard (1997) as a part of IGCP-425 sub-project "Study and protection of Inca cultural heritage on landslide zone at Cusco, Peru".

DPRI/KU team investigated Machu Picchu in March 2000 with support from staffs of INC (Instituto Nacional de Cultura) and INRENA (Instituto Nacional de Recursos Naturales) in Machu Picchu, and installed extensometers in November 2000 (Sassa et al. 2000), and reported those results in the UNESCO/IGCP Symposium on Landslide Risk Mitigation and Protection of Cultural and Natural Heritage in Tokyo, 2001 (Sassa et al. 2001a). Those reports were introduced by *Yomiuri Newspaper* in Japan, "*New Scientists*" in UK (Hadfield 2001), *Civil Engineering* by ASCE in U.S.A. (Wright and Zengarra 2000; Brown 2001) and others (Wright and Wright 1997; Sassa 2005a). The article by "*New Scientists*" was sensational. Articles in *Yomiuri Newspaper* and in *Civil Engineering* were published after the review of the articles and correction of misunderstanding by Sassa. However, the New Scientist article was based only on telephone interview and not reviewed before publication by Sassa. The article was written with serious misunderstandings in some parts. This article caused sensation over the world. Therefore, UNESCO thought the necessity to inform a real state of investigation and published "*Rumbles at Machu Picchu*" article in "*World Heritage Review*," no. 20, May 2001 (Bandarin 2001).

The landslide investigation in Machu Picchu was proposed by K. Sassa of the Disaster Prevention Research Institute, Kyoto University (DPRI/KU) and initially applied for UNESCO using the Japanese Fund in Trust to UNESCO in June 2000. Thereafter, communication and discussion were exchanged in 2000–2001. At the time of the UNESCO-Kyoto University joint symposium "Landslide risk mitigation and protection of cultural and natural heritage" held on 21–25 January 2002, DPRI/KU invited the representatives of the Government of Peru. The government sent three representatives including INC, INRENA and the honorary president of Geological Society of Peru (also representing INGEMMET and IGP) headed by E. Benavente of INC. Six representatives from UNESCO were invited to the symposium. Mrs. Galia Saouma-Forero of the Division of Cultural Heritage was representing UNESCO on the matter of Machu Picchu landslide investigation issue.

During the 2002 UNESCO-Kyoto University joint symposium, the DPRI/KU group headed by Kyoji Sassa, the group of representing the Government of Peru headed by E. Benavente of INC, and Mrs. Galia Saouma-Forero of the Division of Cultural Heritage representing UNESCO Cultural heritage jointly examined on the Machu Picchu landslide investigation based on the results of previous landslide investigation since March 2000. The basic agreement and principle between the Government of Peru and the DPRI/KU was obtained to conduct the landslide investigation in Machu Picchu using the Japanese Fund in Trust to UNESCO.

The International Consortium on Landslides founded on 21 January 2002 (Sassa 2005b) organized the ICL secretariat meeting at Kyoto University on 13–14 March 2002 in prior to the ceremony of UNITWIN cooperation program "Landslide risk mitigation for society and the environment" between UNESCO, Kyoto University and ICL on 18 March 2002 (Sassa 2005b). In the meeting, ICL made a proposal of Landslide investigation in Machu Picchu to the Government of Peru and submitted through the Peruvian Embassy in Tokyo. To respond the proposal, the government of Peru under the initiative of INC and INGEMMET organized an advisory committee for the physical safety of the Machu Picchu archaeological sanctuary on 8 April 2003 in INC. Thereafter, the Government of Peru established a new "*Inter-institutional Commission on the Environmental and Geodynamic Studies at the Urubamba Valley*" to coordinate the environmental and geodynamic studies at Urubamba Valley including landslides in Machu Picchu on 24 May 2003. The commission is composed of:

- Instituto Nacional de Cultura (INC);
- Instituto Nacional de Recursos Naturales (INRENA);
- Instituto Geológico Minero y Metalúrgico (INGEMMET);
- Instituto Geofisico del Peru (IGP);
- Unidad de Gestión del Santuario Histórico Machu Picchu (UGSHMP, Joint Machu Picchu Management Unit); and
- Gobierno Regional Cusco (GR Cusco, Cusco Regional Government).

In the meeting of this commission held on 19 June 2003, the agreement with DPRI/KU obtained during the 2002 Kyoto Symposium was decided to apply for the Machu Picchu investigation to the ICL proposal. A *Technical Inter-institutional Sub-commission of Machu Picchu* was formed under the "Inter-institutional Commission on the Environmental and Geodynamic Studies at the Urubamba Valley". The sub-commission is in charge of coordinating all technical aspects of the evaluation and follow-up of the projects proposed by ICL.

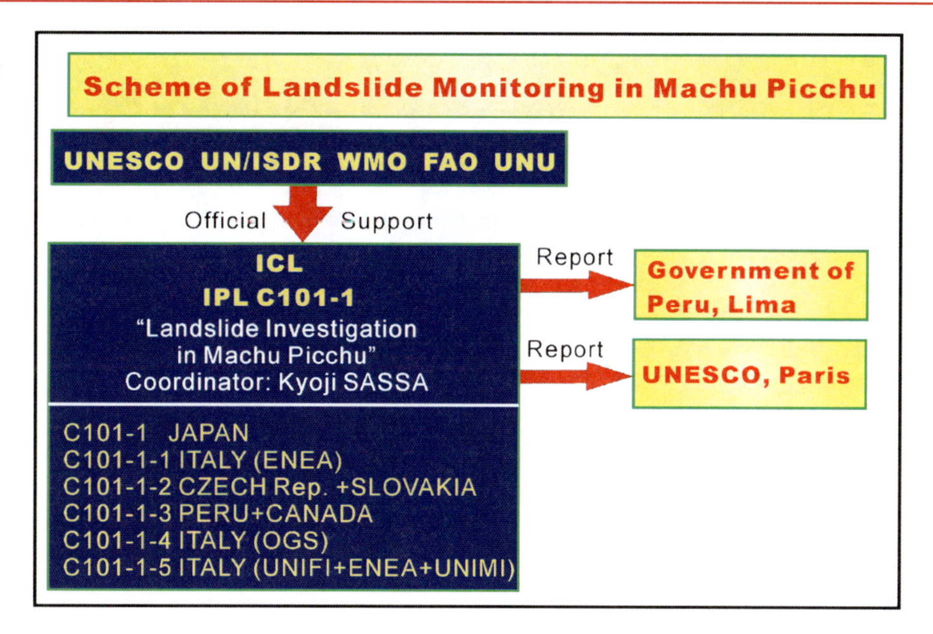

Fig. 2.1. Scheme of IPL C101-1 "Landslide Investigation in Machu Picchu"

At the establishment of this formal cooperation framework crossing different Peruvian agencies involved in Machu Picchu, close and effective cooperation and joint efforts between the Inter-institutional Commission of the Government of Peru and ICL through IPL C101-1 Landslide investigation in Machu Picchu became possible and they have developed very well through reliable partnership. The current investigation teams under IPL C101-1 Landslide investigation in Machu Picchu in 2005 are as follows.

C101-1: Landslide investigation in Machu Picchu (2002–)
Coordinator: Kyoji Sassa
Office: Research Centre on Landslides, DPRI, Kyoto University, Japan
Contact: sassa@scl.kyoto-u.ac.jp

C101-1-1: Low environmental impact technologies for slope monitoring by radar interferometry: application to Machu Picchu site (2002–)
Office: ENEA (Italian Agency for New Technology, Energy and Environment), Italy
Proposer: Claudio Margottini
Contact: margottini@enea.casaccia.it

C101-1-2: Expressions of risky geomorphologic processes in deformations of rock structures at Machu Picchu (2002–)
Office: Research Center of Earth Dynamic, Charles University, Czech Republic
Proposers: Vit Vilimek and Jiri Zvelebil
(This group includes Jan Vlčko and Rudolf Holzer of Comenius University of Slovakia)
Contact: vilimek@natur.cuni.cz

C101-1-3: Shallow geophysics and terrain stability mapping techniques applied to the Urubamba Valley, Peru: Landslide hazard evaluation (2004–)
Office: Instituto Geologico Minero y Metalurgico, Peru (INGEMMET)
Proposers: Romulo Mucho and Peter Bobrowsky
Contact:
rmucho@ingemmet.gob.pe, pbobrows@NRCan.gc.ca

C101-1-4: A proposal for an integrated geophysical study of the Cuzco Region (2004–)
Office: Istituto Nazaionale di Oceanografia e di Geofisica Sperimentale (OGS), Italy
Proposer: Daniel Nieto Yabar
Contact: dnieto@ogs.trieste.it

C101-1-5: Satellite monitoring of Machu Picchu (2005–)
Office: University of Firenze, Italy
Proposers: Paolo Canuti, Claudio Margottini, Fabio Rocca
Contact: canuti@geo.unifi.it

The report of Machu Picchu landslide investigation, risk analysis, and recommendation for disaster management must be comprehensively compiled integrating both results of satellite monitoring and ground monitoring as well as various field investigations, societal impact consideration in Peru and previous experiences and research on landslide risk investigation studies under the coordination of IPL C101-1. The integrated investigation results shall be reported formally through diplomatic channels to the Government of Peru, and also UNESCO as shown in Fig. 2.1.

2.2 Landslides in Machu Picchu

The Inca's world heritage is located in northwest of Cusco, Peru (Fig. 2.2). It was declared a World Heritage of Humanity in terms of both cultural and natural property by UNESCO in 1983. It is also an ecological sanctuary because of its ecological richness. The present style of citadel was

Fig. 2.2.
Location and view of Machu Picchu Inca citadel on the mountain ridge, Cusco, Peru

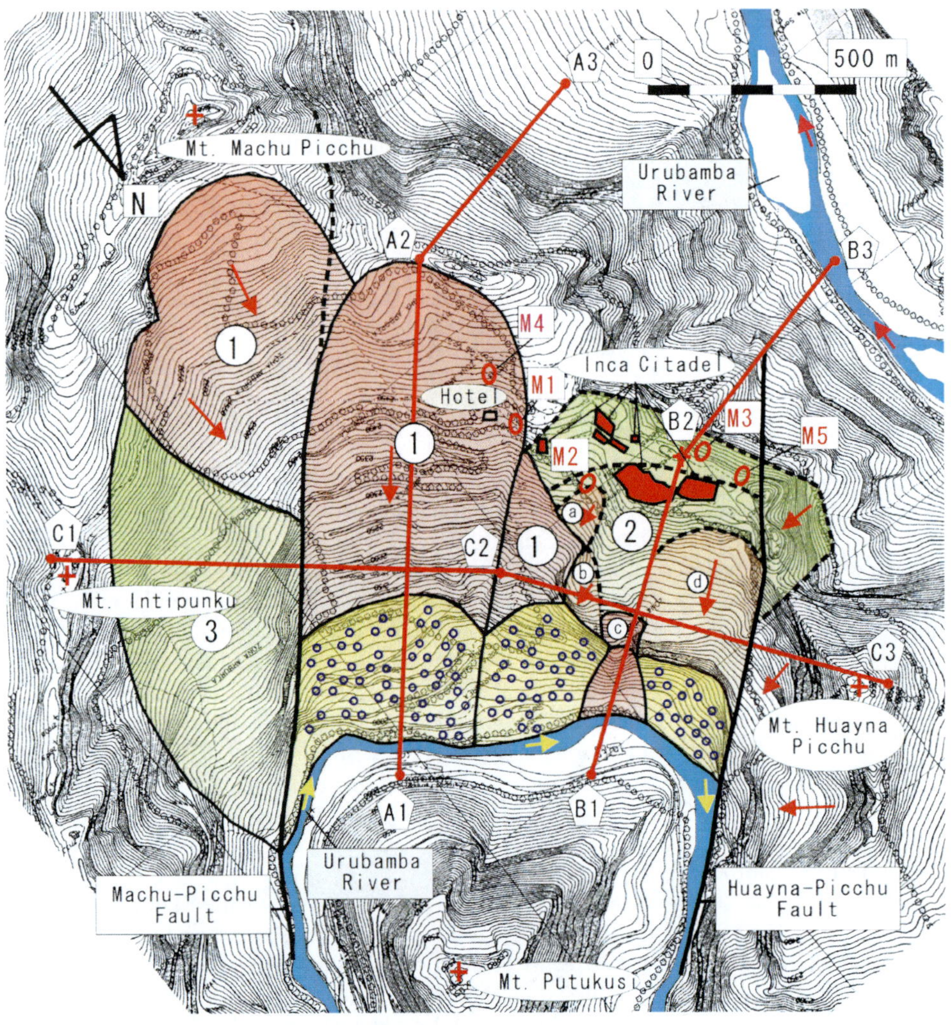

Fig. 2.3.
Working map of landslide blocks 1, 2 and 3 in front slope (Sassa et al. 2002). *A1–A3*, *B1–B3*, *C1–C3:* Lines for sections for block 1, block 2 and cross sections; ①, ②, ③: blocks of landslides or potential landslides; ①: expanding blocks of block 1 to block 2 and 3; ⓐ, ⓑ, ⓒ, ⓓ: sub-blocks of block 2 which seem to move up towards the citadel area from the border (ⓐ: a retrogressive landslide; ⓑ: possible rock topple; ⓒ: a recent landslide; ⓓ: shallow landslides which have not yet slid so much); [symbol] landslide debris at the toe of slope; *M1–M5:* locations of extensometers

probably built by the Incas in the fifteenth century. It remained untouched after the collapse of the Inca Empire in 1540 through the colonial period because of its isolated location on the top of a steep mountain. Machu Picchu became known to the world after its "scientific discovery" by Prof. Hiram Bingham in July 1911. As time passed, it seemed that Machu Picchu Citadel possibly has been affected by landslides. You may see a beautiful citadel constructed on the top of rocky mountain. At the same time, it is obvious that this part of mountain ridge is different from other part of the steep mountain ridge seen behind the citadel. The part called as *Plaza* in which the citadel was constructed has a flat area or an even concave area sandwiched by two ridges (Intiwatana in the left side peak and the residential building area in the right side peak). By the eye of landslide researchers, the Plaza area seems to be re-profiled by filling earth or debris to the cracks between two ridges along the dotted line. As shown in the previous paper (Sassa et al. 2001a), the dot line passed through a zone of broken structure in the extension of right bottom of Fig. 2.2 can be a potential head scarp of landslide. This hypothesis is not yet proved by monitoring, drilling, geophysical exploration or direct excavation or other reliable methods. However, any phenomenon to deny this hypothesis is not yet found. Therefore, what is the cause of Plaza geomorphology, which is apparently different from other parts of mountain ridge in this Machu Picchu area, should be investigated. These investigations and monitoring is on-going under the IPL C101-1 "Landslide Investigation in Machu Picchu" at present.

Sassa et al. presented a landslide distribution map based on chiefly air photo by Geographical Institute, Peru, 1963 and field investigation from a chartered helicopter and walking (Sassa et al. 2001a). The ridge connecting Mt. Machu Picchu and Mt. Huayna Picchu separates the slope as a gentle slope where Hiram Bingham Accessing Road is constructed, and a steep slope in another side. We have called the gentle slope with the accessing road as *Front Slope* including blocks 1, 2 and 3, and the steep slope as *Back Slope*.

The authors investigated Machu Picchu area, especially the Front Slope during 24–28 October 2001, and examined the site for a series of long span extensometers including a span crossing Plaza in the citadel. Based on this investigation, and 1 : 5 000 map we obtained from UGM (Machu Picchu Joint Managing Unit), we made a new landslide distribution map focusing only for the Front Slope as shown in Fig. 2.3. Sites of most information which are referred later in this paper were included in this map, as well.

The frontal view of the Front Slope taken from the tope of the opposite mountain "Mt. Putukusi" was presented in Fig. 2.4 with the location of two sections and the interpretation of landslides. Figure 2.5 presents *(a)* a side view of block 2, *(b)* view of the step (shear band) from the back side, and *(c)* its zoomed up view of the shear band. Comparing photos in Fig. 2.4 and Fig. 2.5 to the interpretation map Fig. 2.3, we may obtain the following overview of the slope from the landslide research aspect.

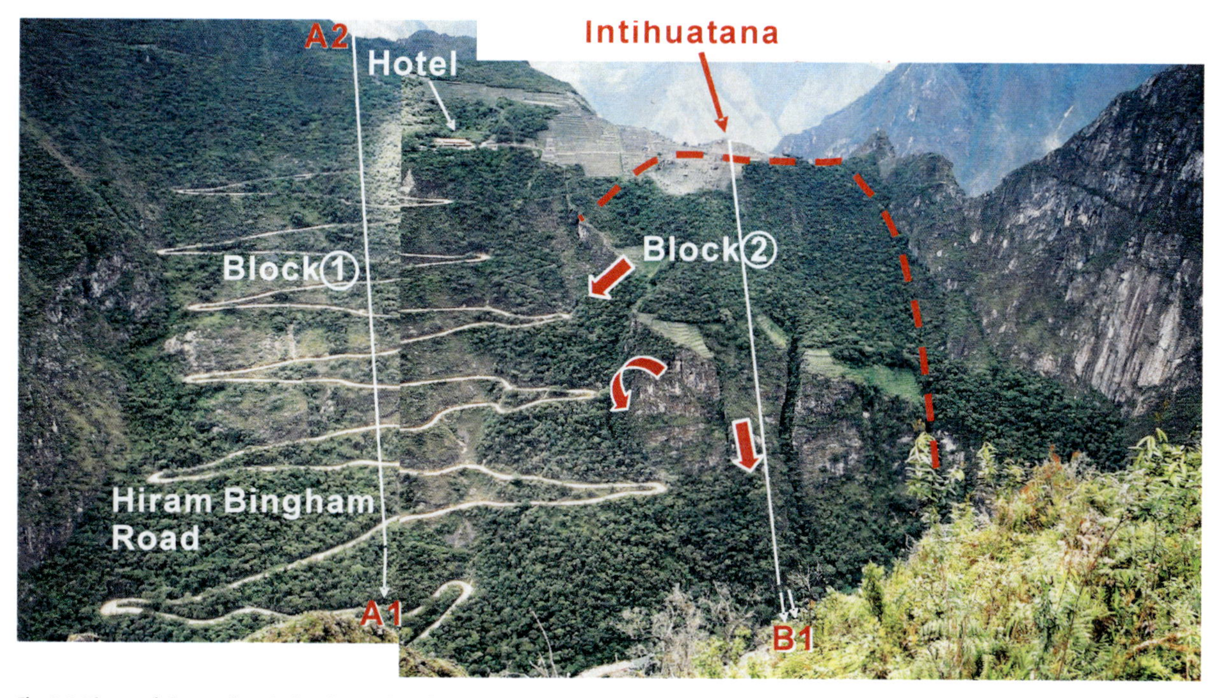

Fig. 2.4. Photo of the Machu Picchu slope taken from Mt. Putsukusi. The smooth ground surface of block 2 would be the sliding surface of past landslides. Block 1 is covered by landslide debris which is currently slowly moving

Fig. 2.5. a Photo of the Machu Picchu Citadel from a chartered helicopter taken by K. Sassa 2000; **b** a photo of point 'A' from the back side and **c** its zoomed up photo of point 'A'. A *dotted red line* indicates a possible head scarp of a precursory stage of landslide. A *dotted yellow line* indicates a potential sliding surface of a precursory stage of landslide. *Arrows 'A'* in these photos (**a**, **b** and **c**) point out the same place from different angles

1. It is obvious that block 1 is currently an active landslide, and the ground level is one step (which is almost 100 m at the position of section C as presented in Fig. 2.9) lower than the block 2 and block 3.
2. Block 2 seems to be a slope of intact hard rock in general. It is NOT an active landslide. However, the slope is being invaded by retrogressive landslides or rock topples from the Urubamba River side and the block 1 side as shown by red arrows in Fig. 2.4.
3. The ground surface of block 2 is very smooth and straight. Those straight surfaces are usually formed along shear bands (shear failure surfaces by tectonic stresses). In the case of shear bands formed by tectonic stress, a set of shear bands are often formed parallel to each other.
4. Air photo taken by Sassa in 2000 suggests that the existence of a possible shear band (yellow line in Fig. 2.5a) in parallel to the present smooth ground surface of block 2. The back side views (Fig. 2.5b and 2.5c) present that the yellow step was a shear band. Another shear band is also visible in the lower height. The slope is composed of hard granitic rock, which were sheared at some shear bands.
5. The shear band indicated by "A" is likely the extension of sliding surface of block 1. In this case, the lower end of this shear band exists within debris deposits near the Urubamba River (probably at a slightly higher than the present level of river bed). It is still a hypothesis which is not yet confirmed by drillings.

2.3 Possible Landslide Process in Machu Picchu

To get the general image of this, an air photo taken from a high altitude will be the best solution. A stereo pair photos in Fig. 2.6 presents the location of Machu Picchu Citadel (a circled part) and its surroundings. The Urubamba River encircled the Machu Picchu area along the strong fault system. You may find the front slope (block no. 1, 2 and 3 in Fig. 2.3) of Machu Picchu seems to have been excavated by previous mass movements. The mountain ridge of the part of citadel seems to have been subjected to landslides, and partly removed to both sides of the Urubamba River. Mt. Putsukusi is not a part of landslide block. It is a stable rock mass as the whole, though surface rock slides are observed. No great flow-mountain as a result of a big landslide is found along the Urubamba River. Therefore, it will be supposed that landslides having excavated the front slope were probably not a single gigantic landslide, but a series of retrogressive landslides so as to be seen in block 1 at present and in the border of block 2.

To assess the landslide risk, the process of landslide development in this slope should be correctly understood by the field investigation and interpretation. But the in-

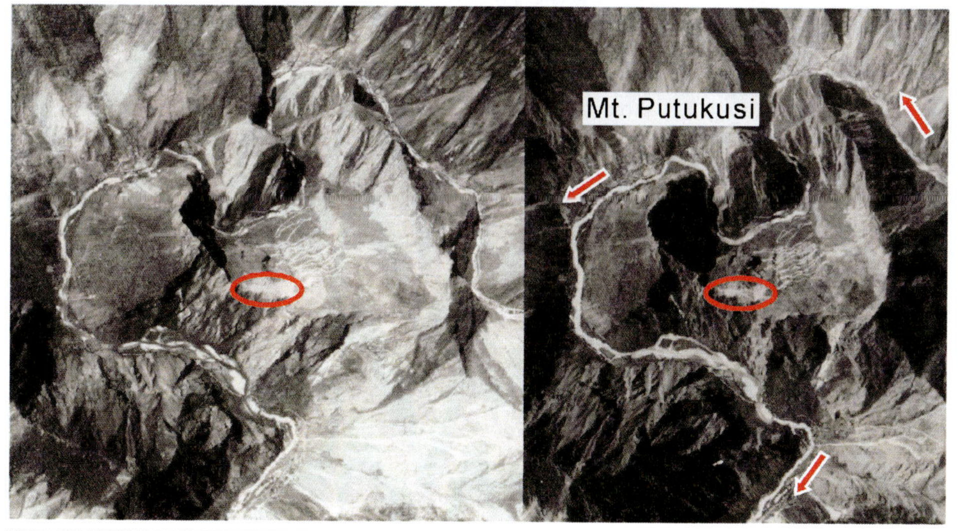

Fig. 2.6.
Airphoto of Machu Picchu (Geographical Institute, Peru)

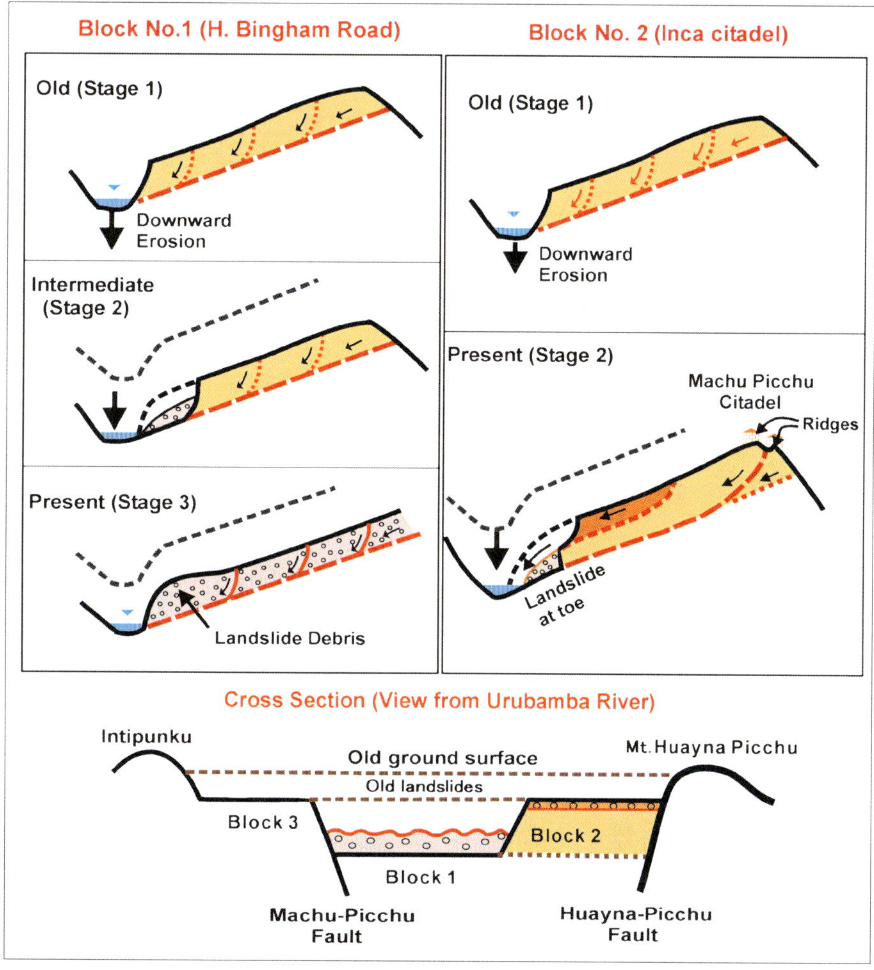

Fig. 2.7.
Hypothesis of landslide process in Machu Picchu

vestigation is not yet enough to conclude the process. However, the draft interpretation was presented in the previous symposia (Sassa et al. 2001a, 2002). Figure 2.7 shows the interpretation presented in 2002. The stages of landslide evolution were illustrated for block 1 and block 2 in a schematic form. In the old time, probably retrogressive landslides occurred in the front slope of Machu Picchu. After the whole landslide debris moved out to the

Urubamba River, the downward erosion proceeded further. The level of Urubamba River was shifted around one hundred meters downward to the present level. Three slopes of block 1, 2, 3 showed the difference in the landslide evolution speed. The slope of block 1 had been most heavily subjected to toe erosion of the Urubamba River as easily understood by the curved path of the river and flow direction in Fig. 2.3. Firstly the slope started to slide as illustrated in the intermediate period (stage 2) of block 1. The initial landslide retrogressively expanded to the upper slope and toward the side slopes, then, the present situation of stage 3, where active landslide debris covers the slope, was formed. You may find in Fig. 2.3 that the Urubamba River was pushed forward by the landslide debris provided by block 1. So this landslide debris has probably worked for the protection of toe erosion of block 2. Because of this protection, the evolution of block 2 was much delayed and it is still in the stage 2. The most delayed block in slope evolution is block 3. After previous landslide debris moved out, no major landslides occurred because there is almost no river erosion as imagined from Fig. 2.3. The landslide evolution in the cross section is illustrated in the bottom of Fig. 2.7. Only block 1 was subjected to major landslides at the present level of Urubamba River and located in the lower elevation which continues to the present level of the river bed. Block 2 is now following the process of block 1. Figure 2.8a shows the longitudinal section along A1–A2 in block 1. The depth of landslides is not known, but the whole slope from the mountain ridge to the river is affected by active landslides and covered by landslide debris. Figure 2.8b presents the longitudinal section along B1–B2 in block 2. The section B1–B2 shows the active landslide at the toe (c in Fig. 2.3), and others are still at the precursor stage. The potential sliding surface will be a shear band found in Fig. 2.5. If we consider that the po-

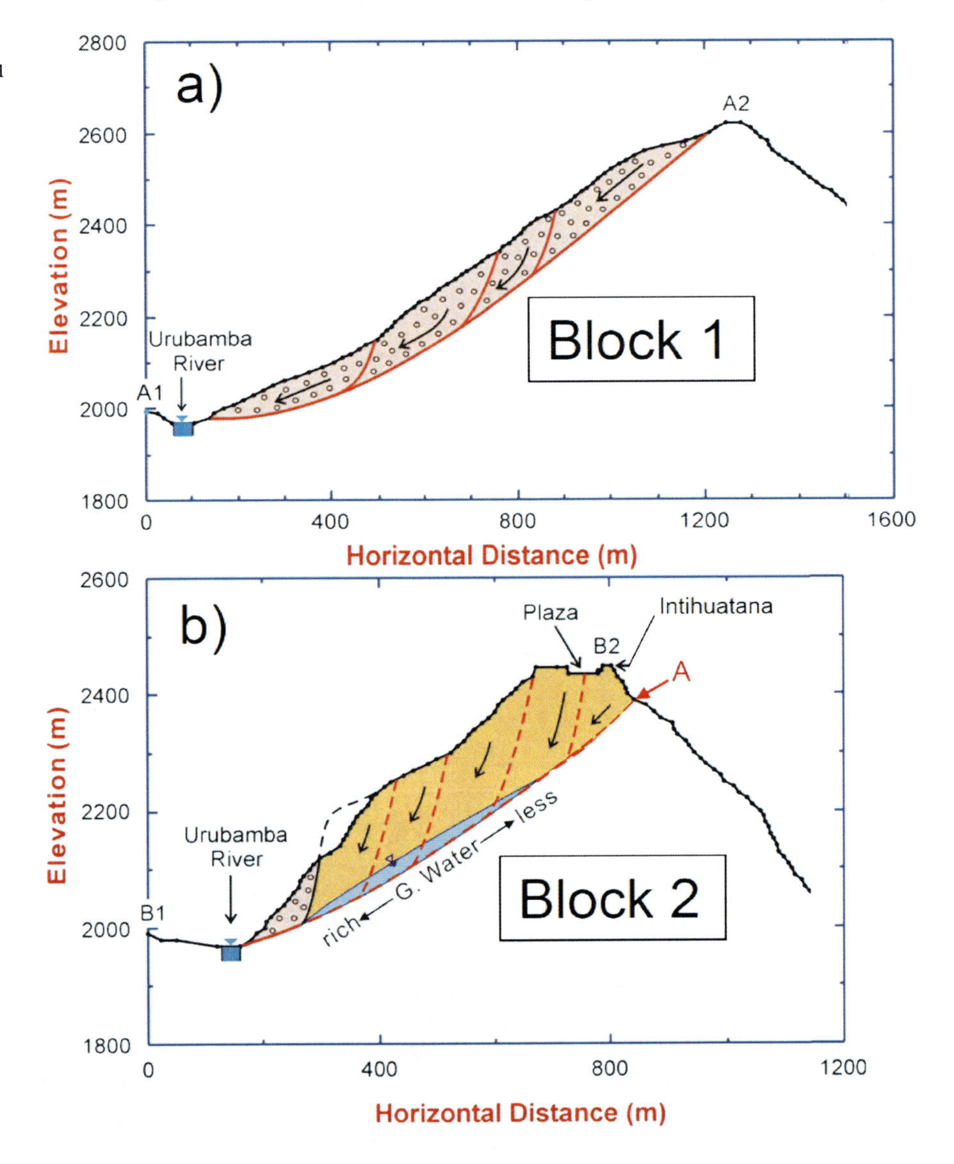

Fig. 2.8. Longitudinal sections of block 1 (*A1–A2*) and block 2 (*B1–B2*)

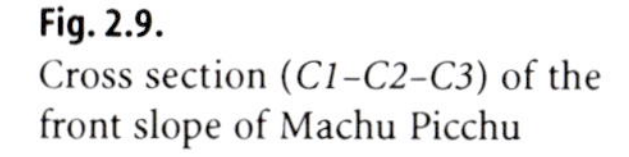

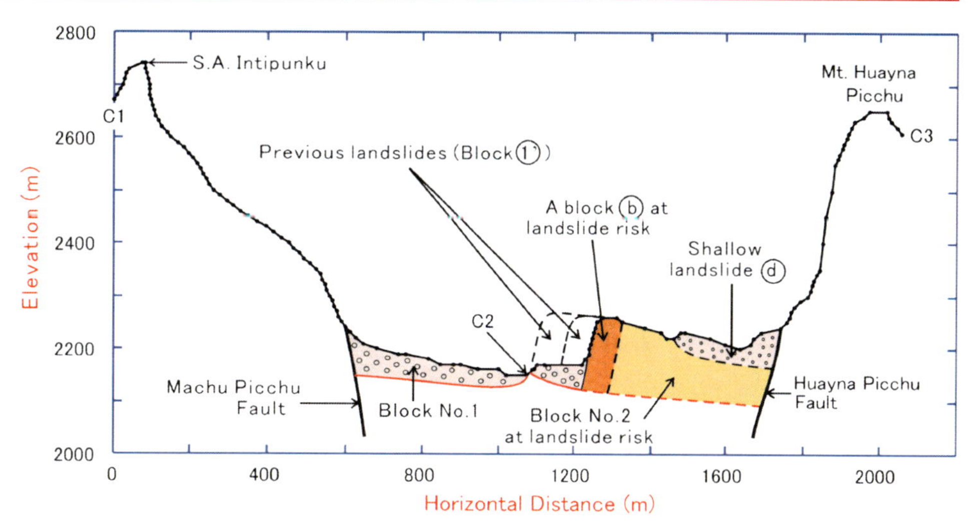

Fig. 2.9. Cross section (*C1–C2–C3*) of the front slope of Machu Picchu

tential sliding surface is located in the level as shown in Fig. 2.8b, the depth of landslide is around 100–150 m.

Figure 2.9 presents the cross section along C1–C2–C3. The potential sliding surface of block 2 may be the extension of the present sliding surface of block 1. In this figure, the sliding surfaces of real landslides are drawn by a real line though the depth is not yet confirmed. The sliding surfaces of potential or precursor stage of landslides are drawn by a dotted line.

There is a *question* for the reason why the sliding surface goes up to the Plaza in block 2 (Fig. 2.2, 2.3, 2.5, and 2.8), which may split Plaza of Inca Citadel. The reason is interpreted below from the data obtained up to now.

The ground water will increase in the lower part of the slope collecting ground water from the upper slope because the shear band is probably having a low permeability and dipping parallel with the front slope. Ground water should increase in the lower slope because of rain fall infiltration is accumulated. Therefore, probably the ground water level and pore water is greater in the lower part of block 2, and almost no ground water near the top of slope. In this case, the top of block 2 (area of Plaza and Intiwatana) is rather stable, so tensile stress should act between the lower instable part and the upper stable part. Accordingly tension cracks may be formed. The concavity in the Plaza likely corresponds to the tension crack, though the concavity was filled possibly by Inca's people for flatter ground suitable for their living, which is not clearly visible on the ground surface at present.

2.4 Monitoring of Extensometers in Machu Picchu

To evaluate landslide risk, monitoring of the ground deformation, the geological drilling, monitoring of shear displacement and ground water level and/or pore water pressure inside drill holes are vitally important and necessary. Without such investigation, neither landslide risk, nor the safety of slope can be reliably evaluated in the convincing way. Without such reliable landslide risk evaluation, neither effective landslide remedial works can be planned, nor high costs of remedial works can be approved.

As the first step of quantitative investigation, two types of simple extensometers were installed in the Machu Picchu slope in 2000. One is a handmade manual reading extensometers, using a pulley and a super invar wire (a special kind of metal with least influence of temperature), movement of the distance between two points is mechanically enlarged by 5 times and indicated on the dial with a pointer. Another type is theoretically the same, movement was also mechanically enlarged by 5 times, and it is recorded on a recording paper continually. The recording drum is rotated by landslide movement, while the recording pen shifts in a steady speed using a dry battery driven mechanical clock.

Extensometers have merits, which are not affected by moisture or atmospheric pressure in the air and cause less trouble and very reliable because of very simple mechanical recording system. They can be easily handled and understood by every body without special knowledge, which are different from sophisticated monitoring systems such as EDM (electronic distance meter) and GPS (global positioning system) and others.

A series of extensometer installation from a part of stable ground crossing over the landslide mass until another stable ground is most desirable to detect landslide blocks. The system is used in Lishan, China (Sassa et al. 2001b), and Zentoku, Japan (Hong et al. 2005). However, the easiest way to check the existence of movement is to set an extensometer crossing a head scarp to monitor the landslide movement. Twelve extensometers were installed in five sites (M1–M5) as shown in Fig. 2.3. Figure 2.10 (*top*) presents the monitoring results of S1 in site M1 near the hotel Sanctuary Lodge in block 1. The extensometer S1 showed gradual extension over a long term corresponding to the rainfall during rainy season. Such slow and gradual movement suggests that it detects a movement

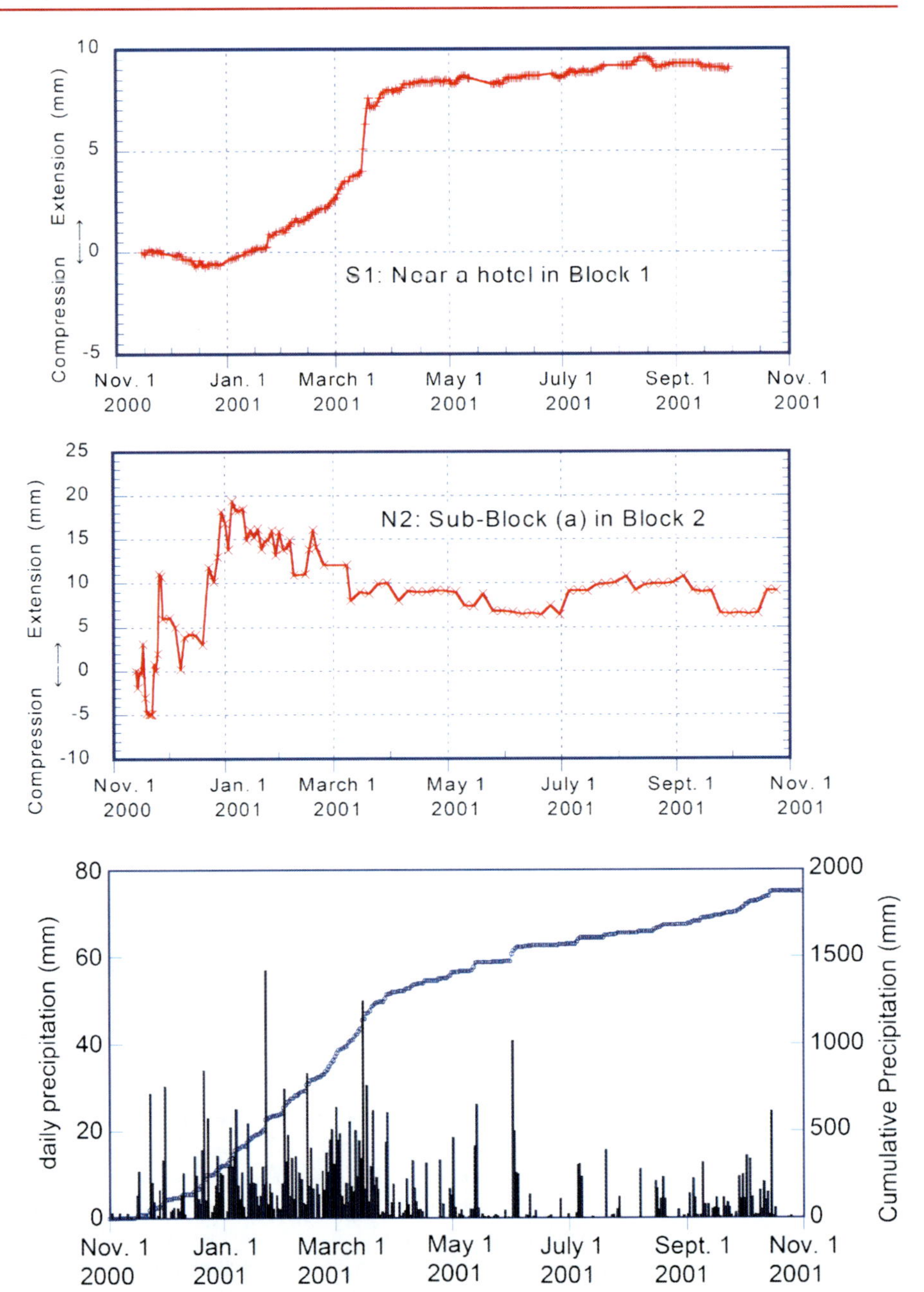

Fig. 2.10. Records of extensometer (*S1*) in the monitoring site (*M1*) near Hotel Sanctuary Lodge in block 1 and extensometer (*N2*) in the monitoring site (*M2*) in block 2

of relatively large and deep seated landslide. Figure 2.10 (*middle*) presents the monitoring of the manual reading extensometer N2 installed in site M2 in the border of sub-block (a) of block 2. The movement is rather great in the order of 10 mm. The response on the rainfall event seems to be a bit quicker than that recorded by the extensometer S1. Probably it detected a relatively shallow landslide. Those extensometers were not installed in a series along a line crossing the landslide block as installed in Lishan, China (Sassa et al. 2001b). It is not possible to estimate the landslide size from single extensometer monitoring, however, it is sure that a landslide movement (probably corresponding to large-scale landslides) responding to rainfall clearly exists in the Machu Picchu slope. Namely rainfall may trigger landslides in this slope, on the contrary, the reduction of rainfall infiltrating to the shear surface by surface drainage or underground drainage may prevent landslide movement.

2.5 Further Landslide Investigation

The results of field investigation and monitoring suggest the possibility of landslides, which possibly affect the Inca's citadel. The exact understanding of slope conditions and landslide risk evaluation need further investigation.

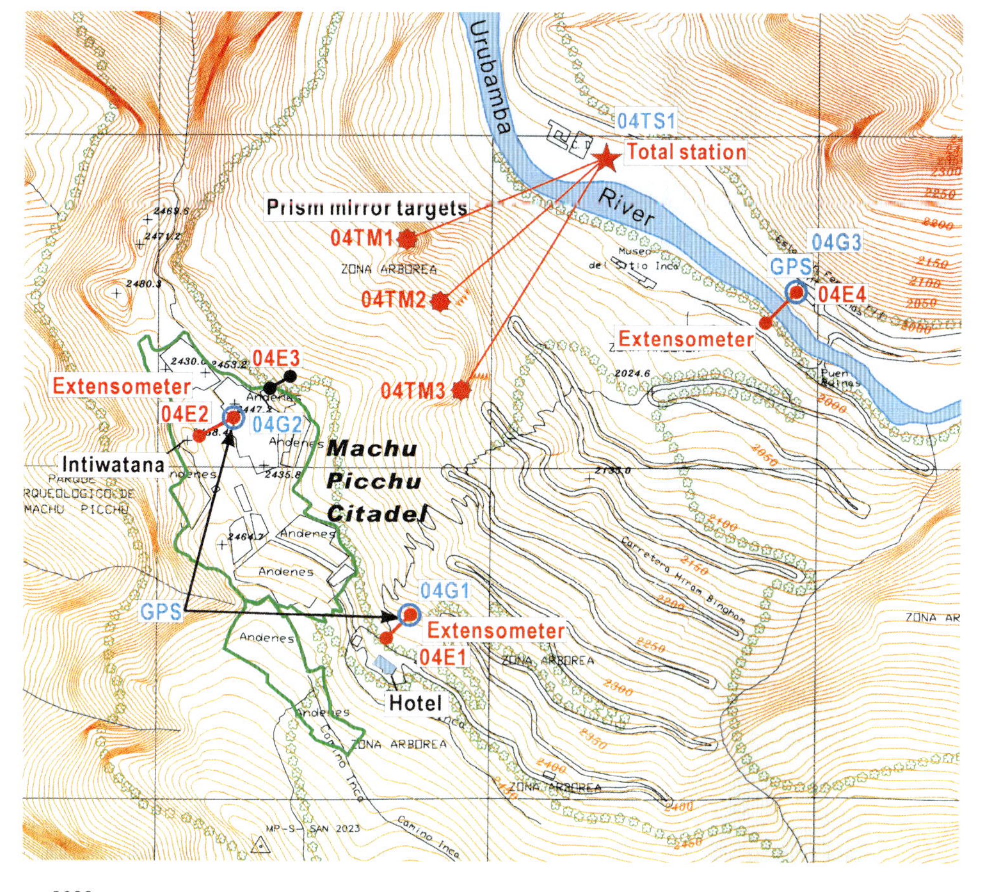

Fig. 2.11.
Monitoring instruments in Machu Picchu installed by Japanese Team in 2004. Long-span extensometers (*04E1, 04E2, 04E3, 04E4*), static GPS receivers (*04G1, 04G2, 04G3*), total station (*04TS1*) and prism mirror targets (*04TM1, 04TM2, 04TM3*) are installed in and around Machu Picchu monument

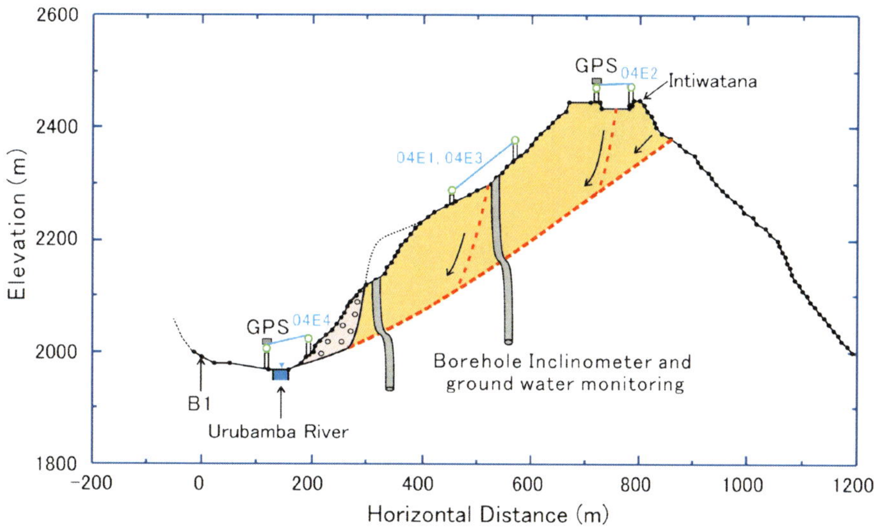

Fig. 2.12.
Illustration of monitoring system for the precise understanding of potential landslides

The Japanese team obtained permission to install extensometers, GPS and the total station in Machu Picchu including Plaza area from the Government of Peru in 2004. Then, we installed 4 sets of long-span extensometers, 3 sets of static GPS receivers, and one total station with 3 prism mirror targets in September, November of 2004 and in March of 2005. Figure 2.11 is the plan map of the location of each instrument.

Figure 2.12 illustrates the location of extensometers and GPS as regards the landslide block. Extensometer 04E2 is installed crossing the Plaza from a big rock below Intiwatana to a stone wall of the Inca's citadel as shown in Fig. 2.13. Super-invar wire position of 04E2 is shown as a red line in Fig. 2.13a and black line in Fig. 2.13b, although the actual wire is so fine and not so visible to tourists.

Fig. 2.13. Long-span extensometer installed to pass over the central part of Machu Picchu "Plaza"

Figures 2.13c,d are the photos of the extensometers of both pen-recording and electronic-recording type (04E2) which is located inside a window. The super-invar wire is hooked on a pulley which is attached to a non-Inca stone plate fixed by clay on the top of the Inca wall, because it is prohibited to drill or use cement on Inca materials inside the citadel. The extensometer records the extension/compression of the span crossing the Plaza on a paper recording sheet by an ink pen as well as store inside memory card in digital format at every one hour.

Extensometer 04E3 is located beneath the citadel in block 2, and the extensometer 04E1 is located near the extensometer S1 in the block 1, but its span is longer than S1. Both extensometers are intended to detect the extension of a potential head scarp of landslides in the middle of the slope in block 1 and block 2. Extensometer 04E4 is located at the toe of block 1 crossing the Urubamba River to monitor the compressive movement of the block. Figure 2.12 presents the section of block 2, while 04E1 and 04E4 are installed in block 1. So the positions of those two extensometers are not real, but projected to block 2.

On top of the pole for extensometers, GPS receivers 04G1, 04G2, 04G3 are installed. Static GPS network has now 3 stations. Reference point is settled at 04G3 in the Peru Rail Machu Picchu station. Relative movement of 04G1 in block 1 and 04G2 in block 2 can be monitored.

Fig. 2.14. Total station and location of prism mirror targets placed on the head of toe cliff of the block no. 2

Total station is located in the point of 04TS1 in Fig. 2.11 in the garden of the dormitory and the office of the Machu Picchu Sanctuary Lodge Hotel at the Urubamba River side. Figure 2.14 presents the view from the monitoring point to the rock cliff. Three prism mirror targets of 04TM1, 04TM2, and 04TM3 are located on the head of the toe cliff of the block 2. Only the central target 04TM2 showed a small down slope displacement of 1 mm along the line of sight of about 350 m for about 4 months from 27 November 2004 to 23 March 2005. Other two targets showed no displacement more than 1 mm, although this observation period includes a rainy season.

Drilling and the borehole inclinometer or alternative methods to detect the sliding surface are very necessary to know the depth of landslide. Ground water monitoring is very essential to evaluate the effect of ground water on this landslide. The illustration of drillings in Fig. 2.12 is a draft plan for that. The movement should be minimal even it can be detected, because the slope is estimated to be at the precursor stage of landslide.

Identification of the sliding surface and ground water (pore water pressure) effect acting the sliding surface (shear band) enables the plan of effective ground water drainage. The prevention of toe landslides as well as the reduction of river erosion are also effective in order to prevent further retrogressive landslides.

2.6 Conclusions

The results of investigation by the Japanese team conducted since March 2000 and the background of IPL C101-1 Landslide Investigation in Machu Picchu were introduced. Major findings of the investigation by the Japanese group are:

1. Retrogressive landslides have probably proceeded in the past and are still on-going in the Machu Picchu slope. The progress of landslide evolution stage is different in block 1, 2, 3. Present stages of each block are informative for the assessment for the future process of landslide activities.
2. A shear band, possibly a potential sliding surface of block 2 was found. It is probably gently dipping to the front slope. Ground water is likely to flow downward possibly along the shear band. The ground water must be poor in the top of block 2 and rich in the lower slope. It may be the cause of possible head scarp which may be splitting Plaza.
3. Ground water along the shear band in block 2 and also retrogressive landslides around block 2 (namely in the border to block 1, in the toe cliff near the Urubamba River) will be major causal factors of block 2. Reduction of those causal factors must be effective landslide remedial measures.

4. Four sets of long-span extensometers, three sets of GPS receivers, a total station with 3 prism mirror targets were installed on the slope in 2004 to start monitoring of the displacement.
5. Reliable understanding of the present state and process of landslides cannot be obtained without reliable monitoring of deformation on the ground surface, geological drillings and preferably geophysical exploration, monitoring of the shear displacement and ground water level and/or pore pressure inside the boreholes. Those investigations of deformation monitoring and geophysical exploration are being conducted by the Italian group, the Peruvian-Canadian Group and the Czech-Slovak group of IPL C101-1.Some of those results may be introduced in this volume.
6. We may conclude that there is a necessity of further investigation to evaluate landslide risk in Machu Picchu, and it will be possible to protect Machu Picchu from landslide risk by reducing landslide causes based on the reliable investigation results though it is not an easy task in all of the scientific, technological, social and financial aspects.

Acknowledgments

The Japanese investigation team was well received by the related agencies in Peru and obtained significant support and cooperation. We acknowledge the following persons and institutes for their cooperation to this investigation: Dr. Romulo Mucho, Vice Minister of Mine, Dr. Victor Benavides, and colleagues of INGEMMET (Peruvian Instituto Geologico Minero y Metalurgico) for their cooperation and coordination of many of Peruvian counterparts.

Prof. Mutsumi Ishitsuka, Director of the Ancon Observatory, Instituto Geofisico del Peru (IGP) has cooperated from the beginning of this Machu Picchu investigation in all aspects. Dr. Raul Carreño of Grudec Ayar, Cusco has cooperated through the whole investigation, and supported our activities. Mr. Cleto Quispe, and other engineers of Machu Picchu office of Instituto Nacional de Cultura (INC) and Instituto National de Recursos Naturales (INRENA) have cooperated with monitoring and field investigation. All of them are deeply appreciated for their kind cooperation. Finally all colleagues of Research Centre on Landslides of the Disaster Prevention Research Institute, Kyoto University are acknowledged for their cooperation to analyze and to present the results of field investigation and monitoring.

References

Bandarin F (editorial director) (2001) Rumbles at Machu Picchu. World Heritage Review, UNESCO, no. 29, 56 p

Brown JL (2001) Landslides may threaten Machu Picchu. Civil Engineering, ASCE, May 2001 issue, p 16

Carreño R, Bonnard C (1997) Rock slide at Machupicchu, Peru. Landslide News 10:15–17

Hadfield P (2001) Slip sliding away. New Scientists, Reed Business Information 169(2281):20

Hong Y, Hiura H, Shino H, Sassa K, Fukuoka H (2005) Quantitative assessment on the influence of heavy rainfall on the crystalline schist landslide by monitoring system-case study on Zentoku landslide, Japan. Landslides 2(1):31–41

Sassa K (1999) Landslides of the world. Kyoto University Press (ISBN4-87698-073-X C3051), pp 311–316

Sassa K, Fukuoka H, Shuzui H (2000) Field investigation of the slope instability at Inca's World Heritage in Machu Picchu, Peru. Landslide News 13:37–41

Sassa K, Fukuoka H, Kamai T, Shuzui H (2001a) Landslide risk at Inca's World Heritage in Machu Picchu, Peru. In: Proc. UNESCO/IGCP Symposium on Landslide Risk Mitigation and Protection of Cultural and Natural Heritage, Tokyo, pp 1–14

Sassa K, Fukuoka H, Wang FW, Furuya G, Wang GH (2001b) Pilot study of landslide hazard assessment in the Imperial Resort Palace (Lishan), Xi'an, China. In: Proc. UNESCO/IGCP Symp on Landslide Risk Mitigation and Protection of Cultural and Natural Heritage, Tokyo, pp 15–34

Sassa K, Fukuoka H, Shuzui H, Hoshino M (2002) Landslide risk evaluation in the Machu Picchu World Heritage, Cusco, Peru. In: Proc. UNESCO/Kyoto University Int Symp on Landslide Risk Mitigation and Protection of Cultural and Natural Heritage, Kyoto, pp 469–488

Sassa K (2005a) Precursory stage of landslides in the Inca World Heritage site at Machu Picchu, Peru. Independent brochure for UNITWIN Programme, Education Sector of UNESCO, 4 p

Sassa K (2005b) ICL history and activities. In: Sassa K, Fukuoka H, Wang F, Wang G (eds) Landslides – Risk analysis and sustainable disaster management. Springer-Verlag, Heidelberg, pp 3–21 (this book)

Wright KR, Wright RM (1997) Machu Picchu: its engineering infrastructure. 37th Annual Meeting of the Institute of Andean Studies, Berkeley, Calfornia. 24 p

Wright KR, Zengarra AV (2000) Machu Picchu – a civil engineering marvel. ASCE Press, ISBN 0-7844-0444-5

Chapter 3

Preliminary Remarks on Monitoring, Geomorphological Evolution and Slope Stability of Inca Citadel of Machu Picchu (C101-1)

P. Canuti · C. Margottini* · R. Mucho · N. Casagli · G. Delmonaco · A. Ferretti · G. Lollino · C. Puglisi · D. Tarchi

Abstract. The geology of Machu Picchu area is characterized by granitoid bodies that had been emplaced in the axial zones of the main rift system. Deformation of the granite, caused by cooling and tectonic phases, originated 4 main joint sets, regularly spaced (few decimeters to meters). Several slope instability phenomena have been identified and classified according to mechanism, material involved and state of activity. They are mainly related to rock falls, debris flows, rock slides and debris slides. Origin of phenomena is kinematically controlled by the structural setting and relationship with slope face (rock falls, rock slide and debris slides); the accumulated materials are the source for debris flows. Geomorphological evidences of deeper deformations are currently under investigation.

A low environmental impact monitoring system has been established on the area with the purpose to minimize equipments usage and, in the mean time, to collect reliable data on surface deformations. The monitoring network is based on a GPS, multi-temporal laser scanner survey, Ground based Radar interferometry (GB-SAR) and Satellite Interferometric Synthetic Aperture Radar (InSAR). The preliminary results are partially confirming the field evidences of slope deformation but, in the mean time, they require a longer period of observations since the sliding processes are relatively slow.

Keywords. Geomorphology, slope instability, GPS, radar, monitoring, Machu Picchu

3.1 Geological Setting

The high Eastern Cordillera of Peru formed as a result of inversion of a Late Permian-Triassic rift system (e.g. Sempéré et al. 2002). The geology of Machu Picchu area reflects the same pattern.

The area (Fig. 3.1) is characterized by granitoid bodies that had been emplaced in the axial zones of the main rift system that are now exposed at the highest altitudes, together with country rocks (Precambrian and Lower Paleozoic metamorphics) originally constituting the rift 'roots'. The Machu Picchu batholith is one of these Permo-Triassic granitoid bodies (a biotite Rb-Sr age of 246 ±10 My by Priem et al. is reported for this intrusion in Lancelot et al. 1978).

The bedrock of the Inca Citadel of Machu Picchu is mainly composed by granite and subordinately granodiorite. This is mainly located in the lower part of the slopes (magmatic layering at the top). Locally, dikes of serpentine and peridotite are outcropping in two main

Fig. 3.1. General view of Machu Picchu area

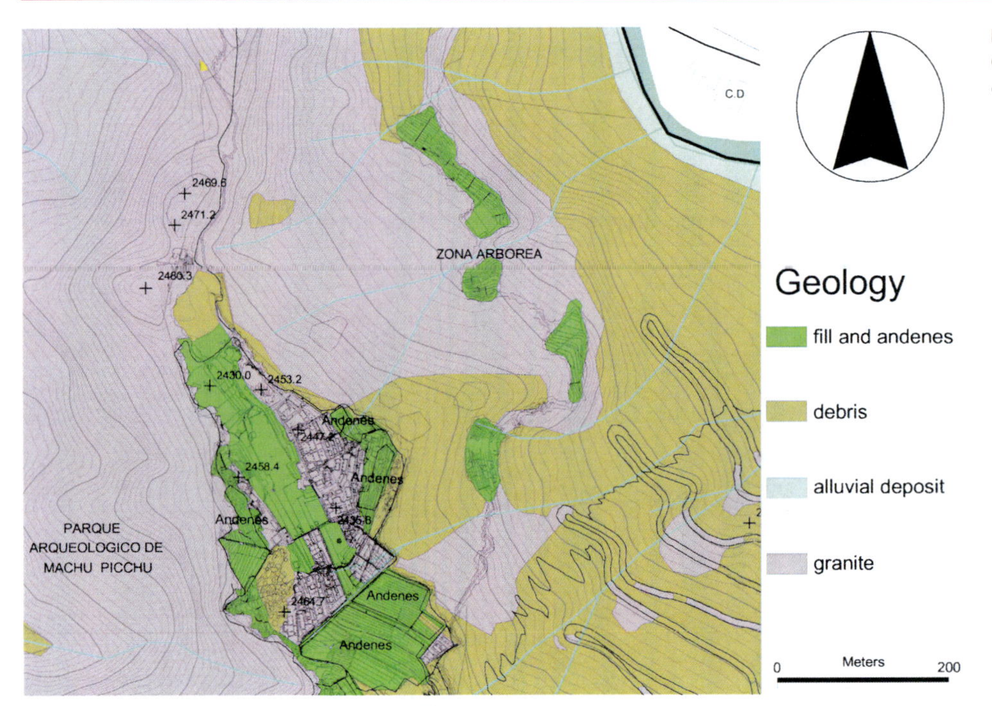

Fig. 3.2. Geological map of the Inca citadel of Machu Picchu

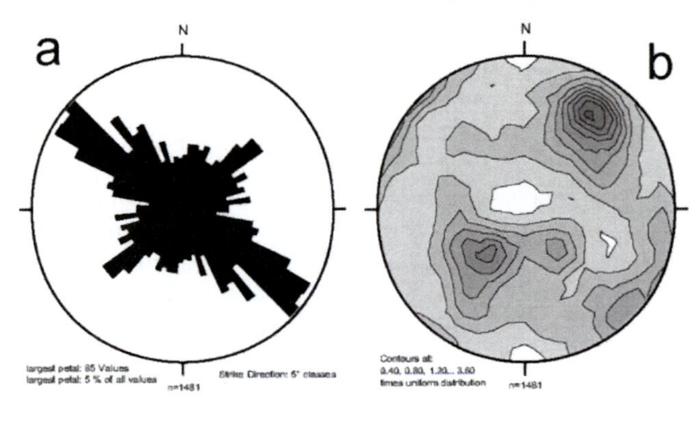

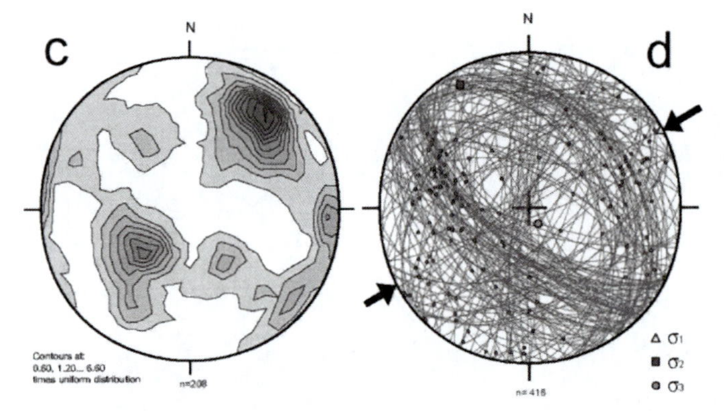

Fig. 3.3. Orientation data. **a** Rose diagram showing strike of planar discontinuities from the batholith. **b** Contour diagram of poles to planar discontinuities (lower hemisphere, equal-area projection). **c** Contour diagram of poles to shear planes (lower hemisphere, equal-area projection). **d** Fault slip data, showing calculated principal stress axes (grid search method) (from Mazzoli et al. 2005)

levels; the former is located along the Inca trail, near Cerro Machu Picchu (vertically dipping), the latter is located along the path toward "Templo de la Luna" in Huayna Picchu relief. Superficially, the granite is jointed in blocks with variable dimensions, promoted by local structural setting. The dimension of single blocks is variable from 10^{-1} to about 3×10 m^3. Soil cover, widely outcropping in the area, is mainly composed by individual blocks and subordinately by coarse materials originated by chemical and physical weathering of minerals. Portions of the slopes exhibit debris accumulation as result of landslide activity. Grain size distributions of landslide accumulation are closely related to movement types and evolution. Talus and talus cones are composed by fine and coarse sediments, depending on local relief energy. Alluvial deposits outcrop along the Urubamba River and its tributaries. They are composed by etherometric and polygenic sediments, that may be in lateral contact with the talus deposits. Anthropogenic fill and terraces, on the top of the Citadel, reflect the work of Inca civilization in the area (Fig. 3.2).

3.2 Structural Setting

Deformation of the granite is highly localized into differently oriented sets of regularly spaced (few decimeters to meters) shear zones. All of these shear zones show well-defined, sharp, fault-like discontinuities, characterized by slickenside surfaces and shear fibres. These indicate consistent reverse to oblique-slip (transpressional), to strike-slip kinematics, depending on fault set attitude. Most of these shear planes show limited measurable displacements (a few centimeters to decimeters being common) and essentially undeformed wall rock (Mazzoli et al. 2005). However, in several instances ductile deformation of the granite has also occurred along the walls of fault-like discontinuities. These deformation zones range from a few centimeters- to decimeters-thick mylonite horizons, showing sharp contacts with surrounding undeformed granite, to more continuous shear zones displaying sigmoidally-shaped S-foliations (progressively decreasing in intensity in wall rock granite) and S-C or composite S-C-shear band fabrics. Mylonite microstructures are characterized by extensive dynamic recrystallization of quartz and grain-size reduction. Chlorite and white mica-dominated mineral assemblages point to lower greenschist-facies conditions for the deformation. Inversion of fault-slip data indicates that the latter occurred as a result of mean N60° E oriented shortening, well compatible with regional data from the nearby Cusco area (Carlotto 1998). Structural analysis clearly suggests that the analyzed shear zones form part of a larger population of regularly spaced surfaces that most probably originated as early (cooling) joints in the igneous rock. Reactivation of differently oriented sets of precursor joints allowed the granite to deform effectively by relatively small displacements occurring on a very large number of shear zones, the strain being therefore localized at the m-scale, but distributed at the km-scale (Mazzoli et al. 2005).

The structural settings of terrains is finally related to the main following dip orientations/dipping:

a 30°/30° in the Machu Picchu Hill
b 30°/60°
c 225°/65°
d 130°/90°

Secondary systems (e.g. 130°/45°, 315°/30° and 310°45°) have been also surveyed in the area, with minor relevance as shown in Fig. 3.3.

Table 3.1. Qualitative and quantitative geomechanical parameters referred to the rock outcropping in the Citadel area

Scanlines	RMR	Q system	GSI	Sigma$_f$ (MPa)	C (kPa)	Phi (°)
101	73.5	9.8	71	36.5	367	41
102	57.8	14.4	57.2	30.4	289	34
105	73.5	8.8	68.9	36.6	367	41
107	68.9	26.5	65	34	345	38
108	65.6	10.2	61.9	30	328	38
109	42.5	0.5	42.5	16.8	212	26
110	49.2	3.5	44.8	31.4	246	30
111	79.6	12.5	74.6	40.7	398	45
113	78	11.2	73.3	27	390	54
116	74.5	8.7	69.5	29.1	372	45
118	73.5	8.6	68.5	39.9	367	42

3.3 Geomechanical Setting

The geomechanical characteristics of the rock mass, mainly granite and granodiorite, have been obtained by implementing a specific geotechnical field investigation (i.e. scanlines for classifications such as RMR, GSI, Q-system, direct measurement of compressive uniaxial strength through Schmidthammer, tilt test) and analysis of technical reports. Finally, the description of materials and general conditions of the rock masses have been summarized through direct observation.

The following Table 3.1 reports briefly the qualitative and quantitative geomechanical parameters referred to the area of the Citadel. The implementation of tilt test analysis on granite rocks has provided values of $\phi_{(b)} + \phi_{(i)}$ for smoothed joint surfaces from 37 to 53. The following parameters in Table 3.2 refer to geotechnical analysis of granite and granodiorite rock mass located at km 122 of the railway track in the area of the Inca sanctuary.

3.4 Geomorphology and Slope Instability

During three joint missions in Machu Picchu in 2003 and 2004, the group performed a geomechanical and geomorphological survey of the entire area. Field observations

Table 3.2. Geotechnical analysis of granite and granodiorite rock mass located at the km 122 of the railway track in the area of the Inca sanctuary

	Density (kN m^{-3})	E_d (MPa)	K_n (MPa)	v	Sigma$_f$ (MPa)	Tensile strength (MPa)	Phi (°)	C (kPa)
Granite	23 – 25	80000 – 90000	20 – 100	0.15	50 – 250	7 – 13	33.5 – 40.5	220 – 355
Granodiorite	23 – 25	100000 – 150000	20 – 100	0.15	60 – 250	7 – 13	35 – 42	270 – 355
Superficial cover	25	–	–	–	–	–	32	270
Talus	18 – 25	–	–	–	–	–	34 – 38	0

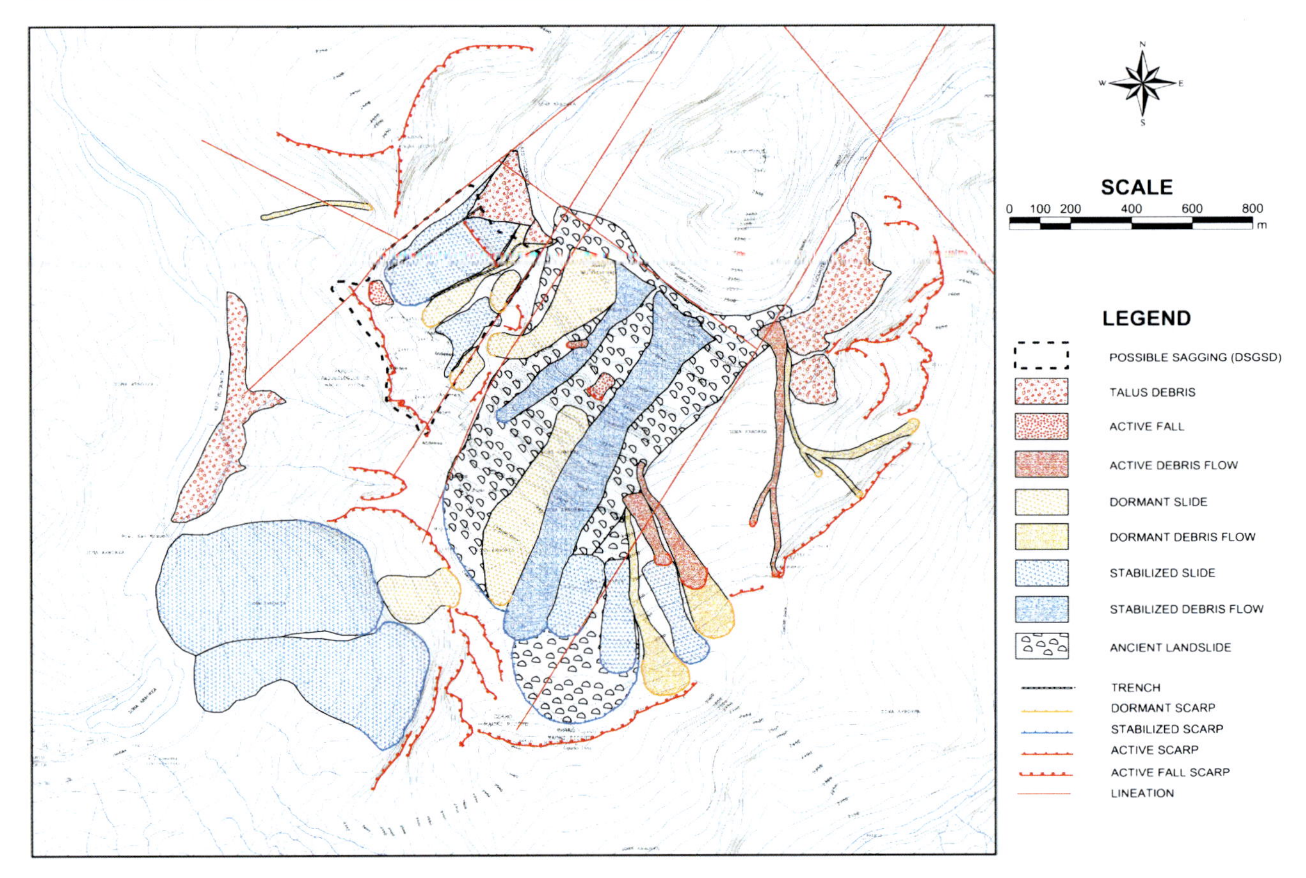

Fig. 3.4. Preliminary geomorphological map

were integrated with interpretation of stereoscopic aerial photos and of two optical very-high-resolution satellite images (Quickbird) dated 18 June 2002 and 18 May 2004, respectively.

The general morphological features of the area are mainly determined by the regional tectonic uplift and structural setting. As consequence, kinematical conditions for landslide type and evolution are closely depending on the above factors.

Several slope instability phenomena have been identified and classified according to mechanism, material involved and state of activity (Fig. 3.4). They are mainly related to the following: rock falls, debris flows, rock slides and debris slides. The area of the Citadel has been interpreted as affected by a deep mass movement (Sassa et al. 2001, 2002) that, if confirmed by the present day monitoring systems, could be referred to a deep-seated gravitational slope deformation (DSGSD), probably of the type of the compound bi-planar sagging (CB) described by Hutchinson (1988). A main trench with NW-SE trend, related to a graben-like structure, is located within the archaeological area and supports this hypothesis. Other trenches are elongated in the dip direction of the slope.

In the SW cliff the local morphological depends on the intersection between the systems 225°/65° and 130°/90° which bounds lateral evolution. This kinematical condition causes high angle rock slide which very often may evolve in rock falls. These are also conditioned by 30°/30° and 30°/60° systems which originate overhanging blocks. The SW slope exhibits some morphological terraces, regularly spaced, the origin of which it is still under investigation (i.e., fluvial erosion, sagging, joint, etc.).

The morphological evolution, in the NE flank below the Inca Citadel, is constrained prevalently by the 30°/30° and 30°/60° systems and marginally the 225°/65° one; the intersection of the first two systems with slope face it is kinematically compatible with the occurrence of planar rock slides; the intersection of the slope face with the 225°/65° system is kinematically compatible with rock falls. Rock slides and rock falls may produce blocks with dimension variable from 10^{-1} to 10^{2} m^3.

Debris produced by rock slides and rock falls, as well as from weathering processes, is periodically mobilized as debris slides and debris flow. The most recent landslide occurred in 1995 and affected part of the "carrettera" Bingham. Debris slides and debris flows are characterized by an undifferentiated structure generally of chaotic blocks immersed on coarse sand matrix. The grain size distribution is mainly depending on the distance from the source areas and slope angle.

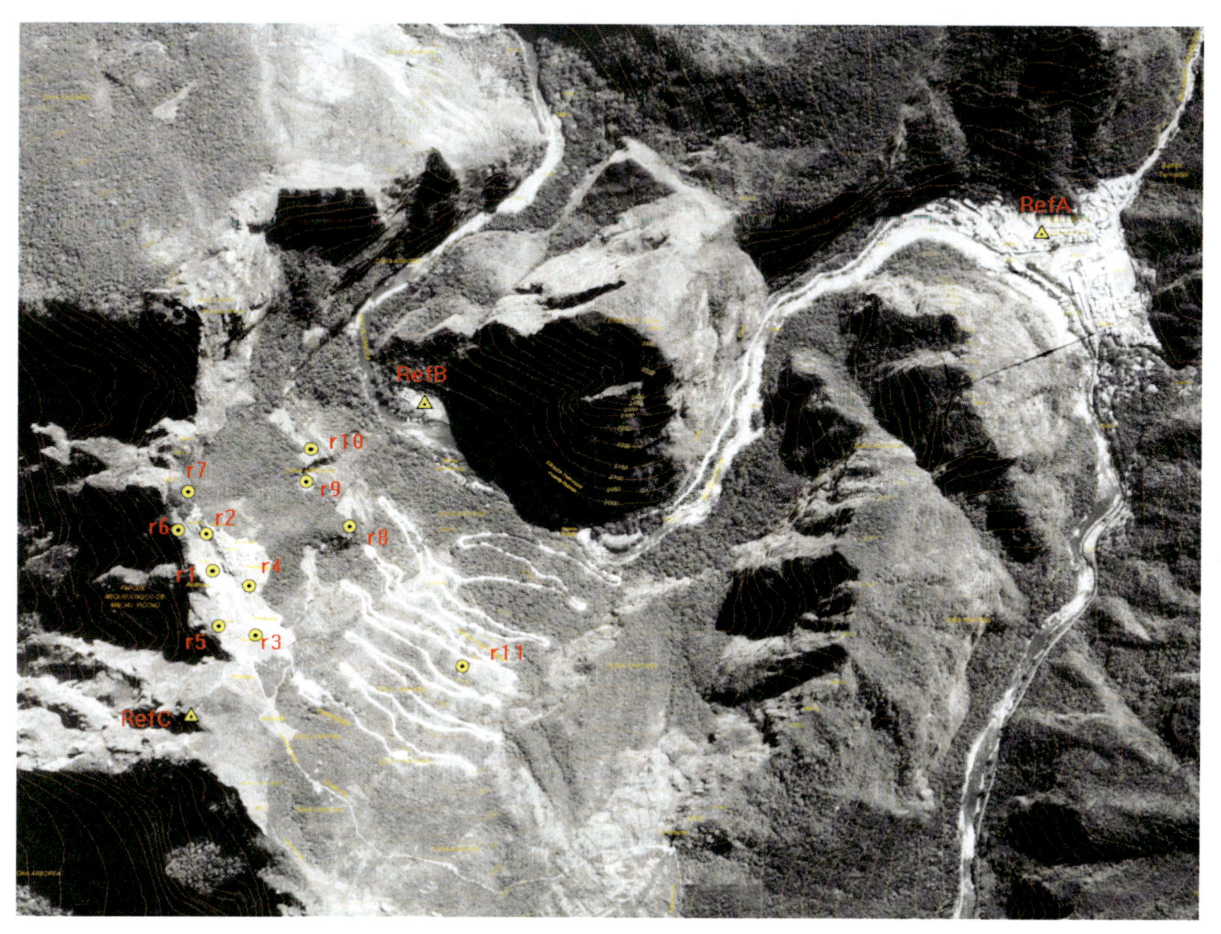

Fig. 3.5. The GPS network of Machu Picchu area

Finally, a worthwhile geomorphological element, on the NE side, is represented by a large debris accumulation, located just below the Citadel, presently being eroded by surrounding dormant slides. The accumulation is probably the result of an ancient landslide now stabilized, still not clear in its original feature. Anyway, the mass movements occurred certainly before the Inca settlement since some of their terraces ("andenes"), are founded over this accumulation area.

3.5 GPS Monitoring Network

A GPS network was installed in September 2003, in the Urubamba Valley and Machu Picchu Citadel (Fig. 3.5). The GPS network is constituted by 3 control points (*a priori* FIXED) and 11 rover points installed inside the Citadel's area.

A first survey was conducted in September 2004, after one year of monitoring. Displacement of rover points of the citadel was calculated using baselines generated from the adjusted geodetic coordinates of reference points. The results show that the points where a displacement was appreciated are (Table 3.3):

Table 3.3. Displacement detected along the three axis (N, E, h) for points 9 and 10. Data in italics are considered reliable

Point	Delta N (m)	Delta E (m)	Delta h (m)
9	*0.012*	–0.002	0.002
10	*0.010*	–0.003	0.009

These results demonstrate how the major displacement vector runs along the North axis; while no significant displacement was detected along the Est and the ellipsoidical height direction.

3.6 Realization of a Three-Dimensional Digital Terrain Model of the INCA Citadel

A survey of the Citadel was acquired using terrestrial LIDAR techniques to obtain a three-dimensional digital model of the area.

This kind of terrestrial laser scanner, also equipped with a digital metric camera, allows to generate terrestrial orthophotos and to link the RGB tone to the single

Fig. 3.6. Laser scanner acquisition in the andenes

Fig. 3.8. The LISA system

Fig. 3.7. Numerical model corresponding on the acquiring phase of the Fig. 3.6

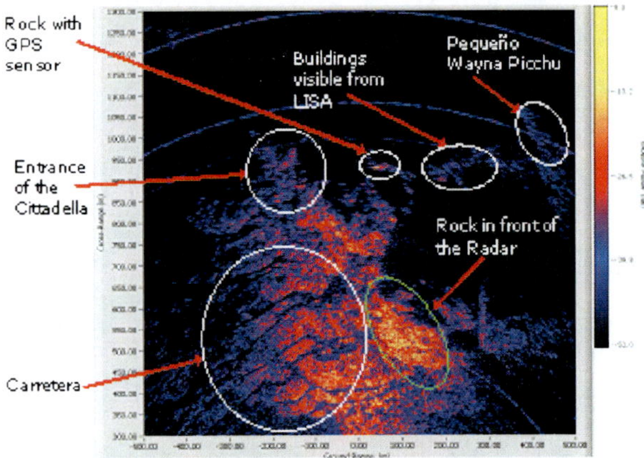

Fig. 3.9. Power image with reference points

laser impulse. Therefore, detailed sector of the Citadel with particular scientific and cultural relevance, like the King's Grave and the principal Temple was also acquired.

This numerical model has a great relevance to continue the investigation of the possible evolution of the area and to evaluate geologic interferences.

These dataset constitute a numerical base to generate detailed and dedicated cartography and to realize numerical modeling (Figs. 3.6 and 3.7).

3.7 Monitoring with JRC GB-SAR

On the 3 October 2004 the LISA Radar system was installed in a small open area which is part of the old train station in Puente Ruinas, at the beginning of the road going up to the archaeological village (Fig. 3.8). From this site (Fig. 3.9) it is possible to see some buildings of the lower part of the "Ciudadela" entrance and to see the whole vegetated zone below, where it is expected to observe the possible sliding movement. According to the data set, that has been continuously acquired up until now, no significant displacements have been detected. Considering the very low displacement rate measured by other monitoring systems, at least a full year of measurement is needed to get reliable data for assessing small displacement that may affect the Inca site.

3.8 Interferometric Synthetic Aperture Radar (InSAR)

Recent advances in optical and radar imagery capabilities, e.g. high spatial resolution, stereoscopic acquisition and high temporal frequency acquisition, the development of new robust techniques based on the interferometric analysis of radar images, such as the Permanent Scatterers Technique (Ferretti et al. 2000, 2001), and the possibility of integrating these data within a Geographical Information System (GIS) have dramatically increased the potential of remote sensing for landslide investigations (Hilley et al. 2004; Colesanti et al. 2004; Farina et al. in review). The resolution is of the order of millimeter.

Due to the very rough topography, steep slopes and local weather conditions the Machu Picchu area is a very challenging test site for the application of satellite radar data. Apart from that, the lack of an historical data-set of ESA-ERS acquisitions prevented – at least in the first part of the project – the application of the POLIMI PS Technique. In fact, the identification of the measurement points and the estimation (and removal) of the atmospheric components can be usually carried out whenever at least 15–20 scenes are available. Unfortunately, just a few ERS scenes were acquired since 1991 for interferometric processing.

Nevertheless, in the framework of the INTERFRASI Project, all satellite radar data acquired by the ESA sensors ERS-1, ERS-2 and Envisat over the area of interest have been processed trying to identify coherent areas where displacement information could be recovered, by applying the conventional approach (DInSAR). However, the coherence level of the interferometric pairs turned out to be to low and no information could be recovered over the AOI.

Due to the lack of ESA-ERS scenes, more than 30 scenes gathered by the Canadian radar sensor RADARSAT in different acquisition modes have been planned and processed. The use of RADARSAT "Fine-Beam" data, characterized by higher spatial resolution with respect to ESA data, turned out to be very important in order to identify good radar targets. Moreover, the shorter repeat-cycle of RADARSAT (24-day rather than 35) allowed the creation of a time series of 16 radar data in two different acquisition modes in about 1 year. The increased resolution allowed the selection of a dozen of "PS Candidates" within the AOI, characterized by a sufficient level of signal-to-clutter ratio (SCR).

An in-depth analysis of the time series of RADARSAT data, however, highlighted severe decorrelation phenomena probably due to microclimatic conditions at the time of the acquisitions and strong phase artifacts due to tropospheric inhomogeneities. Preliminary results have shown evidence of differential motion between the radar targets of about 5 cm (Figs. 3.10 and 3.11), but the estimation, so far, has to be considered not totally reliable using the available data-set. An independent confirmation with more PS is required.

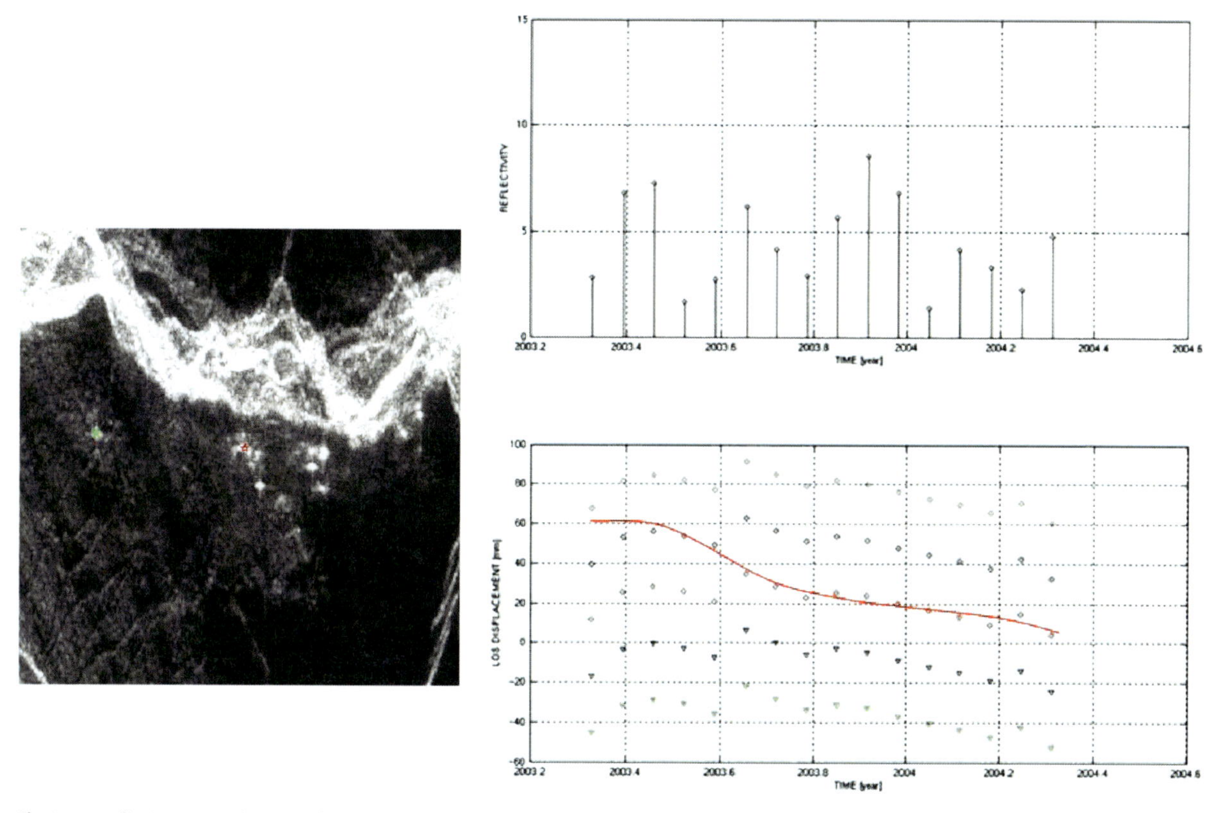

Fig. 3.10. Preliminary results recently obtained by estimating the differential motion between two PS identified in the AOI. Multiple time series show phase ambiguities. Amplitude values of the PS were found to be characterized by high dispersion values

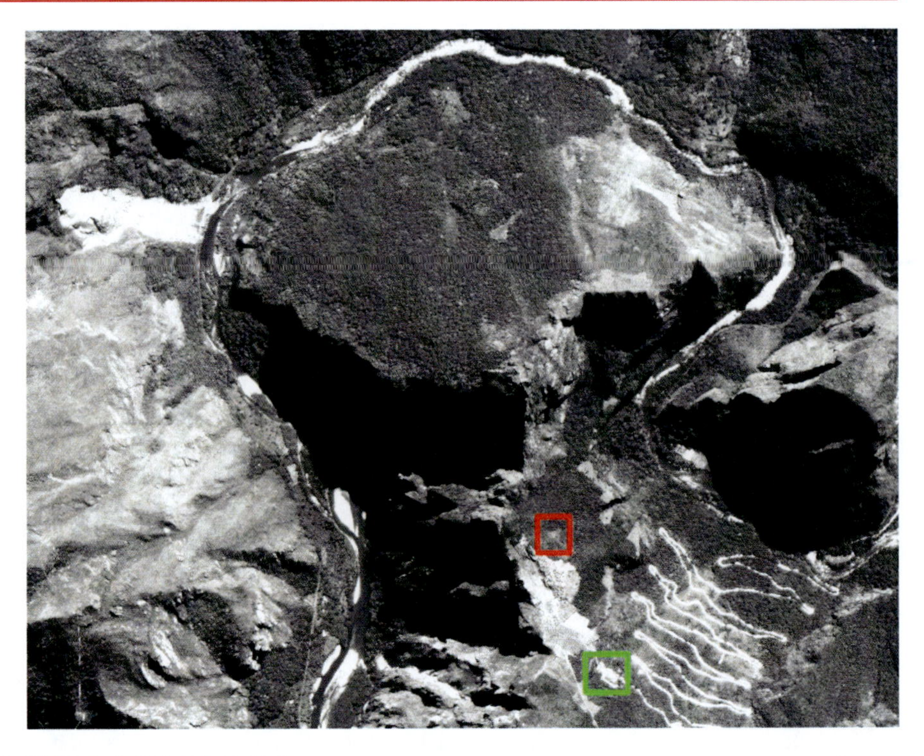

Fig. 3.11.
Identification of the two reported PS: the *green* is considered stable and the detected movement is assigned to the *red* PS

3.9 Conclusion

The geological and geomorphological investigations conducted in the area of Machu Picchu, even if preliminary and to be confirmed by the in progress monitoring network, confirm the general picture of slope instability described by Sassa et al. (2001, 2002), with some new elements and observations. Several slope instability phenomena have been identified and classified according to mechanism, material involved and state of activity. They are mainly related to rock falls, debris flows, rock slides and debris slides. Origin of phenomena is kinematically controlled by structural setting and relationship with slope face (rock falls, rock slide and debris slides); loose terrains generated by ancient or recent landslides are the source for debris flow.

In the area of the Carretera a precise mapping of debris deposits and past debris flows has been carried out, leading to a zonation of processes within the limits of the ancient landslide detected by Sassa et al. (2001). The situation of the slope with the Citadel is more complex due to the strong structural control of the master joints on the slope evolution. In this, planar rock slides are mainly affecting the NE flank while rock falls are predominant on SW cliff.

The analysis of monitoring data are still preliminary and to be better confirmed. Nevertheless it is possible to integrate all the available information in a synthetic map (Fig. 3.12) showing

1. the coincidence of GPS deformation, PS as well as a large concentration of restoration works on the N-E slope of the Citadel. In this area a large debris deposit is accumulated on the cliff and also the geological field analysis has detected some irregular pattern in the local structural setting; on the other hand GPS northward movement has not been detected by ground-based radar, probably due to the short time of observations with respect to GPS monitoring; in addition, the GPS movement is almost parallel to the orientation of the measurement, making not effective the radar distance control;
2. the stability of the upper part of the Citadel where several GPS sensors are not exhibiting any displacement; furthermore, the archaeological structures seem to be relatively undamaged;
3. the continues rock falls in the S-W side of the cliff and related Citadel's border, where also archaeological structure have been damaged by a progressive lateral detensioning; this is probably the area with the most severe short-term conservation problems. Finally, the collected data are beginning to provide a first sketch of the slope evolution of the site. Nevertheless, the analysis of the monitoring data collected from the systems installed by the Italian, Japanese, Czech and Slovak teams, together with data provided by Canadians and Peruvians, will allow a better evaluation of the mechanisms of slope processes and landslides, leading to a complete harmonization amongst the observation of the different research groups involved.

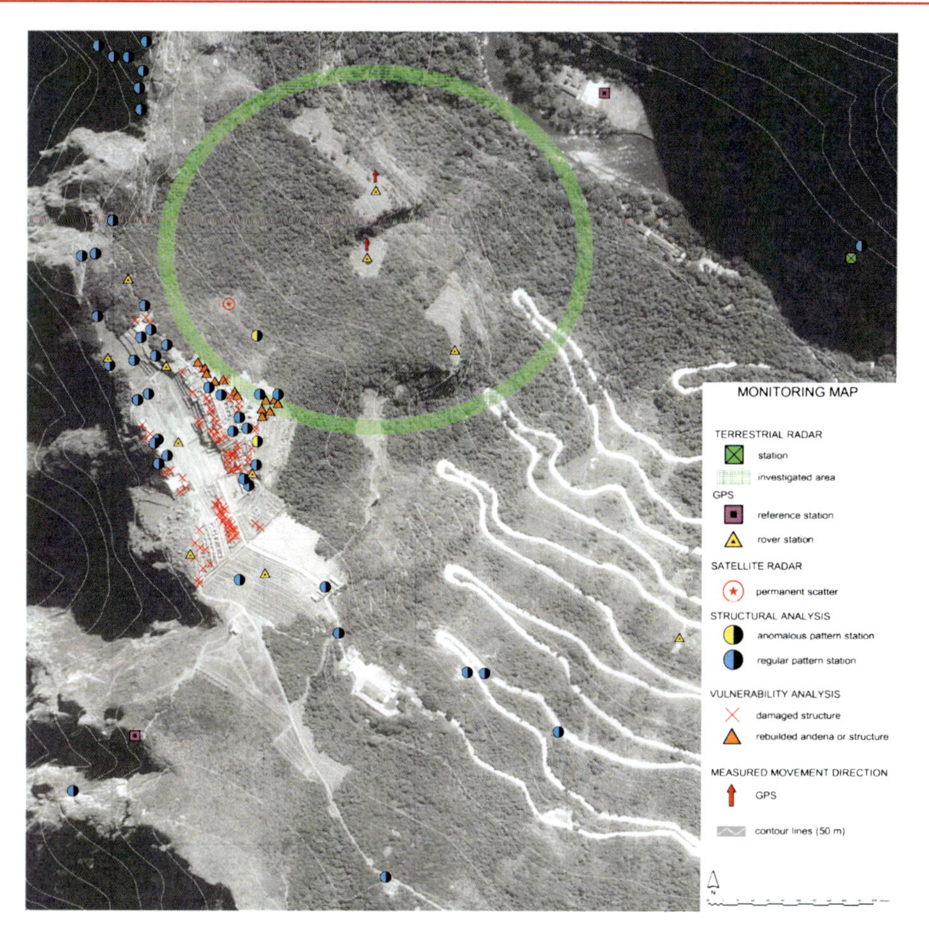

Fig. 3.12.
Integrated map of surface deformation evidences and present monitoring data

Finally, the collected data are beginning to give a first picture of the slope evolution of the site. Nevertheless, the analysis of the monitoring data collected from the systems installed by Italian, Japanese and Czech-Slovak groups, together with data provided by Canadians and Peruvians, will allow a better evaluation of the mechanisms of slope processes and of landslides, leading to a complete harmonization amongst the observation of the different research groups involved.

References

Carlotto V (1998) Mémoire H.S. 39, Université I de Grenoble, France, 203 p

Colesanti C, Wasowski J (2004) Satellite SAR interferometry for wide-area slope hazard detection and site-specific monitoring of slow landslides. Proc. International Landslide Symposium – ISL2004 Rio de Janeiro, Brasil

Farina P, Colombo D, Fumagalli A, Marks F, Moretti S (in review) Remote sensing techniques for landslide risk analysis: outcomes from the ESA-SLAM project. Eng Geol

Ferretti A, Prati C, Rocca F (2000) Non-linear subsidence rate estimation using permanent scatterers in differential SAR interferometry. IEEE T Geosci Remote 38(5):2202–2212

Ferretti A, Prati C, Rocca F (2001) Permanent scatterers in SAR interferometry. IEEE T Geosci Remote 39(1):8–20

Hilley GE, Bürgmann R, Ferretti A, Novali F, Rocca F (2004) Dynamics of slow-moving landslides from permanent scatterer analysis. Science 304(5679):1952–1955

Hutchinson JN (1988) General report: morphological and geotechnical parameters of landslides in relation to geology and hydrogeology. Proc. 5th Int. Symp. on Landslides, July 1988, Lausanne, 1, pp 3–36

Lancelot JR, Laubacher G, Marocco R, Renaud U (1978) U/Pb radiochronology of two granitic plutons from the eastern Cordillera (Peru). Extent of Permian magmatic activity and consequences. Geol Rundsch 67:236–243

Mazzoli S, Delmonaco G, Margottini C (2005) Role of precursor joints in the contractional deformation of the granite pluton, Machu Picchu batholith, Eastern Cordillera, Peru. Rend Soc Geol It, I, Nuova Serie, 3 ff (in print)

Sassa K, Fukuoka H, Kamai T, Shuzui H (2001) Landslide risk at Inca's World Heritage in Machu Picchu, Peru. In: Proceedings UNESCO/IGCP Symposium on Landslide Risk Mitigation and Protection of Cultural and Natural Heritage, Tokyo, pp 1–14

Sassa K, Fukuoka H, Shuzui H, Hoshino M (2002) Landslide risk evaluation in Machu Picchu World Heritage, Cusco, Peru. In: Proceedings UNESCO/IGCP Symposium on Landslide Risk Mitigation and Protection of Cultural and Natural Heritage, Kyoto, pp 1–20

Sempéré T, Carlier G, Soler P, Fornari M, Carlotto V, Jacay J, Arispe O, Neraudeau P, Cardenas J, Rosas S, Jimenez N (2002) Late Permian–middle Jurassic lithospheric thinning in Peru and Bolivia, and its bearing on Andean-age tectonics. Tectonophysics 345:153–181

Chapter 4

Geomorphological Investigations at Machu Picchu, Peru (C101-1)

Vít Vilímek* · Jiří Zvelebil · Jan Klimeš · Jan Vlcko · Fernándo V. Astete

Abstract. The landslide hazards analysis was the principal motivation to start geomorphological investigations in the area of Machu Picchu Sanctuary. But very soon, the need of a broader research was revealed, because the landscape evolution of the Urubamba River meander, where the archaeological site is located, is rather complex. Besides slope movements, also deepwards erosion, and selective mass wasting by weathering, suffusion etc. has been active there. Majority of them have been following predisposition by tectonical structure. Large-scale slope deformations have seriously affected mountain morphology in the area. The actual activity of those deformations is not well known yet, inspite that systematic monitoring has provided information as about irreversible movements on open cracks of rock outcrops within the archaeological site up to 1 mm yr^{-1} (lengths of time series is mostly 3 years), as about movement across the Main Plaza up to 6 mm yr^{-1}. There still are other possibilities to explain those movements by ground deformation due to underground erosion along tectonically shattered zones, or by settlements of heavy stone buildings on water more saturated grounds.

In any case, even under present state of knowledge, the occurrence of a large-scale catastrophically rapid slope collapse is improbable within the short time interval of months or several years, if present-day conditions are maintained (in the absence of a strong earthquake). The question of present day activity of large scale landslide bodies could be solved only by as multilateral research with using of all useful overlapping of geomorphology with hydrogeology, geophysics structural geology and engineering geology.

Keywords. Landslides, monitoring, geomorphology, Machu Picchu, Peru

4.1 Introduction

Geomorphology, in comparison with geological studies, had been at the periphery of scientific interests in the Machu Picchu Sanctuary for a long time since its discovery in 1911. However, in papers by Carreño et al. (1996) and Carreño and Bonnard (1997) questions had been arisen as to the safety of the archaeological World Heritage Monument in case of catastrophic landslide events. Therefore, after preliminary investigations of Sassa et al. (2000); Sassa (2001); Vilímek and Zvelebil (2002), the site has been promoted into UNESCO/IGS, IGSP 425 and then into ICL/IPL Projects of complex assessment of landslide risk.

Geomorphological research in this area was carried out within regionally oriented geological investigations (e.g. Ponce 1999; Carlotto et al. 1999). In these studies, geomorphology was presented mostly in a descriptive way without synthesis and deduction about landscape evolution. On a local level there exists one study directly devoted to the geomorphology of the archaeological site written by Manrique et al. (1998).

An important task in the landslide risk assessment is to estimate the actual rates of displacements deformations in landslide susceptible areas. A monitoring network was recommended already by Carreño and Bonnard (1997), but the first steps to its realization were taken by Sassa et al. (2000). Twelve sites were instrumented by wire extensometers. Two of them were equipped by automatic registration. The results of those monitoring measurements (Sassa 2001; Sassa et al. 2001) suggested existence of a permanent displacement of the order of mm per month or greater.

Our geomorphological investigations in this area started in 2001, when a preliminary field survey was done and a new basic strategy for monitoring network of sites

Fig. 4.1. Photo of Machu Picchu archaeological site with localization of dilatometric measurements. *C:* Cave (measuring site C1); *T:* Principal Temple (T1, T2, T3); *I:* Intiwatana (I1, I2, I3); *P:* Plaza (P1); *W:* Huayna Picchu (W1, W2, W3); *A:* Acllawasi (A1); *M:* Wairana o Mirador (M1, M2, M3, M4); *R:* Rodadero (R1, R2, R3); *Q:* Qhata (Q1). Qualitative evaluation of trends of movements. *Red color:* irreversible movements; *green color:* without any irreversible movements; *light blue:* cannot be still interpreted

Fig. 4.2. Photo of the archaeological site taken from the top of Mt. Huaynapicchu with localization of two extensometric profiles. *Red color:* opening; *magenta:* closing

for measurements by a portable dilatometer was set up upon its results. Due to its site-friendly nature, this type of instrumentation was authorized to be introduced under supervision of Peruvian authorities already in 2001 (Fig. 4.1). In 2002, two extensometric profiles have been added to the network. The reason was to test the hypothesis of ongoing large deep-seated slope movements producing mountain ridge spreading, which should take part just across the Main Square site (Fig. 4.2).

4.2 Study Site

Machu Picchu Sanctuary is located in Cordillera Oriental, about 15 km westward from Nevado Veronica (5 682 m a.s.l.) and 18 km northward from Nevado Salcantay (6 271 m a.s.l.). The archaeological site is situated on the mountain ridge forming the central part of a meander of Urubamba River.

Up-stream from Machu Picchu, the river valley of Urubamba is called Torontoy Canyon. This canyon represents – apart from the spring area – the steepest section in the longitudinal profile of the whole river (Peñaherrera 1986). The morphological and geological comparison of the Machu Picchu meander with the other ones in its surrounding reveals that the former is the most tectonically dissected one and has the most flat saddle-shaped form in its upper parts. In fact, this meander was the most appropriate building area for Inca people. The saddle form on which the urban area is placed is limited by two morphologically well-defined mountain peaks – Huayna Picchu (2 640 m a.s.l.) and Machu Picchu (3 072 m a.s.l.). The urban area is located mostly between 2 400 and 2 500 m a.s.l. and the Urubamba River is running around the meander at the altitude of 2 000 m a.s.l.

From the point of view of geology, Machu Picchu is situated on a 40 km^2 portion of the Vilcabamba Batholith complex of Permian/Triassic age in the Cordillera Oriental between the High Plateau and Subandine Zones of the Peruvian Andes (Marocco 1977). White to gray-colored granite characterized by its abundance of quartz, feldspar and mica (predominantly biotite) is the predominant rock type. Apart from granites, there may be found tonalites, granodiorites and serpentinites.

The prevailing structural elements in the Machu Picchu Massif are closely connected with magmatic intrusion and the origin of the batholithic structure. Deep running joints and fractures of different orientations can be observed. The tectonic pattern of first order is represented by three regional faults: Huaynapicchu fault, Machu Picchu fault (NE-SW direction) and Urubamba fault (NW-SE direction). Local joints and fractures with prevalent E-W direction and N-S direction can be observed directly at the site. The dip of these structures is generally higher than 50–60°. Along these high-angle reverse faults a graben (subsided structural block) had developed on which the ancient Inca people built their city (Wright et al. 1997).

Hydrogeology – especially from the point of view of paleohydrology and civil engineering – was described by Wright et al. (1997, 2000). According to them, the Machu Picchu fault formed hydrogeologic setting for the spring, which the Inca people enhanced by their spring collection system. Preliminary results of our most recent hydrogeological studies, which would be completed by the end of 2005, are indicating the key role of hydrogeology for ground stability conditions.

Besides the study area, various types of slope movements were also described in its broader surroundings in past few years. They vary in their typology as well as in their volumes. Recently, the most significant was a debris flow that destroyed the village of Santa Teresa in 1998 and debris flows in valleys of Aguas Calientes and Alcamayo in 1950, 1970, 1995 (Carlotto et al. 1999). Carreño and Bonnard (1997) described the rockslide from Mt. Putucusi (2 579 m a.s.l.) of December 1995. A rockfall from the same mountain above the southern portal of a railway tunnel north from Aguas Calientes occurred in 2002. The most recent events – debris flows – occurred in 2004 (Alcamayo) and in 2005 (Mandor).

4.3 Methods

General geomorphological survey and mapping were carried out to obtain information on sculpture, genesis and dynamics as well as conditions of modeling (Demek and Embleton 1978). To understand more in detail the relation between landforms and geological structure, morphostructural analysis was elaborated. Major landforms in tectonically predisposed areas are related to tectonics.

Both passive structural forms and active tectonics are modified at varying degrees by denudation or accumulation processes (Demek and Embleton 1978). In accordance with Ollier (1981), neotectonic activity was studied when searching for direct signs but also for indirect ones manifesting themselves by abnormal activation of exogenic processes, like enormous erosion or sedimentation. Nevertheless such "indirect" manifestations should be very carefully proved in the context of a much larger area, because they frequently could be also explained without the influence of active tectonics (e.g. Vilímek 2004).

For aerial photo interpretation, black and white aerial photos (1:50 000 scale) from National Airphoto Service in Lima were used. Major tectonically based valleys were distinguished and compared with geological maps and structural field measurements. Structural mapping of joint orientation was done in the extent of 250 measurements of strike and dip. A special attention was paid to tectonic mirrors and to the orientation of their striations.

The monitoring strategy tried to be as environmentally friendly as possible with full installation restriction for any archaeological structure and with restriction for permanent installations that would be clearly visible on rock surfaces. At the same time, a high precision method was needed because of our hypothesis on small magnitudes of ground deformations within the preserved archaeological area.

As the basic method, measurement of relative displacements across open fissures in rock outcrops in basement of important Inca structures by portable dilatometer was chosen. This type of measurement is able – under fully favorable measuring conditions – to detect irreversible movements of order of 0.1 mm per year within minimally 3 years lasting time series (e.g. Zvelebil and Stemberk 2000). Its requirements on permanent installations are restricted to a pair of tiny metal bolts fixed opposite across the measured joint.

Extensometry by a portable extensometric type was chosen to span across many meters wide space (25–35 m) of the tension zone of the hypothetical deep-seated slope failure, as it was necessary to install permanently on rock outcrops only extensometric bolts. Under fully favorable measuring conditions, the accuracy of this method is 0.5 mm per year in minimally 3 year time series.

The accuracy listed above was considered to be sufficient to obtain reliable data about the actual deformation activity on selected sites. Both dilatometric as well as extensometric measurements were performed by our Peruvian colleagues of the Administration of Machu Picchu Heritage Site.

4.4 Results

In the studied area, erosion forms dominate in the relief (Fig. 4.3), but strongly pronounced structural predisposition of landscape evolution were also found there. Intensive deep erosion of Urubamba River has created a canyon-like form without river terraces from older stages of the landscape evolution which enables to establish the chronostratigraphy of the local valley development. River erosion followed the main tectonic lines/zones: of the Machu Picchu, Huayna Picchu and Urubamba faults, as it was documented by structural measurements (Fig. 4.4). Rapids in the riverbed of Urubamba can be found just at places where those main faults are crossing the valley. The longitudinal profile of the river is still not adjusted to the neotectonic uplift of the wider region of the Vilcabamba Batholith.

Among the other accumulation forms of relief, dejection cones or terraces of probably Quaternary age are dominating. Their dating is only according to their relative position; no absolute dating is available at this mo-

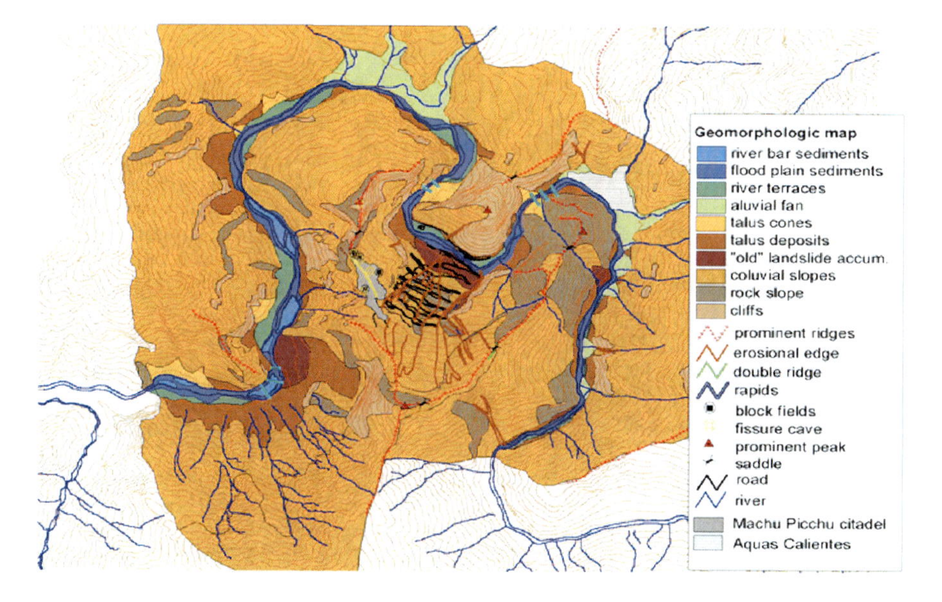

Fig. 4.3. Geomorphological map of the surrounding of Machu Picchu archaeological site (the Urubamba River meander). The map is oriented to the North

ment. A massive input of sediments caused by intensive erosion in the middle and upper parts of watersheds resulted in formation of significant alluvial fans. Their punctuated evolution is controlled by occasional debris flows (e.g. Alcamayo and Mandor).

Lateral slopes in local watersheds are very steep (60–70°) and still in disequilibrium stage which can be documented by various forms of recent slope failures by landsliding (in a very broad sense which includes sliding and flows in soils, various types of rock falls and rock slides). Beside the recent predominantly small-scale landslide forms, also forms reflecting a possible activity of large-scale slope movements have been mapped (Fig. 4.5).

Fig. 4.4. Structural measurements in different areas of Machu Picchu. Ridge: *F:* NE slope outside the Bingham Road (front slope); *BR:* slope of the Bingham Road; *B:* SW slope (back slope); *G:* general summarizing tectonogram

The landslide forms on both slopes of the Machu Picchu ridge can be ranked into a hierarchically ordered system. The first order forms are remnants of a possible large-scale movement, which activated the Machu Picchu ridge in its full height. That slide contributed significantly to the development of the saddle-form (a flat area in the most upper part of the ridge between Machu Picchu and Huayna Picchu Mts.). As this saddle was the building site of the Inca settlement, this slide has to be of pre-Inca age. Inside the boundary of the first order slide, there are forms of lower hierarchic orders formed by activation of shallower parts of the slope mass. They appear to be younger – some of them contemporaneous with Inca occupation and some even younger, quite recent (e.g. Carreño and Bonnard 1997).

On the east slope of Machu Picchu ridge, these forms seem to be produced by a sequence of several huge slides. Their slide origin is strongly supported by a structural discrepancy between the areas along the Bingham road and side slopes of Machu Picchu and Huayna Picchu Mts. (see Fig. 4.4). On the western slope, they have the form of large rock falls.

Due to the fact that three year time series of measurements will be completed in November 2005, the monitoring results are presented only qualitatively in three categories (sites with non-reversible movements, without any movements, cannot be still interpreted, see Fig. 4.1). Only as a preliminary result, it can be stated that the movements registered on the dilatometric sites have magnitudes up to a millimeter per year. An example of non-

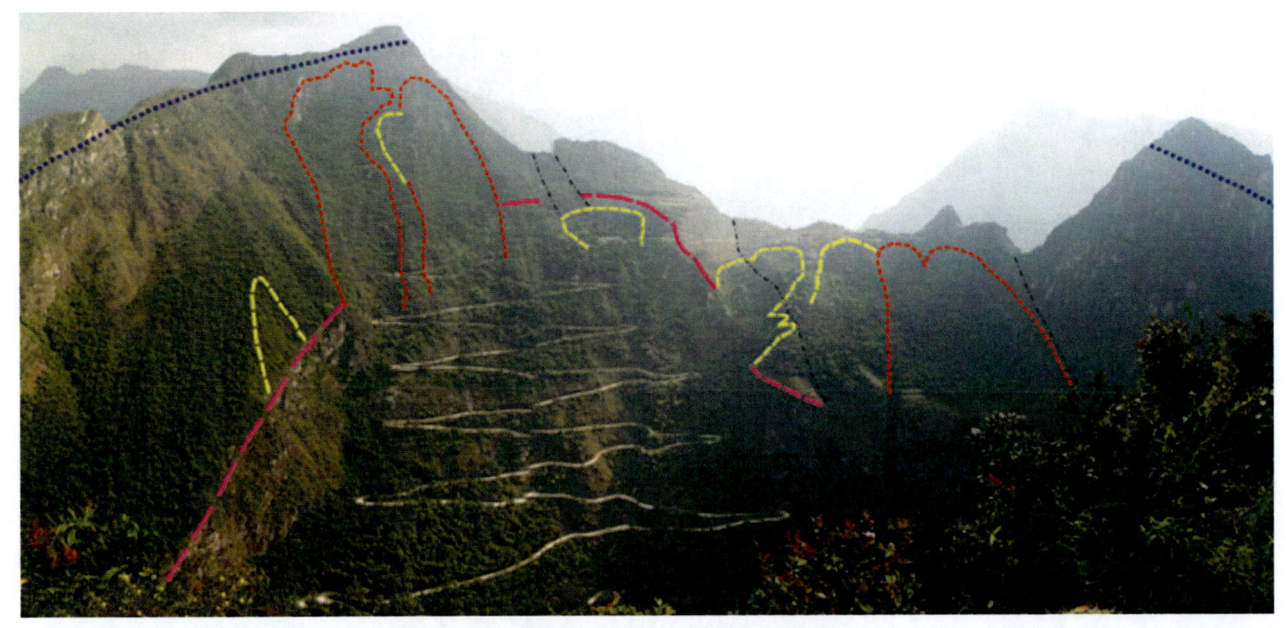

Fig. 4.5. Morphologically apparent boundaries of individual slope deformation bodies depicted in the photo of Machu Picchu archaeological site taken from Mt. Putucusi. *Blue color:* largest deformation of the first hierarchical order (pre-Inca age); *magenta:* large deformation bodies of the second hierarchical order (highly probably of pre-Inca age); *red and yellow:* bodies of the third and fourth hierarchical orders (only according their spatial scales); *black:* morphologically well pronounced structural lines

reversible movement registered about 2 m northwards from deformed masonry of the Principal Temple is depicted in Fig. 4.6.

The plot of extensometric measurements from one of the two key profiles across the Plaza is presented in Fig. 4.7. Regarding the relatively high value of measuring error, it is too early for any rigorous interpretation. Nevertheless, if we consider only the control measurements, a closing trend of 6 mm per 2 yr is apparent.

4.5 Discussion

Even if no recent tectonic movements were described in this area by other authors and no data about present seismicity were published up to now, we can find in the local geomorphology a rather significant exhibition of neotectonic movements. Even the existence of the largest scale prehistoric movement which had affected areas where are now the archaeological site and Bingham road located and which is conditioned by Machu Picchu and Huaynapicchu faults could be related to the neotectonic activity. In fact, such huge failure does not have any other equivalent in the adjacent meanders in the Torontoy Canyon and the river erosion itself was probably not a sufficient triggering factor for such failure. At least a passive influence of tectonics on the failure evolution is evident.

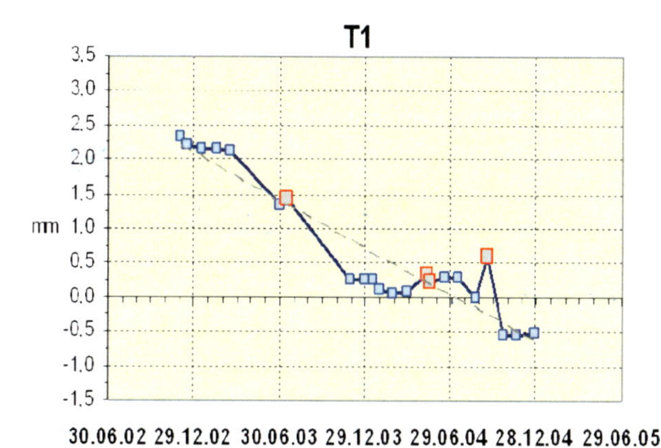

Fig. 4.6. An example of irreversible movement by dilatometric measurements (site Temple I). *Blue dots:* Current measurements; *red-gray dots:* special control measurements

The last 500 year period is only a part of a long-term history of the complex slope development on the Machu Picchu Ridge. In addition to slope movements, there are other geodynamic processes, which could have damaged the archaeological site and be linked to the landslide activity (e.g. weathering, subsidence, changes in infiltration). Therefore without a complex study, it is impossible to assess the present-day geodynamical risk for the valuable World Heritage Site. The first steps of such study include a longer monitoring program to detect displacements in the rock mass. A full catalogue of historical, archaeological, geological and civil engineering analyses of deformations is needed. Geophysical surveys for identification of possible underground openings are recommended along with other case studies (weathering, rock strength, etc.).

4.6 Conclusions

The main processes responsible for the landscape evolution in the study area are: *(i)* neotectonic uplift which enhanced exogenic processes, *(ii)* intensive river erosion

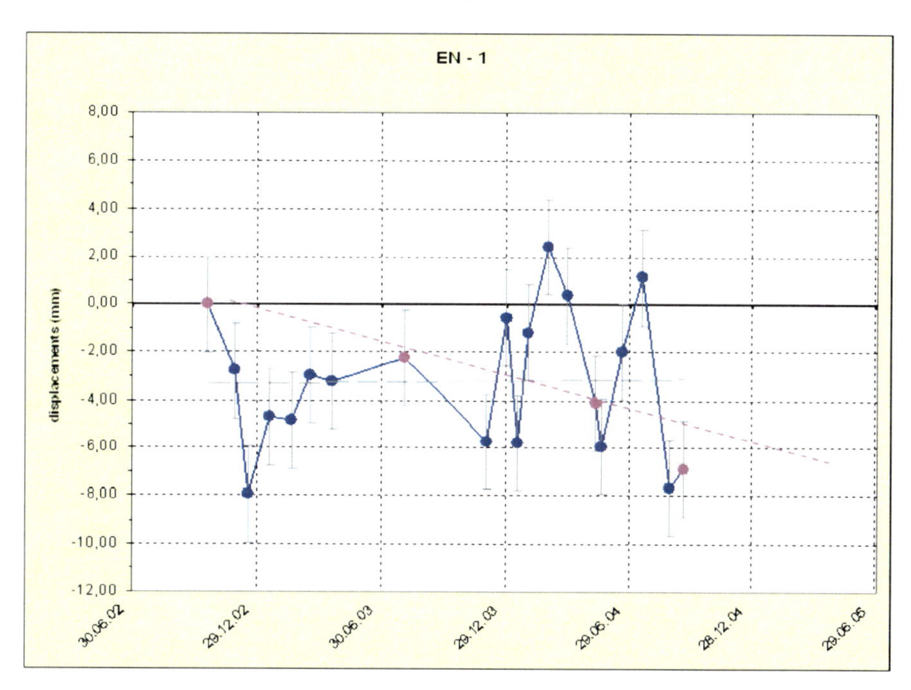

Fig. 4.7. An example of irreversible movement by extensometric measurements (*EN-1*). *Blue dots:* Current measurements; *magenta dots:* special control measurements

in a highly tectonically disturbed zone in places of great faults crossings; *(iii)* various types of slope movements; *(iv)* intensive weathering along the broken zones.

Slope movements have strong structural predisposition. The main Machu Picchu and Huaynapicchu fault zones conditioned the occurrence of the largest first-order slide of pre-Inca age. The eastward dipping planes of granites caused kinematical differentiation of slope movements; sliding on the east slope is in contrast to toppling rockfalls on the western slope. The youngest generation of slope movements is represented by small-scale predominantly surface movements of the lowest hierarchical order.

Our interpretation of monitoring results has a preliminary character due to the uncompleted time span needed. Displacements on joints within rock outcrops situated inside the archaeological area reach maximally one millimeter per year. At the present state of knowledge, we cannot exclude the influence of other processes involved beside possible slope movements, as ground deformations due to underground erosion and irregular subsidence on the tectonically highly jointed ground rock mass. Results of extensometric measurements cannot be interpreted yet.

The occurrence of a large-scale catastrophically rapid slope collapse is highly improbable within a short time interval of months or few years, if the present-day conditions remain (especially without any strong earthquake).

Acknowledgments

We would like to express our thanks to Instituto Nacional de Cultura (INC), Instituto Nacional de Recursos Naturales (INRENA), Instituto Geológico Minero y Metalúrgico (INGEMMET) and Instituto Geofísico del Perú (IGP). We would also like to thank to the Ministry of Education of Slovak Republic. The research was supported by the Ministry of Education, Youth and Sports of the Czech Republic (Projects INGO, LA 157 and MSM 00216 20831) and by the ICL Project C101-1-2.

References

Carlotto V, Cárdenas J, Romero D, Valdivia W, Tintaya D (1999) Geología de los cuadrángulos de Quillabamba y Machupicchu. Hojas: 26-q y 27-q. INGEMMET, Lima

Carreño R, Bonnard C (1997) Rock slide at Machu Picchu, Peru. Landslide News 10:15–17

Carreño R, Lopéz R, Tupayachi L (1996) Informe preliminary sobre el derrube de Machupicchu del 26.12.95 y situación del cerro Putucusi. Nota técnica, MS PROEPTI, Cusco

Demek J, Embleton C (1978) Guide to medium-scale geomorphological mapping. IGU Commission on geomorphological survey and mapping, Brno

Manrique D, Quispesivano L, Moncayo O (1998) Estudio Geomorfologico-estructural del parque arqeologico de Machu Picchu. INGEMET, Peru

Marocco R (1977) Geologie des Andes Peruviennes: Un segment W.E. de la Chaine des Andes Peruviennes; la Deflexion D'Abancy. Etude Geologique de la Cordillere Orientale et des Hauts Plateaux Entre Cusco et San Miguel. These, Academie de Montpellier Universite des Sciences et Techniques de Languedoc

Olliver CD (1981) Tectonics and landforms. Longmann, London, New York

Peñaherrera C (1986) Gran Geografía del Perú, vol. I, Barcelona

Ponce CF (1999) Plan Maestro del Santuario Histórico de Machupicchu. INRENA, Peru

Sassa K (2001) The second investigation report on the slope instability in Inca's World Heritage, in Machupicchu, Cusco, Peru. Unpublished report for INRENA, MS DPRI Kyoto University

Sassa K, Fukuoka H, Shuzui H (2000) Field investigation of the slope instability at Inca's World Heritage, in Machupicchu, Peru. Landslide News 13:37–41

Sassa K, Fukuoka H, Kamai T, Shuzui H (2001) Landslide risk at Inca's World Heritage in Machu Picchu, Peru. In: Sassa K (ed) Landslide risk mitigation and protection of cultural and natural heritage. Proceedings of of the UNESCO/IGCP Symposium, Tokyo, Japan

Vilímek V (2004) Morphotectonic effects in fault zones. Acta Univ Carol Geogr 39(1):47–58

Vilímek V, Zvelebil J (2002) Slope instability at Machu Picchu: ideas and questions. Acta Montana 19:75–89

Wright KR, Witt GD, Zegarra AV (1997) Hydrogeology and paleohydrology of ancient Machu Picchu. Ground Water 35(4):660–666

Wright KR, Zegarra AV, Wright RM, McEvan G (2000) Machu Picchu, a civil engineering marvel. ASCE Press, Reston, Virginia

Zvelebil J, Stemberk J (2000) Slope monitoring applied to rock fall management in NW Bohemia. In: Bromhead E, et al. (eds) Landslides in research, theory and practice. Proceedings 8th International Symposium on Landslides, London 3:1659–1664

Chapter 5

The Application of Ground Penetrating Radar (GPR) at Machu Picchu, Peru (C101-1)

Romulo Mucho Mamani · Victor Carlotto Caillaux · Walter Pari Pinto · Martin Jhonathan Oviedo · Marten Douma · Mel Best Peter Bobrowsky*

Abstract. A ground penetrating radar survey was conducted in the spring of 2005 at the archaeological site of Machu Picchu, Peru by INGEMMET and the Geological Survey of Canada. The aim of the study was to evaluate the nature and characteristics of the shallow surficial sediments and uppermost bedrock at the citadel. Results of the survey permit a high level of resolution indicating that the surficial deposits consist of two separate unconsolidated facies. The uppermost part of the bedrock surface was also captured during the survey imaging. There is no evidence to support the presence of an extensive north-south trending fracture, fault or failure plane crossing the citadel.

Keywords. Machu Picchu, shallow geophysics, ground penetrating radar, landslide hazard

5.1 Introduction

The famous UNESCO World Heritage Site of Machu Picchu, Peru was the royal estate for the Inca ruler Pachacuti, who occupied the site with several hundred other individuals from about A.D. 1450 to A.D. 1562 (Wright and Valencia Zegarra 2000). Discovered early in the last century, the site is now host to some 1 million tourists per year. Moderate modern construction directly adjacent to the site and extensive urbanization at the nearby town of Aguas Calientes has drawn international attention following a series of shallow translational landslides, rock falls and debris torrents in the area. The impact of the failures includes 11 fatalities in 2004 and economic concerns involving the closing of the only access road to the site (Hiram Bingham Road in 1996).

The additional potential threat for large-scale landslide events affecting the archaeological site itself prompted INGEMMET to initiate an evaluation of the sub-surface conditions in the area. INGEMMET in cooperation with the Geological Survey of Canada began a multi-year multi-parameter shallow geophysical assessment program at Machu Picchu in 2003. The data collected will be used to enhance other data on structural geology, engineering properties and geomorphology of the area as collected by collaborative international scientific and engineering teams.

The purpose of this paper is to provide comments on the results obtained from the application of one shallow geophysical method applied at Machu Picchu. We briefly describe the background theory, applications and interpretations of a ground penetrating radar study conducted in June 2005. Results of the other shallow geophysical techniques applied during the past few years will be presented elsewhere.

5.2 Setting

The Inca archaeological site of Machu Picchu is located approximately 100 km north of the city of Cusco, Peru (Fig. 5.1). Perched at an elevation nearly 2 500 m a.s.l., Machu Picchu rests on a saddle ridge stretching between two prominent peaks: Machu Picchu to the south and Huayna Picchu to the north. The Rio Vilcanota, a headwater stream of the Amazon River, winds its way around the site on the east, north and west sides some 500 m below. The unique and steep high relief topography in the

Fig. 5.1. Location map of Machu Picchu, Peru study area

region is underlain by part of the Vilcabamba Batholith a white to gray colored granitic complex dated by Rb/Sr to about 246 ±10 Ma (Carlotto et al. 1999). This granitic complex is cut by several large faults and is characterized by an extensive jointing pattern. Both faults and joints are thought to be the primary contributors to the slope instability in the area.

5.3 Background Theory

Ground penetrating radar (GPR) systems are uniquely suited for shallow subsurface investigations of archaeological sites (Sternberg and McGill 1995; Leckebusch 2000; Carrara et al. 2001; Goodman et al. 1995, 2004) and landslide areas (Hack 2000; Bichler et al. 2004). GPR systems are capable of generating high-resolution images of the shallow subsurface. GPR depth penetration is controlled by the conductivity of the near-surface material. Depth penetration of 30 m or more is possible in resistive ground whereas in conductive ground the GPR signal is attenuated within the first meter or so.

A GPR system consists of transmitting and receiving antennas and associated electronics (Fig. 5.2). The transmitting antenna emits a very high frequency (between 25 and 200 MHz for the applications being discussed in this paper) electromagnetic pulse into the ground that propagates as a wave at or near the speed of light. When this emitted wave encounters changes in electrical (dielectric constant) properties of the ground a reflected wave is generated, similar to seismic reflections. The time required for the reflected wave to reach the receiving antenna and the amplitude of this wave is recorded and used to determine the properties of the shallow subsurface.

The standard operational mode for GPR reflection surveys is shown in Fig. 5.3a. The antenna dipoles are placed perpendicular to the direction of the survey line. A constant separation between the two antenna dipoles is maintained throughout the survey. The transmitter and receiver positions are moved in increments along the survey line as shown in the figure. Figure 5.3a illustrates schematically the ray paths at several locations along this example survey line. Note the reflections from the finite body as well as at the boundary between overburden and bedrock.

Fig. 5.2. This figure shows the sensors and software 200 MHz transmitter and receiver antenna assemblies. Not shown is the Pulse-EKKO 100 console and the recording laptop computer

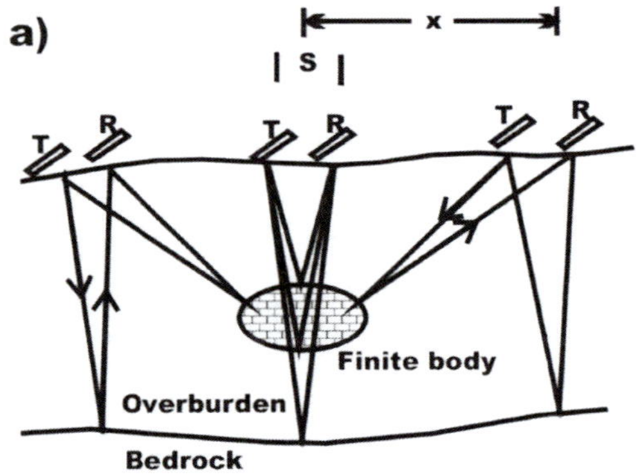

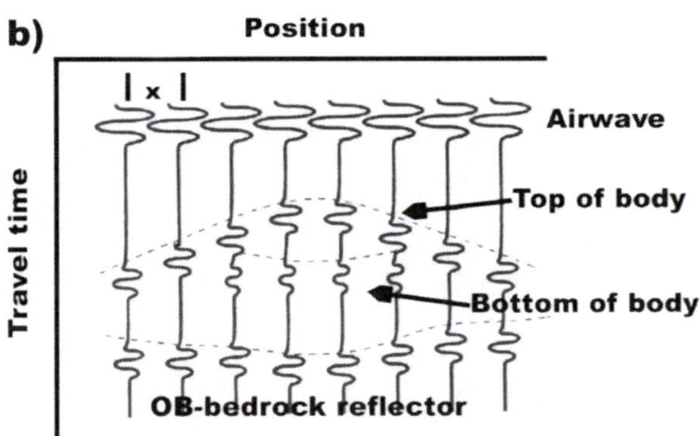

Fig. 5.3. a A schematic diagram illustrating GPR reflection profiling. The antenna dipoles are placed perpendicular to the direction of the survey line. *S* is the constant separation between the two dipoles and *x* is the incremental distance the GPR system is moved along the survey line. b The GPR cross-section for the geological situation given in a. The time axis is measured in nanoseconds since the velocity of light is 0.3 m ns^{-1}. Note the reflections from the top and bottom of the finite body

Figure 5.3b is a simplified diagram showing the corresponding GPR cross section for the geological example shown in Fig. 5.3a. The vertical axis is the time required for the propagating waves to reach the receiving antenna measured in nanoseconds (ns). The horizontal axis is a plot of the position of the mid-point of the transmitter-receiver array. The cross section illustrates how a horizontal boundary (in this case between overburden and bedrock) and a finite body will appear on a GPR cross section, assuming of course these features have different values of dielectric constant than the surrounding material. Other (refracted and diffracted) waves are also generated so real GPR cross sections are generally more complex than this example.

5.4 Methods and Results

One of the first targets of the GPR survey was the Main Plaza area, separating the Intywatana and Western Urban Sector from the Eastern Urban Sector of the citadel (Fig. 5.4). Of primary interest was the subsurface configuration of the bedrock beneath the plaza. Outcrops of granite occur on both east and west sides of the plaza, and it is easy to visualize that the surface of the bedrock could lie at considerable depth, especially given the relief of the adjacent outcrops.

Survey flags placed at 5 m intervals provided fiducial marks for the radar survey. Since one of the survey objectives was to determine the presence or absence of a north-south trending slip-face or failure plane, the first profiles were conducted across the short axis (W-E) of the plaza.

The ground penetrating radar used in the study was a PulseEkko 100, operating at 100 MHz and a pulser voltage of 400 V. Radar velocity was set at 0.1 m ns^{-1}, and the antennae were set 1 m apart, and moved in 25 cm steps. Figure 5.4 shows the location of the Principal Plaza in the background, and the orientation of the survey lines across the plaza. Survey lines were not evenly spaced, but selected, as the data were collected, to make the best use of the time available and generate sufficient coverage to visualize the subsurface configuration of the plaza.

Three primary facies were identified in the radar sections in this area (Fig. 5.5). The top 0.5 to 1 m of the sections shows a strong reflection, sometimes accompanied by one or two fainter, but parallel reflections directly above. Excavations and laboratory test results show that much of the topsoil in the terraces and plaza areas of Machu Picchu is typically some 0.5 m thick (Wright and Valencia Zegarra 1999), therefore it seems reasonable to propose that the strong reflection observed in the transects represents the base of the topsoil. This facies is shown in red in Fig. 5.5, and was observed in all the radar sections conducted on terraces and plazas. Although not discussed

Fig. 5.4.
Location of the radar profiles on the two plazas, viewed from the south. The Intywatana is the hilltop structure on the left (west) side, whereas the Eastern Urban Sector is on the right

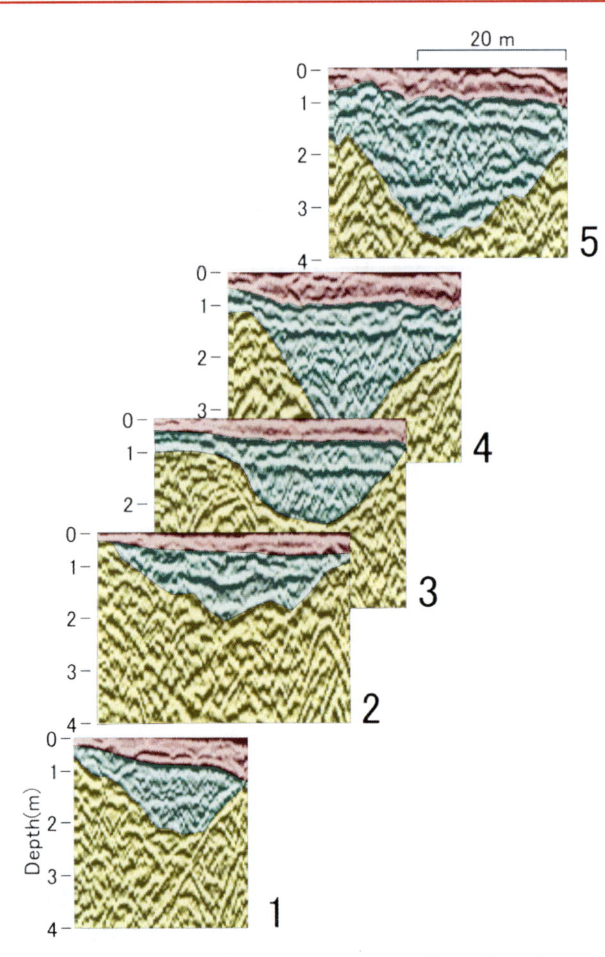

Fig. 5.5. Interpretation of the parallel radar profiles collected across the Principal Plaza, Machu Picchu. See text for a description of the radar facies

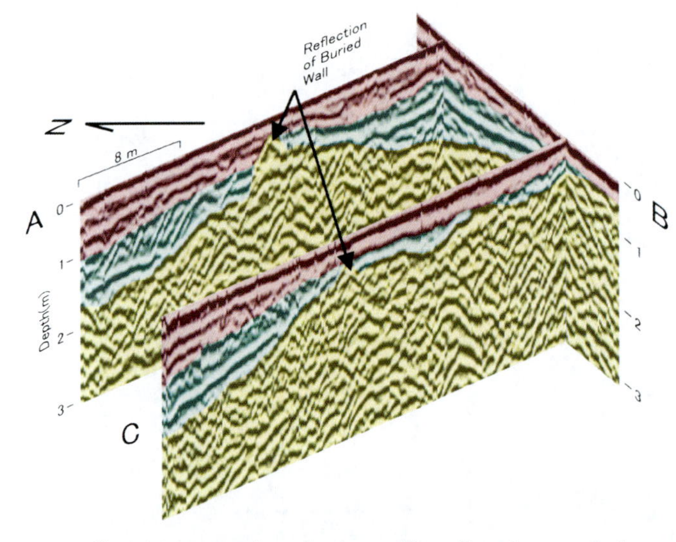

Fig. 5.6. Interpretation of radar profiles collected across the lower plaza located directly south of the Principal Plaza. See text for a description of the radar facies

and illustrated in this paper, radar profiles collected along certain temple floors and a temple plaza show markedly different near surface characteristics, indicating that construction in such areas was not used for purposes such as retaining walls which are capped with clay to form hard, level surfaces (Alfredo Valencia Zegarra, pers. com. 2005).

Below the topsoil facies lies a unit of generally less organized reflections, interrupted in places by a few sub-horizontal reflections. This second unit is seen to extend down to almost 4 m in places, and represents a sub-soil unit composed primarily of boulders, rocks, rock chips, and other by-products of the Inca's quarrying and stone-masonry efforts. The lack of coherence in many of the reflections represents the scattering of radar waves by the rubble. The occasional sub-horizontal reflectors observed in the survey may represent past cultural efforts to level the ground surface. The overall impression of the series of west to east sections across the plaza is one of a relatively significant depression that had been back-filled during construction of the citadel.

Underlying the soil facies, indicating the limits of the radar's ability to penetrate, is a zone of poorly organized or occasionally interwoven hyperbolic reflections. This type of appearance is consistent with either bedrock or a lithology in which larger boulders are a major component. The hyperbolae are merely artifacts of the manner in which radar reflections are returned from point reflectors that do not lie directly below the source-receiver equipment.

Figure 5.6 shows three radar profiles for a major plaza located directly south of the Principal Plaza, as distinguished by the presence of a substantial tree at its center (Fig. 5.4). This plaza was examined at the request of the on-site archaeologists, who had, in 1996, exhumed a previously unsuspected stone-wall, buried to a depth of about 1 m. Whether the wall was built as a temporary retaining wall for construction purposes, or abandoned as a result of a change in construction plans remains unknown.

The radar facies in this second plaza are consistent with the facies identified and described earlier, although the sub-soil facies shows more and better developed, sub-horizontal reflections (Fig. 5.6). The reflection of the wall appears as a poorly developed hyperbolic reflection on a ridge of the lowermost boulder/bedrock facies described earlier.

5.5 Conclusions

The application of ground penetrating radar at the Inca archaeological site of Machu Picchu, Peru has provided a good "snapshot" of the subsurface conditions at this important cultural heritage site. As discussed in this paper, our study shows a consistent pattern of surficial or overburden sediment accumulation over the citadel consisting of easily discernible and readily interpreted geophysical facies consisting of top-soil, back-fill and most likely bedrock. Although evidence for an archaeological feature was obtained during this survey, we were not able to recognize any prominent or distinguishable evidence for a fault plane or pre-existing failure plane at the site. Speculation by other researchers that a north-south plane of detachment could be present below the plaza area was not substantiated with the ground penetrating radar survey conducted in 2005. Although not discussed in detail here, the results of the full GPR survey across other portions of the site also confirm the absence of such a proposed failure feature.

Acknowledgments

This paper is a contribution to the Multinational Andean Project: Geoscience for Andean Communities directed by Dr. Catherine Hickson. We appreciate the financial support of CIDA (Canadian International Development Agency), INGEMMET (Geological Survey of Peru) and the GSC (Geological Survey of Canada) towards the completion of this project. We appreciate the constructive internal review of this manuscript provided by Dr. Steve Wolfe and Andree Blais-Stevens (GSC, Ottawa).

References

Bichler A, Bobrowsky PT, Best ME, Douma M, Hunter J, Calvert T, Burns R (2004) Three-dimensional mapping of a landslide using a multi-geophysical approach: the Quesnel Forks landslide. Landslides 1:29–40

Carlotto V, Cardenas J, Romero D, Valdivia W, Tintaya D (1999) Geologia de los Cuadrangulos de Quillabamba y Machupicchu. Boletin no. 127, Serie A: Carat Geologica Nacional, INGEMMET, Instituto Geologico Minero y Metalurgico

Carrara E, Carrozzo MT, Fedi M, Florio G, Negri S, Paoletti V, Paolillo G, Quarta T, Rapolla A, Roberti N (2001) Resistivity and radar surveys at the archeological site of Ercolano. J Environ Eng Geophys 6:123–132

Goodman D, Nishimura Y, Rogers JD (1995) GPR time slides in archaeological prospection. Archaeol Prospect 2:85–89

Goodman D, Piro S, Nishimura Y, Patterson H, Gaffney V (2004) Discovery of a 1st century AD Roman amphitheater and other structures at the Forum Novum by GPR. J Environ Eng Geophys 9:35–42

Hack R (2000) Geophysics for slope stability. Surv Geophys 21: 323–338

Leckebusch J (2000) Two and three-dimensional ground penetrating radar surveys across a medieval choir: a case study in archaeology. Archaeol Prospect 7:189–200

Sternberg B, McGill J (1995) Archaeology studies in southern Arizona using ground penetrating radar. J Appl Geophys 33:209–225

Wright KR, Valencia Zegarra A (1999) Ancient Machu Picchu drainage engineering. J Irrig Drain E-ASCE 125:360–369

Wright KR, Valencia Zegarra A (2000) Machu Picchu: a civil engineering marvel. ASCE Press

Chapter 6

Assessing the Capabilities of VHR Satellite Data for Debris Flow Mapping in the Machu Picchu Area (C101-1)

Nicola Casagli · Riccardo Fanti* · Massimiliano Nocentini · Gaia Righini

Abstract. Machu Picchu is an ancient Inca city located on a narrow ridge, within the Andes, approximately 80 km north-west of Cusco, Peru. This site of exceptional cultural heritage and its related infrastructure are being undermined by rapid debris flows, that are related to the presence of thick debris deposits produced by granite weathering, past slides and climatic conditions. On 26 December 1995 a rock fall/debris flow occurred on the road that leads to the citadel (Carretera Hiram Bingham) interrupting the traffic coming from the railway station of Aguas Calientes, and on 10 April 2004 a major debris flow, channeled in the Alcamayo stream, devastated the village of Aguas Calientes, causing 11 casualties and damaging the railway. Within the framework of the International Consortium on Landslides (ICL) a program of monitoring the instability conditions at this site was undertaken. In this work the preliminary results of the field survey and the analysis of some very high resolution (VHR) satellite images are presented. A multi-temporal analysis of Quickbird satellite (from Digitalglobe®) panchromatic and multispectral data was carried out: an archive image dated 18 June 2002 was available while a new acquisition with a good image was obtained on 18 May 2004. The main purpose of the analysis was the reconnaissance of debris flows using remote sensing techniques. The remote sensing data analysis was integrated with a field survey, carried out in September 2004. This allowed us to confirm the interpretation of the images, to produce a detailed geomorphological map of the area around the Carretera Hiram Bingham and to assess the thickness of debris deposits on the slopes. The results constitute a first step towards a complete debris flow hazard assessment in the area, where the interactions between slope instability and land use can produce very critical conditions.

Keywords. Debris flows, VHR satellite images, landslide hazard assessment, geomorphology, field survey, remote sensing

6.1 Introduction

Landslide identification, mapping and monitoring are basic tools for landslide risk and hazard assessment. They are traditionally carried out through field surveys, geotechnical and geophysical techniques and the analysis of aerial photographs of different dates and scales to produce detailed thematic maps. Optical satellite remote sensing technology has recently been exploited for landslide identification, since it is capable of providing reliable, cost-effective and repetitive information over wide areas. In fact remote sensing techniques offer an additional tool from which information can be extracted concerning landslide causes and occurrences, aiding investigations, on both a local and regional scale. Although they do not replace fieldwork, satellite images can provide useful information for steep terrain or areas covered by forest where access is difficult. Furthermore, interdisciplinary research strategies can use remote sensing data for testing the reliability of landslide prediction models.

Earth observation optical systems are passive sensors, measuring the sun reflectivity originating from a target on the Earth surface and/or from the atmosphere, in a range of wavelengths varying between 0.4–0.7 μm (visible spectrum), 0.8–0.9 μm (near-infrared) and 1.5–1.8 μm (medium infrared). Environmental missions such as Landsat, TERRA-ASTER and other environmental optical sensors have been not widely used for individual landslide mapping due to the insufficient spatial resolution. However, they are useful for indirect mapping methods, when the distribution of slope instability factors, such as geomorphology, lithology, land use, may be identified on these satellite images (Mantovani et al. 1996). In this sense, medium resolution data have been used for mass movement detection (Scanvic and Girault 1989; Nagarajan et al. 1998; Liu et al. 2003). The most important characteristic of optical sensors is the spatial resolution, which represents the detail discernible in an image and refers to the size of the smallest possible picture element (pixel) that can be detected. As a general statement, medium resolution (MR) refers to a pixel size of 30 to 15 m, high resolution (HR) refers to a pixel size of 10 to 5 m and very high resolution (VHR) refers to a pixel size of less than 5 m.

Exploitation of HR data is growing, especially with the integration of traditional instruments, and may sometimes even replace the interpretation of stereoscopic airborne images (Haeberlin et al. 2004). New generation very high resolution (VHR) satellite imagery (IKONOS, Quickbird) can provide a powerful tool for a quick reproduction of a regional map, up to a scale of 1 : 2 000, with a relatively low cost/benefit ratio due to the fact that these satellites have global coverage and the acquisition cycle is over a short-time period, making the images readily available. In hazard assessment, risk, emergencies and disaster management applications, with essential requirements such as: high spatial resolution of information, quick delivery of data, reliable interpretation and short revisiting time, these new instruments represent a viable tool in many fields including landslides (Hervas et al. 2003; Chadwick et al. 2005), floods, water management and land cover changes (Davis and Wang 2002; Sawaya et al. 2003).

Traditionally, optical satellite data represented a valuable tool for environmental monitoring (Crosta and Moore 1989; Vanverstaeten and Trefois 1993; Alves et al. 1999; Catani et al. 2002) due to the multi-spectral capability, the synoptic view, the high revisiting time and the medium spatial resolution. In new generation instruments two of these characteristics have been substantially improved:

- the concept of "multi-temporal resolution" has been modified by the possibility of programming the data acquisition, allowing information to be obtained shortly after an event, although acquisition still depending on the capability of satellite and cloud cover;
- with the introduction of the panchromatic band, the spatial resolution has increased to approximately 0.70 cm, decisively improving the possibility of deriving detailed ground observations of small-scale geomorphological processes.

Although in VHR data the multi-spectral characteristic is limited to the visible and the short infrared range of the electromagnetic spectrum, color composites and digital image processing allows the detection of features with a different spectral signature and variability in texture. For landslide interpretation from remote sensing images good brightness/contrast balance and high spatial resolution are fundamental: contrast enhances spectral differences between landslides and their surroundings while the landslide itself can not be identified if the resolution is too low (Mantovani 2000).

In this paper VHR images have been used to produce a debris flow inventory of the Machu Picchu area (Peru). Quickbird (from Digital Globe®) images were acquired in order to monitor frequent fast-moving, rock falls and debris flows triggered by intense rainfall, which have caused extensive damage in recent years.

6.2 Description of the Study Area

Machu Picchu represents the main monument of the Inca civilization. It stands 2 430 m a.s.l. in the middle of a tropical mountain forest in the eastern slopes of the Andes, overhanging the Urubamba River, which is a tributary of the Amazon River (Fig. 6.1). The citadel (Fig. 6.2) was revealed to the modern world after the 1911 Hiram Bingham expedition. The site was included in the UNESCO World Heritage List in 1983, for its natural and cultural relevance and has since become one of the main destinations for international tourism. The direct and indirect income

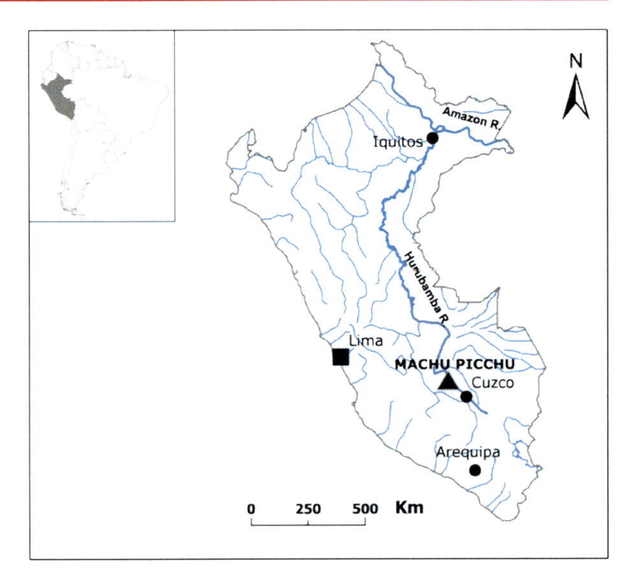

Fig. 6.1. Location map

derived from tourism constitutes a significant component of the Peruvian economy, but the relationship between the cultural and natural heritage, land use and visitor pressure are very precarious.

A symbol of this problem is the town of Aguas Calientes, located at the end of the railway from Cusco, which is linked to the archaeological area via a 8 km road (*Carretera Hiram Bingham*) running on the left bank of the Urubamba River. The village was built without urban planning on a fan along the Urubamba River at the base of some granitic faces, in a site of very high risk in terms of flash flooding and rockfall hazard. On 10 April 2004 a major debris flow, channeled in the Alcamayo stream, devastated the village, causing 11 fatalities and damaging the railway. More than 1 500 tourists remained isolated in the village and had to be rescued by helicopter.

The citadel is also affected by slope instability processes, with extensive deep-seated slow deformations and frequent shallow debris flows. On 26 December 1995 a rock fall/debris flow occurred on the road that leads to the citadel (Carretera Hiram Bingham) interrupting the traffic from the railway station of Aguas Calientes (Carreño and Bonnard 1997). Recent studies (Carreño and Bonnard 1997; Sassa et al. 2001, 2002) were focused on the deep-seated, slow landslides affecting the citadel and the Carretera, while this work focuses only on the shallow debris instability.

The main geological element of the region is the *Machu Picchu Batholith* (also known as the *Vilcabamba Batholith*), a large intrusive body formed of white-gray granites and granodiorites, dating from 246 ±10 Ma B.P. (Carlotto et al. 1999). It outcrops widely in the citadel and surrounding area, constituting the highest relief, such as the Cerro Machu Picchu (3 066 m a.s.l.), the Huayna Picchu (2 678 m a.s.l.) and the Putucusi (2 560 m a.s.l.),

Fig. 6.2. Machu Picchu archaeological site

which are the three peaks surrounding the archaeological site.

The batholith has a complex structural history, due to the cooling processes and superimposed tectonic phases that have determined the present structure, with a NE-SW major joint system, including the Huayna Picchu and the Machu Picchu faults (Carlotto et al. 1999). The intersection of this system with a regional NW-SW trend of master joints creates a regular network that controls the course of the Urubamba River. At a more local level, other joint sets become relevant and the most important dip NE, parallel to the slope below the citadel.

The rock mass is highly affected by chemical weathering, as a consequence of the feldspar sericitization and, more frequently, of the limonitization. These chemical processes, added to the physical weathering, caused the local formation of variable thickness debris sheets, that represent the source material for shallow landsliding. Processes of debris instability in granite, related to weathering and chemical alteration are well known in the scientific literature (Durgin 1977; Lee and De Freitas 1989; Zhao et al. 1994; Calcaterra et al. 1996; Chigira 2001; Palacios et al. 2003). Shallow landslides also occur due to older landslide debris deposits.

6.3 Methodology

The Quickbird satellite, launched in October 2001, acquires panchromatic (black and white) images with a resolution of 70 cm in the range 0.45–0.9 µm of the electromagnetic spectrum, as well as multi-spectral images (4 bands) with a resolution of 2.44 m in the visible (bands 1, 2 and 3) and near infrared (band 4) covering a minimum surface area of 16.5 × 16.5 km. In this study a multi-temporal analysis of Quickbird panchromatic and multi-spectral data was carried out: an archive image dated 18 June 2002 was available, while a new acquisition was scheduled for the middle of April 2004 with a good image being obtained on 18 May 2004.

Figure 6.3 shows the multi-spectral Quickbird image dated 18 May 2004 and printed in true color composite for the whole of the study area: ridge, valleys, rivers and urban areas are evident, while information on vegetation, outcrops and bare lands are also visible. The village of Aguas Calientes is clearly shown along the Urubamba River, while the Machu Picchu ruins can be identified on the left-hand side of the image together with the Carretera running up the hill.

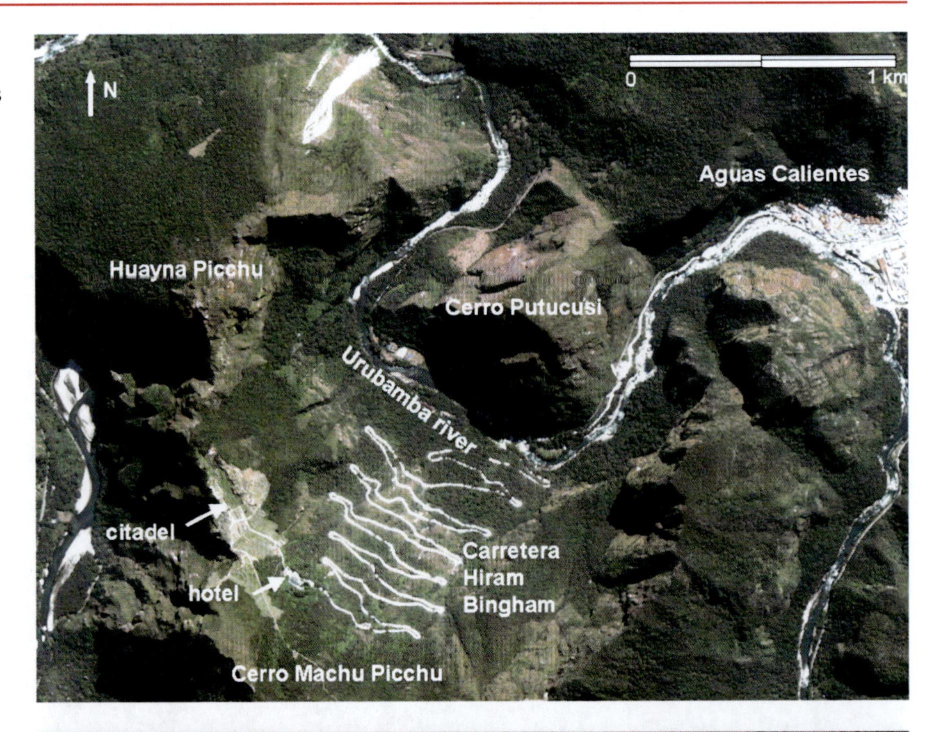

Fig. 6.3.
Quickbird satellite image dated 18 May 2004, bands 3-2-1 in RGB colors

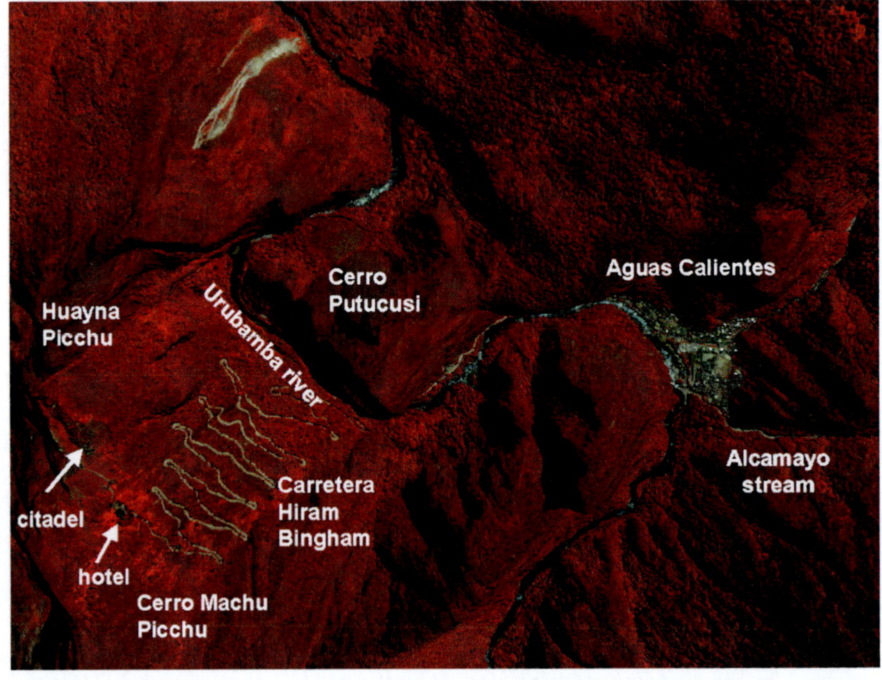

Fig. 6.4.
Quickbird satellite image dated 18 May 2004, bands 4-3-2 in RGB colors

The following areas were studied in detail in order to identify the debris flows that took place between the two satellite acquisitions:

a the basin of the Alcamayo stream and the village of Aguas Calientes;
b the northern slope of the Huayna Picchu peak;
c the northern slope of Cerro Machu Picchu and the Carretera Hiram Bingham.

Debris flow reconnaissance was the main purpose of the analysis and interpretation of the images. This involved important aspects such as: the size of the features, their texture in the image, the variety of forms and the contrast in terms of the difference in spectral characteristics between the landslides and the surroundings.

Images were geocoded in UTM projection zone 18 South Datum WGS84, and orthorectified through the

Rapid Polynomial Coefficients (RPC) process which combines several sets of input data to place each pixel in the correct ground location: RPC were available in the original data sets and elevation information was derived from a 10 m resolution Digital Elevation Model (DEM) previously acquired from contour line digitization.

Radiometric enhancement was carried out on both panchromatic and multi-spectral images in order to develop the most suitable product for a visual interpretation; the color composite of bands 4, 3 and 2 in red, green and blue with special contrast enhancement led to the discrimination of certain features and gives evidence of the main changes which occurred between the two acquisitions. Forest appears in light red, rock outcrops and bare areas from green to cyan while the debris deposits are shown in white or very bright cyan (Fig. 6.4).

In late September 2004 a field survey was conducted in the investigated area in order to validate the interpretation of the satellite images. Field observations were particularly focused in the areas of Aguas Calientes and of the Carretera Hiram Bingham, due to their high risk conditions.

An accurate field analysis focused on the textural differences observed in the Quickbird images. Each spot was investigated, with the aims of identifying the real geomorphological elements matching the image irregularities and detecting the evolution of slope processes between 2002 and 2004. Quantitative measurements of the thickness of the debris cover were carried out in order to asses the potential instability of the source areas of future debris flows.

6.4 Discussion of Results

6.4.1 The Alcamayo Basin and Aguas Calientes

Figure 6.5 shows the comparison of the satellite images of 2002 and 2004, using Bands 4-3-2 in RGB colors. The debris flow deposits along the Alcamayo stream related to the event of 10 April 2004 are evident on the image of 18 May 2004.

The panchromatic band allows a better detection of the size and distribution of the debris flow deposits along the stream (Fig. 6.6). This image has been processed by applying suitable spatial high-pass and directional filters, enhancing the structures with specific wavelengths and trends, showing that the material has accumulated along the stream over a distance of hundreds of meters, within which there are boulders up to some meters in diameter.

Table 6.1. Measurements of the debris deposits within the Alcamayo channel identified on the satellite images

No.	Area (m^2)	Perimeter (m)
1	682.27	107
2	301.58	76.2
3	1 077	306.8
4	1 111	219.4
5	687.59	175.6
6	939.04	194.7

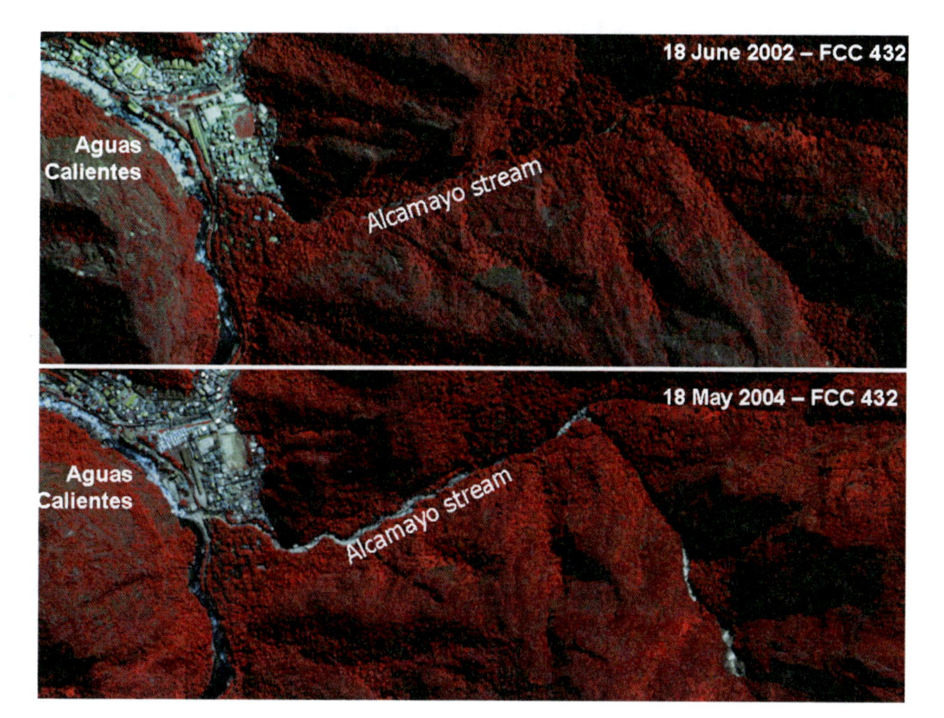

Fig. 6.5. Alcamayo stream and Aguas Calientes: comparison between the two Quickbird satellite images dated 18 June 2002 and 18 May 2004, respectively. Bands 4-3-2 in RGB colors

Six debris deposits were left in the channel of the Alcamayo after the devastating event of 10 April 2004. Of these, the area labeled as number 4 attracts particular attention, since its position is just downstream of a confluence. In fact, this may have been the decisive factor concerning the damage in Aguas Calientes, as it may have produced an ephemeral dam that subsequently caused a flash flood after its sudden collapse.

Table 6.1 shows the area and the perimeter of the debris deposits as measured on the Quickbird image.

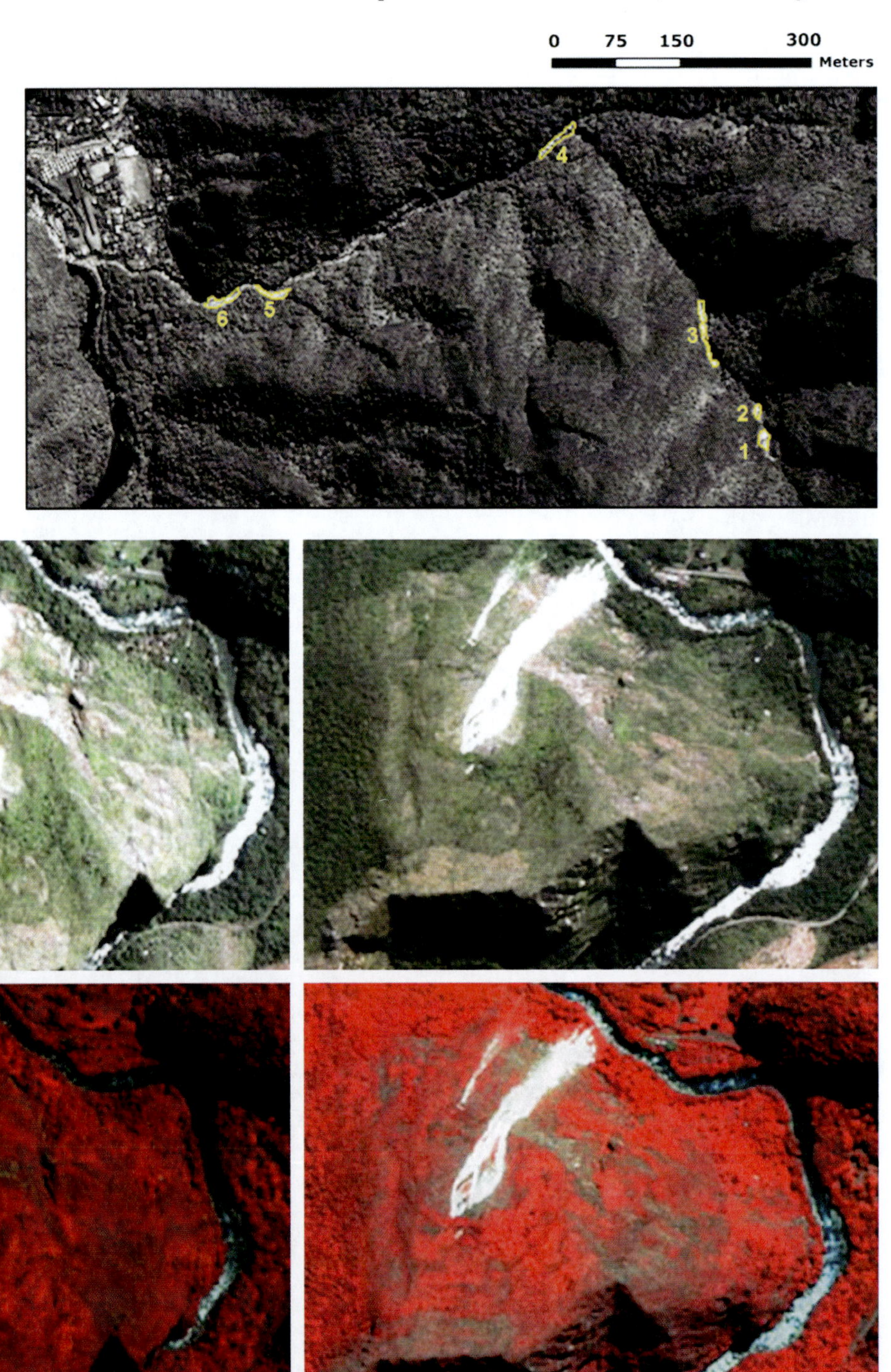

Fig. 6.6. Alcamayo stream and Aguas Calientes: Quickbird satellite image dated 18 May 2004, panchromatic band with debris deposits along the stream

Fig. 6.7. Northern slope of Huayna Picchu: Quickbird satellite image, bands 3-2-1 in RGB colors (above) and bands 4-3-2 in RGB colors (below). *Left side:* 18 June 2002; *right side:* 18 May 2004

6.4.2 The Northern Slope of the Huanya Picchu Peak

On the northern slope of the Huayna Picchu peak a major debris flow occurred within the period under investigation; as shown in Fig. 6.7 the initiation of the landslide is hardly detectable in the 2002 image while in 2004 a debris flow of more than 400 m long is evident along the slope. The texture and development of the debris flow is clearly detected in the panchromatic image processed by contrast stretch and convolution filtering with a sharpen kernel (Fig. 6.8). The debris flow deposits cover an area of 38 509 m^2, with a perimeter of 1 429 m. A minor debris flow is evident just to the north of the main one, covering an area of 5 255 m^2 with a perimeter of 601 m.

These two events are probably not related to the rainfall event which caused the Alcamayo flows and flood in April 2004. Some information gathered on site and the analysis of a few photos taken in a field survey in September 2003, allow us to date the landslides to the 2002–2003 austral summer. The comparison between the Quickbird images, the photos and the field evidence also leads us to the conclusion that the debris flows occurred as single events, without further reactivations.

6.4.3 The Northern Slope of Cerro Machu Picchu and the Carretera Hiram Bingham

The northern slope of Cerro Machu Picchu was the object of a detailed study, since the presence of the Carretera Hiram Bingham, with the repeated transit of tourist buses, is related to very high risk conditions. As mentioned in the previous sections, a major rock fall/debris flow occurred in this area on 26 Dec. 1995, creating a road block which continued for several months (Carreño and Bonnard 1997; Copesco 1997).

Figure 6.9 shows a detail of the site on the satellite image dated 18 May 2004 in panchromatic band. Several scars of rock falls and debris flows, of small scale, are evident in correspondence with the road cuts.

The main sources of debris flow hazard correspond to the sectors of the slope covered with thick debris deposits. For this reason, during the field survey in September 2004, the debris sheets were characterized with over 80 measurements (Fig. 6.10), aimed at assessing debris thickness, grain size and permeability.

The reconstruction of the debris thickness isopaches, shown in Fig. 6.11, was the first outcome of the field survey. The SE sector of the Carretera contains debris deposits up to 5 m thick and is the area of highest hazard for new debris flow initiation.

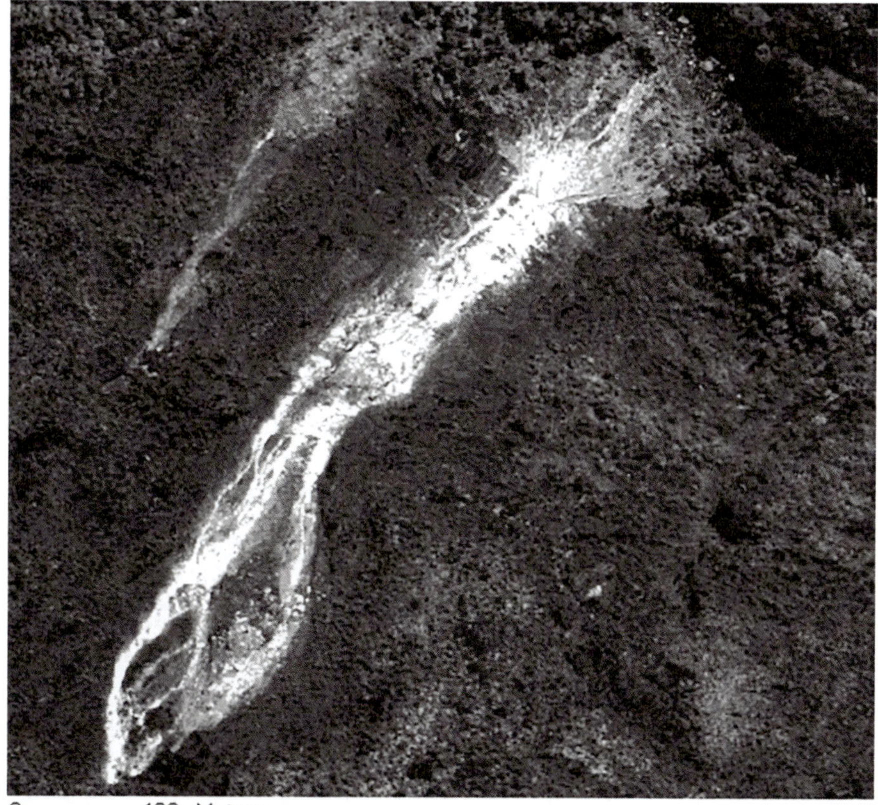

Fig. 6.8. Northern slope of Huayna Picchu: details of the debris flows on the Quickbird satellite image dated 18 May 2004, panchromatic band

Fig. 6.9.
Northern slope of Cerro Machu Picchu: Quickbird satellite image dated 18 May 2004, panchromatic band. Detail of the debris flows along the Carretera Hiram Bingham

Fig. 6.10.
Field survey of debris along the Carretera Hiram Bingham

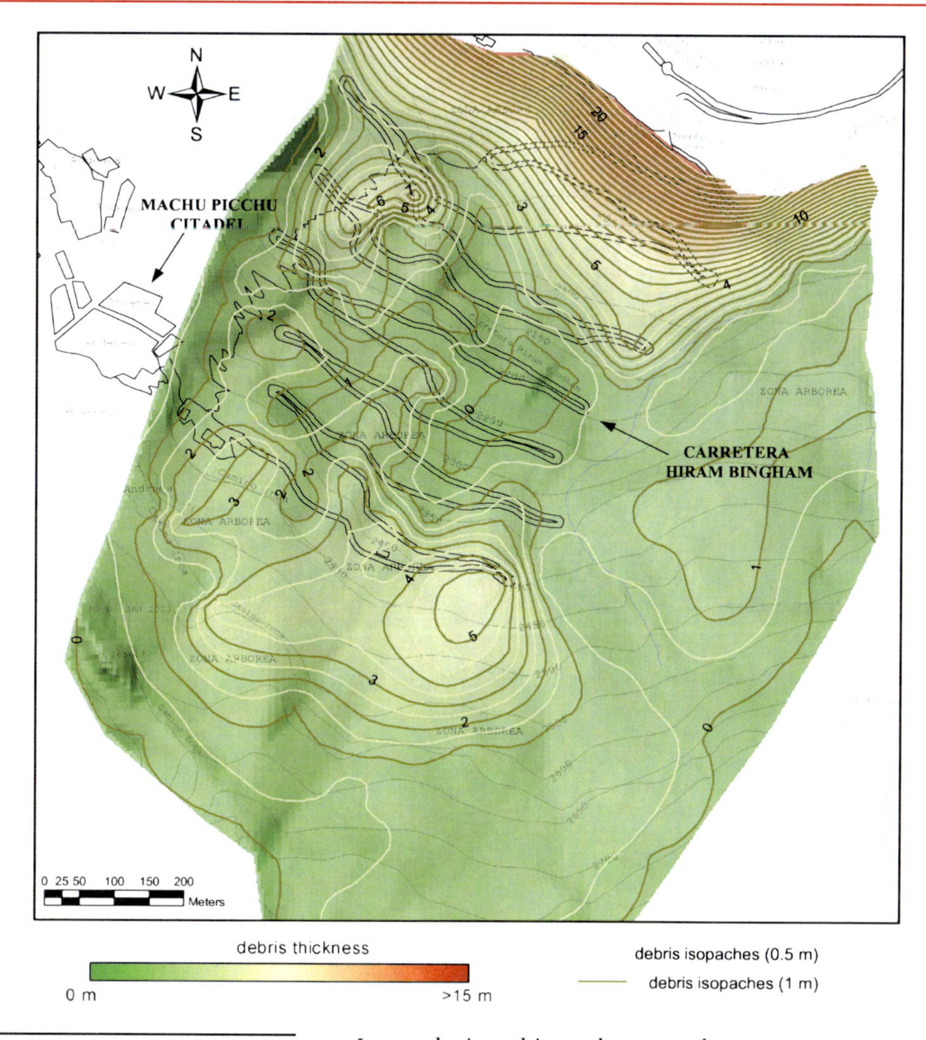

Fig. 6.11. Map of the debris thickness in the northern slope of Cerro Machu Picchu

6.5 Conclusions

The results of the analyses confirm that the VHR satellite data is capable of detecting superficial landslides, even of small scale, and is very useful for supporting traditional field surveys. The comparison between the 2002 and 2004 Quickbird images allowed us to determine the main instability processes that occurred over this period, specifically within the Rio Alcamayo Basin, on the northern slope of the Huayna Picchu and on the northern slope of Cerro Machu Picchu.

The integration of remote sensing data and field observations was also employed to produce detailed maps of the debris cover and to qualitatively assess the instability potential of the future debris flow source areas. This is particularly important in a region, such as the one under consideration, where rainfall has a very high spatial variability and is influenced by the slope aspect. Oral reports of residents, after the 10 April 2004 disaster, confirm a discontinuous rainfall pattern in space and in time. The presence of debris sheets over the slopes and within channels explains the different response of similar sites to the same rainfall event.

In conclusion, this work proves that remote sensing imagery should be regarded as an important source of information able to improve the assessment of natural hazards, bearing in mind that the meaning and value of the results requires skilled interpretation used in conjunction with conventionally mapped information and field-collected data.

Acknowledgments

Machu Picchu and the surrounding zones are the test sites of some ongoing projects under the co-ordination of the International Consortium on Landslides (ICL). The authors are involved in the general ICL project, coordinated by K. Sassa (DPRI, University of Kyoto) and titled "Landslide Investigation in Machu Picchu", and in its sub-project, funded by the Italian Ministry of Education, University and Research, devoted to the integration of remote sensing techniques for slope instability monitoring and mapping (INTERFRASI Project, Coordinators: P. Canuti, University of Firenze and C. Margottini, ENEA Rome).

References

Alves DS, Pereira JLG, De Sousa CL, Soares JV, Yamaguchi F (1999) Caracterizing landscape changes in central Rondonia using Landsat TM imagery. Int J Remote Sens 20(14):2877–2882

Calcaterra D, Parise M, Dattola L (1996) Debris flows in deeply weathered granitoids (Serre Massif, Calabria, Southern Italy). In: Senneset K (ed) Proceedings of the Seventh International Symposium on Landslides, Balkema, Trondheim, pp 171–176

Carlotto V, Càrdenas J, Romero D, Valdivia W, Tintaya D (1999) Geologìa de los quadràngulos de Quillabamba y Machu Picchu. Boletìn no. 127, serie A: Carta Geologica Nacional, Lima, pp 321

Carreño R, Bonnard C (1997) Rock slide at Machupicchu, Peru. Landslide News 10:15–17

Catani F, Righini G, Moretti S, Dessena MA, Rodolfi G (2002) Remote sensing and GIS as tools for the hydro geomorphological modelling of soil erosion in semi-arid Mediterranean regions. MIS2002 Management Information Systems, WIT Press, 2002. pp 43–52

Chadwick J, Thackray G, Dorsch S (2005) Landslide surveillance: new tools for an old problem. EOS 86(11)

Chigira M (2001) Micro-sheeting of granite and its relationship with landsliding specifically after the heavy rainstorm in June 1999, Hiroshima Prefecture, Japan. Eng Geol 59:219–231

Copesco (1997) Mehoramento acceso Puente Ruinas conjunto arqueologico de Machupicchu. Projecto Especial Regional Plan Copesco, Cusco, pp 147

Crosta AP, Moore JMcM (1989) Geological mapping using Landsat Thematic Mapper imagery in Almeria Province, Southeast Spain. Int J Remote Sens 10(3):505–514

Davis CH, Wang X (2002) Urban land cover classification from high resolution multi-spectral IKONOS imagery. Proceedings of IGARSS 24–28 June 2002, Toronto, Canada

Durgin PB (1977) Landslides and the weathering of granitic rocks. Rev Eng Geol 3:127–131

Haeberlin Y, Turberg P, Retiere A, Senegas O, Parriaux A (2004) Validation of SPOT-5 satellite imagery for geological hazard identification and risk assessment for landslides, mud and debris flows in Maragalpa, Nicaragua. XXth ISPRS Congress, XXXV, part B1, Int. Soc. for Photogramm. and Remote Sensing, Istanbul, Turkey, 12–13 July 2004

Hervas J, Barredo JI, Rosin PL, Pasuto A, Mantovani F, Silvano S (2003) Monitoring landslides from optical remotely sensed imagery: the case history of Tessina landslide, Italy. Geomorphology 54:63–75

Lee SG, De Freitas MH (1989) A revision of the description and classification of weathered granite and its application to granite in Korea. Q J Eng Geol 22:31–48

Liu JG, Mason PJ, Clerici N, Chen S, Davis AM, Miao F, Deng H, Lieng L (2003) Landslide hazard assessment in the Three Gorges area of the Yangtze River using ASTER imagery. Proceedings of IGARSS 2003, 21–25 July 2003

Mantovani F (2000) Remote sensing techniques on landslide detection and monitoring. Natural Hazard on built-up areas Proceedings of CERG Intensive course, 25–30 September 2000

Mantovani F, Soeters R, Van Westen CJ (1996) Remote sensing techniques for landslide studies and hazard zonation in Europe. Geomorphology 15(3–4):213–225

Nagarajan R, Mukherjee A, Roy A, Khire MV (1998) Temporal remote sensing data and GIS application in landslide hazard zonation of part of Western Ghat, India. Int J Remote Sens 19(4):573–585

Palacios D, Garcia R, Rubio V, Vigil R (2003) Debris flows in a weathered granitic massif: Sierra de Gredos, Spain. Catena 51:115–140

Sassa K, Fukuoka H, Kamai T, Shuzui H (2001) Landslide risk at Inca's World Heritage in Machu Picchu, Peru. Proceedings UNESCO/IGCP Symposium on Landslide Risk Mitigation and Protection of Cultural and Natural Heritage, Tokyo, pp 1–14

Sassa K, Fukuoka H, Shuzui H, Hoshino M (2002) Landslide risk evaluation in Machu Picchu World Heritage, Cusco, Peru. Proceedings UNESCO/IGCP Symposium on Landslide Risk Mitigation and Protection of Cultural and Natural Heritage, Kyoto, pp 1–20

Sawaya KE, Olmanson LG, Heinert NJ, Brezonik PL, Bauer ME (2003) Extending satellite remote sensing to local scales: land and water resource monitoring using high-resolution imagery. Remote Sens Environ 88(1–2):144–156

Scanvic JY, Girault F (1989) Imagerie SPOT-1 et inventaire des mouvements de terrain: l'exemple de La Paz (Bolivie). Photointerpretation 89-2(1):1–20

Vanverstaeten M, Trefois P (1993) Detectability of land systems by classification from Landsat Thematic Mapper data of Virunga National Park (Zaire). Int J Remote Sens 14(14):2857–2873

Zhao J, Broms BB, Zhou Y, Choa V (1994) A study of the weathering of the Bukit Timah Granite. Bull Int Ass Eng Geoll 50:97–111

Chapter 7

Influence of Thermal Expansion on Slope Displacements (C101-2)

Jan Vlcko* · Michal Jezny · Zuzana Pagacova

Abstract. Extremely slow deformations are a frequent phenomenon in the territory of Western Carpathians. These slope movements are in common generated by *internal* (geological, structural, morphological, geomechanical, hydrogeological, etc.) and *external* factors (climate factors as temperature, precipitation, air pressure etc.). One of several possibilities how to get more detailed information about kinematics of the rock displacements is monitoring. Within the monitoring records indicating rock slope movement besides real displacements (length/time) climate factors mentioned above are included. In our study, from among several external factors, we tried to estimate the daily temperature fluctuations influence on slope movement – i.e. its kinematics and dynamics as well as on monitoring recording sets. This paper describes the thermally induced influences on the rock mass behavior (expansion, contraction) as well as the possibility of data records filtering and thus to bring representative and essentially correct data into slope activity estimations, geotechnical calculations as well as time prediction for potential failure.

The temperature monitoring we performed at Spiš Castle (Eastern Slovakia), a monument included in the UNESCO World Heritage List.

Keywords. Extremely slow deformations, monitoring, thermal expansion, unitary expansion – relative dilation, linear thermal expansion coefficient, thermal residual strain, Spiš Castle

7.1 Introduction

A slope surface is affected by both daily and seasonal temperature fluctuations caused by air temperature changes, wind cooling and solar radiation. These cyclic fluctuations are partially transmitted to the interior of the rock mass by means of conduction and induce mechanical changes relative to thermal expansion (Gunzburger et al. 2004). Thus thermal expansion (compression or contraction) is causes physical rock deterioration (rock loosening) and rock mass disintegration and thus in many cases acts as a trigger for various exogenous phenomena starting with weathering up to various forms of slope movements.

The reliability and accuracy of monitored data recorded by the help of any crack gauge is on one hand based on the material temperature correction of the used device, thermal properties of the monitored rock body are generally not considered. For that reason we included into our work the study on thermal characteristics of rocks and thus to bring the realistic knowledge concerning monitoring data sets as well as the information if the data presented are essentially correct. The test site for our research was selected the Spiš Castle (Fig. 7.1).

In order to estimate the representative rate of displacement two approaches were adopted:

- monitoring carried out by the mechanical-optical crack gauge type TM-71,
- and determination (laboratory and in situ) of thermal expansion characteristics of travertine rocks.

7.2 Study Site

From a geological point of view, the Spiš Castle is built on a travertine mound, which is underlain by Paleogene soft rocks formed by claystone and sandstone strata (flysh-like formation). Lateral spreading caused by the subsidence of the strong upper travertine into the soft underlying claystone has fractured and separated the castle rock into several cliffs. The differential movement (rate of displacement, direction) of individual cliffs is the phenomenon influencing the stability of the monument.

Fig. 7.1. Spiš Castle (www.grafika.cz)

The history of monitoring at Spiš Castle goes back to 1980 when three TM-71 devices were installed. Six additional devices were installed later on following our survey carried out in 1992. It should be noted that only four of nine crack gauges operate, since unexpected visitors demolished the rest of them (Vlčko 2004). We focused our attention to so called Perun's rock, a southern part of the castle rock where three monitoring devices of crack-gauge type TM-71 (Fig. 7.2) were installed and the rate of displacement 1–2 mm per year was recorded.

As can be seen from data records (Fig. 7.3) the temperature plays the key role in interpretation of monitoring time series. With the rising of the temperature, the distance between opposite crack faces is reduced (rock expansion) and reversal. This phenomenon can be explained by the thermal expansion of the particular rock or rock body.

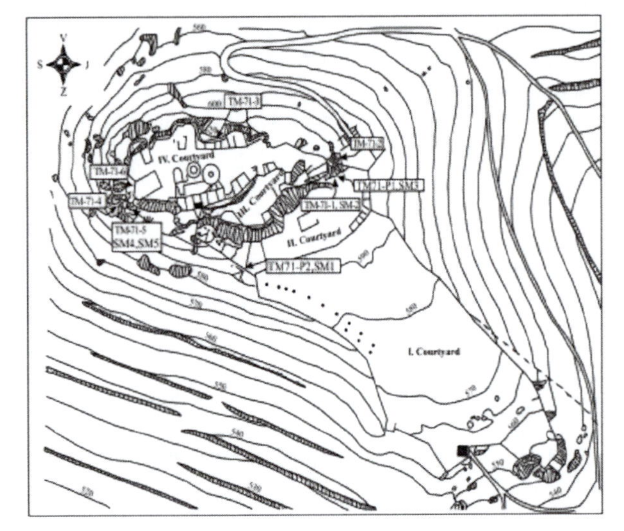

Fig. 7.2. Location of monitoring devices at Spiš Castle

To learn more about thermal properties of travertine, the dominating rock type forming the subgrade of the monument, we undertook several laboratory and in situ experiments.

As far as the laboratory methodology concerns we followed the European Standards Proposal – "Natural stone test methods: Determination of the thermal expansion coefficient" which specifies methods to determine the linear thermal expansion coefficient of natural stone based on mechanical length-change measurements.

The linear thermal characteristics are:

- Unitary expansion – relative dilation, ε
- Linear thermal expansion coefficient, α
- Thermal residual strain, L_r

Unitary expansion – relative dilation (ε) expresses relative length change within certain temperature range and is expressed by the formula:

$$\varepsilon = \Delta L / L_i \ [\text{mm m}^{-1}]$$

ΔL is the change of length of the sample between initial temperatures T_i and final experimental temperature T_{fin}. L_i: initial sample length at the temperature T_i 20 °C).

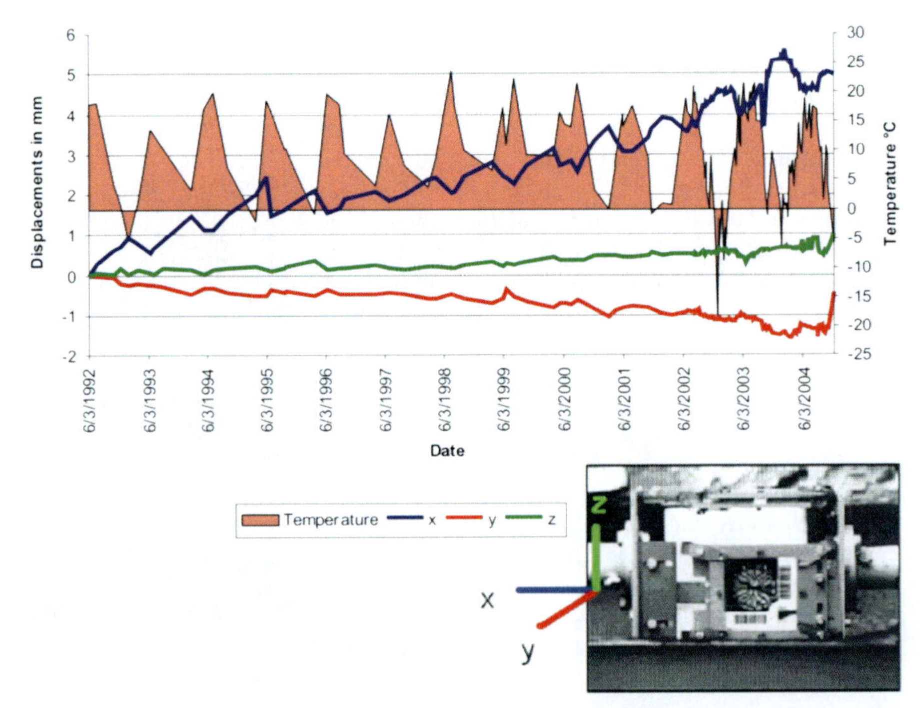

Fig. 7.3. Correlation between temperature and monitoring records at Perun's rock (Spiš Castle); *x*-axis corresponds to distance variation (compression, extension) between opposite crack faces, *y*: horizontal shear displacement along a crack and *z*: vertical shear displacement

Linear thermal expansion coefficient (α) of rocks can be defined as unitary expansion of the sample – relative dilation (ε) related to the temperature interval $\Delta T = T_{fin} - T_i$,

$$\alpha = \frac{\varepsilon}{\Delta T}$$

and can also be expressed as

$$\alpha = \frac{\Delta L}{L_i \Delta T}$$

Temperature *change* ΔT corresponds to the temperature range $\Delta T = T_{fin} - T_i$.

Apart from the temperature the thermal rock expansion properties depend on mineralogical composition, texture, structure and physical properties.

Thermal residual strain, ΔL_r, represents a measurable quantity corresponding to the destruction of the sample after heating/cooling treatment (Siegesmund et al. 2000, 2001; Kirschner et al. 2003). Thermal residual strain corresponds to the length increment of the sample length before (L_i, initial sample length) and after experiment until in the heater/cooler chamber initial temperature (T_i) is fixed. Actually it represents a non-reversible sample length.

Thermal expansion tests were performed on travertine using a thermodilatometer *VLAP 01* especially constructed for these purposes with heater/cooler chamber capable of raising the temperature from −20 ±0.2 °C to +80 ±0.2 °C at a rate of 0.1 °C min^{-1} and maintaining temperatures within that range at least two hours with an accuracy of at least ±0.5 °C. The apparatus is linked with data acquisition system in order to record the temperature and length change of tested sample and to calculate linear thermal characteristics. Thermodilatometer VLAP 01 is capable to measure thermal expansion on two samples simultaneously. The rock samples were of cylindrical shape 25 × 50 mm in size cut in three x-, y-, and z-axis orientations. A dummy sample corresponding to the same rock type with a thermocouple in its center regulates the temperature and secures the identical experimental conditions for both rock samples.

In order to simulate in laboratory the temperature conditions comparable with the natural ones four thermocouples at four different depths (7, 25, 40 and 80 cm) inside the creeping and monitored travertine cliff (Peruns rock) were installed and since January 2004 continually recording the temperature inside the rock mass (Vlčko et al. 2005). Based on the temperature recordings three thermo cycles corresponding to real seasonal temperature conditions were distinguished and in laboratory tests using apparatus VLAP 01 were performed. These are as follows:

- Summer cycle with the temperature range from +20 °C to +50 °C,
- Winter cycle with the temperature range from −5 °C to −20 °C,
- Spring/autumn cycle with the temperature range from +20 °C to −5 °C.

7.3 Results

The experiments were undertaken on 72 samples and the results we gained is related to directional dependence and maybe summarized as follows: the highest value of linear thermal expansion coefficient was observed on the samples corresponding to the y-axis direction with the mean α value reaching 15.060×10^{-6} °C^{-1}, then in the x-axis direction with the mean α value reaching 13.310×10^{-6} °C^{-1} and finally the lowest mean α value was determined on samples with z-axis direction reaching 13.299×10^{-6} °C^{-1}.

The highest value of thermal residual strain (in some papers assigned as residual dilation) was observed in direction of y-axis with mean value $\Delta L_r = 3.9 \times 10^{-6}$ m, following direction in z-axis with mean value $\Delta L_r = 1.5 \times 10^{-6}$ m and the lowest one was determined in direction of x-axis with mean value $\Delta L_r = 1.23 \times 10^{-6}$ m. The observed differences in thermal residual strain can be explained by means of the thermal anisotropy of the calcite grains in connection with the natural preferred orientation in travertine rocks.

In order to get representative and essentially correct data from readings taken once a week from each crack gauge installed at Spiš Castle the data were corrected in accordance with thermal properties tests. As a "correction factor" we applied mean L_{fin} value (Fig. 7.4) determined on rock samples cut in x-axis corresponding to the crack-gauge orientation (across the tension crack) for each of tested thermo cycles. Weekly monitoring data records were re-

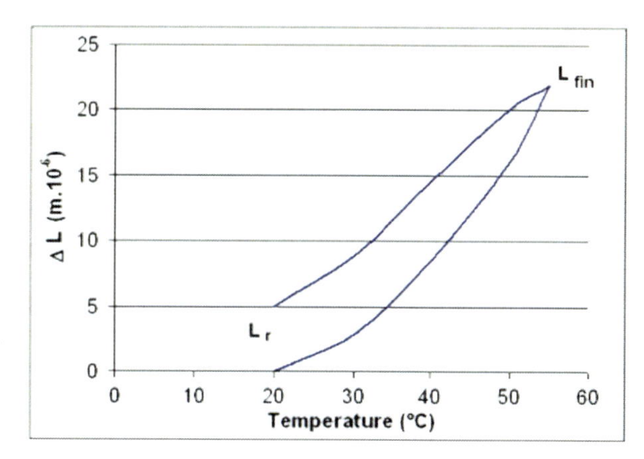

Fig. 7.4. Typical plot showing the change of length (ΔL) of the sample between initial temperatures T_i (20 °C) and final experimental temperature T_{fin} (55 °C). ΔL_r corresponds to the length increment before and after experiment until in the heater/cooler chamber initial temperature (T_i) is fixed. ΔL_r corresponds to the value 5×10^{-6} m

duced by $2L_{fin}$ (there is the same rock type along both sides of the monitored tension crack with the same L_{fin} mean value) when mean L_{fin} values for each thermo cycle were determined by laboratory tests as follows:

- Spring/autumn cycle 0.027 mm
- Summer cycle 0.017 mm
- Winter cycle 0.032 mm (Fig. 7.5)

According the procedure mentioned above the final plot-portraying rate of displacements at TM-71-1 (Fig. 7.6) shows the constant trend of slope movement. As far as the monitoring records concerns the correction has no essential effect on the rate of displacement, particularly when long-term monitoring records exist, this only gives the user objective information about monitored phenomenon.

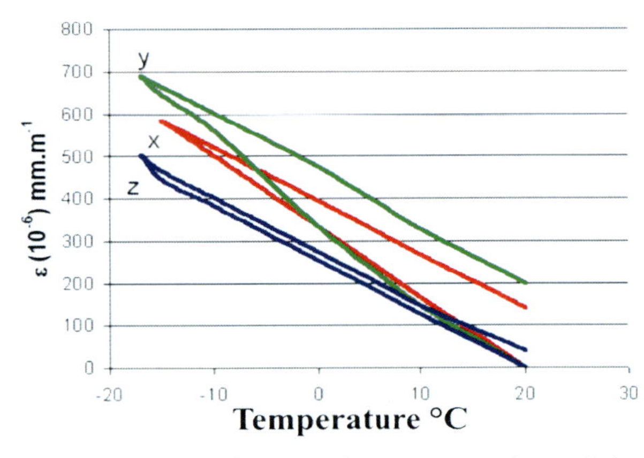

Fig. 7.5. Typical plot of winter cycle unitary expansion – relative dilation (ε) vs. temperature. *1:* Samples cut in y-axis orientation; *2:* samples cut in x-axis; *3:* samples cut in z-axis

Acknowledgments

This paper was prepared as part of the framework of Partial monitoring system of geological factors of the environment in Slovakia, VEGA grant project no. 1/1028/04, IPL Project no. 101-2 OF and a bilateral project between Kyoto University, Faculty of Natural Sciences Jap/Slov/JSPS/40. The author is thankful to the Ministry of the Environment of the Slovak Republic, Ministry of Education of the Slovak Republic, grant agency VEGA and the International Consortium on Landslides for their kind support.

References

Gunzburger Y, Soukatchoff VM, Guglielmi Y (2004) Influence of daily surface temperature fluctuations on rock slope stability: case study of the Rochers de Valabres slope (France). International Journal of Rock Mech and Mining Sciences

Kirschner D, Ondrasina J, Siegesmund S (2003) Freeze-thaw cycles. In: Natural stone weathering phenomena, conservation strategies and case studies. Z Deut Geol Ges

Siegesmund S, Weiss T, Tschegg E (2000) Control of marble weathering by thermal expansion and rock fabrics. In: 9th International Congress on Deterioration and Conservation of Stone, Venice

Siegesmund S, Rasolofosan PN, Weiss T (2001) Thermal micro cracking in Carrara marble. Z Deut Geol Ges

Vlčko J (2004) Extremely slow slope movements influencing the stability of Spiš Castle, UNESCO site. In: Landslides, vol. 1, no. 1, Springer-Verlag, Berlin Heidelberg, pp 67–71

Vlčko J, Jezny M, Pagáčová Z (2005) Thermal expansion effect on slope deformation recordings at Spiš Castle. 15th Conference on Engineering Geology, Erlangen, Germany

Fig. 7.6.
Typical plot of displacement records at Spiš-Castle from the crack gauge TM-71-1

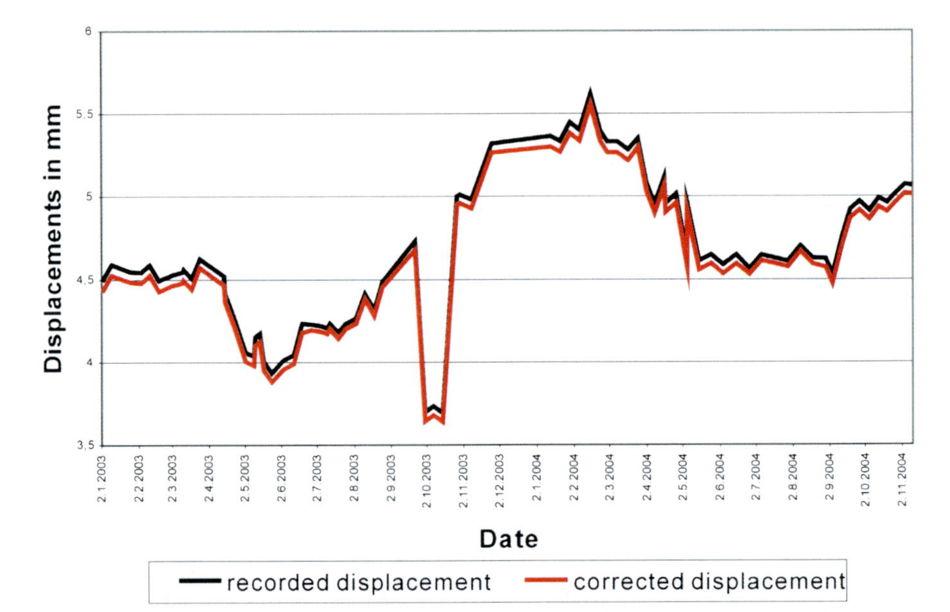

Chapter 8

Emergency Intervention for the Geo-Mechanical Conservation of the Niches of Bamiyan Buddhas (Northern Afghanistan) (C101-3)

Claudio Margottini* · Vittorio Colombini · Carlo Crippa · Gedeone Tonoli

Abstract. The historical site of Bamiyan is affected by geomorphological deformation processes which were enhanced during the Talibans' bombing in March 2001, when the two standing Buddhas, dating back to sixth century A.D. were destroyed. An invaluable cultural heritage was irremediably lost and the consequences of the explosions, as well as the collapse of the giant statues, added greatly to the geological instability of the area. Traces of rocks recently slid and fallen are relevant proofs of the deterioration of its stability conditions and most parts appear prone to collapse in the near future.

Under the coordination of the UNESCO, a global project to assess the feasibility conditions for the site's restoration was developed; field data were collected and a mechanism for the potential cliff and niches' evolution was provided. In the mean time some consolidation works were carried out in the most critical rock fall-prone areas to avoid any further collapse in the coming winter season, but also to enable archaeologists the safe cataloguing and recovering of the Buddha statues' remains, still laying on the floor of the niches. The emergency activities started in October 2003 and included: the installation of a monitoring system, the realization of temporary supports for the unstable blocks, the stabilization of the upper-eastern and upper-western parts of the small Buddha niche, the minimization of the environmental impact of the actions taken. Consolidation works were mainly implemented by professional climbers, directly operating on the cliff.

Keywords. Rock fall, mitigation works, Buddha statues, Bamiyan (Northern Afghanistan)

8.1 Introduction

In the great valley of Bamiyan, 200 km NW of Kabul, central Afghanistan, two big standing Buddha statues appear to visitors, carved out of the sedimentary rock of the region, at 2 500 m of altitude. Following the tradition, this remarkable work was done by some descendants of Greek artists who went to Afghanistan with Alexander the Great, probably around sixth century A.D.

Under the worldwide astonishment, the two statues were demolished on March 2001 by the Talibans, using mortars, dynamite, anti-aircraft weapons and rockets. The Buddhists as well as the world community, UN and UNESCO failed to convince the Talibans to avoid the destruction of this cultural heritage. Nevertheless, since 2002 UNESCO is coordinating a large international effort for the protection of the World Heritage Site of Bamiyan and the future development of the area.

8.2 General Features of the Area

The investigations performed in the Buddha niches and surrounding cliff in the Bamiyan Valley (northern Afghanistan) highlight the following general features (Margottini 2003, 2004):

- the area is located in mountainous central Afghanistan in a dry part of the world that experiences extremes of climate and weather. Winters are cold and snowy, and summers hot and dry. Mean annual precipitation in Bamiyan is about 163 mm and mean annual temperature, 7.4 °C.
- the rocks outcropping in the area are mainly conglomerate, with some strata of siltstone that largely slake under wet conditions. The lower part of the cliff is predominantly siltstone, with two main set of discontinuities spaced every 20–40 cm. The central part of the cliff is mainly conglomerate, well cemented and with a limited number of vertical discontinuities mainly paralleling the profile of the slope.

Major geomorphological processes include water infiltration, gully erosion, progressive opening of discontinuities in the outer parts of the cliff, weathering and slaking of siltstone levels, toppling of large external portions as well isolated blocks along the cliff face, occurrence of mud flows probably when the siltstone is saturated, sliding of a large portion of the slope, accumulation of debris at the toe.

The explosion of March 2001, besides the destruction of the statues, reduced the stability of the slope, mainly in the outer parts of the niches.

In the small Buddha niche, as well as the collapse of statue, three minor rock falls from the top of the niche occured. The blasting also degraded the upper-eastern part of the niche where a stairway is located inside the cliff, and the wall between the stairs and the niche is quite thin (about 30–50 cm). This part is presently the most critically unstable site. The western side, as consequence of an existing buttress, suffered less damage. Nevertheless, a rock fall occurred and some instabilities are now also evident only in the eastern part.

Fig. 8.1. The effect of destruction in the large (*left*) and small (*right*) Buddha niches

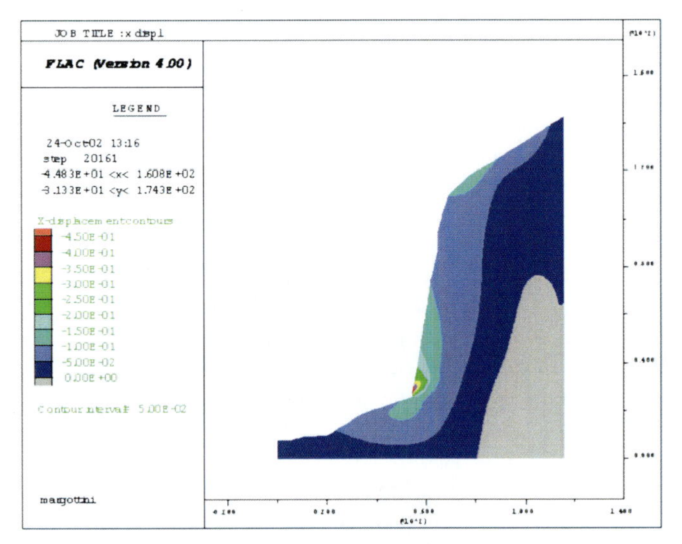

Fig. 8.2. Instability condition of the cliff in case of a discontinuity reaching the lower siltstone and weathering it. This condition was probably the cause for two large mass movements, detected close to each niche

Major effects in the great Buddha niche were the collapse of the statue and the consequent instability of the rear of the niche. A small rock fall occurred from left side of the top of the niche. Probably, the strength of the greater thickness of wall between the stairway going up into the cliff and the niche (about 1 m), reduced the effects of blasting and resulted in less severe damage (Fig. 8.1). Investigation of the possible stability conditions of cliff were computed using the explicit-difference-finite code, FLAC (ITASCA Consulting Group 2000). Considering the Hoek and Brown (1980) shear strength criteria for conglomerate and siltstone, and with a major discontinuity ranging from the middle of cliff till the middle of the niche (only friction value for shear strength) the deformation of the cliff is relatively low and it seems to be now in condition of stability. Assuming the fracture in conglomerate reaching the lower sandstone formation and decreasing gradually the cohesion of siltstone due to fracturing/weathering, the cliff is become unstable when the cohesion is near to nil. Under this situation maximum displacement and vector are at the base of the niche (Fig. 8.2).

8.3 Long-Term Conservation Strategy for the Geo-Mechanical Preservation of the Site

Considering the processes affecting the site, any long term stabilization strategy has to solve the causes of active processes and not to focus only in mitigating the effects. Different typology of stabilization measures have been investigating, considering also the need of implementing the work in an area of high cultural value but low, at the moment, technological support. Furthermore the consolidation works should be addressed to the entire area besides the restoration of, especially for those processes that are affecting the entire cliff (e.g. rainwater drainage).

Fig. 8.3. Unstable blocks in the upper-Eastern part of small Buddha niche and pattern of existing discontinuities

The final proposed long term solution includes the following typologies of stabilization techniques for both small and large Buddha statues:

- the protection from water circulation and infiltration from the upper part of the cliff (Mastropietro 1996);
- stabilization of cliff and potential rock falls by means of passive anchors (stainless steel) and bolts, grouted with cement with low water release. This intervention can be restricted to the Buddhas site and surroundings;
- In the back side of both Buddha niches a very fragile area has been surveyed. This is the place where original plaster is still existing. These areas should be safeguarded by a restorer as soon as possible. Removal of as minimum as possible unstable blocks of rock (barring down) using hand tools and installation of a net is required: (1) to allow the protection and restoration of the few remains of Buddha Statues; (2) to allow archaeologists the beginning of recovering and restoring of fragments.

8.4 Emergency Measures

After the general strategy for stabilization a follow up of activities was performed in September 2003 and still in progress, aimed at the identification of potential negative evolution of cliff and niches during the following months. The result of a UNESCO mission suggested an immediate response to the upper Eastern part of the small Buddha where existing large fissures were widening and the risk of an immediate rock fall was estimated to be very high. This collapse could involve large part of the upper Eastern part of the cliff and, as a consequence, to partially bury the niche (Fig. 8.3).

Fig. 8.4. The beam (*left*) and steel rope (*right*) for the temporary support of upper Eastern part of small Buddha

The field work started in October 2003 and included four different steps:

- the installation of a monitoring system;
- the realization of temporary supports for the unstable blocks (Fig. 8.4);
- the stabilization of the upper-eastern part of the small Buddha niche (superficial stainless steel nails, deep stainless steel anchors and the grouting of discontinuities to avoid water infiltration and increase shear strength);
- the minimization of the environmental impact of the actions taken.

Fig. 8.5. Final consolidation with the use of professional climbers

Since the winter season was approaching soon, speediness was to be given priority and climbers had to be sent to directly operate on the cliff. They were also supported by ground staff when working at consolidating the cavities from the inside to the outside (Fig. 8.5).

High attention was posed to the methodology for consolidation. To avoid any induced vibration firstly surficial bolts were drilled by means of 3.6 cm diamond head. Cooling water was reduced at minimum by means of a combined mixture with compressed air. Deep anchors were drilled from the further part of the unstable blocks, towards the most critical one. Grouting was made with cement added with superplasticizer to avoid any water release, capable to interfere with the slaking siltstone. The demonstration of correct realization of grouting is given from the anchor suitability tests, performed to understand the bounding capacity of anchors in both siltstone and conglomerate. The design strength of passive anchors was assumed in 20 t, for a bounded length, after the major discontinuity, at least of 5 m (about 4 t per linear meter); the anchor suitability was performed for 1 m length, till 40 t, close to yield capacity of steel. Up to this value no remarkable permanent elongation was detected, to demonstrate the correct bounding effect between siltstone and conglomerate and the anchors (Fig. 8.6).

8.5 Conclusion

The destruction of the two statues was a dramatic example of people living in conflict with history and

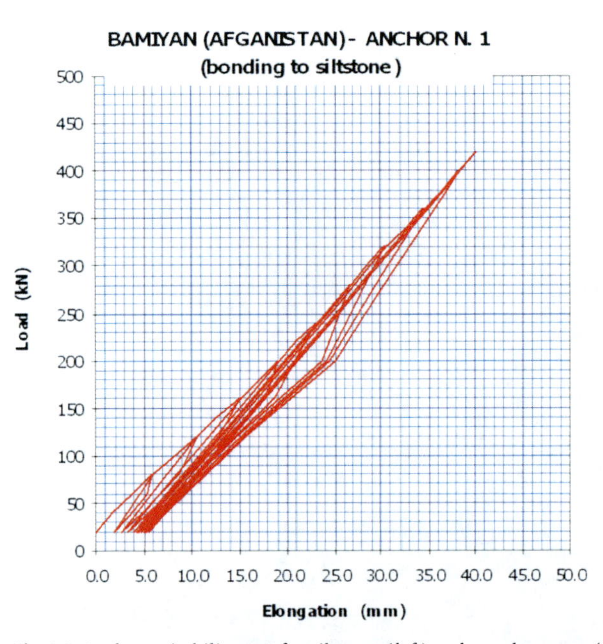

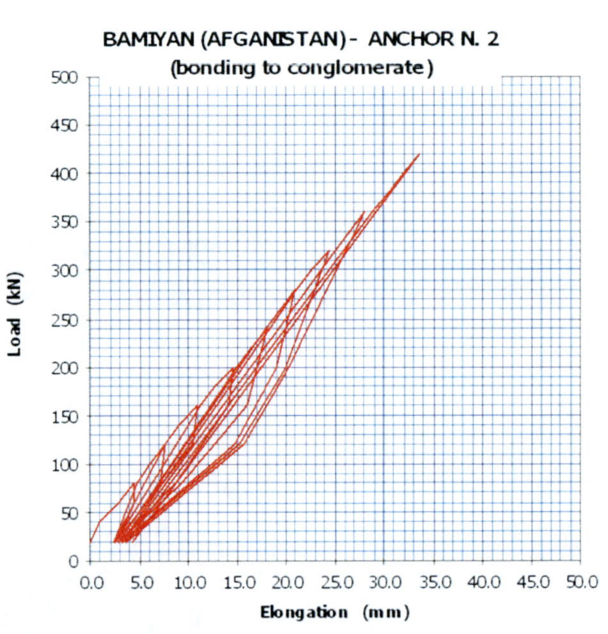

Fig. 8.6. Anchor suitability test for siltstone (*left*) and conglomerate (*right*)

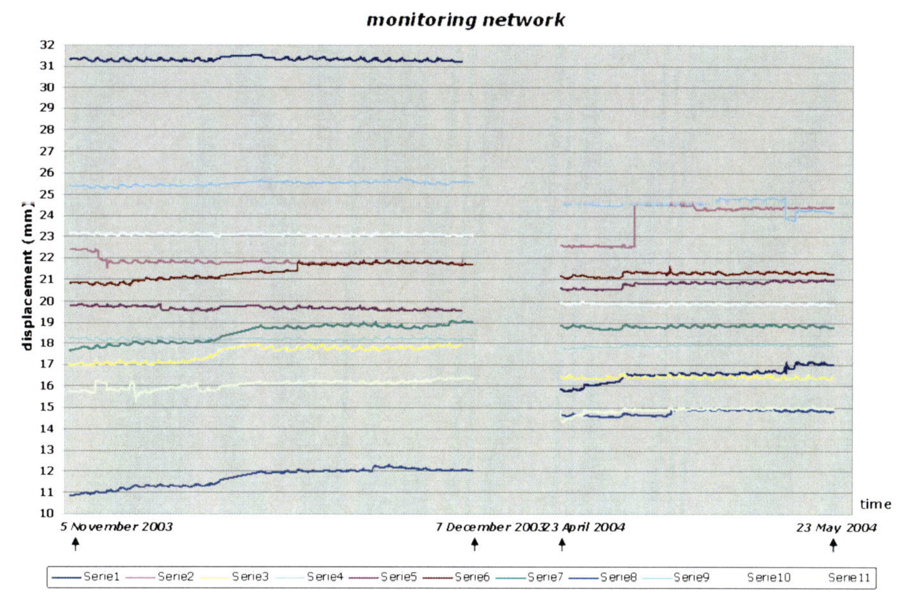

Fig. 8.7. The result of the crack gauge monitoring system. The few small sudden steps are related to external fault

tradition. The preservation of the few remains is a mandate to which our answer has to be the most rapid and appropriate.

The solution and the techniques adopted as well as the four steps improvement of activities proved quite satisfactory since the monitoring system did not record any remarkable deformation in the unstable blocks through the working period (Fig. 8.7).

After the intervention of October–December 2003 and, partially in March 2004, the cliff and niche of the small Buddha (the most critical part) it is now more stable and the risk of collapse almost avoided. Further work will be needed in the future but, al least, the major risk to have a collapse involving also the few remains not destroyed by Talibans is now turned away.

References

Hoek E, Brown LW (1980) Empirical strength criterion for rock masses. J Geotech Eng-ASCE 106(GT9):1013–1035

ITASCA Consulting Group (2000) Fast Lagrangian analysis of Continua, ver. 4.0. Minneapolis, Minnesota, U.S.A.

Margottini C (2003) The Buddha niches in the Bamiyan Valley (Central Afghanistan): instability problems and restoration plans in the UNESCO intervention. Journal of the Japan Landslide Society 40(3):246–249 (in Japanese)

Margottini C (2004) Instability and geotechnical problems of the Buddha niches and surrounding cliff in Bamiyan Valley, Central Afghanistan. Landslides 1(1):41–51

Mastropietro M (ed) (1996) Restoration and behind. Architecture from conservation to conversion. Projects and works by Andrea Bruno (1960–1995). Edizioni Lybra immagine, Torino

Chapter 9

Landslide Risk Assessment and Disaster Management in the Imperial Resort Palace of Lishan, Xian, China (C101-4)

Hiroshi Fukuoka* · Kyoji Sassa · Gonghui Wang · Fawu Wang · Yong Wang · Yongjin Tian

Abstract. Landslide risk assessment and disaster management are challenging problems in nowadays along with the progress of societal development towards mountainous area. Through a Japan-China joint research project for the landslide hazard assessment in the ancient imperial resort palace of Lishan, Xian, China, detailed investigation has been performed on the Lishan slope to assess its landslide risk, and then offer suitable preparedness method. It was clarified that the Lishan slope is deforming, showing the characteristics of precursor landslide movement. A short-span and two lines of thirteen long-span extensometers installed over high trees proved the effectiveness of the monitoring method. They gave an evidential deforming data showing the landslide risk which made the decision by the Chinese national and municipal governments to install landslide prevention measures. The development of semi real-time monitoring system enables us to give warning of landslide disaster in advance.

Keywords. Lishan slope, extensometer monitoring, landslide risk assessment, precursor landslide, landslide risk preparedness

9.1 Introduction

Landslide hazard assessment is to estimate the probability of landslide occurrence and the area, volume, speed and travel distance. Sassa and other 18 Japanese landslide researchers investigated the Lishan slope behind Huaqing Palace in the suburb of Xian City of China (Fig. 9.1) together with 50 Chinese landslide researchers at the Japan-China Joint Field Workshop on Landslides that was organized in Xian and Lanzhou, October 1987. Loess blocks of this slope already failed and slope deformation was obvious in other parts of the slope. Opinions of researchers were diverse regarding the depth and the area of landslides (shallow and small superficial soil creep to deep and large slides) and the present risk of this slope. The surface soil of this slope is mainly loess deposits; the bedrock is formed by Precambrian gneiss, which are outcropping. In the 1987 workshop, one thought that hard Precambrian gneiss rock mass is very unlikely to be subjected to landslides, another thought that this slope is the cliff of large active fault and the rock mass should be sheared and fractured which can move in this steep slope over 35 degrees. But dominant opinions in both countries were the former. The fact that this rock slope was stable since Tang Dynasty (A.D. 618–907) era though strong earthquakes repeatedly attacked this area (Lin 1997; Xian Seismological Bureau 1991) supported this opinion. However, members agreed the existence of surface deformation and the necessity of further investigation.

Almost in every country, landslide prevention measures are usually implemented by obtaining its budget initially after remarkable economic damage or death of people occurred. It is very difficult to obtain the necessary budget to prevent landslides prior to their occurrence. This is due to the fact that high capital investment is required, but no returned profit, and also due to the lack of reliable prediction method of landslide hazards. However, the most economical measure to reduce landslide disaster should be the reliable landslide risk prediction (assessment), followed by relocation of people, houses and other facilities or temporary evacuation before the landslide occurs. We believe that this can be the best case study in assessing a large scale landslide hazard before event, affords to obtain the budgets necessary for investigation because of the significance of the Imperial Resort Palace, especially for the famous Lady Yang-Que-Fe Palace.

After some efforts, Japanese group obtained budgets from Ministry of Education (later reorganized to Ministry of Education, Culture, Sports, Science and Technology (MEXT)) for the Japan-China Joint Research on the Assessment of Landslide Hazards in Lishan, Xian since 1991. In 1999, this project became the first sub-project of the UNESCO/IUGS joint project International Geoscience Programme (IGCP, former called as International Geological Correlation Programme) no. 425 "Landslide Hazard Assessment and Mitigation for Cultural Heritage Sites and Other Locations of High Societal Value" (1998–2003), which was proposed and headed by Sassa.

9.2 Objects at Risk in Lishan

The Imperial Resort Palace in Lishan was constructed in an area surrounded by the active fault system, where great earthquakes occurred several times in the past. The greatest one was the Huaxian earthquake with a Richter magnitude of 8 on 23 January 1556. However, no seismic ac-

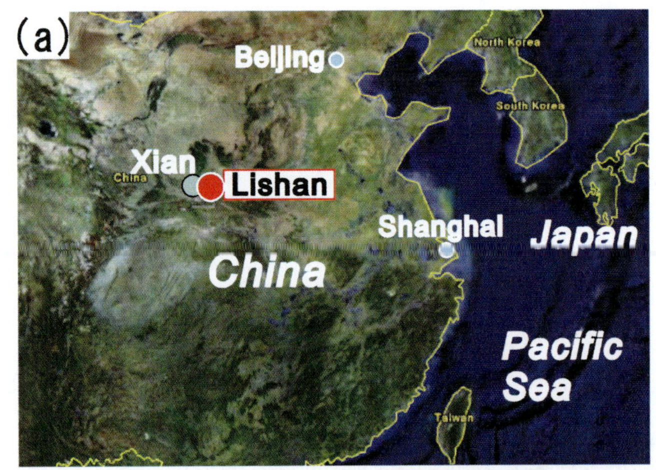

Fig. 9.1.
a Location of Lishan, China; **b** view of the Lishan slope from Huaqing Palace

tivity was reported in twentieth century. The mountain behind the Palace is a great cliff of the fault system. Although it is not volcanic area, hot water comes through the deep fault and such hot springs are used in the Huaqing Palace. This palace is well known for the Emperor Xuan Zong (A.D. 685–762) of Tang Dynasty era who enjoyed the hot spring bath with Lady Yang-Que-Fe. The baths still exist in the Palace. The Palace buildings have been reconstructed in twentieth century for the purpose of tourism. At present, this site attracts 3 million tourists annually from not only China but also from all over the world. Figure 9.2 is the view of the center of the Lintong Town at the foot of the Lishan slope. About 200 000 people lives in this area. The failure of this slope should cause a serious disaster without any doubt.

9.3 Potential Landslide Zoning

The Chinese group extensively investigated this slope. The conducted works include the topographical mapping of 1 : 500 scale, the geological mapping of 1 : 1 000 scale, more than 100 geological drillings, two investigation tunnels and monitoring by leveling and laser EDM, etc. The Japanese side provided advanced monitoring equipments (long-span and short-span extensometers, borehole inclinometers, the Three Dimensional Shear Displacement Meters, GPS receivers, Total Station with prism mirror targets, seismograph, data analysis system, rain gauge, etc.) and cooperated on monitoring of slopes as well as soil mechanical tests to study the dynamics of landslide.

Fig. 9.2. View of Huaqing Palace and the center of Lintong Town from the Lishan slope

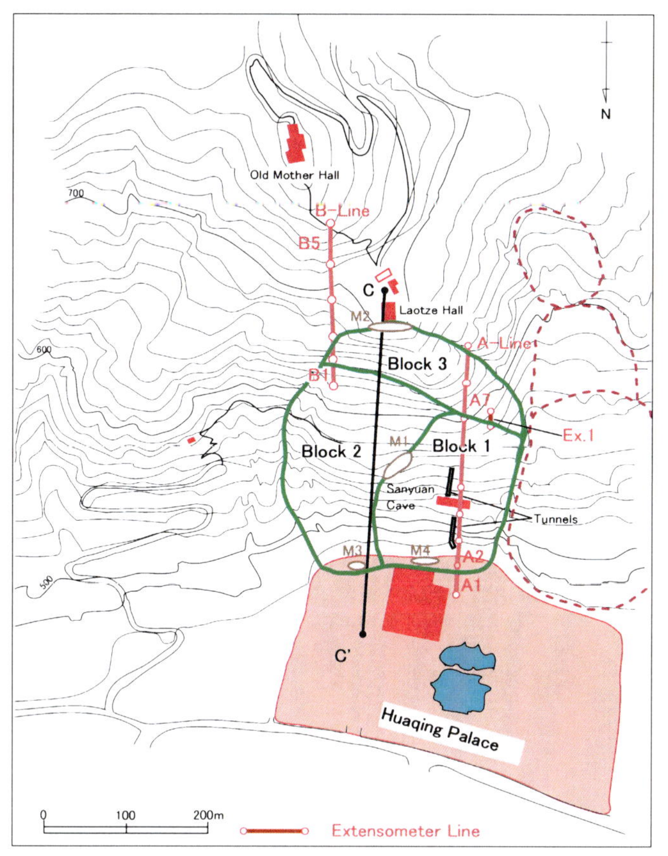

Fig. 9.3. Plan map and potential landslide blocks of the Lishan slope

Figure 9.3 is the plan map of Lishan slope and surrounding area and the currently estimated dangerous landslide blocks (block 1, 2, 3). In the right side of figure, clear landslide topography is delineated by dashed lines. This is a landslide block in thick loess deposit that already slid and now being at the residual state. Therefore, this part has low possibility of rapid and long runout landslide. In the left part of this figure, there is a gentle slope where a road is constructed in it, and a creep movement was detected through EDM monitoring by the Xian Seismological Bureau (Fukuoka et al. 1994). However, this slope is gentle and is estimated that it is of low possibility to develop as a rapid landslide. Deformation seems to be limited in the surface loess layer in the left side.

Potential landslides that may cause great disasters are the blocks in the Precambrian gneiss rock mass (block 1, 2, 3 in Fig. 9.3). The slope on this area is very steep and did not slide yet. However, once the slop starts to slide, it will have a high possibility to become rapid slide due to its peak-strength-slide fact (Sassa 1985, 1989). The shear-strength decreasing from peak strength to the residual strength will accelerate the landslide movement. Block 1 is the most active part, and its toe and surrounding area has been showing rather active ground subsidence for years (Lin 1997) possibly due to pumping up hot water for tourism. Probably affected by this ground subsidence, retrogressive landslides is induced in the loess layer as well as in the gneiss rock mass. Extension cracks were observed in the upper border of block 1, so the first extensometer on Lishan slope was installed at Ex.1 shown in Fig. 9.3. This extensometer recorded a repeated landslide movement. Meanwhile, the monitoring of a series of long span extensometers installed in A-line and B-line (Fig. 9.3) in 1996 showed a clear compression in A1 and the corresponding extension in A7. The monitoring results will be introduced later in details. These monitoring results indicate the existing of a large-scale rockslide as block 1. Extension deformation had also been observed along B1. The movement of block 2 is relatively small, but the topographical feature and shear disturbance in rocks suggested the activity of this landslide block. Block 3 could be a retrogressive landslide affected by the movement of block 1 and 2 or a landslide along a steep fault. The marked part as M1 is the shear zone between block 1 and 2, where the surface feature as the border of block 1 had been recognized. A linear subsidence and a sink-hole were observed at M2, suggesting the extension of ground deformation along the upper boundary of block 3 (Sassa et al. 1994). M3, M4 are locations showing the compressive failure or deformation of concrete and brick structures at the toe of slope.

9.4 Depth of the Lishan Landslide

Figure 9.4 shows the central section C–C' of Fig. 9.3. It was estimated from the outcrop of gneiss rock mass, geological borings and two horizontal investigation tunnels. Figure 9.5 shows the geological section of the lower tunnel. In this tunnel, a gently dipping fault was found. At the end of the tunnel, a very hard intact gneiss rock was exposed and then excavation was stopped. The location of lower tunnel is projected to Fig. 9.4, which is situated around the toe of potential sliding surface. Hence, it is referred that the present creeping landslide may use the gently dipping fault as a part of its sliding surface. Figure 9.6 presents a close-up view of the slickenside within the gently dipping fault. Some rounded gravels are found in this shear zone, suggesting that the experienced shear displacement was not very short.

Fig. 9.6. Close-up view of the slickenside within the gently dipping fault (potential sliding surface) found in the lower tunnel

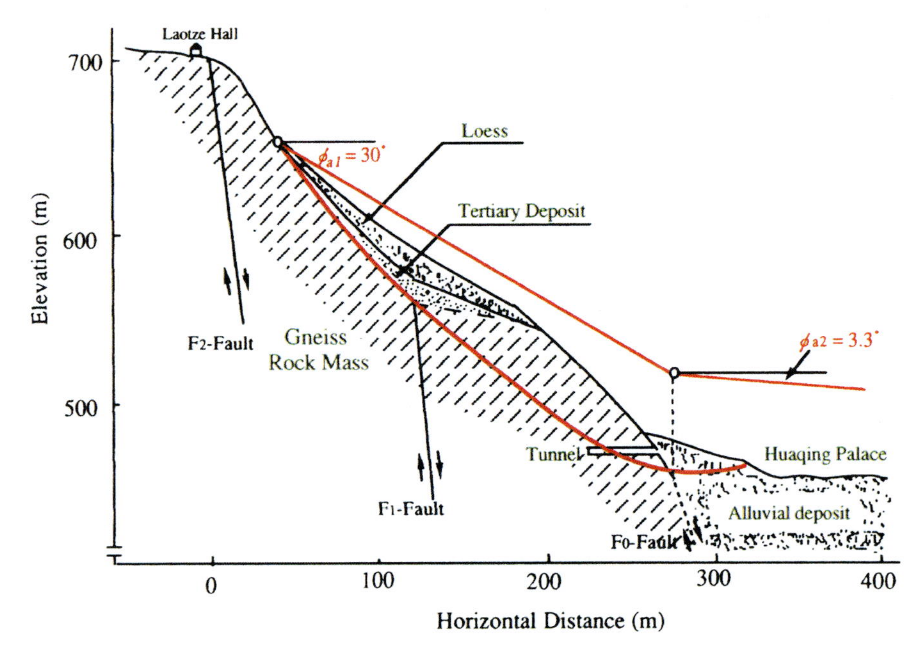

Fig. 9.4. Central section of the Lishan slope and location of potential sliding surface

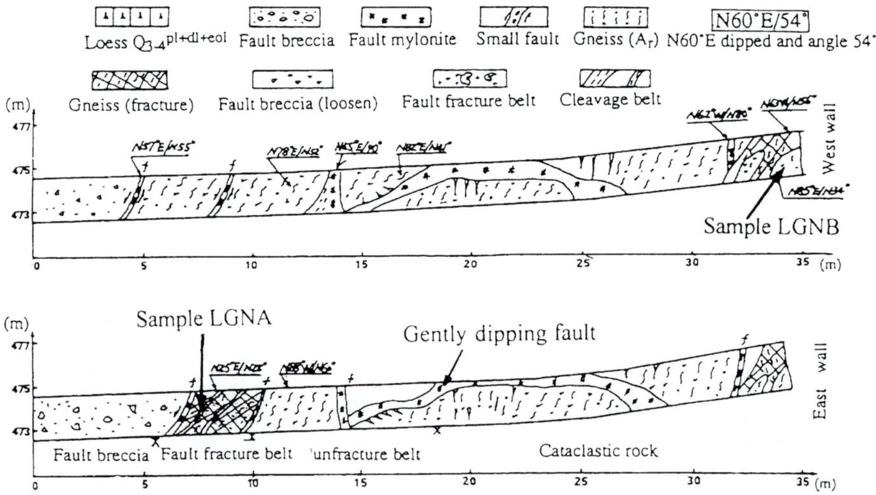

Fig. 9.5. Geological section of the lower tunnel (Xie et al. 1994)

A series of three dimensional shear displacement meters (developed by Sassa et al. 1975) were installed inside the tunnels in 1996. The detected shear displacement from 1996 to 2000 was 11 mm between 25–30 m of Fig. 9.5. If the location of sliding surface shown in Fig. 9.4 is correct, the depth of landslide is estimated to be about 50 m. Although its deformation is not clear yet, block 3 may extend from the Laotze Hall downward.

9.5 Travel Distance of the Landslide

9.5.1 Undrained Loading and Landslide Classification

For landslide hazard assessment, the most important factor is the travel distance. Some landslides slide slowly with limited runout distance, while some landslides move rapidly and travel long distance. Sassa (1985, 1989) proposed the geotechnical classification of landslides to classify landslides by its initiation mechanism, and he added another mechanism of sliding surface liquefaction (Sassa and Fukuoka 1995), based on the research on landslides triggered by the Hyogoken-Nambu earthquake (Sassa 1996).

The peak strength slide, mass liquefaction slide, and the sliding surface liquefaction slide are rapid landslides. According to this classification, Lishan landslide will be the peak strength slide because the material is gneiss rock mass. Rock mass has great difference between its peak shear strength at failure and its shear resistance during motion, and then may suffer rapid landsliding after failure. Usually rapid landslide travels long distance. Nevertheless, very long travel distance can be caused by one of the following two cases of mechanisms; case 1: Mass liquefaction of landslide mass corresponding to debris flow, and case 2: Undrained loading onto a saturated ground (Sassa 1988).

The precipitation in Lishan is not so great, and the slope is very steep, so case 1 is unlikely to be here. Case 2 is very common when large landslide mass moves onto the alluvial deposit, and also the most possible case here (Sassa et al. 2001). When the landslide mass moves onto the saturated ground, excess pore pressure will be generated on the saturated soil layer of the ground, and then the shear resistance will be reduced consequently. Furthermore, if the alluvial deposit is prone to suffer grain crushing during shear, sliding surface liquefaction (liquefaction only in the shear zone due to volume shrinkage by grain crushing) will take place, and it will further reduce the apparent friction.

9.5.2 Travel Distance in the Lishan Landslide

Undrained ring shear tests were performed on the sample taken from the alluvial deposit (fine silt from loess and others) at the toe of Lishan slope from the depth of water level (2.0–2.3 m deep).

The estimated apparent friction angle for the landslide mass of 40–50 m thickness loading onto the alluvial deposit based on the undrained ring shear tests results (Sassa et al. 1994) will be 3.3 degrees. Although the apparent friction angle in the gneiss rock slope is not known, three cases of 25, 30, and a bit higher angle of 35 degrees could be considered, referring the case of Frank slide (Sassa et al. 1998). Corresponding to these three values, the estimated runout distances from Laotze Hall should be 2 430, 1 530, and 930 m, respectively (as shown in Fig. 9.7. Details on the estimation of runout distance can be obtained in Sassa et al. 2001). From Fig. 9.7 it is seen that given the minimum runout distance being 930 m, once landslide was triggered, Huaqing Palace should be destroyed and the densely populated central part of Lintong County will be heavily damaged by the landslide mass.

9.6 Deformation Monitoring of the Lishan Landslide

To monitor the creep movement and the precursor phenomenon possibly relating to a catastrophic rock slide, various monitoring instruments were installed in Lishan; extensometers, inclinometers, three dimensional shear displacement meters inside two investigation tunnels (Sassa 1997). Piles and points for GPS, Total Station and level monitoring were also installed, and periodically monitoring has been carried out. Among those methods, short-span and long-span extensometer monitoring is the most simple, reliable and durable, accordingly they are most successful and having provided good data for prediction of landslides. Then, some of monitoring data are introduced.

The short-span extensometer (the span length is about 10 m) was installed at Ex.1 of Fig. 9.3 in the position crossing the upper boundary of the block 1 and also crossing an active fault, in order to monitor the block 1 movement. Its span is installed underground to guard against thieves (Fig. 9.8a). The extension in response to precipitation continued to about 40 mm within first four years (Fig. 9.8b)

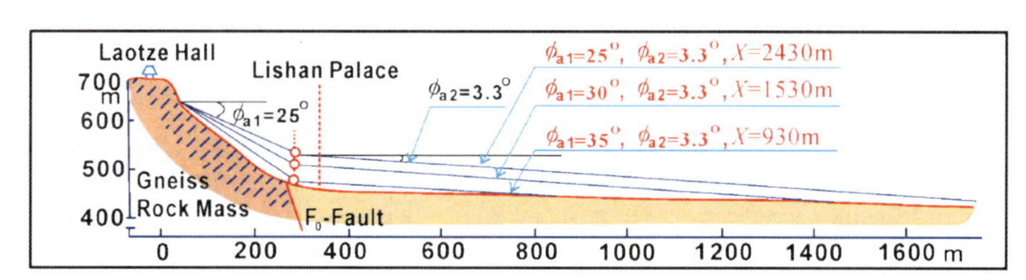

Fig. 9.7. Energy lines of the potential landslide blocks in case of undrained loading of landslide mass onto a saturated alluvial deposit based on undrained loading ring shear tests of the alluvial deposit in Huaqing Palace

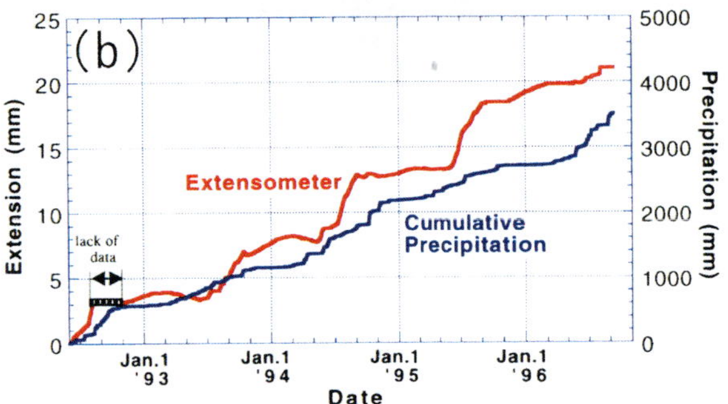

Fig. 9.8. a Photo of short-span extensometer EX-1 crossing the upper border of block no. 1; **b** EX-1 record and the cumulative precipitation during 1992–1996

(Fukuoka et al. 1997; Sassa et al. 1997). It suggested that this slope is affected by creeping. However, the monitored value may be affected by a superficial landslide. In order to detect the movement of possible deep landslide, long-span extensometers (Fig. 9.9) were continually installed in two lines as shown in Fig. 9.3; 8 extensometers in the A-line and 5 extensometers in the B-line (Fig. 9.3). Figure 9.9a is the about 12 m high pole and anti-thief metal box of the long-span extensometer. With use of this high pole, continuous installation of 5 and 8 extensometers were possible on the slope, and tourists can not touch the super-invar wires and pass over roads on the slope. Figure 9.9b is the pen-recording type extensometer and linear electronic sensor located inside the metal box. Figure 9.9c is the facilities for measuring and recording located in a tunnel at the foot of the Lishan slope.

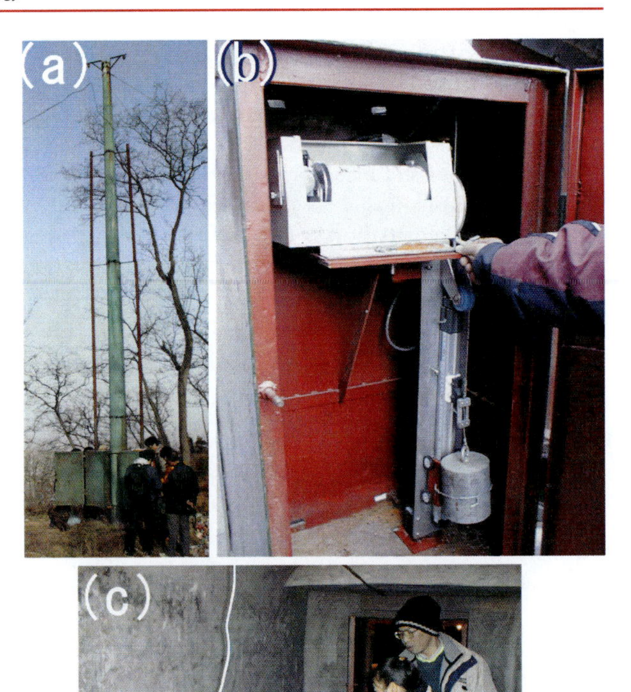

Fig. 9.9. a Tall pole of a long-span extensometer in the B-line; **b** pen-recording type extensometer and linear electronic sensor for the long-span extensometer system; **c** facilities for measuring and recording of linear electronic sensor data located in a tunnel at the foot of the Lishan slope

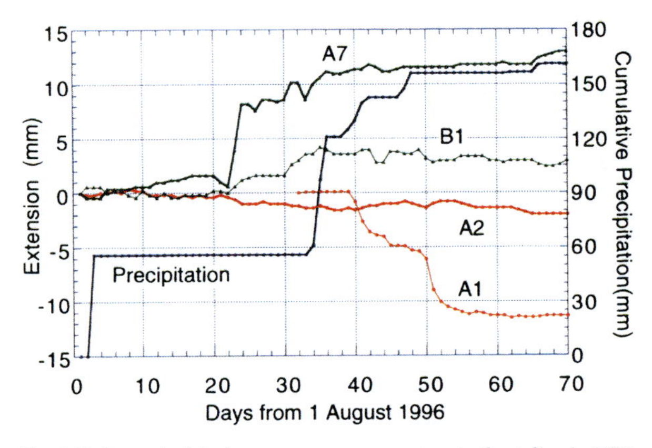

Fig. 9.10. Record of the long-span extensometers in the A-line in 1996 just after installation

9.7 The Results of Extensometer Monitoring

Figure 9.10 shows the clear precursor landslide movement monitored in the A-line found just after installation. Extensometer A7 moved in 20 days after about 60 mm precipitation, and A1 also moved after next precipitation of about 100 mm. Extensometer A1 is installed inside the almost flat garden of the Lishan Palace, therefore, no superficial landslide is expected there. And superficial landslide moves during rainfall, not delayed so long. Therefore, the movement of A1 and A7 must be the movement of deep seated landslide. A7 crosses the upper border of estimated landslide block 1, A1 crosses the lower border of block 1. From the time deviation appeared in different extensometers, it is inferred that the sliding surface has not been well developed yet. Both movements had close relation to precipitation, and both movements are in the same magnitude of around 10 mm. It is very likely to have detected a precursor landslide movement of block 1. This estimation was further supported by observation of the potential sliding surface inside the investigation tunnel which is located along a possible long and deep sliding surface connecting somewhere in A1 and A7 spans. Note that Fig. 9.10 was probably the decisive evidence for the Chinese government to fund about U.S.$3 million and to start a series of expensive landslide countermeasure works in block 1 from 1997, including the constructions of surface drainage, piles on the toe and ground anchors on the lower part of the slope. These countermeasures have been mainly finished in 1998. Figure 9.11a shows the photos of the surface drainage to collect rain water and prevent infiltration into ground and Fig. 9.11b and c show the photos of ground anchors on the lower part of the slope to enhance the strength of the ground against slide. They were designed and installed by Chinese engineers.

Figure 9.12 present the monitoring results along A-line and B-line from August 1996 to June 2000.

Fig. 9.11. Countermeasures constructed in block 1 of Lishan slope. **a** Surface drainage system installed on the Lishan slope to prevent rain water infiltration into the ground; **b, c** ground anchors on the lower part of the slope

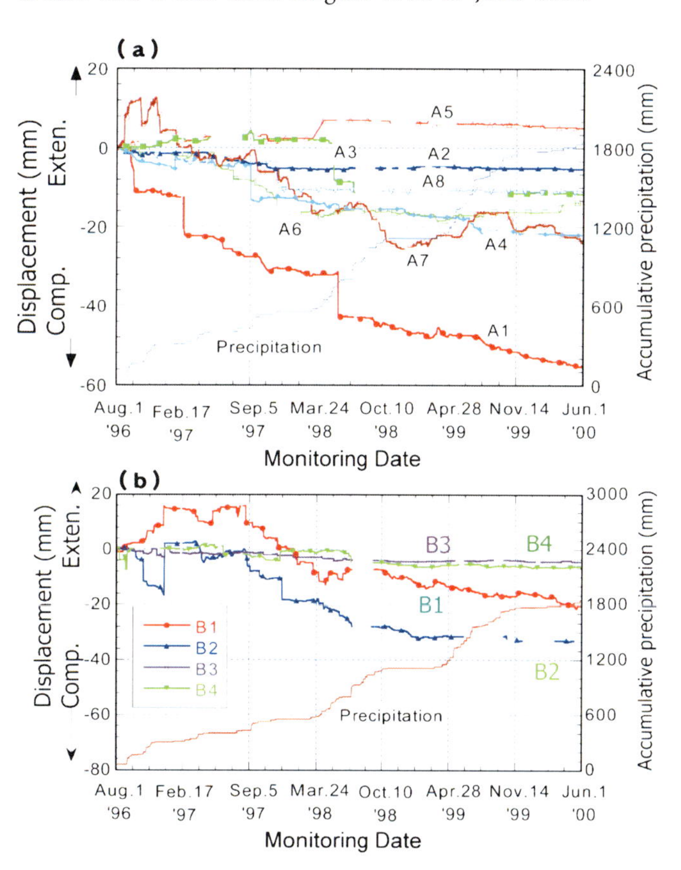

Fig. 9.12. Records of the long-span extensometers until 1 June 2000. Most of large deformation events took place in rain seasons. **a** A-line; **b** B-line

The compression in A1 is very outstanding. Repeated and continual movement has been monitored. It could be noticed that rapid movement of about 10 mm was observed three times before the summer of 1998. However, such big movement seems to have stopped. The remedial measures in block 1 have been completed, and the effect of works might appear in the movement of A1. Extensometer B2 and B1 also showed compression although B1 detected extension in 1996–1997. Compression monitored in extensometer means the downslope movement of the upper slope. Then, this record suggests the upper slope of block 3 is moving. Although we inferred that the area between B3 and B5 probably moved in the past, but it may be stable at present. So we delineated the border of landslide block as shown in Fig. 9.3. However, concerning this, further investigation is needed. Another characteristic of the monitored date is the overall trend of compression. This may be the result of superficial creep of loess; it is likely that the superficial layer is creeping gradually. Then, the monitored data is probably the combination of superficial loess movement and the precursor motion of deep seated rock slide.

Finally, it is noted that although those landslide countermeasures retarded the slope, it seems that the slope has not been stabilized completely. It can be seen from the data monitored on Ex.1 that are plotted in Fig. 9.13 together with the rainfall. Those data were from January 2000 to the end of December 2004. It is seen that extension deformation continued throughout the whole monitoring period although the displacement rate is smaller than Fig. 9.8b. Also at point "R", a relative compression tendency was observed. This was induced probably because the block 3 had downslope movement, and the difference between the downslope movements of block 1 and block 3 resulted in the reduction of recorded compressional displacement at point "R". Analysis on the recorded extensometer data along A-line and B-line after January 2000 are in the process, and further more detailed studies on block 2 and block 3 have been conducted.

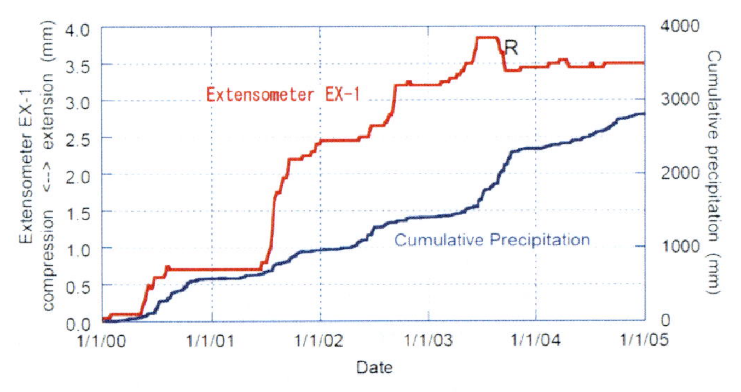

Fig. 9.13. Records of short-span extensometer Ex-1 and precipitation data from 1 January 2000 to 30 December 2004

9.8 Landslide Risk Preparedness

It is almost clear that Lishan slope is in danger of large-scale landslides, although the apparent friction during motion in the gneiss rock is not yet well estimated and information of landslide depth distribution is not sufficient. However, the present investigation results are enough to examine the landslide risk preparedness. We should proceed for disaster mitigation. Two major ways of risk preparedness are: (1) to know the state of slope instability which may be approaching to the critical state. It is useful for the time prediction of catastrophic failure and warning in advance, (2) to stabilize the slope through extensive protective works.

To monitor the slope deformation in real-time from remote places, and then to issue the warning in advance in case of facing signs of imminent danger of disastrous occurrence, an electronic extensometer monitoring system along line-A and point Ex.1 on the Lishan slope has been set up (Fig. 9.9b and c). This system includes nine electronic extensometer devices, a data logger, a data acquisition PC, and cables for data transmission. It is also worth noting here that extensive protective works are necessary, although some measures, such as piles on the toe of the slope and anchors on the lower parts of the slope, had been performed in 1998. The monitored extensometer data after 1998 indicated that the slope was still deforming, although the value was becoming smaller.

9.9 Summary

Landslide risk assessment is a challenging topic to both geo-engineers and governmental administrative officers. The joint research presented in this paper has started since 1991 as a contribution to the International Decade for Natural Disaster Reduction (IDNDR 1990–2000) and also the International Strategy for Disaster Risk Reduction (ISDR 2001–) of the United Nations. This initiative inspired enthusiasm of many Japanese, Chinese and others to work for landslide hazard assessment, and great academic and technological achievements were obtained. Through this research, a new undrained dynamic loading ring shear apparatus was developed, which enables to simulate the undrained loading process during long runout landslide motion. One short-span extensometer and two lines of the long span extensometers installed over high trees (which can not be cut for environmental protection) in the Lishan slope have proved that these monitoring methods are very effective. The monitoring results of this potential landslide thus provided good basis for the first case of installation of prevention works prior to landslide occurrence. The semi real-time monitoring system had also been set up along A-line for the purpose of landslide warning in advance. It has been proved that this is an effective method to monitor and issue warning as well.

To develop these achievements and to apply them for natural disaster mitigation, landslide risk preparedness should be more studied. It could be concluded that landslide risk preparedness for protection of the environment and cultural heritage as well as communities and industrial infrastructures will be getting significance in the society of twenty-first century.

Acknowledgments

This Japan-China Joint Research on Landslide Hazard Assessment in Lishan (Yang-Que-Fe Palace) was supported by many of Japanese and Chinese Agencies, such as the Ministry of Education, Science, Culture and Sports (MEXT) of Japanese Government, the Japan International Cooperation Agency, the Japan Society for the Promotion of Science, the People's Government of Xian Municipality, the People's Government of Shaanxi Province, China, the State Planning Commission, China, the State Science and Technology Commission, China, the Ministry of Geology and Mineral Resources. Thanks go to many colleagues in Japanese universities and institutes who were involved in this joint research since 1991.

References

Fukuoka H, Sassa K, Hiura H, Yang Q, Lin Z (1994) Monitoring of slope deformation by leveling, extensometers inclinometers, EDM and GPS in the Lishan landslide. Proc. Special Session in the International Workshop on Prediction of Rapid Landslide Motion, "Prediction of Landslides in Lishan", pp 93–118

Fukuoka H, Sassa K, Yang QJ, Song BE (1997) Extensometer monitoring in the Lishan Landslide, Xian, China. Landslide News, 10:23–25

Lin Z (1997) Huaqing Palace, Xian, China and landslide hazard. Proc. International Symposium on Natural Disaster Prevention and Mitigation, Kyoto, published by Disaster Prevention Research Institute, Kyoto University, pp 299–307

Sassa K (1985) The geotechnical classification of landslides. Proc. 4th International Conference and Field Workshop on Landslides, pp 31–40

Sassa K (1988) Geotechnical model for the motion of landslides, Special lecture for 5th International Symposium on Landslides. "Landslides", Balkema, 1:37–55

Sassa K (1989) Geotechnical classification of landslides. Landslide News 3:21–24

Sassa K (1996) Prediction of earthquake induced landslides. Special lecture for 7th International Symposium on Landslides, "Landslides", Balkema, 1:115–132

Sassa K (ed) (1997) Proceedings of International Symposium on Landslide Hazard Assessment (ISBN4-9900618-0-2 C3051), IUGS Working Group on Landslide (WGL/RLM), 421 p

Sassa K, Fukuoka H (1995) Prediction of rapid landslide motion. Proc. XX IUFRO World Congress, Technical Session on Natural Disasters in Mountainous Areas, pp 71–82

Sassa K, Nakano K, Takei A (1975) Monitoring of horizontal and vertical displacement of a landslide in the fractured rock area – from 64 sets of newly developed shear displacement meters installed along a crossing line. Kyoto University Forest 47:98–111

Sassa K, Fukuoka H, Lee JH, Shoaei Z, Zhang D, Xie Z, Zeng SW, Cao B (1994) Prediction of landslide motion based on the measurement of geotechnical parameters. Proc. Special Session "Prediction of Landslides in Lishan", the International Workshop on Prediction of Rapid Landslide Motion, pp 13–47

Sassa K, Fukuoka H, Yang QJ (1997) Cultural heritage and Lishan landslide hazard assessment, Xian, China. Annuals Disas Prev Res Inst, Kyoto University, no. 40, IDNDR Special Issue, pp 119–138 (in Japanese)

Sassa K, Fukuoka H, Wang FW (1998) Sliding surface liquefaction and undrained loading mechanism in rapid landslides in Japan and Canada. Panel Report for 8th IAEG World Congress, "Engineering Geology", Balkema, 3:1923–1930

Sassa K, Fukuoka H, Wang FW, Furuya G, Wang G (2001) Pilot study of landslide hazard assessment in the Imperial Resort Palace (Lishan), Xian, China. In: "Landslide risk mitigation and protection of cultural and natural heritage". Proc. UNESCO/IGCP Symp, January 2001, Tokyo, pp 15–34

Xian Seismological Bureau (1991) History of earthquakes in Xian. 134 p

Xie Z, Chen Y, Lin Z, Li T (1994) Prediction of landslides in Lishan. In: Proc. Special Session in the International Workshop on Prediction of Rapid Landslide Motion, pp 49–91

Chapter 10

Formation Conditions and Risk Evaluation of Debris Flow in Tianchi Lake Area of Changbai Mountains Natural Protection Area, China (C101-5)

Binglan Cao* · Xiaoyu Zheng · Hui Wang

Abstract. Tianchi Lake area of Changbai Mountains is an area with frequent and dense debris flow disasters due to the special volcano geology and landform. In this area the three basic conditions inducing debris flow developed. The widely distributed accumulation of collapse and landslide provide the main solid sources of debris flow; the high and steep slopes along the Valley of Edaobai River take the landform of steps with a big gradient, to provide the moving way; and the concentrated rainstorm in summer provide debris flow for the force to move. The mount-slope type is the main type of debris flow occurred in the area, and it can be classified into further two sub-types according to the formation condition and activity, namely frequently active debris flow and relative stable debris flow. Based on analysis of controlling factors of debris flow, to evaluate the risk degree of 8 debris flow galleries. The result shows a good correspond with the practical situation, i.e. this method has a valuable application foreground.

Keywords. Debris flow, formation conditions, risk degree, fuzzy evaluation method, Changbai Mounts

10.1 Introduction

Changbai Mountains is located in Jilin province in the northeast China, adjoining with DPRK, with a range length of 200 km in W-E direction and 310 km in S-N direction. It is also the famous international natural protected biotic division and a tourist site with beautiful natural landscape (Figs. 10.1 and 10.2). The area of Tianchi Lake is about 20 km^2 as the highest and the biggest lake of volcanic vent in China.

In research area the geological background is determined by special volcanic geological environment (Liu 1999; Xie et al. 1993), so debris flow happened densely and frequently as a key geological disaster. At the both sides of Erdaobai River to the north of the waterfall, more than 15 gullies of mount-type debris flow can be seen with more or less accu-

Fig. 10.1. Tianchi Lake of Changbai Mountains

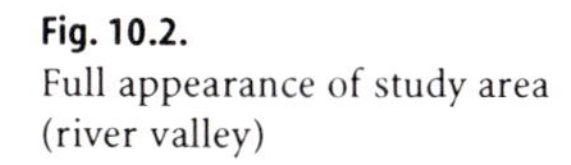

Fig. 10.2. Full appearance of study area (river valley)

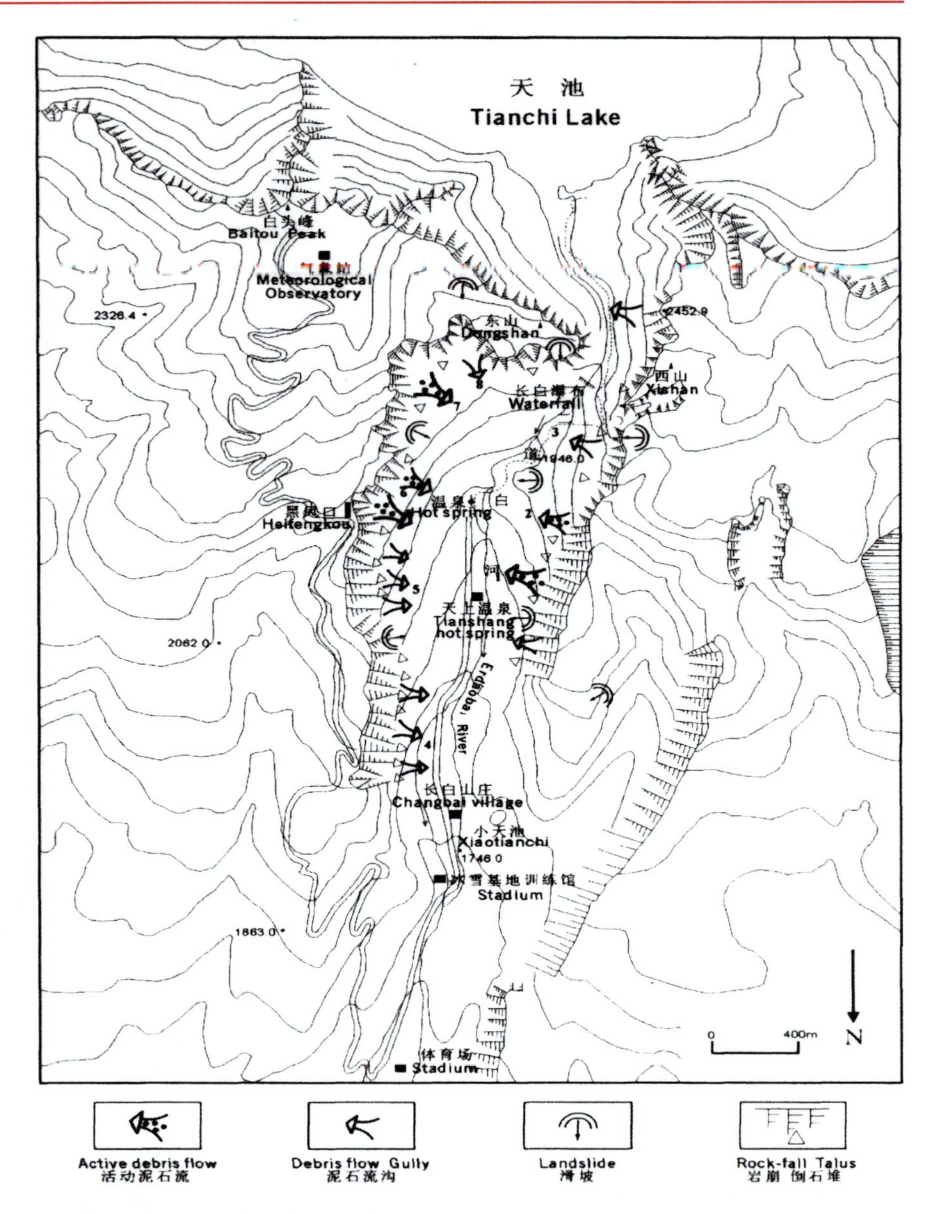

Fig. 10.3.
Distribution of debris flow in Tianchi Lake Area, China (modified from Cao et al. 1999)

mulation (Fig. 10.3). Among these gullies the Heifengkou gully and Tianshang Hot Spring gully of debris flow are worthy to study deeply because of their taking place both frequently and intensity.

10.2 Formation Condition of Debris Flow

As an outcome of the co-action of geology, atmosphere etc., the formation of debris flow is bound to the three basic conditions, namely the loose solid material, high and steep mount landform and heavy and sudden precipitation, which are all provided in the area.

The stratum and structure in the area are available for the formation of loose solid material. The principal part of rock emerged in the area is deeply gray trachyte and yellow volcanoclastic rock as well as the weathering mantle of volcanic ash (Bureau of Geological and Mineral Resources of Jilin Province 1988). The stratum assembly of alternate weak and hard layers offers the condition to the destruction of rock body (Fig. 10.4). The weathering mantle of ash formed in the intermission of volcano has loose texture, which becomes the origin material of accumulation of falling directly, and in another way fail to support upper hard rock and driving it in the state of easy sliding. The developed original prismatic joints in the hard trachyte cut the rock body into broken dollops (rock-fall) (Fig. 10.5). Meanwhile the nearly horizontal weak plane and the gently declining joint are easy to produce landslide. The above provided a rich solid material sources for the debris flow (Cao et al. 1999, 2001).

The two banks alongside the Erdaobai River valley near the waterfall represent in the shape of steps with steep

Fig. 10.4.
Alternate layer of weak and hard rock and formation area of debris flow

Fig. 10.5.
Rock-fall of hard rock with joints and formation area of debris flow

and high landform (Fig. 10.2), their altitude above sea level being more than 2 000 m. The comparative altitude gap surpasses 300 m with developed gullies, which provide a movement condition for the debris flow.

The special meteorological condition in the area represents as a long winter and with abundant precipitation, snow season reaches 7 months and in summer the snow melts into water. The concentration of rain occurs in July and August and achieves certain limit value rapidly to induce the debris flow. According to the survey the daily precipitation amounted to 50–80 mm. Figure 10.7 shows the situation of Heifengkou debris flow taking place in 23 August 2003, which induce a threat to the traffic and even the environment of traveling in the area (Fig. 10.6).

10.3 The Types and Distribution of Debris Flow in Study Area

From the landform of movement the debris flow in this area is concluded into two types, i.e. valley-type debris flow and mount slope-type debris flow.

Valley-type debris flow usually happened in the area with wide space in the river valley. In Erdaobai River this type is rare to see and only can be presumed based on the historical record, which shows about one thousand years ago a large scale debris flow occurred here accompanying the happening of old volcano. Mount slope-type debris flow is the main type happened in research area, with dense distribution and frequent happening. This kind of

Fig. 10.6.
Heifengkou debris flow induced by rainstorm

Fig. 10.7.
Accumulation of frequently active debris flow at Heifengkou

debris flow occurs in the slope of mount with small catchment moving mainly along the gully. On the basis of detailed investigation, according to the different geological conditions and activities the authors classified the mount-slop-type debris flows into the following two kinds.

10.3.1 Frequently Active Debris Flow

This kind of debris flow distributes near the south of the Tianshang Hot Spring, the west bank of valley, as well as in the south of Heifengkou, the east bank of valley (Figs. 10.3, 10.7, and 10.8). In the two places the mount were constituted of alternant weak layer and hard layer so it is easy to produce avalanche and landslide and accumulation such as talus widely distributes. Once strong and rapid rain especially rainstorm falls it is possible to lead to debris flow.

10.3.2 Relatively Stable Debris Flow

This kind of debris flow occurred in the north side to Tianshang Hot Spring at west bank of Erdaobai River as well as north side to Hefengkou at the east bank of Edaobai River (Fig. 10.9). The debris flows pass through the district of the steep slope with multi-steps and constituting of hard rock with joints. At the higher part of slope accumulation of falling becomes the main source of solid material. This kind of debris flow has a small area of movement. Once the fragmental material is rushed out the debris flow will end.

Fig. 10.8. Full appearance of frequently active debris flow at Tianshang Hot Spring

Fig. 10.9. Relatively stable debris flow in high and steep slopes at Changbai Village

10.4 Risk Degree Evaluation of Debris Flow

There are two aspects need to analyze in evaluating the debris flow risk in a certain area, which are the hazard historically happened and the potential hazard (Liu and Tang 1995). In this paper the study is made on evaluate the risk of potential debris flow based on the current formation condition of debris flow gully. It is the fuzzy method that provide a suitable disposal measure to this complicated problem due to the multi-factors in forming debris flow (Luo and Chen 2000). The follows describe the evaluation process of 8 selected debris flow gullies in Tianchi Lake area, which are represented by number 1 to 8.

10.4.1 Mathematical Model of Fuzzy Method

Factor set $U = \{u_1, u_2, \dots, u_n\}$ and evaluation degree set $V = \{v_1, v_2, \dots, v_m\}$ are set to known, meanwhile the $\tilde{A} = (a_1, a_2, \dots, a_n)$ represents the distribution of each factor upon the U, i.e. a_i correspond to the weight of u_i, and

$$\sum_{n=1}^{n} a_i = 1$$

For the factor i if the judge matrix $\tilde{R} = (r_{ij})_{n\times m}$, the judge result of objective by fuzzy method embody as the fuzzy set upon V: $\tilde{B} = \tilde{A} \circ \tilde{R}$, where "$\circ$" express certain kind of computation process.

10.4.2 Determination of Factors and Objective to Study

To choose the controlling factors in validity, ascertain the weight by determining key and subordinated factors.

To choose the factors that bring about the hazard of debris flow directly or importantly into analysis and discard other unimportant or subordinated factors. As for precipitation, one of the most important affecting factors, because of the nearly same condition upon everywhere in the study area, it is discarded to simplify the computation. By means of *in situ* investigation, 10 are chosen as the judge factors to make the set $U = \{u_1, u_2, \ldots, u_n\}$, they are u_1: rock property, u_2: longitudinal inclination of slope of gully (°), u_3: catchment area (km^2), u_4: relative gap of altitude (m), u_5: average gradient of slope (°), u_6: average width of valley (m), u_7: vegetation, u_8: hazard degree, u_9: maximum particle *size* (m), u_{10}: condition of joints (numbers per square meter).

Among the chosen factors the rock property is the most important one that play a key role in the formation of debris flow. In the study area the mount body developed consist of alternate weak and hard rock layers and developed with joints and cracks, so it not only provide the solid material namely the accumulation of collapse and slide but also contribute to the development of high and steep gallery for debris flow to move. Then the weight of rock property is selected as the largest factor and is taken as the base for calculation. Table 10.1 shows the original data for analysis.

10.4.3 Evaluation of Risk Degree

The risk degree of debris flow are divided into 4 grades, then evaluation set $V = \{v_1, v_2, v_3, v_4\}$ = {grade I, grade II, grade III, grade IV} = {safety, weak risk, medium risk, serious risk}.

By calculate the weight matrix of judge factors and the judge matrix of single factor to get the result matrix of fuzzy judge, namely $\tilde{B} = \tilde{A} \circ \tilde{R}$,where "∘" represent a certain composite computation. Then according to the principle of maximum attaching degree, the grade of risk for each objective is given. In this chapter 5 compound computations are used to analyze and according to the method of weighted average of "$\vee, \cdot$ "and"$\oplus, \cdot$ "the risk degree of Tianchi Lake area are obtained as in Table 10.2.

From Table 10.2 it can be seen that the risk degree of both no. 1 of west slope (Tianshang Hot Spring) debris flow gallery and no. 6 of east slope (Heifengkou) debris flow gallery is grade IV, belonging to serious debris flow. The judgment got by means of fuzzy method evaluation shows that the calculation correspond excellently to the practical situation and then provide a effective analysis method for prevention of the hazard of debris flow at near situation.

10.5 Conclusion

1. In the area three fundamental conditions to produce debris flow are all serious, that is, solid material

Table 10.1. Original data of chosen debris flow galleries

Gully number	Affecting factors									
	U_1	U_2	U_3	U_4	U_5	U_6	U_7	U_8	U_9	U_{10}
1	Alternated weak and hard layer, weak rock is more	20	0.03	300	40	5.0	Covered slightly	Heavily dangerous	0.8	16
2	Alternated weak and hard layer	14	0.02	250	32	4.5	Covered	Slightly dangerous	0.5	10
3	Alternated weak and hard layer	19	0.018	260	35	4	Covered	Clearly dangerous	0.7	15
4	Hard rock with developed joints	25	0.015	100	38	3	Covered slightly	Non-dangerous	0.4	10
5	Hard rock with developed joints	21	0.08	300	45	8	Covered slightly	Clearly dangerous	0.6	12
6	Alternated weak and hard layer, weak rock is more	19	0.14	320	40	13	Uncovered	Heavily dangerous	1.0	18
7	Hard rock with developed joints	15	0.12	300	45	4.5	Covered slightly	Clearly dangerous	0.9	16
8	Hard rock with developed joints	14	0.06	180	30	5	Uncovered	Slightly dangerous	0.7	14

Table 10.2.
The evaluation result and risk degree obtained by Fuzzy method

Debris flow gully	Evaluation result by Fuzzy method				Risk
No. 1 of west slope (Tianshang Hot spring)	0.118	0.205	0.067	0.364	IV
No. 2 of west slope	0.178	0.265	0.179	0.067	II
No. 3 of west slope	0.185	0.129	0.304	0.056	III
No. 4 of east slope	0.242	0.163	0.188	0.110	I
No. 5 of east slope	0.032	0.189	0.302	0.022	III
No. 6 of east slope (Heifengkou)	0	0.022	0.243	0.394	IV
No. 7 of east slope	0.090	0.230	0.300	0.074	III
No. 8 of east slope	0.132	0.264	0.074	0.120	II

sources, high and steep landform as well as rich rainstorm. The development of debris flow possesses the special property of volcanic rock area. The mount-slope-type debris flow distributes widely and frequently.

2. The stratum assembly of alternate weak and hard layers, together with the weak rock body with dense joints and cracks, constitute the main controlling factors to determine the development and distribution regularity of debris flow in the research area. The mount-slope-type debris flow in this area can be classified into two kinds, their formation areas situate at different places namely weak or alternate weak and hard rock layer, as well as hard rocks with joints and cracks developed respectively.
3. On the foundation of *in situ* investigation and analysis as well as in door testing of the main controlling factors, the fuzzy method are applied to evaluate the risk degree of debris flow in the area, with a valid result suitable to the practical situation.

References

Bureau of Geological and Mineral Resources of Jilin Province (1988) Regional geology record of Jilin Province. Geological Publishing House, Bejing

Cao Binglan, Wang Gangcheng, Liu Zhenghua (1999) The disaster features of rock avalanche and landslide in Tianchi Lake tourist area of changbai mountain, China. IGCP-425 Meeting at Bonvin Building, UNESCO, Paris

Cao Binglan, Sun Ping, Zheng Xiaoyu, Wang Gonghui (2001) The assessment and protection plan to the rock avalanche and landslide in Tianchi Lake area of Changbai Mountains natural protection area. China, Proceedings of UNESCO/IGCP Symposium, Tokyo

Liu Jiaqi (1999) China Volcano. Science Publishing House, Bejing

Liu Xilin, Tang Chuan (1995) Risk assessment of debris flow. Science Publishing House, Beijing

Luo Yuanhua, Chen Chongxi (2000) Numerical simulation to accumulation of debris flow and risk evaluation of debris flow disaster. Geological Publishing House, Beijing

Xie Yuping, Liu Xiang, Xiang Tianyuan (1993) The researches on Cenozoic volcanoes and volcanic rocks in the middle distric of Northeast China. Northeast Normal University Press, Changchun

Chapter 11

Aerial Prediction of Earthquake and Rain Induced Rapid and Long-Traveling Flow Phenomena (APERITIF) (M101)

Kyoji Sassa* · Hiroshi Fukuoka · Hirotaka Ochiai · Fawu Wang · Gonghui Wang

Abstract. The major research achievements of the Aerial Prediction of Earthquake and Rain Induced Flow Phenomena (APERITIF) project (IPL M101) are introduced in this paper in three parts. (1) Detection of landslide prone slope using airborne laser scanner, (2) Full-scale landslide flume experiments and an artificial rainfall-induced landslide experiment on a natural slope to understand the fluidization mechanism of rainfall induced landslides; (3) Landslide risk evaluation and hazard zoning in urban development areas. In part (3) as a comprehensive research of this project, application studies were conducted in two test sites in the Tama residential area near Tokyo. A series of field and laboratory investigations including laser scanner, geological drilling and ring shear tests were carried out and the results show that there is a risk of sliding surface liquefaction triggered by coming earthquakes for both sites. Developed in this project, a geotechnical computer simulation of landslide (Rapid/LS code) using parameters obtained from the undrained dynamic-loading ring shear tests makes urban landslide hazard zoning possible even at individual street level.

Keywords. Risk analysis, hazard zoning, fluidization, airborne laser scanning, full-scale flume experiment, natural slope failure by artificial rainfall, numerical simulation

11.1 Introduction

The national imperative towards safety and human security is increasing in countries suffering from natural disasters. However, landslide disasters are generally less recognized in the statistics of the United Nations and many governments than events such as earthquakes, volcanic eruption and meteorological disasters, because most landslide disasters often occur in association with earthquakes, typhoons or hurricanes and volcanic activities and are thus classified as earthquake disasters, meteorological disasters or volcanic disasters. Nonetheless, the total number of deaths in Japan owing to landslides during the past 30 years from 1967–1998 was 3 152, while the number owing to earthquakes was 6 254, including the 1995 Kobe earthquake. Taking an even longer period, landslide disasters have caused a greater number of deaths than earthquakes.

Some landslides move slowly, on the order of centimeters per year, whilst others move at meters per day, and still others move rapidly at velocities over 50 m s^{-1}: high velocities from which people cannot always escape. Within this kind of rapid landslides, some move only a short distance, whereas others travel long distances. As illustrated in Fig. 11.1, landslides can be classified into four types according to velocity and travel distance. They are rapid long-travel landslides, slow long-travel landslides, rapid short-moving landslides, and slow short-moving landslides. The first type of rapid and long-travel landslides is clearly the most dangerous since the rapid motion does

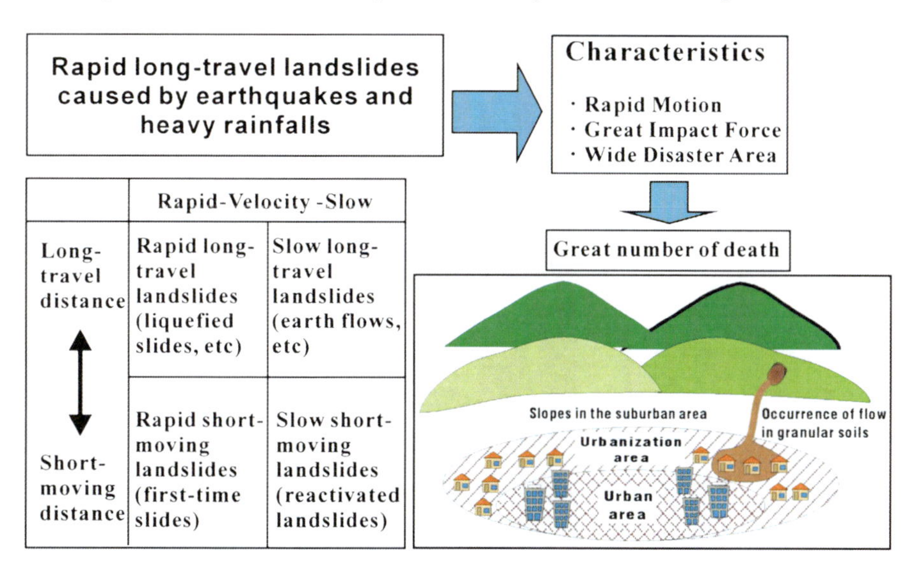

Fig. 11.1. Types of landslides in terms of velocities and travel distances

not allow for evacuation, and a great impact force can destroy houses and disastrously affect a large area. Houses that have been even constructed away from steep slopes, or often on gentle slopes can be destroyed by this phenomenon.

Figure 11.2a presents a concept of landslide mobility, which is expressed by the apparent friction (H/L) and the apparent friction angle (ϕ_a). Low apparent friction means high mobility. Figure 11.2b presents a statistic of H/L and ϕ_a for the landslides occurred in the period of 1972–1995 in Japan. Most landslides (more that 97%) had an H/L value of more than 0.5 or an apparent friction angle of more than 26°. Less than 3% showed high mobility in the debris slides and rockfalls in Japan (locally called Gake-kuzure, steep slope failures). Although the 3% is quite a small percentage in the landslide occurrence, the damages caused by the 3% landslide are quite huge.

(a)

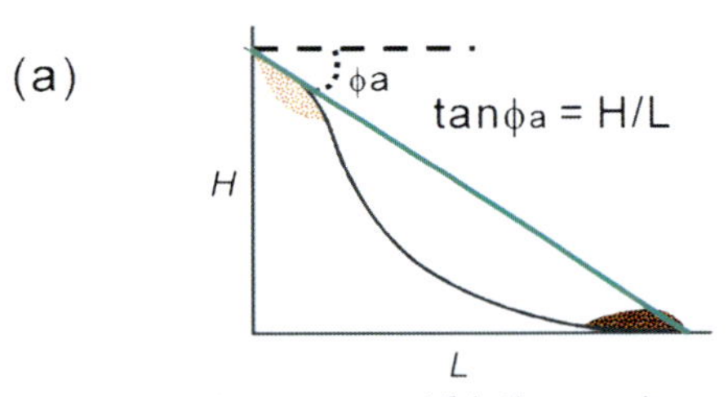

(b)

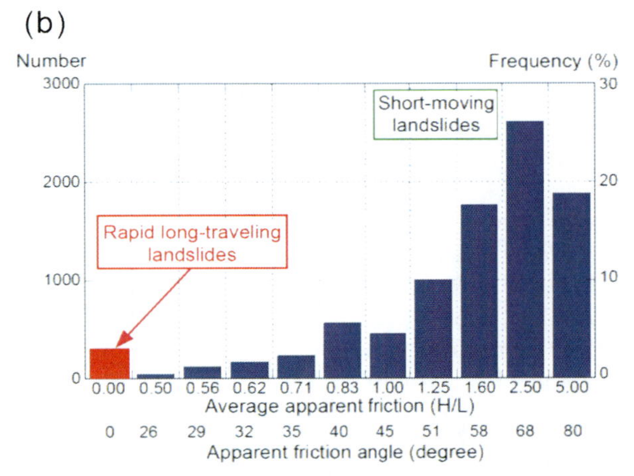

Fig. 11.2. Index to express mobility and frequency of debris slides and falls in Japan. **a** Illustration of apparent friction angle. **b** Statistical chart of life loss (by the Ministry of Construction of Japan 1997)

In Fig. 11.3, some examples of recent major landslide disasters are listed in terms of fatalities and the apparent friction angles, chiefly in Japan, also including recent large disasters in other countries. It is understandable that most of these had high mobility, with apparent friction angles of around 10 degrees. It is clear that large disasters have been caused by landslides with exceptionally high mobility.

Both Fig. 11.2 and Fig. 11.3 visualize the significance of research on rapid long-travel landslides with high mobility. A project called Areal Prediction of Earthquake and Rain Induced Rapid and Long-traveling Flow Phenomena (APERITIF) was proposed for a Special Coordinating Fund for Promoting Science and Technology of the Ministry of Education, Culture, Sports, Science and Technology of Japan (MEXT) under this understanding, and as a project in the group category of Social Infrastructure for a 3-year period from 2001 to the end of March 2004. This project was also approved as one of projects of the International Programme on Landslides (IPL M101) coordinated by the International Consortium on Landslides (ICL).

The APERITIF project consists of four topics: (1) mechanism of rapid and long-travel flow phenomena, (2) development of micro-topographical survey technology to extract dangerous slopes, (3) development of technology to assess hazardous area, and (4) integrated research for urban landslide hazard zoning. Topic (1) includes the de-

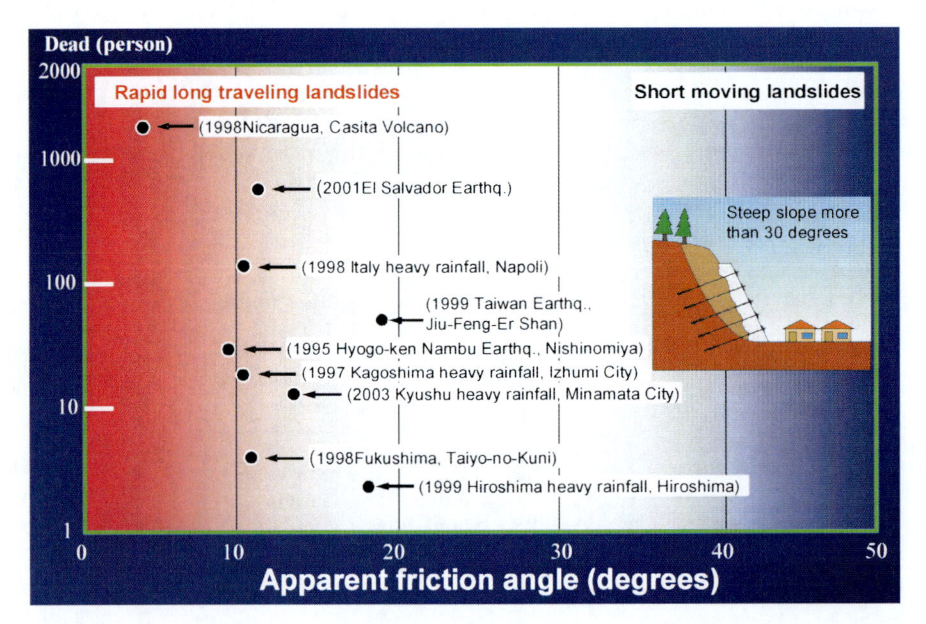

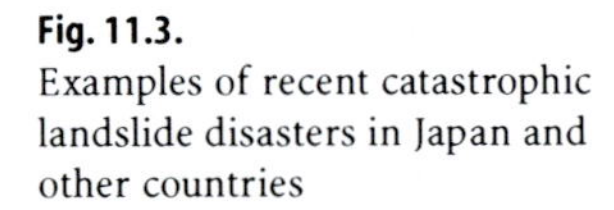

Fig. 11.3. Examples of recent catastrophic landslide disasters in Japan and other countries

velopment of dynamic loading ring shear apparatus with a transparent shear box (Sassa et al. 2004a); topic (2) includes the airborne laser scanner technology to measure micro-topography under forests (Sekiguchi and Sato 2004; Chigira et al. 2004); topic (3) includes large-scale (similar to the real scale of landslides) flume tests to reproduce landslides (Moriwaki et al. 2004), and a field experiment to reproduce the rapid and long-travel flow phenomena by artificial rainfall onto a natural slope (Ochiai et al. 2004); and topic (4) includes risk analysis of rapid and long-travel flow phenomena which may be triggered by earthquake and rainfall (Sassa et al. 2004b). This paper summarizes major research achievement of this project.

11.2 Mapping of Micro Topography Using Airborne Laser Scanning

Airborne laser scanning has recently been used in terms of landslide identification (Aleotti et al. 2000; Sekiguchi et al. 2003). It is expected that airborne laser scanning will enhance not only high-precision mapping, but also landform analysis (Ackermann 1999; Wehr and Lohr 1999). Using the acquired data, high-precision topographic maps, inclination classification maps and shading maps can be produced more efficiently. Furthermore, through micro topographic mapping, analysis for landslide simulation and unstable slope identification may be actively performed (Sekiguchi et al. 2003).

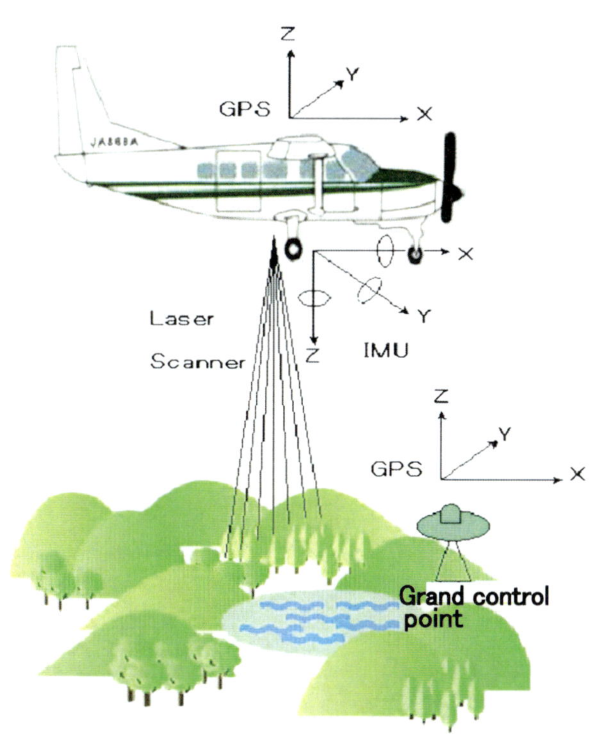

Fig. 11.4. Measuring principle of airborne laser scanning

11.2.1 Principle of Airborne Laser Scanning

Airborne laser scanning uses an active sensor that measures the distance from the sensor to the ground where the laser beam is reflected (Fig. 11.4). Aircraft positions are calculated using a combination of Global Positioning System (GPS) data, both on the aircraft and on the ground, aircraft acceleration and three-axial attitude (ω, φ, κ) data measured by an Inertial Measurement Unit (IMU). Furthermore, the direction data of the laser beams are measured by a sensor onboard the aircraft. These data are combined to calculate the three-dimensional position (X, Y, Z) on the ground (Sekiguchi and Sato 2004).

11.2.2 Application Results

Chigira et al. (2004) applied this technology to a area where many landslides were triggered by rainstorm in 1998 in Fukushima Prefecture, Japan. The aerial photograph (Fig. 11.5a) was taken in September after the landslide disaster, but it is not easy to recognize whether previous landslide scars were present or not. The laser scanner map (Fig. 11.5b), however, clearly shows previous

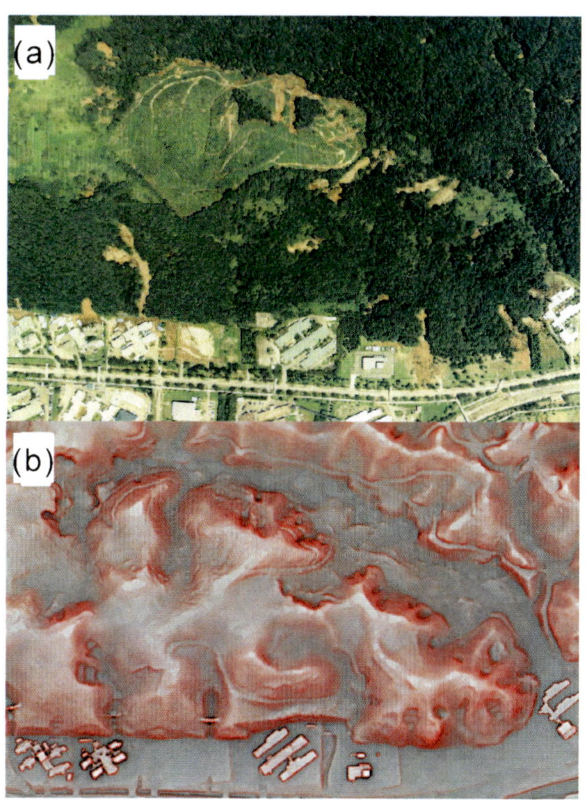

Fig. 11.5. a Aerial photograph; **b** red three-dimensional image (by Asia Survey, Co.). Aerial photograph was taken by Kokusai Co. (Chigira et al. 2004)

landslide scars as well as those during that event. Based on this map, one hundred and two previous landslide scars were identified (Chigira et al. 2004).

11.3 Full-Scale Landslide Flume Experiments and an Artificial Rainfall-Induced Landslide on a Natural Slope

Recently, great efforts have been paid to study the self-fluidization process of landslides. Because it is very difficult to get on-site data through field observations on actual landslide fluidization, model tests have been conducted with a focus on the relationship between pore-water pressure and landslide failure process. Iverson and his colleagues (Iverson and LaHusen 1989; Iverson 1997; Iverson et al. 2000) performed a series of full-scale landslide flume experiments, and found that the pore pressure responses during sliding are significantly dependent on the initial soil porosity, and that rapid fluidized landsliding involves partial or complete liquefaction of the mass by high pore-fluid pressure. Some model experiments covering "self-fluidization" have also been performed (Eckersley 1990; Moriwaki 1993; Spence and Guymer 1997; Wang and Sassa 2002).

Although these above-mentioned results sound reasonable and interesting, some were based on small-sized flume tests. As well known, small-sized experiments have problems with scale effects, similarity relations, and the disruptive effects of sensors and their cables. Therefore, large scale model test or field test at natural slope is desirable in order to realistically reproduce a landslide phenomenon.

11.3.1 Full-Scale Landslide Flume Experiment Using a Rainfall Simulator

The three-stage steel flume used in the landslide experiment was 23 m long, 7.8 m high, 3 m wide, and 1.6 m deep as shown in Figs. 11.6a and 11.6b. The main slope was a 10 m long section inclined 30 degrees, with a 6 m long section inclined 10 degrees connected at the lower end, and a 1 m long horizontal section connected at the upper

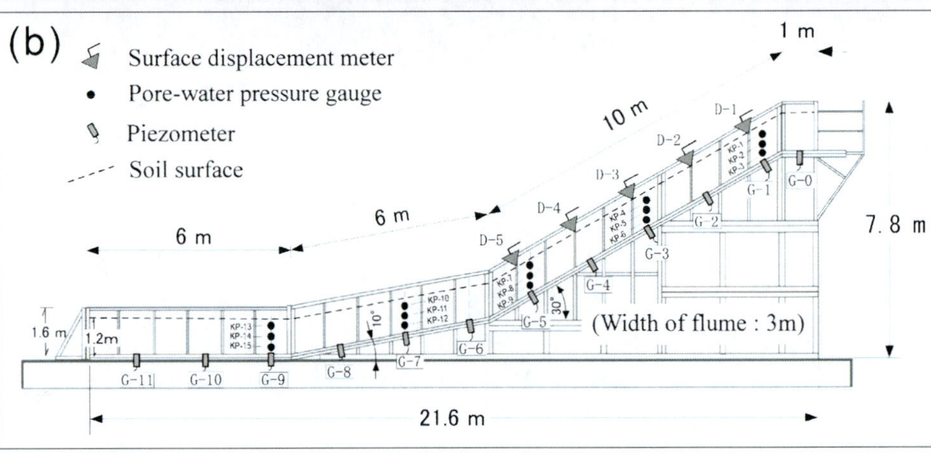

Fig. 11.6. Model slope experiment. **a** Photo before test; **b** configuration and location of sensors (Moriwaki et al. 2004)

end. A 6 m long horizontal extension, with the same width and depth as the sloping flume, was connected to the flume at the downslope end of the 10-degree slope. One wall of the entire flume was transparent reinforced glass to enable direct observation of soil deformation.

Sakuragawa sand (specific gravity: 2.69, uniformity coefficient: 3.43, mean grain size: 0.39 mm), a fine river sand was used in the flume test. The rainfall simulator in the National Research Institute for Earth Science and Disaster Prevention (NIED), Japan was employed. It is 75 m long, 50 m wide and 22 m high. The height of the nozzles above the ground was 16 m, far enough for raindrop to reach the terminal velocity. In this experiment, we sprinkled at a constant intensity of 100 mm h^{-1}.

The following items were monitored. (1) Surface displacement: extensometers with wire attachments were installed on the 30-degree slope; (2) basal piezometric levels: sealed diaphragm-type pressure transducers fixed to the slope bed at a regular interval of 2 m. Transducers could measure up to 3 m of pressure head; (3) internal pore-water pressures: three strain-gauge type meters were installed at regular interval of 4 m in the soil layer. These gauges were buried 30 cm, 60 cm, and 90 cm deep, respectively, at each location; (4) slip surface and deformation: colored-sand indices with a width of 6 cm were inserted vertically between the soil and the glass wall at a regular interval of 1 m; (5) landslide motion: high-speed and video cameras and digital cameras were used.

This test shows that when a slope consists of loose soil with high water content, the pore-water pressure can increase remarkably during failure and can simultaneously decrease the shear strength of the slope (Moriwaki et al. 2004).

11.3.2 Fluidized Landslide Experiment on a Natural Slope by Artificial Rainfall

An artificial rainfall-induced landslide experiment on a natural slope was conducted in this project. This experiment has more complex and heterogeneous characteristics than the flume models, in an attempt to investigate the dynamic movements of the soil surface, the formation of the sliding surface, and hydrological characteristics, based on the results of the laboratory flume testing.

The purpose of this experiment is to produce hopefully a fluidized landslide on a natural slope by artificial rain fall. A natural slope in the Koido National Forest at Mt. Kaba-san, Yamato Village, 25 km north of Tsukuba City, Ibaraki Prefecture, Japan was selected for the controlled experiment on landslide and possible fluidization in cooperation with the Forestry Agency of Japan.

The selected portion of hillslope was 30 m long, with an average gradient of 33 degrees (maximum 35 degrees). The soil was 1 to 3 m deep. A 5 m wide experimental slope was isolated from its surroundings by driving thin steel plates about 1 m deep into the soil. These plates prevented lateral diffusion of infiltrated rain water and cut the lateral tree root network that imparts resistance within the soil layer. The surface of the slope was covered by straw matting to prevent surface erosion and promote rainfall infiltration. Surface material on the slope consisted of fine weathered disintegrated granite sand, called "Masa" in Japan. Loamy soil blanketed the upper portion of the regolith to a depth of about 1 m; this soil mainly originated from tephra of Mt. Fuji, Mt. Akagi, and other volcanoes located west of Mt. Kaba-san.

During experiment, soil-surface movements were monitored by using stereo photogrammetry (5 stereo pairs of CCD video cameras). White-colored targets were placed on the experimental slopes and their movements were traced by video camera. To detect the formation of the sliding surface, soil-strain probes were inserted into the soil to 2 m depth at deepest. Tensiometers with porous ceramic cups were set into the slope at middle of the slope to measure changes in pore-water pressures within the soil. Artificial rain at the rate of 78 mm h^{-1} was applied to the slope segment during the experiment. The rainfall simulator consisted of a framework of steel pipes with 24 sprinkling nozzles arranged 2 m above the soil surface.

11.3.3 Overview of the Triggered Landslide Movement

On the 12 November 2003, artificial rainfall was given to the slope at a rainfall intensity of 78 mm h^{-1} for four hours and a half until sunset. No slope movement was observed. The second experiment was conducted on 14 November 2003. Artificial rainfall was started from 9:13 at a rainfall intensity of 78 mm h^{-1}, the slope deformation was detected from around 15:00, and a clear movement was observed to start at 16:03. The initiated landslide was a type of an expected fluidized landslide, the landslide mass moved rapidly and traveled long.

Some images from the digital video camera are presented in Fig. 11.7a–c, and d. As soil surface movement increased, a tension crack became visible at the head (Fig. 11.7a), and a compressive bulge resulting from downslope movement was observed 5 m above the base of the slope (Fig. 11.7b). The bulge enlarged (Fig. 11.7c) before the main landslide mass began to undulate and rapidly enter the stream (Fig. 11.7d). The compressive bulge was observed only in the left part of the landslide.

Figure 11.8 shows the landslide deposit one day after the experiment. The straw matting, the cover of the tensiometers, and the white-colored targets were conveyed to the toe of the fluidized landslide.

Changes in pore water pressure including suction monitored by the tensiometers at *T* are presented in Fig. 11.9. The cover of the tensiometers started to incline

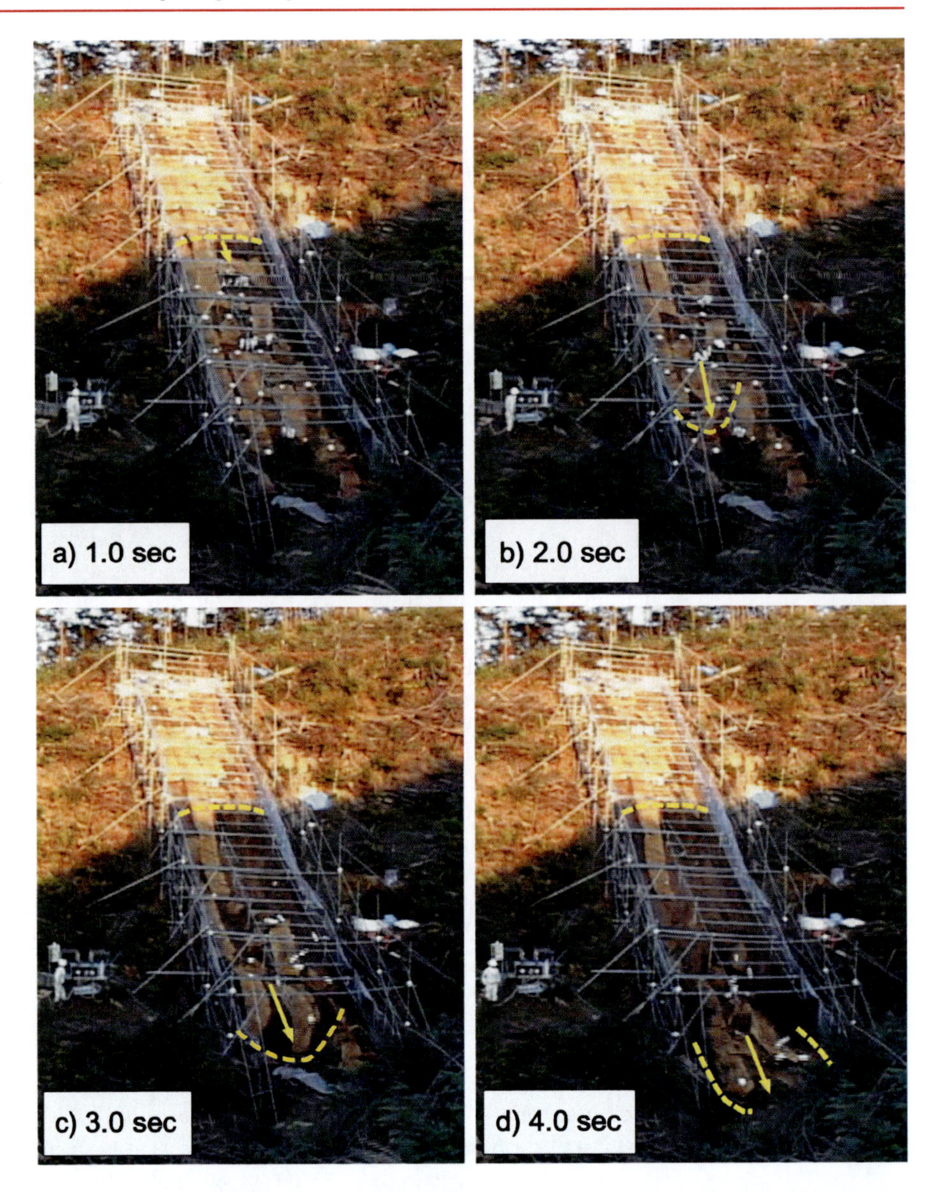

Fig. 11.7.
Views of the landslide fluidization between one second and four seconds after failure. **a** At 24628.5 s (one second after failure initiation); **b** at 24629.5 s (two seconds); **c** at 24630.5 s (three seconds); **d** 24631.5 s (four seconds)

downslope at 24627.5 s (410 min 27.5 s) after sprinkling commenced. We interpret this as indicating that slope failure initiated at 24627.5 s. The tensiometers at *T* were placed at depths of 50, 100, 150, 200, 250, and 290 cm. All tensiometers showed negative pore pressures at the start of sprinkling, indicating that the soils at all depths were unsaturated or partly saturated. When the wetting front passed, the tensiometers showed increases in pore pressure in sequence of the depths. At 410 min, when the failure took place, all of the tensiometers showed positive pore pressures. The pore pressure of the deepest tensiometers (290 cm) rapidly increased its values from about 290 min. Hence, it can be deduced that general slope instability increased from 290 min, before final failure at 410 min.

This test obtained valuable information for further qualitative interpretation of the shape of the landslide and definition of failure initiation and deposition times. As monitored by the tensiometers, a rapid increase in pore-water pressure after about 290 min from the start of sprinkling, and it almost coincided with the time when the strain was first observed on the sliding surface (Ochiai et al. 2004).

11.4 Landslide Risk Evaluation and Hazard Zoning for Rapid and Long-Traveling Landslides in Urban Development Areas

As a comprehensive study part of the APERITIF project, we try to supply an integrated method for landslide risk evaluation and hazard zoning for rapid and long-traveling landslides in urban development areas (Sassa et al. 2004b). Two fields in Japan were selected to conduct this study. One is the upper slope connecting to the site of 1995 Nikawa landslide (34 fatalities) triggered by the Hyogoken-

Fig. 11.8. The landslide deposit one day after the experiment

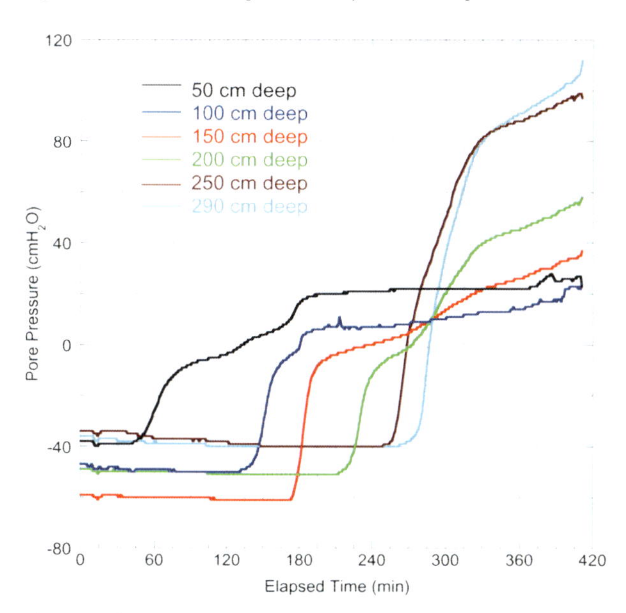

Fig. 11.9. Changes in soil-water pressure in tensiometers at *T*

Nambu earthquake in Nishinomiya City between Osaka and Kobe; and the other site corresponds to slopes inside or adjacent to a large scale residential development in the Tama Hills near Tokyo. Here, only the assessment method for rapid and long-travel landslides which will be triggered by future huge earthquake in a densely populated residential town near Tokyo is reported.

11.4.1 Field Study for the Development Area near Tokyo

Figure 11.10 shows a study area in the hilly Tama area near Tokyo. It is located at the border of Hino and Hachioji Cities. The area is covered by an almost flat sedimentary sandy layer (Kazusa Group soil), which was formed during the Pliocene and Pleistocene. Two sites, A and B, were selected in this area. Site A is a large infilled valley within the residential area. Site B is a projecting ridge toward the residential area. Those areas are shown in the rectangular boxes in Fig. 11.10.

Sedimentary sandy layers are distributed through almost horizontal bedding, so this saturated sandy layer should extend below the extrusive ridge. To investigate this situation, four drillings were planned. However, due to non-cooperation of a development company in this area, two drillings could not be conducted along the central section. So they were drilled some distance apart at a higher elevation. The blue color part indicates an unsaturated silty layer. Probably this layer cannot experience rapid and long-travel landslides. However, if the saturated sandy layer will be liquefied or semi-liquefied during earthquake loading, there is a high possibility of landslides along the sliding surfaces because most resisting parts in these expected landslides are parts passing inside the saturated sandy layer.

Two boreholes (A and B) were drilled at the Hino site. Samples were taken from the filled materials in the infilled valley. An undrained cyclic shear stress loading test was performed to assess the acceleration required to cause failure. An initial normal stress conditions to reproduce the potential sliding surface of 15 m deep in a 12–13 degrees slope were applied. The initial ground water table was 2.5 m from the surface. Therefore, an initial pore-water pressure of about 120 kPa (corresponding to the water table above the sliding surface) was imposed. A control signal to produce the required cyclic shear stress by increasing its value step by step up to around 600 kPa ensured the occurrence of failure. The mobilized maximum shear stress was about 115 kPa (which corresponds to 220 gal) in shear stress increment to cause failure from the following relation.

The steady state shear resistance reached 16.3 kPa. The apparent friction angle was only 3.5 degrees. It is a typical sliding surface liquefaction (Sassa et al. 1996). Rapid and long-travel landslides should result from such stress condition.

Then, a naturally drained cyclic loading test using the same sample, control signal testing procedure with the undrained test was performed. These are similar to the undrained test, although the excess pore pressure could not be monitored because the pore pressure transducer inlet is located 2 mm above the shear surface. Excess pore pressure was generated within the shear zone (there, grains are crushed and causing a decrease in permeability). The influence of open valve drainage is much great. However, the shear resistance decreased until 29 kPa and shear displacement accelerated. It is apparently the sliding surface lique-

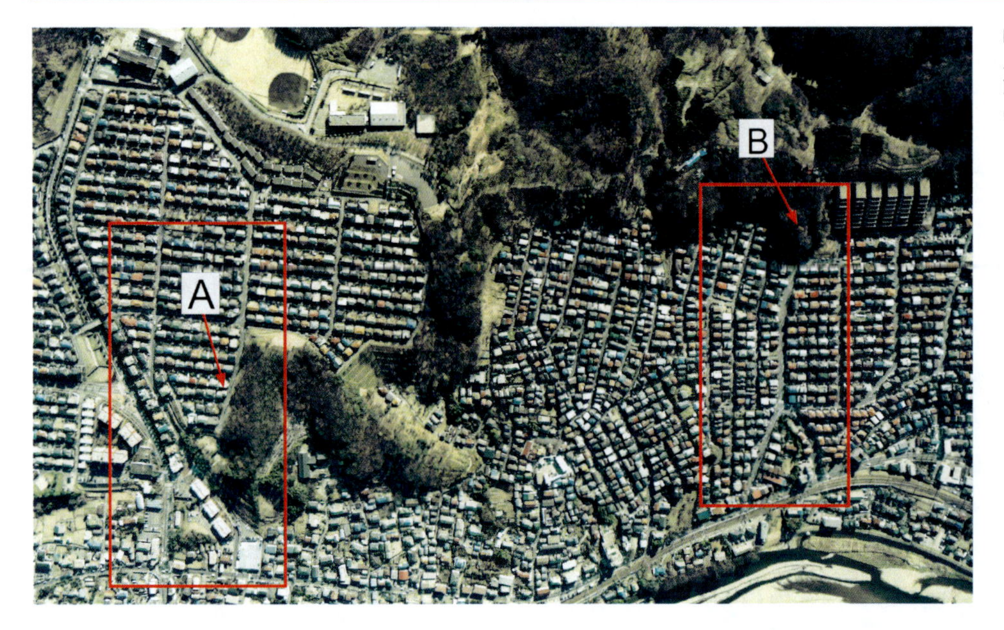

Fig. 11.10. Air photo of the densely populated developed area near Tokyo (*A:* Hino site; *B:* Hachioji site)

faction phenomena occurred. The steady state shear resistance was around 1.8 times greater than in the undrained test. The difference can be explained by pore pressure dissipation. The real soil response should fall in between these extremes.

Similar tests were conducted on a sample taken from sands in the Hachioji area, only with the different initial condition. Initial normal stress and shear stress corresponding to a depth of 20 m in a 10 degrees slope without initial pore water pressure were reproduced. The slope is in a projecting ridge, therefore a high ground water level is not expected in ordinary period without rainfall, while the bottom of ridge (sandy layer) is expected to be saturated. In the simulating experiment, failure took place for a shear stress around 151 kPa, corresponding to a critical seismic acceleration of 270 gal. The steady state shear resistance was 35 kPa, and the apparent friction angle was 4.5 degrees. This value is low enough to suggest the possibility that long-travel landslides will be a potential hazard in this area.

11.4.2 Computer Simulation in Investigated Areas

A geotechnical landslide simulation model was proposed by Sassa (1988). It was improved to a computer code for general use during APERITIF project especially in the data input and three dimensional presentation of output. This new computer code for rapid landslides (Rapid/LS) was applied to reproduce the two cases using the measured values of the steady state shear resistance obtained through undrained and naturally drained ring shear tests (Sassa et al. 2003). In addition to the steady state shear resistance, another parameter for geotechnical simulation, the lateral pressure ratio K (σ_h / σ_v) was introduced to express "softness" or "potential for lateral spreading" of the moving mass.

Figure 11.11 presents three-dimensional visual output of the result. Figure 11.11a shows the initial state; Fig. 11.11b shows a intermediate stage during motion; Fig. 11.11c shows the final deposition state; and Fig. 11.11d presents the assessed hazard area superimposed to the three dimensional view of the area obtained by the laser scanner technology by Japanese Geographical Survey Institute. In Fig. 11.11d, the larger area corresponds to the data obtained by the undrained ring shear test (using 16 kPa as the steady state shear resistance), and the smaller area corresponds to the naturally drained ring shear test (using 29 kPa as the steady state shear resistance). Since the source area includes many houses, the assessed risk is quite large even for the naturally drained case.

Figure 11.12 shows the results for the case of 20 m deep landslide mass onto the infilled valley in the Hachioji area. Figure 11.12a presents the initial state before landslide in which some steps at possible head scarps are visible; Fig. 11.12b and Fig. 11.12c show intermediate states during motion; Fig. 11.12d is the final situation after deposition in which the border of deposited landslide mass is delineated. The landslide mass covered a considerably large area along the infilled valley.

The study shows that an integrated method has formed with the combination of the airborne laser scanning method for detailed topography, field investigation for detailed soil layer condition including hydrogeological condition, geotechnical testing like undrained and drained ring shear tests for soil behavior at different possible loading conditions, and the landslide simulation program to apply all of the relevant information of targeted site and give a visual output, and it is possible to conduct landslide risk evaluation and hazard zoning for rapid and long-traveling landslides in urban development areas in a street precisions.

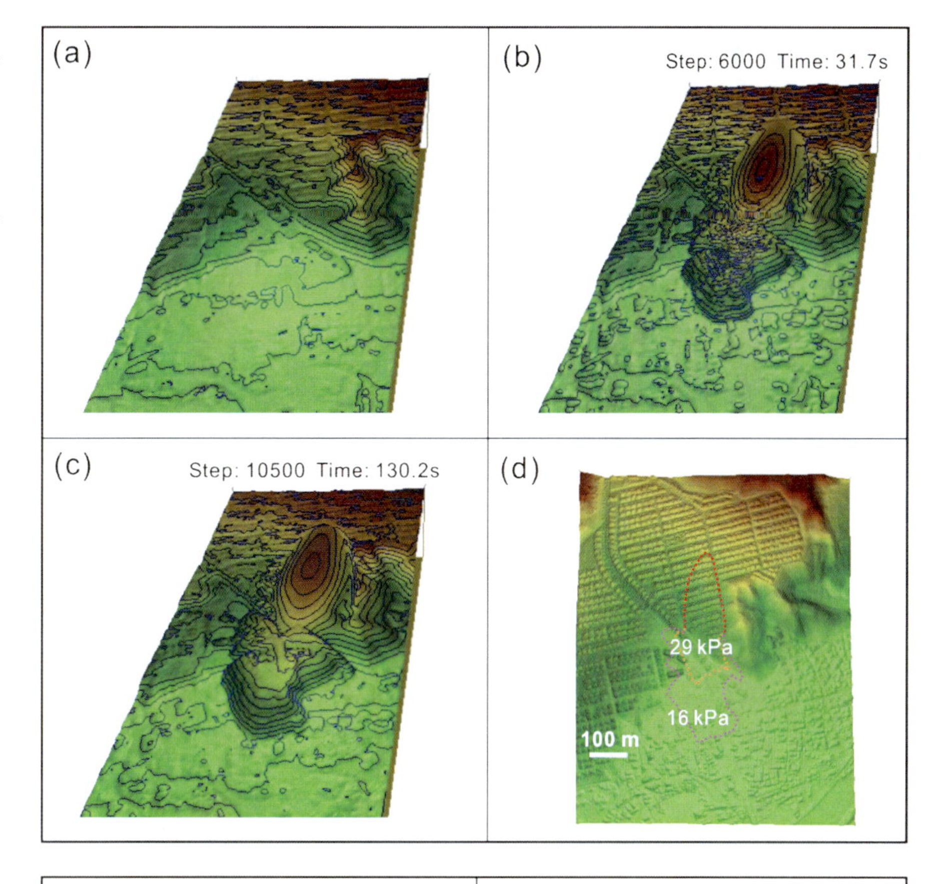

Fig. 11.11. Results of simulation results for Hino area (shear strength at steady state: 16 kPa). **a** Original topography; **b** and **c** during motion; **d** deposited areas corresponding to the steady-state-shear strength of 16 and 29 kPa, respectively. Lateral pressure ratio $K = 0.65–0.80$. The three dimensional view was made by the Geographical Survey Institute of Japan

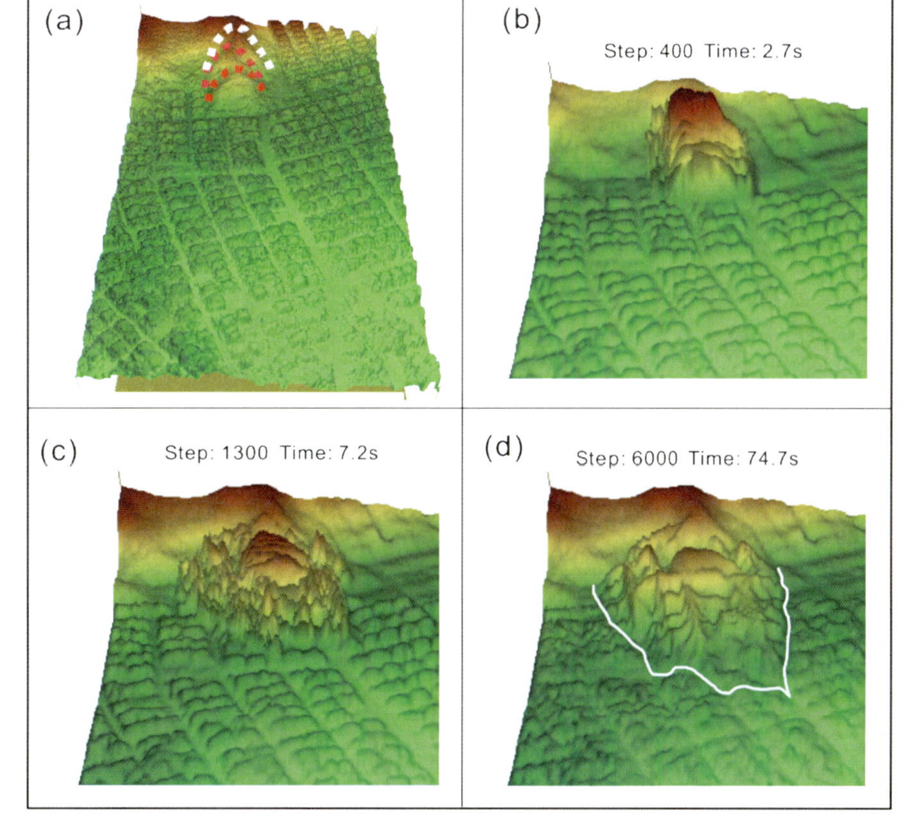

Fig. 11.12. Simulation results for Hachioji area (shear strength at steady state: 16 kPa). **a** Original topography; **b** and **c** during motion; **d** deposited area. The three-dimensional view was made by the Geographical Survey Institute of Japan using the laser scanner data for this APERITIF project. Lateral pressure ratio $K = 0.50–0.65$

11.5 Conclusions

The following conclusions can be drawn:

1. Airborne laser scanning method is developed for landslide risk analysis, and it is approved that that laser contour maps are more precisely than photogrammetric contour maps.
2. Full-scale landslide experiment and artificial rainfall-induced landslide focusing on the mechanism of self-fluidization were successfully conducted, and some insights were obtained for understanding of the fluidization mechanism of rapid landslide.
3. An integrated method for landslide risk analysis and landslide hazard zoning, which includes precise topography measurements, detailed field investigation, suitable geotechnical testing methods, and landslide simulation program, was formed. It is a reliable and convincing approach of urban hazard zoning based on the topographical, geological and geotechnical parameters at street-level scale from two case studies.

Acknowledgments

This research was conducted as a part of the areal prediction of earthquake and rain induced rapid and long-traveling flow phenomena (APERITIF) project of the International Programme on Landslides (IPL M101). The financial support of the Special Coordinating Fund for Promoting Science and Technology of the Ministry of Education, Culture, Sports, Science and Technology (MEXT) in 2001–2004 is highly appreciated.

References

Ackermann F (1999) Airborne laser scanning present status and future expectations. ISPRS J Photogramm 54:64–67

Aleotti P, Canuti P, Iotti A, Polloni G (2000) Debris flow hazard and risk assessment using a new airborne laser terrain mapping technique (ALTM). In Proc 8th Int Symp on Landslides, Cardiff, June 2000, 1, pp 19–26

Chigira M, Duan FJ, Yagi H, Furuya T (2004) Using an airborne laser scanner for the identification of shallow landslides and susceptibility assessment in an area of ignimbrite overlain by permeable pyroclastics. Landslides 1(3):203–209

Eckersley JD (1990) Instrumented laboratory flowslides. Geotechnique 40(3):489–502

Iverson RM (1997) The physics of debris flows. Rev Geophys 35(3): 245–296

Iverson RM, LaHusen RG (1989) Dynamic pore-pressure fluctuations in rapidly shearing granular materials. Science 246:769–799

Iverson RM, Reid ME, Iverson NR, LaHusen RG, Logan M, Mann JE, Brien DL (2000) Acute sensitivity of landslide rates to initial soil porosity. Science 290:513–516

Moriwaki H (1993) Behavior of pore-water pressure at slope failure. In: Proc 7th Int Conf and Field Workshop on Landslides in Czech and Slovak Republics, "Landslides", Balkema, Rotterdam, pp 263–268

Moriwaki H, Inokuchi T, Hattanji T, Sassa K, Ochiai H, Wang G (2004) Failure processes in a full-scale landslide experiment using a rainfall simulator. Landslides 1(4):277–288

Ochiai H, Okada Y, Furuya G, Okura Y, Matsui T, Sammori T, Terajima T, Sassa K (2004) A fluidized landslide on a natural slope by artificial rainfall. Landslides 1(3):211–219

Sassa K (1988) Special Lecture: The geotechnical model for the motion of landslides. In: Proc 5th Int Symp on Landslides, Lausanne, vol. 1, pp 33–52

Sassa K, Fukuoka H, Scarascia-Mugnozza G, Evans SG (1996) Earthquake-induced-landslides: Distribution, motion and mechanisms. Soils and Foundations, Special Issue, pp 53–64

Sassa K, Wang G, Fukuoka H (2003) Performing undrained shear tests on saturated sands in a new intelligent type of ring shear apparatus. Geotech Test J 26(3):257–265

Sassa K, Fukuoka H, Wang G, Ishikawa N (2004a) Undrained dynamic-loading ring-shear apparatus and application for landslide dynamics. Landslides 1(1):7–19

Sassa K, Wang G, Fukuoka H, Wang FW, Ochiai T, Sugiyama M, Sekiguchi T (2004b) Landslide risk evaluation and hazard zoning for rapid and long-travel landslides in urban development areas. Landslides 1(3):221–235

Sekiguchi T, Sato HP (2004) Mapping of micro topography using airborne laser scanning. Landslides 1(3):195–202

Sekiguchi T, Sato HP, Ichikawa S, Kojiroi R (2003) Mapping of micro topography on hill slopes using airborne laser scanning. Bulletin of Geographical Survey Institute 49:47–57

Spence KJ, Guymer I (1997) Small-scale laboratory flowslides. Geotechnique 47(5):915–932

Wang G, Sassa K (2002) Pore pressure generation and motion of rainfall-induced landslides in laboratory flume tests. In: Proc Int Symp on Landslide Risk Mitigation and Protection of Cultural and Natural Heritage, UNESCO and Kyoto University, pp 45–60

Wehr A, Lohr U (1999) Airborne laser scanning an introduction and overview. ISPRS J Photogramm 54:68–82

Chapter 12

Investigating Rock-Slope Failures in the Tien Shan: State-of-the-Art and Perspectives of International Cooperation (M111)

Alexander L. Strom* · Oliver Korup · Kanatbek E. Abdrakhmatov · Hans-Balder Havenith

Abstract. The Tien Shan is an intracontinental mountain system ~1 500 km long and up to 500 km wide that formed between the Tarim Basin and the Kazakh Shield due to the India-Asian collision. It is shared by the five nations of Kyrgyzstan, Kazakhstan, Uzbekistan, Tajikistan, and China. As one of the highest and most seismotectonically active parts of the Central Asian Mountain Belt, it is extremely prone to large rock-slope failures. At least nine rock slope failures >1 km^3 in volume, two of which involved ~10 km^3, have been identified in the Tien Shan. Thanks to an arid climate many of these formerly river-blocking rockslides and long-runout rock avalanches are well preserved, and both their morphology and internal structure may be readily studied in detail. Here we briefly describe the state-of-the-art and planned future international collaborations in research on rock-slope failures in the Tien Shan.

Keywords. Rockslide, rock avalanche, landslide dam, Tien Shan, Kyrgyzstan, landslide hazard and risk

12.1 Introduction

The Tien Shan ("Sky Mountains") is one of the highest and most seismically active parts of the Central Asian Mountain Belt (Fig. 12.1). It is shared by the five nations of Kyrgyzstan, Kazakhstan, Uzbekistan, Tajikistan, and China. Numerous landslides are known here, yet only a small part of them have been described in publications and unpublished technical reports. Landslides in some parts of the Tien Shan have been regularly studied during Soviet times (Fedorenko 1988; Zolotariov 1990; Niazov et al. 2002), but even these data are poorly known by landslide researchers outside the former USSR. We expect a similar situation with landslide research in the eastern part of the mountain belt, i.e. the Chinese Tian Shan.

Recently, international cooperation in landslide research started developing in Kyrgyzstan. These activities have been largely stipulated by the significant environmental hazard posed by large active landslides endangering the stability of numerous radioactive mine tailings in the Mailuu-Suu Valley (Aleshin et al. 2002; Torgoev et al. 2002). These investigations, along with some other projects in the densely populated eastern rim of the Fergana Basin, are focused on monitoring and early warning of selected landslides and landslide-prone slopes, which mainly occur in unconsolidated Quaternary or poorly lithified Neogene and Mesozoic sediments.

Also, large rockslides with potential of long runout and forming large natural dams in mountain rivers are wide-

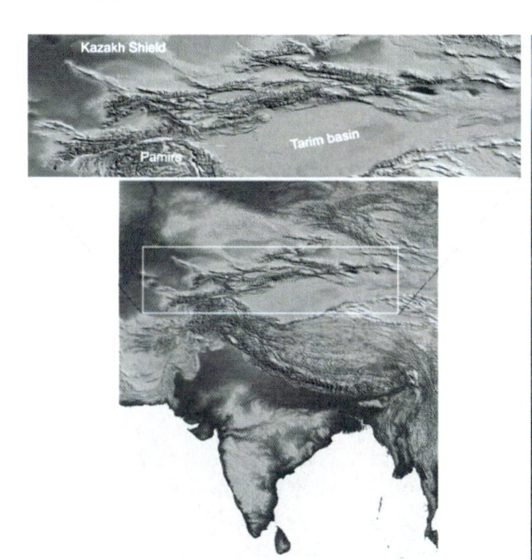

Fig. 12.1. *Left:* Location of the Tien Shan in the Central Asian region; *right:* participants of the NATO Advanced Research Workshop on "The Security of Natural and Artificial Rockslide Dams", Kyrgyzstan, June 2004, in front of traditional jurts

spread in the Tien Shan. They have been studied during the 1960s to 1980s e.g. in the Zeravshan River Basin in the SW (Tajik) Tien Shan (Fedorenko 1988), the central Naryn River Basin in connection with the Toktogul hydropower project (Zolotariov 1990), and the southern Tien Shan, where the catastrophic coseismic Khait rock avalanche killed thousands of people in 1949 (Leonov 1960). Two rockslide dams prone to devastating outburst were studied in northern (Kazakh) part of Tien Shan, and several catastrophic breaches of rockslide-dammed lakes occurred in the 1960s. The most well-known cases were at Issyk Lake east of Almaty, Kazahkstan, and Jashinkul Lake in the southern Tien Shan. More recently, various rockslides were studied in the Naryn Valley upstream from the Toktogul Reservoir, in the Kokomeren Valley, and other parts of the Kyrgyz Tien Shan with special focus on the mechanics of their formation and motion (Strom 1996, 1998).

However, there is as yet no comprehensive landslide inventory comparable to e.g. that of the European Alps (Heim 1932; Abele 1974). Hence, no detailed analysis of the magnitude and frequency, and resulting hazard and risk from landslides in the region is possible up to date.

12.2 Completed and Ongoing Activities

One of the first international studies concerned with rockslides was the INCO-COPERNICUS Project PL96-3202 "Landslides triggered by earthquakes in Kyrgyzstan, Tien Shan" in 1997–2000 (Delvaux et al. 2001; Havenith et al. 2002, 2003). Detailed geophysical and seismological studies were applied to the Ananevo and Kaindy rockslides ($>0.5 \times 10^7$ m^3) triggered by the 1911 M 8.2 Kemin earthquake, to investigate why only so few slopes experienced large-scale instability despite strong ground shaking. Surface ruptures of the earthquake were also mapped and studied with geophysical and palaeoseismological methods (Delvaux et al. 2001; Havenith et al. 2000).

A NATO Advanced Research Workshop "*Security of Natural and Artificial Rockslide Dams*" held in Bishkek, Kyrgyzstan, in June 2004, was the next step in order to introduce the Tien Shan to the international community of rockslide researchers. The meeting convened 48 participants from Austria, Belgium, Canada, China, Germany, Italy, Kyrgyzstan, Mexico, New Zealand, Russia, Switzerland, Tajikistan, UK, and the U.S.A., and included a one-day field trip to the Chon-Kemin Valley, where a cluster of four large prehistoric rockslides had mobilized ~1.2 km^3 of rock. Fifteen researchers from Austria, Canada, Germany, Italy, Kyrgyzstan, New Zealand, Russia, Switzerland, and the UK participated in a five-day post-conference field trip during which several large to giant rockslides and rock avalanches were visited and discussed. Most of the rockslide deposits visited are deeply incised by erosion and thus, participants had the opportunity to study their internal structure, lithology, and grain-size composition. Much of the preparatory work for the workshop and field trips was supported by IPL2002 M-111 Project "Detail study of the internal structure of large rockslide dams in the Tien Shan and the International field mission – Internal structure of dissected rockslide dams in Kyrgyzstan". Two field guidebooks were prepared (Strom and Abdrakhmatov 2004a,b) to compile an overview of the most prominent rockslide sites.

12.3 Aims and Goals of Further Studies

The NATO Advanced Research Workshop strongly contributed to outlining future key objectives for studying large rock-slope failures and slope-instability in the Tien Shan. The first objective is to systematically compile a regional GIS inventory of rockslides, landslides, and related phenomena (sackungen) across the entire Tien Shan, irrespective of political boundaries. This is the main goal of IPL2004 M-127 Project "Compilation of uniform landslide inventory for the Tien Shan Mountain system" approved by the 2004 ICL BOR meeting in Bratislava. Mapping will be based on the analysis of high-resolution (<15 m) space images such as KFA-1000, KFA-3000, Corona, SPOT, ASTER, and IRS, allowing detection of most large-scale slope failures with areas >0.1 km^2, and volumes of roughly $>10^6$ m^3. Similar studies were done for the Suusamyr Basin and surrounding ranges (Havenith et al. 2003), where several hundreds of landslides have been identified on satellite images, forming the database for slope stability analyses. The largest events can be also identified with satellite imagery draped over 3" SRTM digital elevation data. So far, nine "gigalandslides" ($>10^9$ m^3) have been identified in the Kyrgyz Tien Shan (Strom and Korup submitted; Korup et al. in press).

Another very important topic of future studies is the absolute dating of rockslides and other geomorphic features such as fault scarps and river terraces that can shed light on the recurrence of rockslide events. This will allow the establishment of magnitude-frequency curves, and, together with limit equilibrium and back analyses, test our currently favored hypothesis that most of the large rock-slope failures in the Tien Shan would be of seismic origin. We have submitted several funding proposals to the European Commission, the Swiss National Science Foundation, and INTAS, for using radiometric (^{14}C, OSL, and cosmogenic isotopes), dendrochronologic, and lichenometric dating for this purpose.

Process-related research is highly promising, given that excellent exposures along 200–400 m deep river gorges cut into rockslide debris allow studies of landslide internal structure, including grain-size analysis, and mechanical properties of rockslide debris. Another important theme in future studies is the identification of potentially unstable rock-slopes prone to catastrophic failure. The phenomenon of rockslide clustering is widely developed in the Tien Shan and shows that even along active fault zones some local areas are subjected to larger and, likely, more frequent bedrock slope failures. These "focal points" will serve as starting points for regional slope-stability analyses.

All these objectives and research issues strongly depend on successful international collaboration. Joint efforts of landslide researchers from countries adjoining the Tien Shan should be directed to allow using harmonized criteria of rockslide selection and classification, as well as exchange of data and expertise. An overall aim will be the promotion of rockslide and landslide hazard and risk assessment across the entire mountain system. In close connection with the above investigations, we are planning an annual international summer school on rockslides and related phenomena for students and young landslide researchers in the Kokomeren Valley, Kyrgyzstan, known for its clustering of several rockslides and rock avalanches (Fig. 12.2). Participants of the summer school will be trained, among others, in methods of bedrock slope failure mapping, absolute dating, detail study of their internal structure, grain-size composition, and geomorphic evidence of rockslide-dam failure.

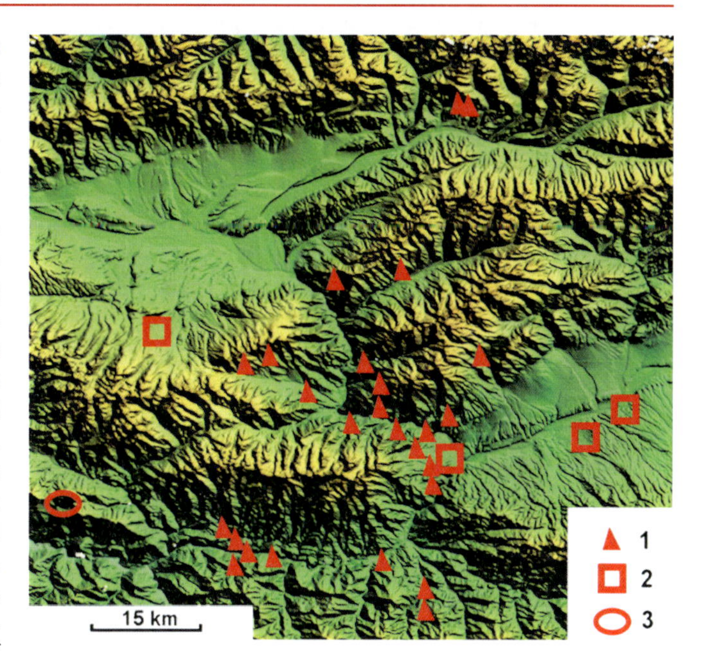

Fig. 12.2. Largest rock slope failures and landslides in the Kokomeren River Basin. *1:* Rockslides and rock avalanches $>10^6$ m^3; *2:* largest landslides in Neogene and Quaternary deposits; *3:* Caldera-like Kyzylkiol collapse

12.4 Conclusions

The Tien Shan provides high potential for studying catastrophic rock-slope failures and their hazard and risk implications before a trans-national background. The arid climate and dense road network in the Tien Shan makes many spectacular landslides sites accessible, that easily rival in size those reported from other mountain belts. Comparison of landslide characteristics and distributions in the Tien Shan with those in seismically less active mountain belts such as the European Alps will also increase our understanding of the role of seismicity in triggering large rock-slope failures.

References

Abele G (1974) Bergstuerze in den Alpen; ihre Verbreitung, Morphologie und Folgeerscheinungen. Translated title: Alpine landslides; their distribution, morphology, and consequences. Wissenschaftliche Alpenvereinshefte. 25, Deutscher Alpenverein, Munich, 230 pp

Aleshin YG, Torgoev IA, Shmidt G (2002) Environmental risk management at uranium tailing ponds in Mailuu-Suu, Kyrgyzstan. In: Merkel B, Planer-Friedrich B, Wolkersdorfer C (eds) Uranium in the aquatic environment. Springer-Verlag, Berlin Heidelberg, pp 881–888

Delvaux D, Abdrakhmatov KE, Lemzin IN, Strom AL (2001) Landslides and surface breaks of the 1911, Ms 8.2 Kemin earthquake, Kyrgyzstan. Russ Geol Geophys 42:1583–1592

Fedorenko VS (1988) Landslides and rockfalls in mountains, and their forecasting. Moscow University Press, Moscow

Havenith H-B, Jongmans D, Abdrakhmatov K, Torgoev I, Delvaux D, Strom A, Trefois P (2000) Application of geophysical methods to seismic landslides and fault scarps in Kyrgyzstan (Central Asia). In: Proc Kyrgyz National Academy of Sciences, Institute of Rock Physics and Mechanics, Bishkek, Kyrgyzstan, pp 173–182

Havenith H-B, Jongmans D, Faccioli E, Abdrakhmatov, K, Bard P-Y (2002) Site effects analysis around the seismically induced Ananevo rockslide, Kyrgyzstan. Bull Seismol Soc Am 92: 3190–3209

Havenith H-B, Strom A, Jongmans D, Abdrakhmatov K, Delvaux D, Tréfois P (2003) Seismic triggering of landslides, part A: Field evidence from the Northern Tien Shan. Natural Hazards and Earth System Sciences 3:135–149

Heim A (1932) Bergsturz und Menschenleben. Beiblatt zur Vierteljahrsschrift der Naturforschenden Gesellschaft in Zürich, vol.77, pp 1–217. Translated by N. Skermer under the title "Landslides and Human Lives", BiTech Publishers, Vancouver, British Columbia, Canada, 1989, 195 pp

Korup O, Strom AL, Weidinger JT (in press) Fluvial response to large rock-slope failures – examples from the Himalayas, the Tien Shan, and the New Zealand Southern Alps. Geomorphology

Leonov NN (1960) Khait earthquake of 1949 and the geological conditions of its occurrence. Geology and Geophysics, Geophysical Series 3:48–56 (in Russian)

Niyazov RA, Nurtaev BS, Minchenko VD (2002) Hazard assessment of geological processes in loess rock of the mountain regions of Uz-bekistan. In: Proc 9th IAEG Congress, Westville (South Africa), p 9

Strom AL (1996) Some morphological types of long-runout rockslides: effect of the relief on their mechanism and on the rockslide deposits distribution. In: Senneset K (ed) Proc 7th International Symposium on Landslides, 1996, Trondheim, Norway, Rotterdam. Balkema, pp 1977–1982

Strom AL (1998) Giant ancient rockslides and rock avalanches in the Tien Shan Mountains, Kyrgyzstan. Landslide News 11:20–23

Strom AL, Abdrakhmatov KE (2004b) NATO ARW guidebook. Rock avalanches and rockslide dams of the Northern Kyrgyzstan (Kyrgyz Range and Chon-Kemin River Valley)

Strom AL, Abdrakhmatov KE (2004c) Post-ARW field trip guidebook. Rockslides and rock avalanches of the Kokomeren River Basin and Karakudjur rockslide

Strom AL, Korup O (submitted) Extremely large rockslides and rock avalanches in the Tien Shan, Kyrgyzstan. Landslides

Torgoev IA, Alioshin YG, Aitmatov IT (2002) Danger and risk of natural and man-caused disasters in mountains of Kyrgyzstan. In Aidaraliev et al. (eds) Mountains of Kyrgyzstan. Bishkek Publishing House "Technology", Bishkek, pp 157–166

Zolotarev GS (1990) Methodology of engineering-geological investigations. Moscow State University Publishers. 384 p

Chapter 13

Multi-Temporal and Quantitative Geomorphological Analysis on the Large Landslide of Craco Village (M118)

Giuseppe Delmonaco* · Luca Falconi · Gabriele Leoni · Claudio Margottini · Claudio Puglisi · Daniele Spizzichino

Abstract. The village of Craco (Basilicata, Italy), is being affected by severe landslide phenomena mainly due to the geological and geomorphological setting of the area. The village has been interested by a progressive abandon of the population after the occurrence in the time of landslides and earthquakes that caused the disruption of large portions of the urban settlement. Several landslide typologies can be recognized in the area: rock-falls in the upper part of the hill, rotational and translational earth slides, earth-flows, rock lateral spreading. The main purpose of the paper is to reconstruct the evolution of the geological and morphological dynamics acting on the southern slope of Craco, where the largest landslides occurred in the past.

The analysis of deep landslide phenomena has been carried out with a multi-temporal geomorphological approach using analogical and digital photogrammetry on 4 different set of aerial photos (1954–1999). The analysis has stressed a progressive retrogression of the crown area in the upper portions of the hill that caused, in the recent past, the degradation or, in some cases, the complete disruption of part of the historical village, where most of cultural heritage was located.

The reconstruction of the landslide evolution acting in this area can be very useful as model to transfer in other areas characterized by similar geological and morphological setting that may result in large landslide phenomena. The suggestion of correct mitigation strategies may help to prevent environmental and, consequently, social degradation of the territory.

Keywords. Landslide intensity, multi-temporal analysis, digital stereoscopy, cultural heritage

13.1 Introduction

The paper reports a multi-temporal analysis of the large landslide of Craco (S Italy) implemented through the application of digital stereoscopy coupled with GIS techniques. This work is part of the Project IPL M-118 *Development of an expert DSS for assessing landslide impact mitigation work for Cultural Heritage at risk (VIP Project)*.

The multi-temporal analysis through aerial photos is a powerful method for assessing landslide evolution. In the last decade, the development of stereoscopy and *GIS* techniques and tools has improved the capability of recognition of morphometry, geomorphological features and quantitative parameters (i.e. mobilized volumes) of a landslide. The evaluation of the intensity of a landslide is a fundamental task in landslide hazard assessment; this can be generally defined by two parameters: volume of the mobilized mass and velocity of displacement and/or their combination.

This study has been addressed to quantify, following the best possible resolution, the dimension of the mobilized mass of by the main landslide that affects the historical village of Craco (Basilicata Region, S Italy). In order to test the opportunity of digital photogrammetry coupled with GIS techniques, a multi-temporal analysis, based on 4 aerial photo taken in the period 1954–1999, has been carried out. The analysis has stressed the plano-altimetric differences caused by the geomorphological evolution of the large landslide of Craco providing an estimation of the mobilized volumes at different time windows.

13.2 Study Site

The village of Craco (Basilicata, Italy), is being affected by severe landslide phenomena mainly due to the geological and geomorphological characteristics of the area. The tectonic-stratigraphical setting of Craco area is characterized by Cretaceous-Oligocene and Miocene sediments (Sicilide Clays), mainly clayey, lying in transgressive contact with a Pliocene clayey-sandy complex (Del Prete and Petley 1982; Carbone et al. 1991). Four distinct landslide typologies (Cruden and Varnes 1996) can be recognized in the study area. The largest phenomena, as involved areas and volumes, develop along the southern slope: rotational slides with very deep failure surfaces (>50 m) in the upper part of the slope, and earth-flows that involve the middle-low portions of the slopes. The crown areas can be detected in the contact between clayey units and conglomerate, although a retrogressive activity, testified by tension cracks, is now involving also areas where the conglomerate outcrops. Most of the accumulation areas, produced by rotational slides, constitute the alimentation zone of slow earth-flows that can affect the entire slopes from the top to the river valleys. These complex landslides, whose activity has been progressively increased in time, caused severe damage to the village that has been gradually abandoned by the population. The southern slope has been mostly affected by instability during the re-activations in 1959, 1965 and 1971 (Beneo 1969; Brugner 1964;

Verstappen 1977; Del Prete and Petley 1982; Del Prete 1990; Gisotti et al. 1996). These movements have severely damaged the buildings and infrastructures sited inside or in the vicinity of the crown area. The great landslide of the southern slope shows typical features of a retrogressive and multi-directional slide. The main scarp has been progressively moved upward of about 50 m, between 1955 and 1972 affecting large areas of the ancient town while a

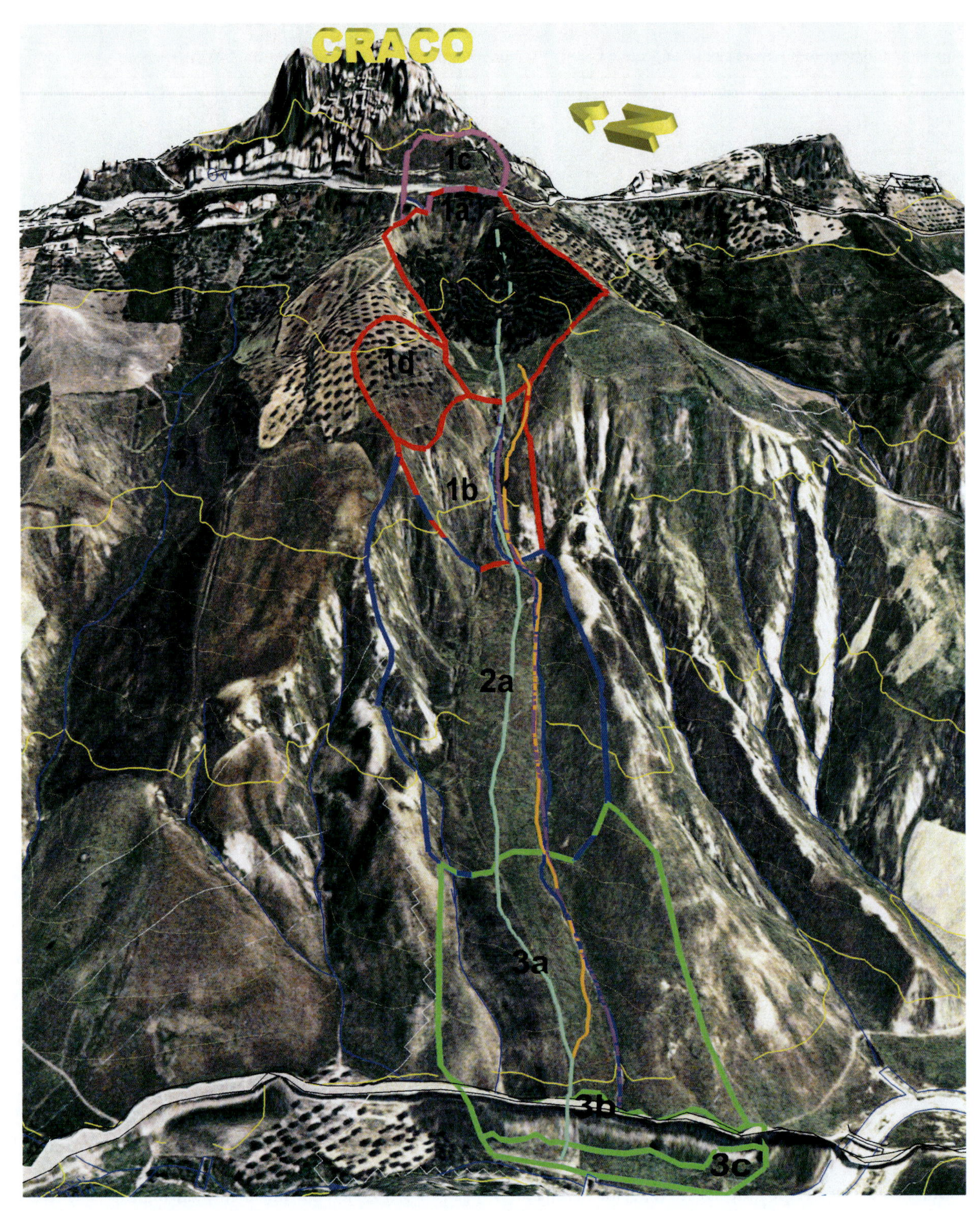

Fig. 13.1. 3D model with landslide sectors interpretation

secondary landslide developed backward at NE involving the Convent of San Pietro. In this area two distinct movements have been detected in both slopes involving the saddle where the structure is sited.

13.3 Methods

A multi-temporal analysis, through digital and analogical stereoscopy and field survey, has been implemented in order to assess the evolution of the large landslide of Craco. The analysis is based on four sets of aerial photos taken in 1955, 1972, 1989 and 1999 (Table 13.1) and Digital Elevation Models processed by the images. The detection and mapping of the various geomorphological sectors and landslide activity distribution has been done through analogical stereoscopic analysis coupled with field survey.

The digital stereoscopic analysis has enabled to assess quantitatively the plano-altimetric differences and, therefore, the changes in volume of the landslide occurred in the periods 1955–1972, 1972–1989, 1989–1999 and 1955–1999. The analysis has been focused on distinct landslide morphological sectors interpreted through analogical photogrammetry of the 1999 photo set (Fig. 13.1). In particular, from the top to the bottom of the slope, the following areas have been recognized:

- Area with retrogressive evolution;
- Upper crown area;
- Lower crown area;
- Track area;
- Area of accumulation.

The upper sector is located upslope the crown area of 1971 and affects partially the urban area. The crown area has been distinguished in two sectors divided by the morphological threshold represented by the boundary Pliocene Clays/Oligocene-Miocene Clays. This plane may represent the original deep failure surface of the landslide.

In the above sectors, the eroded or deposited materials have been calculated by means of the difference between high-resolution DEMs, implemented with a dedicated software of digital stereoscopy.

The correct superimposition of DEMs, for each photo set, has been done comparing a minimum grid composed

Table 13.1. Characteristics of photos used in the analysis

Year	Scale	Type	Dpi	Scanner type
1955	1:33 000	Panchromatic	1 800	Photogrammetric
1972	1:29 000	Panchromatic	1 600	Photogrammetric
1989	1:70 000	Panchromatic	1 800	Photogrammetric
1999	1:40 000	RGB	1 200	Photogrammetric

Table 13.2. Average errors of geo-referencing

Year	Parallaxis (pixels)	*X* (m)	*Y* (m)	*Z* (m)
1955	1.66	1.88	1.41	1.71
1972	0.557	2.280	1.490	0.544
1989	0.348	0.630	1.264	0.653
1999	0.352	1.230	1.438	0.182

Table 13.3. Characteristics of DEMs

ELEV_UNITS	UNIT_METERS
NUM_POSTS_XY	2032 3010
SPACING_XY	0.500000 0.500000
SPACING_UNITS	UNIT_METERS
DTM_FORMAT	DTM_GRID
SMOOTHING	HIGH_SMOOTHING
PRECISION	HIGH_PRECISION

Table 13.4. Average errors during DEM generation

	1955	1972	1989	1999
CIRCULAR_ERROR	1.950541	0.552988	1.371233	0.993244
LINEAR_ERROR	4.290094	1.436584	2.850768	2.347502

of 9×9 common points of elevation, selected in geomorphological stable areas.

The stage of absolute orientation has produced planimetric errors with a range of 1.71–2.28 m (see Table 13.2). The extraction of DEMs has been done following the characteristics, results and errors reported in Tables 13.3 and 13.4, inside a polyline including a sampling step of 0.5×0.5 m.

The photos taken in 1989 and 1999, show a viaduct located in the river valley and the development of a forest in the crown area. In order to avoid the computation of other elements (i.e. forest area) some opportune modifications have been done in the images, calibrating the respective DEMs to the ground level.

13.4 Results

The large landslide of Craco shows, during its evolution, the typical features of retrogressive and multi-directional activities. The main landslide scarp has moved upslope progressively, affecting, in the time, large historical urban areas, while a minor scarp has developed at NE. The large debris body located downslope the village, developed after the large failure of 1971, exhibits a remarkable development towards the river valley.

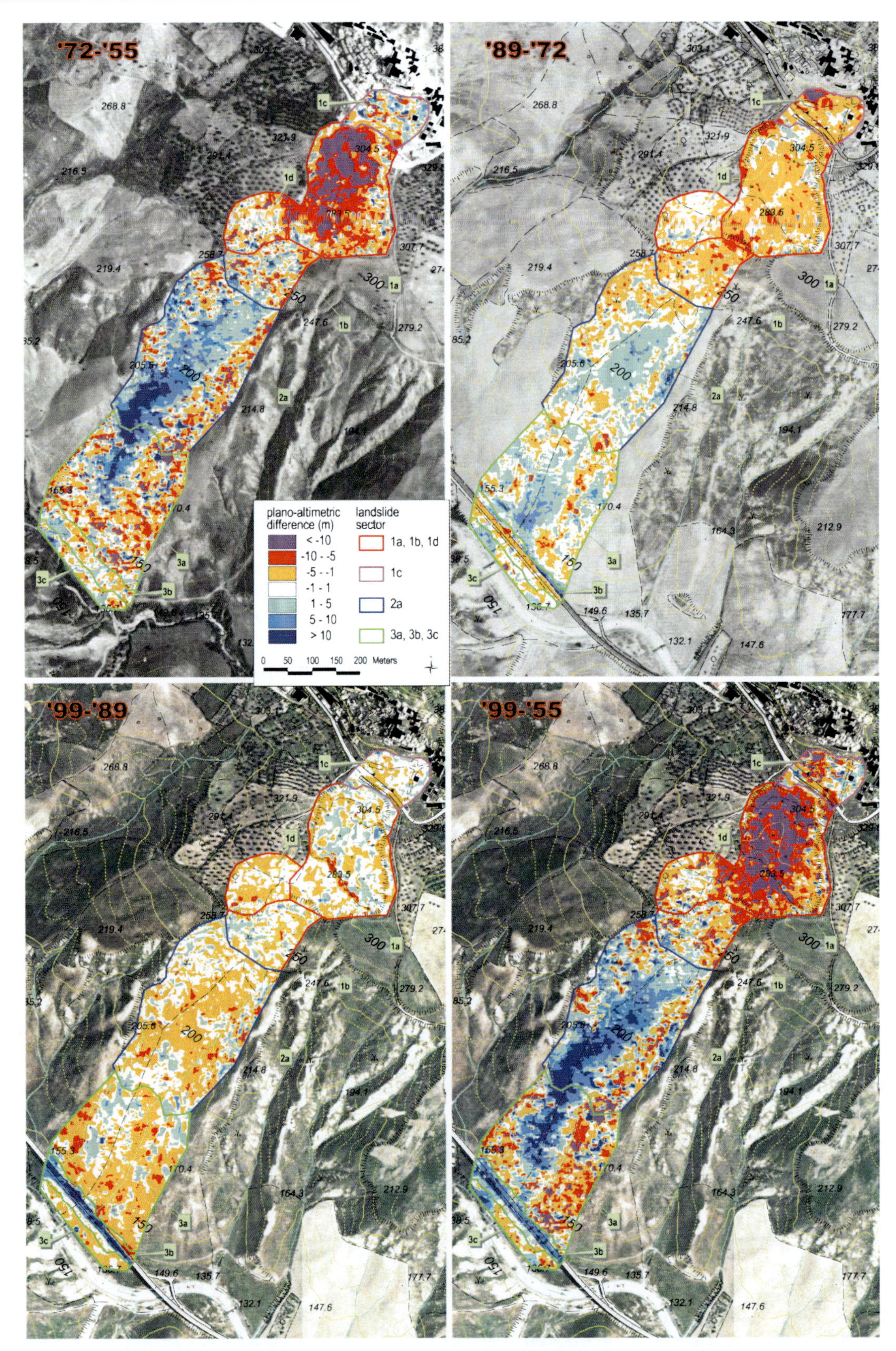

Fig. 13.2. Map of plano-altimetric differences in the period 1955–1972 (*upper left*), 1972–1989 (*upper right*), 1989–1999 (*lower left*) and 1955–1999 (*lower right*)

The images of 1972, 1989 and 1999 enable the assessment of landslide dynamics after the parossistic event of 1971 and, therefore, mainly show the evolution of the remoulded body mobilized by slow earth-flow processes and erosion rather than by rotational slides. The comparison with the photo taken in 1954, permits to analyze the slope characteristics before the large landslide event of 1971.

From analogical and digital comparison and analysis of the available aerial photos, a multi-temporal study of the large landslide of Craco has been carried out. The study has focused the state and distribution of activity as well as the induced morphological changes of the landslide (Figs. 13.1 and 13.2).

The *sector 1a* exhibits, between 1955 and 1972, a general trend to a retrogressive evolution, especially in the western portion. In the interval 1972–1989, the displaced mass has been completely eroded while, during 1989–1999, a general stabilization of the phenomenon occurs. In this period, very small re-activations in the crown area have been recognized, whose accumulations tend to fill up the internal depression of the main scarp.

The *sector 1b* indicates that this depressed area has been between 1955 and 1972 the accumulation area and, overall, the zone of transition for the displaced mass coming from upslope and, in particular, from the toe of the main failure surface of the rotational slide that formed the sector 1a. In the following period (1972–1989) this area exhibits a substantial activity of the displaced mass while in the interval 1989–1999 is mainly inactive with a trend to erosion.

The *sector 1d* represents, between 1955 and 1972, the area of a successive rotational or translational slide promoted by the depression of the sector 1b; in the same interval, in the left flank, an area with gully erosion has developed. From 1972 on, this sector exhibit a substantial stability with a slight erosion of the displaced mass.

The *sector 2a*, that represents the track area of the slow earth-flow generated by the rotational slides in the sectors 1a and 1b, shows between 1955 and 1972 a large accumulation of materials that has caused the displacement of the impluvium of ca. 40 m to E, with a consequent triggering of minor rotational slides along the left flank. In the period 1972–1989 the same material has been translated to the river valley, while the upslope area becomes a track area of further displaced mass. Between 1989 and 1999 the accumulation area shows the development of minor channels due to runoff.

The *sector 3a* represents the accumulation area of the earth-flow. This zone before 1954 was an inactive toe of an old landslide that caused the displacement at S of the river flood plain. During the period 1955–1972, small rotational slides caused the progressive dismantling of the old landslide toe while, upslope, the accumulation of the displaced mass, coming from the large track area, is clearly visible. Between 1972 and 1989 the accumulation area is generally reactivated and characterized by an advancing distribution of the activity. Between 1989 and 1999 the toe has been progressively eroded by runoff due to the lack of alimentation.

Finally, the multi-temporal analysis assessed through a GIS and digital stereoscopy coupled approach, has enabled to estimate the intensity and the activity of the large landslide of Craco measured as volumes of the mobilized materials (Table 13.5). Most of the activity can be attributed to the large phenomena occurred in 1965 and 1971. Presently, the landslide activity shows a typical retrogressive trend that, without the implementation of risk mitigation strategies, may affect in the future the southern sector of the urban area of Craco.

References

Beneo E (1969) Relazione sulle condizioni di stabilità dell'abitato di Craco. Relazione Tecnica, pp 1–18

Brugner W (1964) Sulle condizioni di stabilità dell'abitato di Craco (Provincia di Matera). Relazione Tecnica

Carbone S, Catalano S, Lazzari S, Lentini F, Monaco C (1991) Presentazione della carta geologica del bacino del fiume Agri (Basilicata). Mem Soc Geol It 47:129–143

Cruden DM, Varnes DJ (1996) Landslides types and processes. In: Turner AK, Schuster RL (eds) Landslides: investigation and mitigation. Transportation Research Board Special Report 247. National Academy Press, WA, pp 36–75

Del Prete M (1990) La difesa dei centri storici minacciati dalle frane: Craco un'esperienza da non ripetere. Sc Dir Econ Amb 7–8: 38–41

Del Prete M, Petley DJ (1959) Case history of the main landslide at Craco, Basilicata, South Italy, in Cotecchia V. Il dissesto idrogeologico nella provincia di Matera, in "Annali della Facoltà di Ingegneria", vol. 3, Bari

Del Prete M, Petley DJ (1982) Case history of the main landslide at Craco, Basilicata, south Italy. Idro Appl & Idrog 17:291–304

Gisotti G, Spilotro G, Gisotti N (1996) Note sul dissesto idrogeologico che ha colpito l'abitato di Craco. Relazione Tecnica

Verstappen H (1977) Orthophotos in applied geomorphology: the mudflow hazard at Craco, Italy. ITC Journal, 4 Special

Table 13.5. Differences of volumes (m^3) for each landslide sector, calculated for different periods of observation

	Sector	1955–1972	1972–1989	1989–1999	1955–1999
1a	Upper detachment	−258176	−73702	5218	−326203
1b	Lower detachment	6145	−25711	5505	−13468
1c	Retrogressive evolution	7322	−32090	4113	−20581
1d	Lateral scarp	−21300	−1337	−6545	−29251
2a	Track area	230948	72951	−37717	264996
3	Accumulation area	15233	93601	−10059	98914

Chapter 14

Tools for Rock Fall Risk Integrated Management in Sandstone Landscape of the Bohemian Switzerland National Park, Czech Republic (M121)

Jiří Zvelebil* · Zuzana Vařilová · Milan Paluš

Abstract. There are 327 monitored rock objects with more than 900 measuring sites on the territory of the Bohemian Switzerland NP and its nearest neighborhood, and the monitoring nets are ever growing. Therefore a high-tech, scientifically challenging Integrated System (IS) for effective, but nature-friendly management of rock fall risks on the Bohemian Switzerland NP territory has been under construction since 2002, there. Rapid processing and timely, on-line delivery of relevant, easy-to-understand information to an end-user through an information web portal and cellular phone emergency messages should be the highlights of IS. Other highlights are represented by the implementation of complex dynamical systems knowledge and methods to provide more realistic and mathematically more rigorous grasping of very complex dynamics of rock slope stability failure. Those methods also provide a basis for a qualitative step in implementation of computers for future highly automated run of data assessment, modeling and early warning modules of the system. Several successful case-histories have made those new tools very promising for the practical use. Nevertheless, there are some tasks still unfinished. Especially the one enabling to bridge the gap between science, civil protection and general public and by it to enhance effectiveness of utilization of delivered information by its end-users.

Keywords. Rock fall, holistic approach, risk assessment spatial and time domains, monitoring, early warning, forecasting, Czech Republic

14.1 Introduction

Specific sandstone landscape with plateaus, deep canyons, rock cities and labyrinths, which has developed on massive, sub-horizontally stratified sandstones of Cretaceous age, and the rich eco-diversity related with this landscape have been protected within the frame of Bohemian Switzerland National Park in NW Bohemia. Picturesque, tourist attracting, high-energy relief with rock walls on valley sides of deep canyons and on rims of plateau-mountains is conditioning frequent occurrence of rock falls, there. Nowadays, preparation of those rock falls is causally connected with activity of exogenous geological processes, mainly with weathering and erosional removal of materials at rock wall toes (Vařilová 2002; Vařilová and Zvelebil 2005a).

At one hand, those rock falls belong to a contemporary, natural development of sandstone landscape, so they are an integral part of the protected natural environment. On the other hand, there is also a strong need to protect safety of inhabitants and visitors of the National Park. To fulfill both those tasks, a program to establish an integrated system of effective, but nature-friendly management of rock fall risks has been launched on the Bohemian Switzerland NP territory.

14.2 Integration Issue

Geo-risk management is always a multidisciplinary task, which has to fulfill in the same time four different tasks.

On one hand, a multi-disaster approach should be used to tackle as with the multi-causal origin of individual types of disasters together with very rich spectre of spatial-temporal scales, within them those disasters are being prepared and act, as with the possibility of causal chaining of the individual disasters into one spatial-temporal sequence of threads (e.g. meteorological disastrous event – flood and landslide/rock fall events – transport and agriculture infrastructure collapses etc.). Therefore, the scientific part of that management should integrate knowledge and methods from various fields of geosciences.

On the other hand, outcomes of complex, scientific effort should be delivered to their practical end-users – i.e. to National Park administration, municipalities and other executive entities, in an understandable, simple, ready-to-use form. Easy to comprehend, concise, mainly graphic forms should be implemented or developed to fulfill this task.

Timely delivery of information (from data to be analyzed to early warning launching for public) is an essential demand of both the previous management tasks. Information turnout should be as rapid as possible. Therefore, progressive Information Technologies (IT) should be implemented for an automated data acquisition, online transport, multiple analyses and synthesis of data from different sources in the integrated process of their final evaluation, and consequent transformation of the most important scientific results into the simple, end-user fitted forms.

At last but not at least, operation of the system should be, besides all its above listed functions, also reasonably priced.

14.3 System Structure, Methods, Outputs

Scheme of the system is depicted on Fig. 14.1.The system is forwarding the best practices as they had emerged from the forerunning task to secure a traffic corridor going through a deep canyon of the Labe River between Děčín Town and Czech-Germany boundary-crossing point (Zvelebil 1995; Zvelebil and Park 2001).

14.3.1 Regional Rock Fall Risk Zoning in Spatial Domain, 1 : 10 000

Areas at different level prone to preparation of various types of rock falls are identified within the frame of *rock fall hazard zoning*. In the same time, areas of special interest are demarcated within the former ones – i.e. especially populated areas, traffic corridors, and along main tourist trails. Finally, *rock fall risk zoning* is produced combining the rock fall hazards with probable negative consequences of those hazards for those special interest areas (Fig. 14.2).

The rock fall hazard zoning is based on the principle of geomorphologic developmental units. It is more complex – hence more realistic, than the current linear zoning by GIS layers overlying, which is keeping with very simple condition-process models. Regional model of development of sandstone slopes with regards to spatially and mainly temporally differentiated activity of destructive processes, especially of rock falls, within its course (Zvelebil 1989; Vařilová and Zvelebil 2005a) is exploited for the latter task. The zones defining of differentiated occurrence probability for distinct types of rock fall match according with that method, with zones demarcating areas in selected stages of rock walls development, because different developmental stages are characterized by domination or diminishing of different rock fall types among slope shaping, destructive processes.

Resulted rock fall hazard maps are therefore valid for the time-spans characterizing development of sandstone slopes – i.e. minimally for hundreds, but more probably for thousands and more years. This long validity time span of the rock fall hazard maps makes them the up-date-non-demanding ones in the frame of rock fall risk management tasks. The only exception should be made for any enhancement of knowledge of regional rock slope development model and causal associations of rock fall occurrences with developmental stages of that model.

14.3.2 Detailed Risk Zoning in Spatial-Temporal Domains, 1 : 1 000, 1 : 2 000

In areas of high rock fall risk, detailed *engineering-geological mapping of rock slopes with unstable objects inventory* are carried out. For each inventoried object, its actual instability is fixed using the four degree scale (Fig. 14.3). This instability scale already embodies the time aspect expressed in the human time scales. Simple, numerical equilibrium calculations are combined with empirical-phenomenological models of failure development in sandstone slope to that time assessment: *(x)* immediately

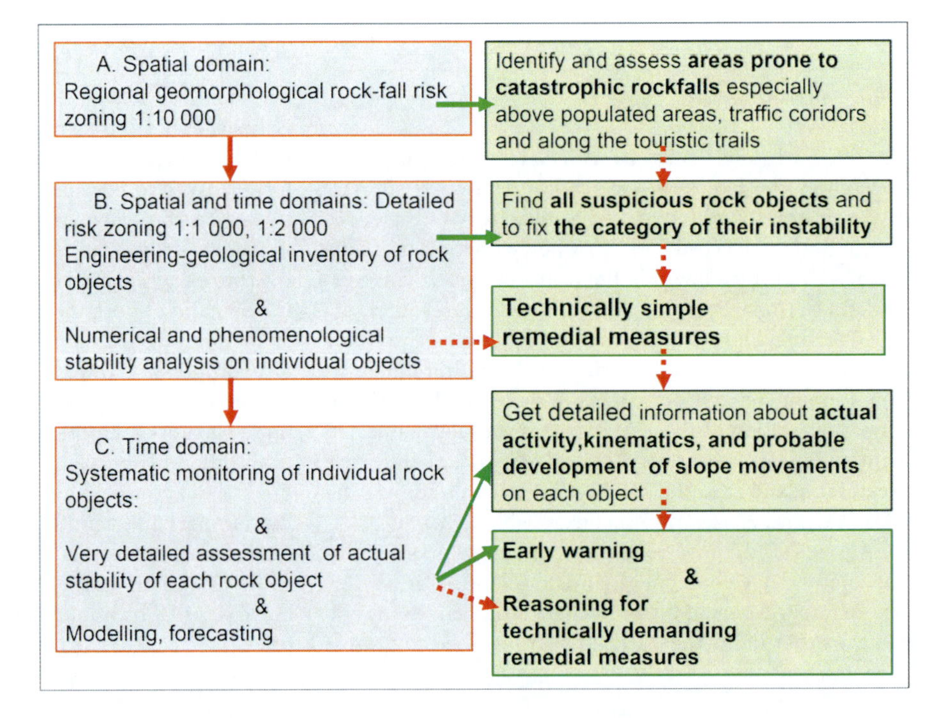

Fig. 14.1. Block function scheme of the Integrated System for Rock Fall Risk Management, which is being implemented in the Bohemian Switzerland National Park

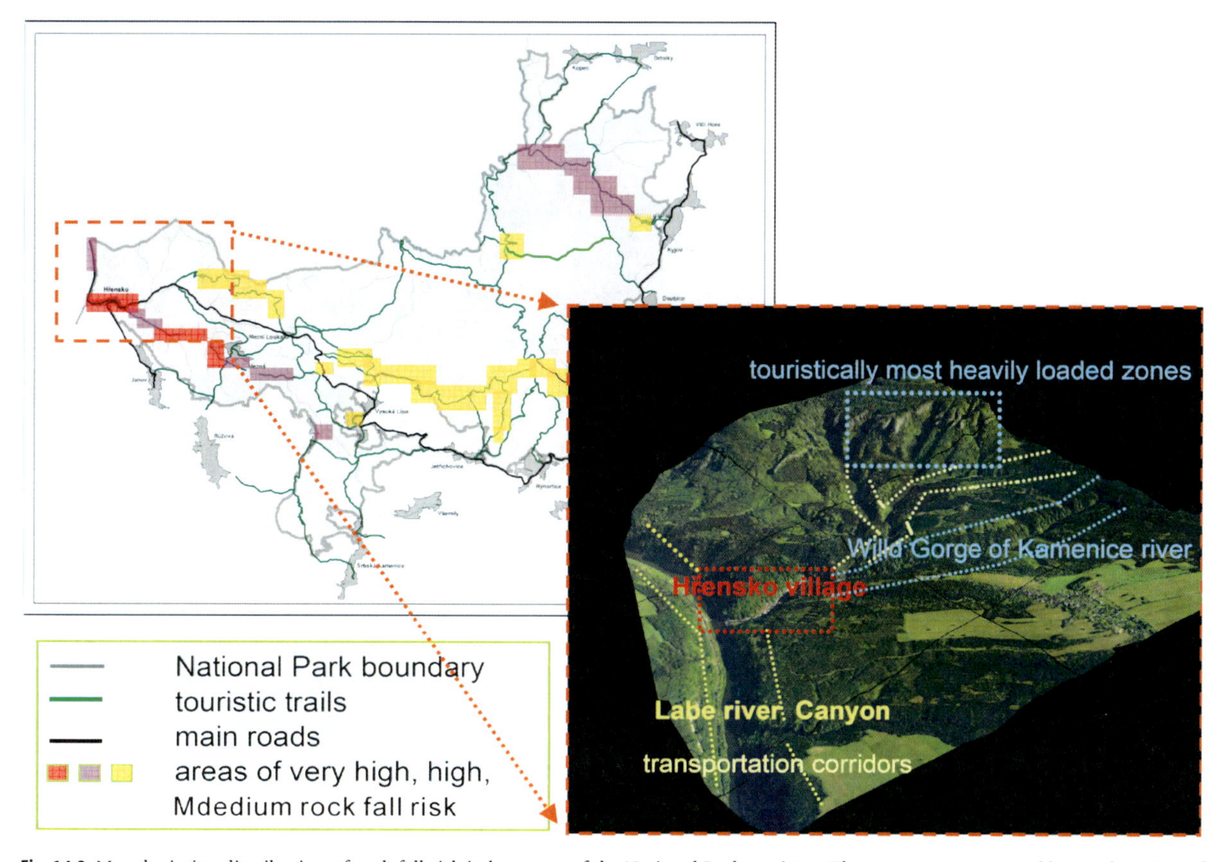

Fig. 14.2. Map depicting distribution of rock fall risk in key areas of the National Park territory. The map was constructed by combination of GIS layer for areas of special interest around settlements, traffic corridors and the main tourist trails, with layers depicting rock fall risks. The *red box* on map and the picture of DEM represent the area of Hřensko Village (cf. Fig. 14.3) and its neighborhood, which is the most important area of thorough rock fall management within the Park territory

unstable objects possess no safety warranty. They could collapse any time within the time span from a few days to a few months. As such objects are classified the ones which posses as the high degree of calculated instability, as clearly visible demonstrations of activity of rock mass destruction and its strength decay (fresh shattering, propagation of new joints, splitting off etc.) within the statically key areas of rock object under evaluation. *(xx)* The second category of objects provides a reasonable degree of security against unexpected rock fall occurrence for time spans of months, maximally of one year. The objects with the same class of numerically fixed instability as in the previous category but without those signs of fresh degradation of rock mass strength represent this second category. *(xxx)* The third category should provide the security for time spans up to 3–4 years. It embodies objects without visible signs of degradation of their rock mass strength and with very small stability reserve. *(xxxx)* The forth category grants security from 4 years up to a few tens of years.

Regarding the time aspects of the listed instability scale, validity of the inventory maps of dangerous rock objects is restricted to 2–3 years. After its expiration, those maps should be up-dated.

14.3.3 Very Detailed Risk Assessment in Time Domain

Only two categories of unstable objects are immediately technically treated – or stabilized or removed, just after the delivery of results of the detailed risk zoning in spatial-temporal domains: *(i)* the unquestionably immediately unstable objects; *(ii)* potentially unstable object of the second and third category on condition that their volumes and positions as provide conditions for cheap and simple technical treatment and, in the same time, their monitoring would or uneconomical, or low reliable one due to the small rock volumes activated by the instability development.

All the other objects are moved to the *group of systematically, by monitoring supervised rock objects* to further refinement of their risk assessment in a time domain. This wide implementation of monitoring brings following multiple practical use. It provides: *(i)* an instant security against an occurrence of unexpected rock falls, even from its very beginning; *(ii)* information as about a real activity of mass disturbation, as about its kinematics, as well as its dynamical patterns

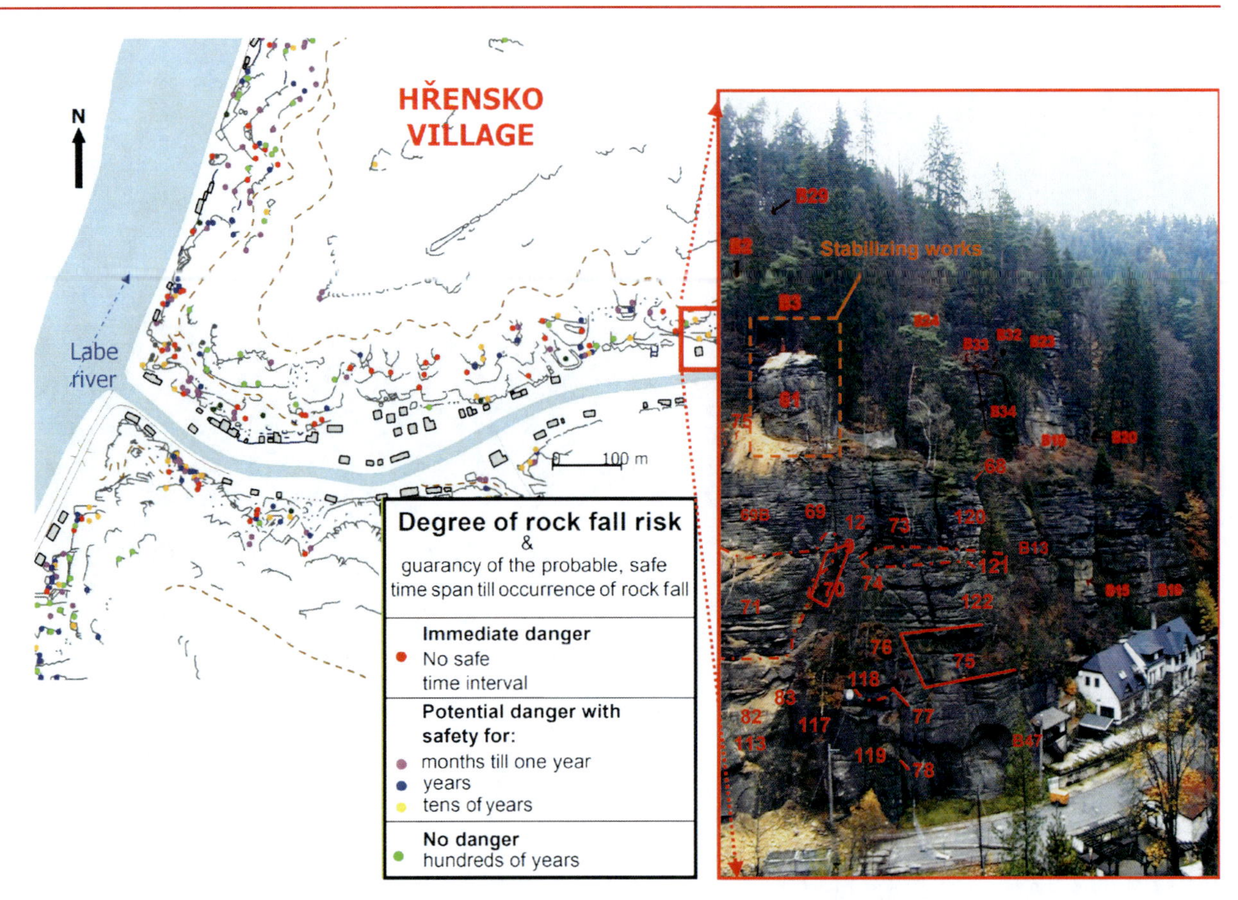

Fig. 14.3. Part of detailed engineering geological map of slopes and unstable rock object inventory in Hřensko Village, 1:1000. The *red box* marks the detail of area presented on the photo. In the *left upper half* of the photo, notice please, a simple remedial measure taking place – the gradual, from *top downwards* advancing removal of immediately dangerous rock tower no. 61 by a group of professional alpinists

including possible triggering factors of its acceleration events; *(iii)* Information according the paragraph *(ii)* enables as further refinement of actual instability and security degree for the monitored objects, as well as provides basic input data to time forecasting of rock fall occurrence *(iv)*. By providing information listed, monitoring also helps to substantially lower uncertainties in input data and to compensate some other drawbacks of geomechanical computing schemes for rock slope stability evaluation, and for planning of optimal remedial measures. During their realizations, monitoring provides not only security for workers, but also for an operational optimization of those works). Finally, it could be used for checking of their stabilizing efficiency.

Monitoring techniques include portable rod dilatometer measurements, which represent the main method, portable tilt-meter and extensometric tape measurements. Recently, an automated, complex system for remote, on-line data acquisition and Internet DB Storage, data processing, and on-line, interactive result visualizations, including the end-practical-user fitted simple graphical forms, has being introduced.

Displacements between rock blocks on selected, kinematically and safety key-sites of rock object are measured together with selected items of micro-climatic characteristics to detect existence of any irreversible deformation and to fix its magnitude and dynamical patterns. Temperature changes are the main environmental influence measured. Temperature has been proved as to be the main driving force for reversible deformation by voluminal changes of rock blocks and thus spoiling the information about intrinsic slope movement dynamics, as to probably taking part in dynamic driving of the intrinsic dynamics of slope failure system in its Near-To-Equilibrium stage (see Paluš et al. 2004).

Dynamical Systems theory and holistic phenomenological models are used as for the assessment of instant instability of rock objects due to detecting the actual developmental stage of rock fall preparation on them, as that process modeling for purpose of its short- and medium-time forecasting. Different developmental stages of rock fall preparation, including the early warning precursors of rock fall occurrence, are detected using our knowledge of differences in slope movement dynamics. Quite *reli-*

able diagnostic of immediate danger state, as well as its short- and medium-time forecasting (from days to 2 years ahead) are at our disposal now. Current phenomenological models, which implantation is rather hand-made, the high skilled personnel demanding work (e.g. Zvelebil 1996; Zvelebil and Moser 2001) have been now accompanied by mathematically rigorous numerical analyses and modeling according the latest challenges of complex dynamical systems theory (Paluš et al. 2004; Zvelebil et al. in print, cf. also "Discussion" section).

Besides the data effective storage, synthesis and evaluation, the GIS systems also provide a basis for an interactive, on-line presentation of the final maps with plots of all monitored objects together with actual monitoring data and their safety evaluation for every object. The first experimental version of specialized web portal has been already launched. The site is accessible only to authorized persons from NP Administrations and Ministry of Environment. For area of Hřensko Village (cf. Fig. 14.3), it provides on-line, hierarchically interactive information about results of detailed rock slope mapping and unstable object inventory together with plots of all monitored objects with actual monitoring results and their safety evaluations. The portal also provides on-line results from 4 automatic data acquisition units installed on other hot spots of NP area.

14.3.4 Remedial Measures

Three groups of remedial measures have been used in accordance with differences in the typology of unstable objects, their actual instability, and levels of risk to inhabited areas, transport routes or tourist trails. *(x) The simple technologies* – which are possible to accomplish by a hand-work, using only technically very simple means. Those works are systematically carried out by the National Park Staff – a special group of alpinists, which is supervised by a geologist *(xx) The most technically, organizationally* (e.g. emergency evacuation of part of a village, medium till long time traffic break off etc.), and, of course, *economically demanding measures* are realized on a commercial platform. An expert team delegated by the National Park Administration and Ministry of Environment supervises specialized geotechnical enterprises during those works. The most dangerous cases are – in accordance with the Emergency Law of CR, treated in collaboration with State Police and Army, and coordinated by the Integrated Rescue Body of Ministry of Internal Affairs. In this case, the supervising team has to embodied experts from all the involved entities. *(xxx)* Special, *medium position* is held by a *long-term – tens of years lasting monitoring* of potentially unstable objects, because it is able to substitute the costly and sometimes even nature-unfriendly technical measures for the majority of objects. In this case, monitoring is used to ensure, that the rock fall preparation process is still remaining in its medium, long-lasting phase. By revealing a shift from that phase towards the last, short time lasting phase, such monitoring also provides a guaranty that the technically very demanding measures will be limited only to the objects, which high danger state had been rigorously proven by hard data from monitoring.

14.4 Discussion

The local – spatially, according the degree of rock fall risk, differentiated approach of successive survey steps had proven its reliability. Since the very beginning of the project in 2002, there had at least occurred one bigger rock fall and several smaller ones within the National Park territory every year (Vařilová and Zvelebil 2005b). But none of them had unexpectedly occurred within the areas of special interest – i.e. above settlements and the main tourist trails. In contrary, even from 2003 till 2005, 12 cases of high instability of rock objects had been detected and treated – 8 cases of the simple, 3 cases of the higher, and 1 case of the highest technological demands (Fig. 14.4), in those areas of special interest before they were able to do substantial harm. The detailed inventory maps of unstable rock object have been also used by the local authorities to regulate the building activity within the area of Hřensko Village.

The latest experience from the 2004/2005 winter could be regarded as a very important lesson. On one hand, it has strengthened our hopes in the ability of the new, on dynamical systems based analytical methods to timely provide the rock fall early warning even from assessment of very short monitoring time series. On the other hand, it has reminded us the possibility of human mistake which is especially dangerous during the survey phase of detailed inventory of unstable rock objects.

A medium size rock fall which had unexpectedly occurred from previously surveyed part of slope above a local road (Stemberk 2004) in winter 2005 provided the first negative exception from the above listed statistics of success. Nevertheless, when the unstable/potentially unstable rock object would be correctly fixed and their monitoring started, chance of such mistake could be substantially decreased.

Relatively very short, only a few of months lasting time series from a suspicious rock pillar with the volume of 30 m^3 which endangered a tourist trail, were not possible to interpret by any regular method due to screening of rock mass irreversible deformations by environmentally driven reversible changes of rock mass volume. Using the new method of displacement/displacement and displacement/temperature correlograms (e.g. Zvelebil et al. in print) those time series were deciphered and unusual ac-

Fig. 14.4. Site of repeated rock fall danger which required technically demanding remedial measures by anchoring in Hřensko Village. **a** Damaged roof and house walls by a 400 m^3 rock fall in 1936. *Yellow box* marks the site of new, detailed photo **b**. **b** Fresh opening of joints by restored preparation activity, of another rock fall in late fall 2002. **c** The same site as on photo **b** under progress of remedial measures. **d** Heavy scaffolding was needed to realize anchoring works during winter 2002/2003

tivity of irreversible deformations shown (Fig. 14.5). Conclusion driven from the latter about the high instability of that pillar was then supplemented by a time forecast of rock fall occurrence, which had been given in December 2004, 3 months before the expected collapse event (Zvelebil 2004). The pillar collapsed on 12 March – just in the predicted, to March 2005 restricted time window, providing an evidence for correctness of our interpretations. Falling rocks hit badly the tourist path bellow, but that path had been closed – in accordance with this stability evaluation, since January (Vařilová and Zvelebil 2005b).

The regional monitoring net has already encompassed all known areas of the highest rock fall risks within the area of Bohemian Switzerland National Park but it still remains expanding gradually in accordance with actual results provided by the unstable object inventory, which is now operating in areas with lower degrees of risk. Up to now, there are 227 monitored rock objects with 511 measuring sites on the territory of Bohemian Switzerland NP. The majority of them – 217 objects with 486 sites, are manually measured by portable dilatometer and extensometric tape. Only 4 objects with 25 sites have been equipped by an automated monitoring system with on-line data transport to a remote, central dispatching. When statistics of the older monitoring net, which spreads over the main road between Děčín Town and Czech-Germany boundary in the Labe Canyon is added, one arrives to the total of 327 rock objects with more than 900 sites. The reason for high degree of automation of IS, especially in its time-series processing and evaluation, which should include an expert module to facilitate the every-day practical use of IS, is clearly visible.

14.5 Conclusions

A high-tech, scientifically challenging Integrated System (IS) for effective, but nature-friendly management of rock fall risks on the Bohemian Switzerland NP territory has been under construction since 2002. Its scientific part has been based on holistic, knowledge and methods as from various fields of geosciences, as from nonlinear dynamics and Dynamical Systems theory integrating approach.

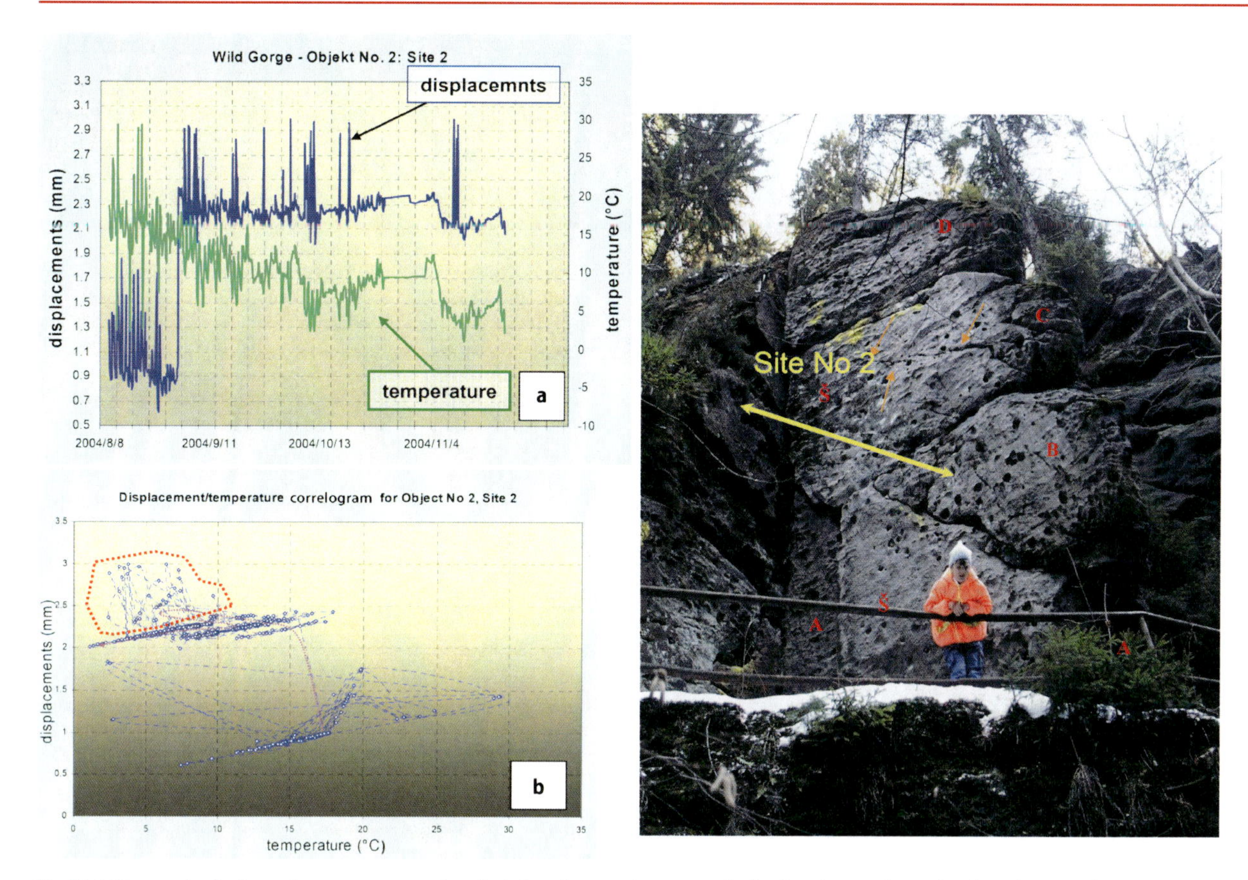

Fig. 14.5. New method of correlograms was used to decipher short, only 4.5 months lasting time series and to launch a rock fall early warning for a rock pillar just above a tourist trail at the bottom of the Wild Gorge of Kamenice River. **a** Current time/displacement plot from the monitoring key site (for its position see photo) is showing only one irreversible jump shortly after the beginning of measurement. **b** Temperature/displacement correlogram: ordered patterns of diagonal, *from* left to *right* elongated *dot clouds* represent well correlated responses of rocks to temperature changes. In contrary to the current plot, many temperature independent micro slip events are clearly visible as dot groups violating the left-right diagonal patterns. The high density area of the slip events is marked by a *red curve*

Its technical part has been using progressive tools of Information Technologies (IT) to make information turn-out as rapid as possible. Rapid processing and timely, on-line delivery of relevant, easy-to-understand information to an end-user through an information web portal and cellular phone emergency messages are the highlights of IS.

The high need of maximal automation of monitoring and mainly of monitoring time series evaluation can be clearly seen from ever-growing number of monitored rock objects. Up to now, there are 227 monitored rock objects with 511 measuring sites on the territory of Bohemian Switzerland NP only, and, together with neighbor, out of NP boundary laying area of Labe River Canyon one arrives to the total of 327 rock objects with more than 900 monitored sites.

GIS system techniques are used as to store data, as to improve gain from their synthesis and integrated evaluation, as well as to present the final results in forms of regional rock fall hazard zoning in spatial domain 1:10 000, and of detailed rock fall risk zoning in spatial and time domains 1:1 000, 1:2 000. The latter has also provided basis for exhibition of actual results of monitoring data evaluation for each monitored object.

Systematic monitoring of displacements on cracks within rock mass of potentially unstable objects has a multipurpose use in IS. To ensure its nature-friendly function, monitoring provides as a immediate guaranty as against an unexpected occurrence of a rock fall, as that the technically very demanding remedial measures will be limited only to those rock objects, which high danger state had been rigorously proven by monitoring hard data. The long-term – tens of years lasting monitoring is able to substitute those technical demanding measures for nearly 90% of unstable rock objects, for which the developmental stage of process of rock fall preparation was fixed as the middle, long time – from years up to hundreds of years, lasting one.

The main qualitative innovation of IS is represented by the implementation of methods from a toolbox of complex dynamical systems. They provide more realistic and mathematically more rigorous grasping of very complex nature of rock slope dynamic. Moreover, due to their explicit, numerical nature, they provide also an opportunity to a wider implementation of rough, big-bulk-of-data-at-once-handling force of computers to analyze and evaluate monitoring time series, and to model development of rock slope failure dynamics for advanced, early warning targeting diagnostics and time forecasting. Several successful cases have made those new tools very promising as to diagnose the dynamical shift from that middle to the last, short time lasting phase of rock fall preparation even from short monitoring time series, as to help to make time forecasts ranging from a few days up to 2 years.

In spite of that partial success, there are two IS modules, which still wait to be substantially enhanced. Methodology of monitoring time series safety evaluation and modeling, which includes data multiple analyses and syntheses of data from different sources is professionally very demanding. Even a specialist – geologist should be supported by an automated expert system module for the current, practical use of the IS. Besides that, not all of the simple and concise forms of IS outputs for their practical end users have been fixed, yet. Still, the demand for bridging the gap between science, civil protection and general public by a constructive dialogue has not been fulfilled.

Acknowledgments

This work has been supported by the ICL Project IPL-M121, and by Academy of Sciences of Czech Republic, Project T110190504.

References

Paluš M, Novotná D, Zvelebil J (2004) Fractal rock slope dynamics anticipating collapse. Phys Rev E 70:36212

Stemberk J (2004) Inventory and evaluation of risky geodynamical phenomena on territory of Bohemiam Switzerland National. Final Report of the Project VaV/610/7/01: 2002–2004, MS, IG Ateliér Ltd. for Ministry of Environment of Czech Republic (in Czech)

Vařilová Z (2002) A review of selected sandstone weathering forms in the Bohemian Switzerland National Park, Czech Republic. In: Přikryl R, Viles H (eds) Understanding and managing of stone decay (SWAPNET 2001). Charles University of Prague, The Karolinum Press, Prague, pp 243–261

Vařilová Z, Zvelebil J (2005a) Catastrophic and episodic events in sand stone landscape: slope movements and weathering issue. In: Cílek V, Härtel H, Herben T (eds) Sandstone landscapes. Academia, Prague (in print)

Vařilová Z, Zvelebil J (2005b) Sandstone relief geohazards and their mitigation: rock fall risk management in NP Bohemian Switzerland. In: Proc Int. Conf. “Sandstones Landscapes in Europe, FERRANTIA – Travaux scientifiques du Musee national d'histoire naturele, Luxembourgh (in print)

Zvelebil J (1989) Engineering geological aspects of rock slope development in Děčín Highland, NW Bohemia. MS, PhD Thesis, Charles University, Prague (in Czech with English summary)

Zvelebil J (1995) Determination of characteristic features of slope movements present day activity by monitoring in thick-bedded sandstones of the Bohemian Cretaceous Basin. Acta Universitatis Carolinae, Geographica, Supp., pp 79–113

Zvelebil J (1996) A conceptional phenomenological model to stability interpretaqtion of dilatometric data from rock slope monitoring. In: Seneset K (ed) Landslides. Proc. VIIth Int. Symo. on Landslides. Balkema, Rotterdam, pp 1473–1480

Zvelebil J (2004) Stability assessment of three unstable rock objects in Wild Gorge of Kamenice River. MS for Administration of Bohemian Switzerland National Park, Krásná Lípa (in Czech)

Zvelebil J, Moser M (2001) Monitoring based time prediction of rock falls: three case-histories. Phys Chem Earth Pt B 26(2):159–67

Zvelebil J, Park HD (2001) Rock slope monitoring for environment – friendly management of rock fall danger. In: Proc. UNESCO/IGCP Symposium “Landslide Risk Mitigation and Protection of Cultural and Natural Heritage”, 15–19 January, Tokyo, pp 199–209

Zvelebil J, Paluš M, Novotná D (in print) Nonlinear science issue in dynamics of unstable rock slopes: new tool for rock fall risk assessment and early warning? In: Cello G, Turcotte DL (eds) Fractal analysis for natural hazards. Special Series Book pf Geological Society, London (in print)

Chapter 15

The Mechanism of Liquefaction of Clayey Soils (M124)

Victor I. Osipov · Ivan B. Gratchev* · Kyoji Sassa

Abstract. An experimental study on the liquefaction of clayey soils was conducted under ICL Project M124 "The influence of clay mineralogy and ground water chemistry on the mechanism of landslides" in order to better understand the mechanism of this phenomenon. The first section of this study deals with artificial mixtures of sand with different clays while the second is concerned with natural soils collected from landslides. The results from the first section are presented in this article. The investigation was conducted by means of a ring-shear apparatus and a scanning electron microscope (SEM). The results obtained for artificial mixtures enabled us to draw a line between liquefiable and non-liquefiable clayey soils and to define a criterion to estimate their liquefaction potential. In addition, the influence of clay content and clay mineralogy on the cyclic behavior of clayey soil was studied. It was found that an increase in clay content as well as the presence of bentonite clay raised the soil resistance to liquefaction. The analysis of microstructures of bentonite-sand mixtures along with the results from ring-shear tests revealed that the soil microstructure is the key factor in determining the dynamic properties of soil. For example, in the microstructures of soils vulnerable to liquefaction, the clay matter was observed to form "clay bridges" between sand grains that were easily destroyed during cyclic loading. In the microstructures of soils resistant to liquefaction, the clay matter seemed to form a matrix that prevented sand grains from liquefaction. The influence of pore water chemistry on the liquefaction potential of artificial mixtures was also studied. It was found that the presence of ions in pore water changed the microstructure of clayey soil, thus making it more vulnerable to liquefaction.

Keywords. Liquefaction, clay, microstructure, pore water chemistry

15.1 Introduction

A great number of earthquake-induced landslides are triggered by liquefaction of soil. The analysis of several landslide accounts revealed the presence of clay fraction in the liquefied soils, a finding that seems to oppose the widely accepted supposition that clayey soils are non-liquefiable. For example, Ishihara et al. (1989) noted the occurrence of liquefaction in silty sand containing clay in the Tokyo Bay area, Japan, during the 1987 Chibaken-Toho-Oki earthquake. Youd et al. (1989) reported the liquefaction of silty sands with as much as 10% clay that occurred at the Kornbloom site in the Imperial Valley, U.S.A., during the 1981 Imperial Valley earthquake. More recently, Miura et al. (1995) noted liquefaction of soils with up to 48% fines and 18% clay fraction due to the 1993 Hokkaido Nansai-Oki earthquake. Liquefaction of fine-grained soils during earthquake has also been observed in China (Perlea et al. 1999) and Tajikistan (Ishihara et al. 1990).

These examples pose many questions, and the most important among them are: where to draw a boundary between the liquefiable and non-liquefiable soils; and what criteria to use for estimating the liquefaction potential of clayey soil. In order to address these questions, an in-depth study on the liquefaction potential of clayey soils has been conducted under ICL Project M124 "The influence of clay mineralogy and ground water chemistry on the mechanism of landslides".

The study is divided into two sections: the first deals with artificial mixtures of sand with different clays while the second is concerned with natural soils from landslides. The results of the first section are presented here, and some parts of it were also published elsewhere (Gratchev et al. 2004, 2006). By mixing sand with clays in different proportions, a few mixtures with different properties were formed. The cyclic behavior of the mixtures were studied by means of a ring-shear apparatus in Kyoto University while the mixtures microstructures were obtained by means of SEM and analyzed at the Institute of Geoscience of the Russian Academy of Science. The obtained results enabled us to understand the mechanism of liquefaction in clayey soils as well as the factors affecting it.

15.2 Tested Soils

The samples were formed by mixing oven-dried sand (S7) with oven-dried commercially available bentonite, kaolin and illite in various proportions in order to obtain soils with different properties and microstructures. S7 is a subangular quartz sand with a specific gravity of 2.65 and minimum and maximum dry densities of 1.23 and 1.57 g cm^{-3}, a mean diameter of 0.14 mm, and a uniformity coefficient of 2.1 ($D_{10} = 0.075$). The bentonite clay used in this research had plastic and liquid limits of 85.7 and 357.9, respectively. The kaolin and illite clays were much less plastic, with plasticity indices of 20.7 and 34.9, respectively.

15.3 Test Procedure

15.3.1 Ring-Shear Tests

To study the undrained cyclic behavior of soil, a ring-shear apparatus (DPRI-4) was used. DPRI-4 is one of a series of intelligent ring shear apparatuses developed and improved at the Disaster Prevention Research Institute, Kyoto University (Sassa et al. 2004). The main features of this apparatus, distinguishing it from other types, are the structure of its undrained shear box and the servo-controlled dynamic loading system which enable cyclic shear and normal stress loading. Before each test, the sample was set into the shear box by dry deposition method and then saturated by means of carbon dioxide and de-aired water. To study the influence of pore water chemistry on the liquefaction potential of soil, samples formed from mixtures of S7 with 11% bentonite were treated with solutions of Sodium Chloride (NaCl) and Calcium Carbonate ($CaCO_3$). The degree of saturation was examined by measuring B_D value, which was defined as the ratio between the increments of generated pore pressure (Δu) and normal stress ($\Delta\sigma$) ($B_D = \Delta u / \Delta\sigma$) (Sassa 1985). The ratio for each test was ensured to be more than 0.95, a value that indicated an approximately full saturation. All samples were normally consolidated under a confining stress of 105 kPa. Then a reversal shear stress with a constant amplitude of about 45 kPa and a loading frequency of 0.5 Hz were applied for 50 cycles. After each test, a cyclic stress ratio (CSR), defined as the ratio between the maximum cyclic shear resistance and the normal stress, was measured for the last (50th) cycle of loading in order to compare the obtained results for different soils. It is noted that soils with CSR_{50} less than 0.1 were considered to have liquefied.

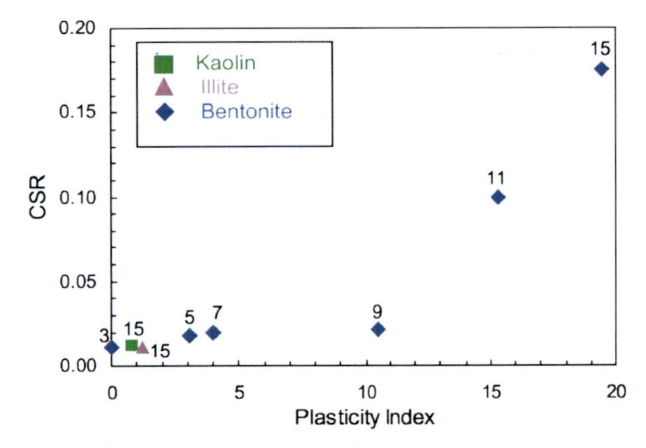

Fig. 15.1. Results of cyclic undrained ring-shear tests on mixtures of sand with kaolin, illite and bentonite plotted in terms of plasticity index (*PI*) and cyclic stress ratio (CSR) (the numbers next to the marks denote clay content of total weight in %)

15.3.2 SEM Analysis

The microstructures of the artificial mixtures were studied by means of a scanning electron microscope (SEM). This part of the research was conducted by the Institute of Geoscience of the Russian Academy of Science in cooperation with Moscow State University. A freeze-drying procedure was used to prepare samples (Osipov and Sokolov 1978). The most interesting peculiarities of microstructure were photographed at a number of magnifications (from ×100 to ×10 000). The obtained images were used to study the interaction between sand grains and clay particles, and the distribution of pores and clay microaggregates.

15.4 Results

15.4.1 Clay Content and Clay Mineralogy

The results from a series of undrained cyclic ring-shear tests plotted in Fig. 15.1 in terms of Plasticity Index (*PI*) against CSR_{50} revealed the strong influence of clay content and clay mineralogy on the liquefaction potential of soil. The data indicated that (1) low plasticity kaolin-, illite- and bentonite-sand mixtures were very vulnerable to liquefaction; (2) an increase in bentonite content raised both the soil plasticity and its resistance to liquefaction; and (3) the bentonite-sand mixtures with $PI > 15$ were resistant to liquefaction. Also, as seen in Fig. 15.1, an increase in *PI* correlated with an increase in soil resistance to liquefaction, suggesting that the *PI* can be used as criterion to estimate the liquefaction potential of soil.

The microscopic examination of the mixtures of sand with different amounts of bentonite indicated that the soil microstructure is the key factor in determining the cyclic behavior of clayey soil, and the soil resistance to liquefaction is strongly related to the state and distribution of clay material. Two examples will be discussed below. Figure 15.2 presents the microstructure of the mixture of sand with 7% bentonite that was found to be vulnerable to liquefaction (liquefied after 5 cycles of loading). The clay material in this microstructure is arranged in a non-uniform pattern and does not produce a continuous matrix (Fig. 15.2a). It generally accumulates on the surface of sand grains or at the contacts of sand particles, forming a bridge-like structure bonding the grains to each other (Fig. 15.2b). The pore space formed by the sand particles (or sand skeleton) is open. The clay matter itself forms large cells with walls made up by microaggregates (Fig. 15.2c). This microstructure (also referred to as "skeleton" after Sergeyev et al. 1980), which seems to be typical for sandy soil with a small amount of clay, yields a low re-

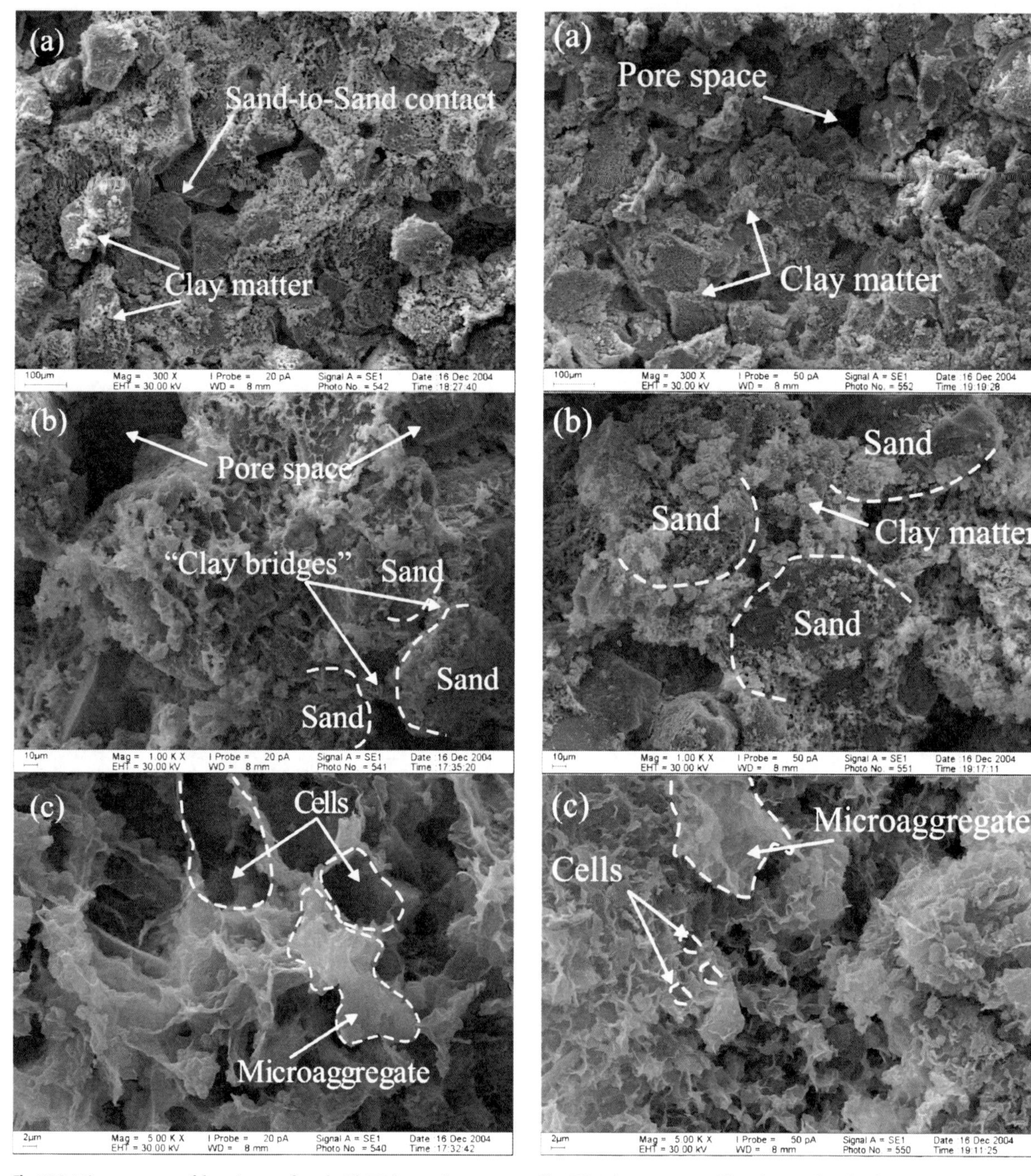

Fig. 15.2. Microstructure of the mixture of sand with 7% bentonite: **a** magnification ×300; **b** magnification ×1 000; and **c** magnification ×5 000

Fig. 15.3. Microstructure of the mixture of sand with 15% bentonite: **a** magnification ×300; **b** magnification ×1 000; and **c** magnification ×5 000

sistance to liquefaction. According to Osipov et al. (1984) and Osipov (1986), cyclic loading easily destroys the low-strength "clay bridges" at the contacts of sand grains and thus destabilizes the sand skeleton, leading to liquefaction.

Increase in clay material leads to a significant change in microstructure of clayey sand. Figure 15.3 presents the microstructure of the mixture of sand with 15% bentonite resistant to liquefaction. Although some pores remain open as shown in Fig. 15.3a, the clay material fills most pore space forming almost a continuous clay matrix (Fig. 15.3b). The clay cells become smaller and thus more isotropic (Fig. 15.3c). This microstructure manifests a higher resistance to liquefaction than the skeleton microstructure (Fig. 15.2).

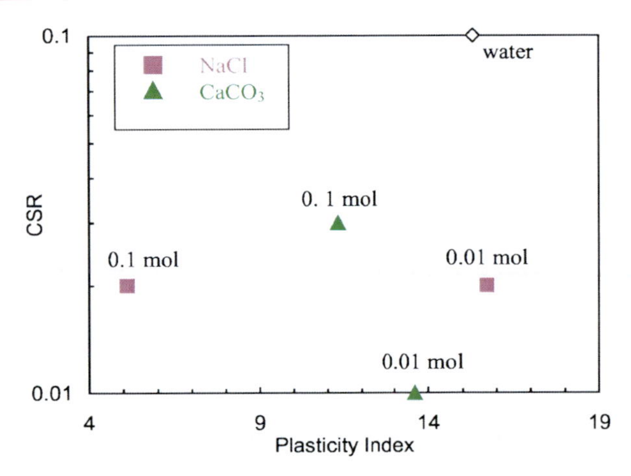

Fig. 15.4. Results of cyclic undrained ring-shear tests on the mixture of sand with 11% bentonite treated with solutions of NaCl and $CaCO_3$ plotted in terms of plasticity index (*PI*) and cyclic stress ratio (CSR) (the numbers next to the marks denote ion concentration in mol)

15.4.2 Pore Water Chemistry

The effect of pore water chemistry on the liquefaction of clayey soil was studied on the mixture of sand with 11% bentonite, which was observed to be resistant to liquefaction ($CSR_{50} \approx 0.1$). The samples formed from this mixture were treated with solutions of NaCl and $CaCO_3$ and then their cyclic behavior as well as their microstructures were studied. The results from a series of cyclic undrained ring-shear tests are presented in Fig. 15.4 in terms of *PI* against CSR. Clearly, the change in pore water chemistry caused the samples to liquefy under cyclic loading (CSR_{50} less than 0.1). To understand why the same mixture underwent such dramatic changes in cyclic behavior, the microstructures of all the samples were studied. The SEM analysis revealed the strong influence of the used ions on the distribution and state of clay matter. In the case of sample saturated with water, the clay matter partly filled the pore space and was spread evenly in the sample resembling the matrix microstructure. Such a resistant-to-liquefaction microstructure was formed due to the honeycomb arrangement of clay matter as shown in Fig. 15.5a. However, the presence of ions caused the formation of big clay aggregates (Fig. 15.5b,c), thus disrupting the clay matrix and leading to liquefaction.

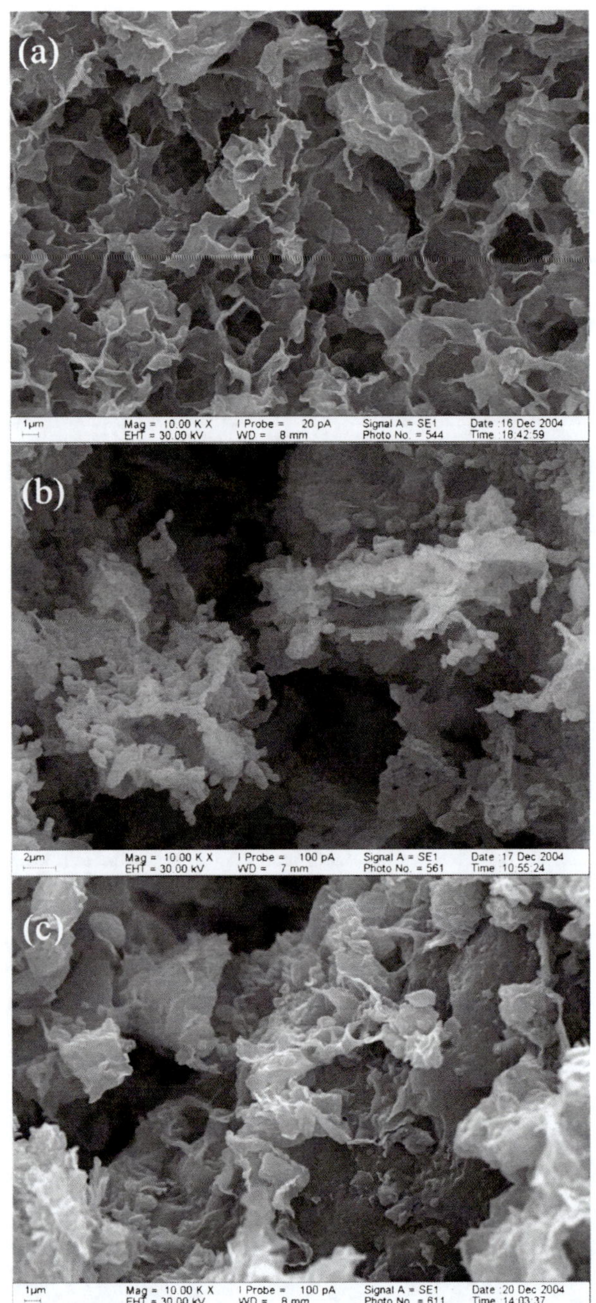

Fig. 15.5. Microstructure of the mixture of sand with 11% bentonite saturated with water (**a**), and treated with solutions of NaCl (**b**), and $CaCO_3$ (**c**) (magnification ×10 000)

15.5 Conclusions

The mechanism of liquefaction of clayey soils and the factors affecting it were studied using mixtures of sand with different clays by means of a ring-shear apparatus and a scanning electron microscope. The following conclusions could be drawn:

- Increase in clay content raised the soil resistance to liquefaction. Also, low plasticity kaolin- and illite-sand mixtures were more vulnerable to liquefaction than that of high plasticity bentonite-sand mixture given the same clay content;
- Plasticity Index (*PI*) can be used as a criterion for estimating the liquefaction potential of clayey soils, based on the observed relation that an increase in *PI* raises the soil resistance to liquefaction, and soil with *PI* > 15 is non-liquefiable;

- The soil microstructure plays an important role in the cyclic behavior of clayey soil. Analysis of the microstructures of different mixtures showed that it was the state and distribution of clay matter that determined the soil resistance to liquefaction;
- Pore water chemistry had a strong influence on the liquefaction potential of clayey soil. The presence of ions in pore water changed the microstructure of the soil, making it vulnerable to liquefaction.

Acknowledgment

The authors wish to thank Prof. V. N. Sokolov (Moscow State University) for his invaluable help with SEM analysis.

References

Gratchev IB, Sassa K, Fukuoka H (2004) Cyclic behavior of clayey sands under different physico-chemical conditions. In: Lacerda W, Ehrlich M, Fontoura S, Sayao A (eds) Proceedings 9th International Symposium on Landslides, Rio de Janeiro, Brazil. Balkema, vol. 1, pp 701–704

Gratchev IB, Sassa K, Fukuoka H (2006) How reliable is the plasticity index for estimating the liquefaction potential of clayey sands? J Geotech Geoenviron (in press)

Ishihara K, Kokusho T, Silver M (1989) Resent developments in evaluating liquefaction characteristics of local soils. In: Proceedings of the 12th International Conference on Soil Mechanics and Foundation Engineering, Rio de Janeiro, vol. IV

Ishihara K, Okusa S, Oyagi N, Ischuk A (1990) Liquefaction induced flow slide in the collapsible loess deposit in Soviet Tadjik. Soils and Foundations 30(4):73–89

Miura S, Kawamura S, Yagi, K (1995) Liquefaction damage of sandy and volcanic grounds in the 1993 Hokkaido Nansei-Oki earthquake. In: Proc. 3rd Int. Conf. on Recent Advances in Geotechnical Earthq. Engg. and Soil Dynamics, St. Louis, MO, vol. 1, pp 193–196

Osipov VI (1986) Liquefaction of soils and dynamic problems. In: Rimoldi HV (ed) Proceeding 5th International Congress International Association of Engineering Geology, Buenos Aires. Balkema, pp 2599–2622

Osipov VI, Sokolov VN (1978) Microstructure of recent clay sediments examined by scanning electron microscopy. In: Whalley WB (eds) Scanning electron microscopy in the study of sediments. Geo Abstracts, Norwich, England, pp 29–40

Osipov VI, Nilolaeva SK, Sokolov VN (1984) Microstructural changes associated with thixotropic phenomena in clay soils. Geotechnique 34(2):293–303

Perlea VG, Koester JP, Prakash S (1999) How liquefiable are cohesive soils. In: Seco e Pinto P (ed) Proceedings of earthquake geotechnical engineering, Lisboa. Balkema, vol. 2, pp 611–618

Sassa K (1985) The mechanism of debris flows. In: Proceedings 11th International Conference on Soil Mech and Foundation Engineering, San Francisco, pp 1173–1176

Sassa K, Fukuoka H, Wang G, Ishikawa N (2004) Undrained dynamic loading ring shear apparatus and application for landslide dynamics. Landslides 1(1):7–21

Sergeyev YM, Grabowska-Olszewska B, Osipov VI, Sokolov VN, Kolomenski YN (1980) The classification of microstructures of clay soils. J Microsc-Oxford 120(3):237–260

Youd TL, Holzer TL, Bennett MJ (1989) Liquefaction lessons learned from the Imperial Valley California. Special volume for Discussion Session on Influence of Local Soils on Seismic Response, 12th ICSMFE, Rio de Janeiro

Chapter 16

On Early Detection and Warning against Rainfall-Induced Landslides (M129)

Ikuo Towhata* · Taro Uchimura · Chaminda Gallage

Abstract. Traditional approaches to prevent rainfall-induced landslides consist of such stabilization of unstable slopes as installation of retaining walls as well as ground anchors. Although having been useful in mitigation of large slope failures, those traditional measures are not very helpful in mitigation of small slope failures which are less significant in scale but numerous in numbers. It is proposed in the present text for people to install slope instability detectors which find precursors of an imminent slope failure and issue warnings so that people may be able to evacuate themselves prior to fatal slope failures. To achieve this goal, model tests as well as laboratory triaxial tests have been conducted in order to understand the behavior of soil prior to failure. Moreover, numerical analyses on ground water percolation and decrease of factor of safety in the course of rainfall were conducted on a sandy slope in order to support findings from model tests. As a whole, a small instrument is proposed for a use of people which can detect minor displacement and change of moisture content prior to failure in a slope and issue warning through internet.

Keywords. Landslide, rainfall, warning, monitoring, model test

16.1 Introduction

There is a long history in prevention and mitigation of rainfall-induced landslides. Typical measures to prevent slope failure are retaining walls and ground anchors which improve factor of safety against failure. These measures have been widely used everywhere in the world and have been effective.

One of the limitations of the traditional measures lies in their cost of installation. Consequently, the traditional measures have been constructed only by governmental money in order to avoid relatively larger landslides such as shown in Fig. 16.1.

In consequence of recent residential developments in hilly area, the risk of smaller landslides has been realized. Figures 16.2 and 16.3 illustrate examples of this kind in which a small slope instability caused by rainfall endangered only a few houses upon heavy rainfall. It is important that such minor slope instability is many in number and difficult to be investigated by conventional engineering. Although efforts are needed to avoid risks of this kind, financial limitations make it difficult to install retaining walls and other conventional measures everywhere.

Early detection of slope failure and quick evacuation are always of significant importance. It has been known empirically that slope failure is preceded by the following precursors:

Fig. 16.1. Site of rainfall-induced landslide at Sakashi-dani in Niigata, Japan, in 2004

1. Sound of cutting tree roots,
2. Falling of stones from a slope which is normally stable,
3. Such distortion as cracking and heaving within or near a slope,
4. Generation of new water spring within or near a slope, and
5. Unusual roaring sound which is probably generated by distortion of ground at depth.

Although the knowledge is very important for disaster mitigation, monitoring of sound, falling stones, and detection of water spring during heavy rainfall is very difficult. It is obvious that meaningful underground sound is erased by the noise of rainfall and possibly thunder storm. Thus, it is desired to find out other precursor which can be monitored by any equipment during rainfall.

16.2 Warning of Rainfall-Induced Landslide

The problem of small landslides during heavy rainfall is now widely acknowledged. In efforts to mitigate this problem, the Meteorological Agency of Japan issues slope instability warnings when heavy rainfall and high moisture content in soil are expected. Although this idea is good, the moisture content is assessed on a regional basis based only on the regional intensity of rainfall, and the issued warning simply indicates the regionally averaged extent of risk. Note that the said "region" stands typically for one third or fourth of prefectures where hundred of thousands of people are living. Certainly, therefore, local topography, soil conditions, and the intensity of rainfall at indi-

Fig. 16.2. Small slope instability in Shizuoka, Japan, in 2004

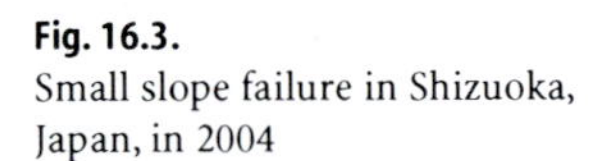

Fig. 16.3. Small slope failure in Shizuoka, Japan, in 2004

vidual slopes are out of scope. Accordingly, risk of any particular slope is not considered and the issued warning can hardly encourage local residents to evacuate.

An alternative approach to mitigate the problem of small landslides is an early warning and quick evacuation based on local monitoring. Not being new, efforts have been made at many potentially unstable slopes to monitor possible movement of slopes and issue warnings by using such transducers as extensometers (Fig. 16.4) and moisture transducers. It seems that there are still limitations in such an approach. Firstly, they are still costly and personal funding cannot afford them. Public fund may be introduced for restoration only after slope failure because everybody is now aware of the risk of repeated slope failure. Secondly, installation of extensometers needs some area so that tensile deformation may be detected. In case the concerned slope is owned by somebody else, an individual effort to install such an equipment is difficult. Thus, there is a need to develop a small and cheap transducer which can monitor the behavior of a slope but does not need much space. Another important issue may be the quick and less expensive transmission and interpretation of monitored data by which warning is issued.

16.3 Laboratory Tests on Effects of Moisture Increase on Shear Failure of Soil

Attempts to develop a methodology to make an early detection of imminent slope instability and to initiate evacuation consist of two factors which are namely the experimental understanding of soil behavior upon wetting due to rainwater percolation and also decision of logic by which the imminent instability is judged based on field monitoring.

Orense et al. (2003a,b, 2004) as well as Farooq et al. (2004) conducted a series of triaxial tests in which the effects of wetting on deformation of soil specimens were investigated. It was therein considered that the situation as occurs in a slope undergoing heavy rainfall was reasonably reproduced by injecting water into a soil specimen while maintaining constant the total stresses. The use of constant stress is supported by the fact that a slope does not change its inclination during rainfall until instability starts. The material properties of Omigawa sand which they used are as specific gravity = 2.67, mean grain size (D_{50}) = 0.49 mm, and fines content = 9.4%.

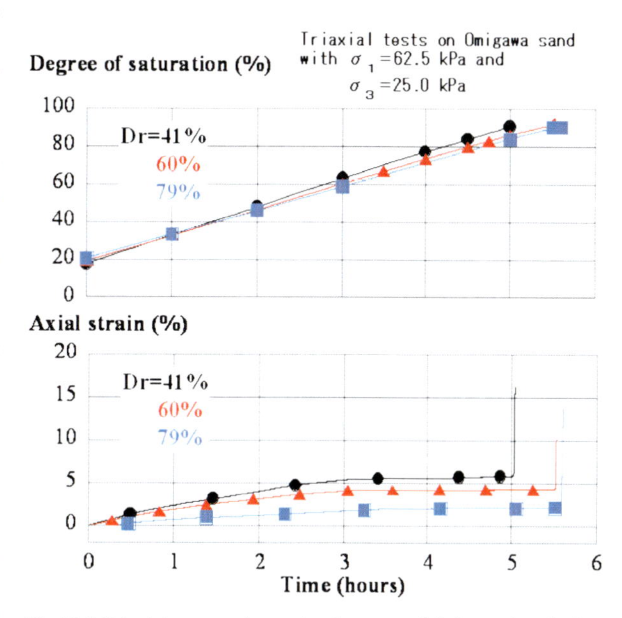

Fig. 16.5. Triaxial test results on development of deformation during increase in degree of saturation (Data by Farooq and Orense)

Fig. 16.4. Monitoring of slope displacement in unstable slope by extensometer (Niigata, Japan)

Fig. 16.6. Initiation of rainfall-induced instability from bottom of slope during model tests (case 9 by Shimoma et al. 2004)

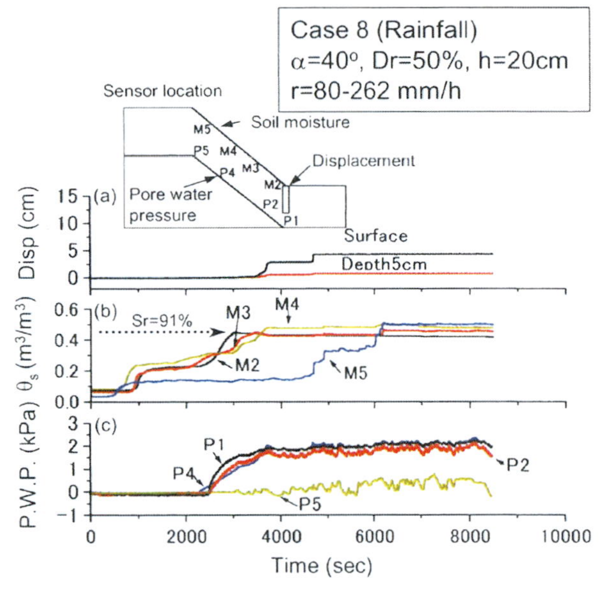

Fig. 16.7. Correlation between rainfall-induced displacement in slope and moisture content both at and near the bottom of slope (model test by Shimoma et al. 2004)

Fig. 16.8. Model test on slope failure with artificial rainfall

Figure 16.5 illustrates the variation of axial strain of three specimens which had different densities. As pore water was injected into specimens, the degree of saturation increased with time, and when it reached more or less 90%, the axial strain started to increase rapidly towards failure. It is noteworthy that strain started to increase at 90% degree of saturation irrespective of density of sand.

To make more insight on precursor of rainfall-induced landslide, Shimoma et al. (2004) conducted model tests in which a sandy slope was subjected to artificial rainfalls. Firstly, Fig. 16.6 indicates that failure was initiated from the bottom of a slope where moisture content takes the highest level within the slope. This suggests that any precursor may be detected most easily at this location. Figure 16.7 shows a correlation between lateral displacement measured at the toe of a model slope and the volumetric moisture contents near the bottom. The volumetric moisture content designates the ratio of volume of pore water and the total volume of soil inclusive of solid, liquid, and gas components. It may be seen that the moisture content near the toe (M2) is a good precursor of displacement towards failure. Furthermore, note that the ultimate displacement at the time 5 000 s was preceded by preliminary displacement at around 4 000 s. It may be reasonable, therefore, that minor displacement could be another precursor to be monitored.

It was further attempted by the present authors to investigate the relationship between rainfall-induced displacement and the factor of safety (FOS) against slope failure. Figure 16.8 shows the model container with rainfall, while Fig. 16.9 the configuration of the model and the location of transducers.

The measured degree of saturation and the lateral displacement are plotted in Fig. 16.10. While the ultimate large displacement was induced after 7 500 s, the degree of saturation at the bottom (M7) was high from the beginning, and that in the lower slope (M5) started at 2 500 s. It is important that minor displacement started at around 2 500 s as well. Thus, moisture content as well

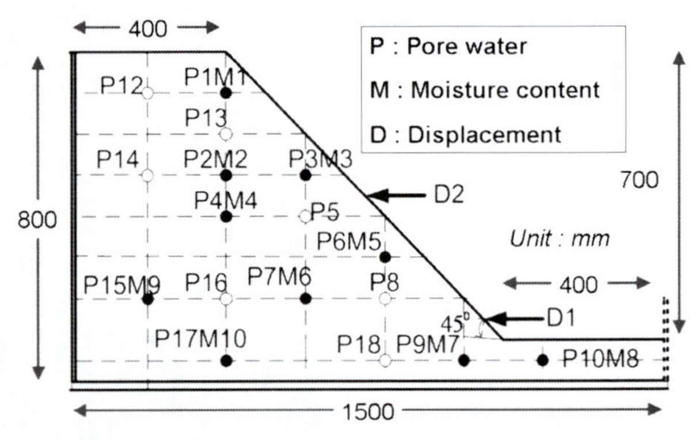

Fig. 16.9. Configuration of model container and location of transducers

as lateral displacement may be good precursors to be monitored. According to the appearance of this slope in Fig. 16.11, some instability occurred at 3 480 s. This further implies that the above mentioned precursors are reasonable. Note in this figure that soil in this test was wet at time zero because there was another rainfall test on the same model without causing significant displacement.

Finally, an attempt was made to reproduce the present test condition by FE seepage analysis. While its detail is not presented in the present text, it should be mentioned here that the calculated pore pressure and effective stress were used to run stability analysis in order to obtain the time history of factor of safety. The result is presented in Fig. 16.12. It may be seen here that the factor of safety became less than unity at 4 500 s at which soil had already been well saturated and minor displacement had started (Fig. 16.10).

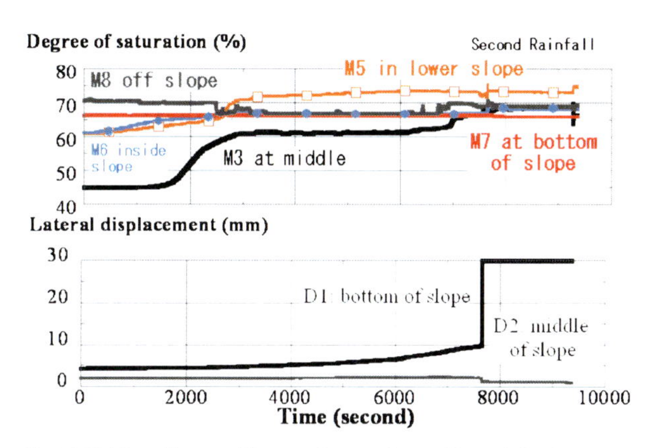

Fig. 16.10. Time history of degree of saturation and lateral displacement

16.4 Proposal of Slope Instability Detector for Personal Use

Laboratory and model tests as stated in previous sections showed the importance of the use of displacement and moisture content at the bottom of a concerned slope. Hence the present study is now aiming to fabricate an inexpensive equipment which can issue a warning signal prior to slope instability. Following points were discussed prior to final specification of this equipment.

1. Since the instability warning equipment is supposed to help people who wish to protect themselves from rainfall-induced slope disaster, the price of the equipment has to be reasonably cheap. Including a possible financial support from a local government, the price should most probably be more or less 40 000 Japanese Yen. It is thus aimed herein to help self efforts of people so that they will be more aware of the importance of natural disasters and mitigation.
2. The present warning system concerns with relatively small slope failures which occur due to water infiltra-

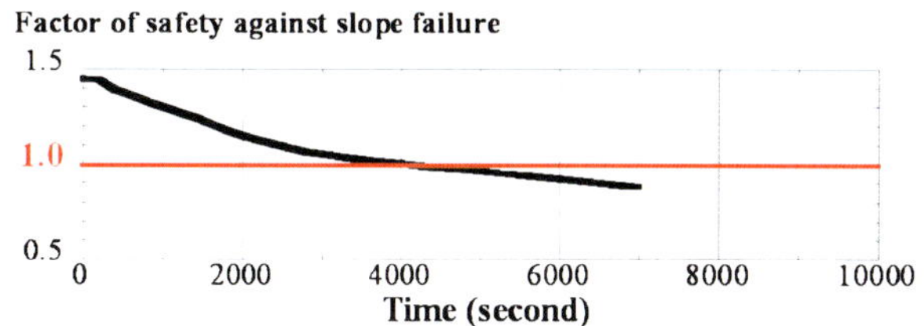

Fig. 16.12. Time history of factor of safety as calculated by seepage analysis

at 3 480 seconds

at 6 300 seconds

at 7 860 seconds

Fig. 16.11. Appearance of model slope during rainfall test tion in surface weathered soil. It is reasonable to say that the thickness of the surface weathered soil scarcely exceeds one meter. Thus, there is no need to monitor displacement and/or moisture content at deeper elevations.

3. It is very likely that a neighboring and potentially unstable slope is owned by somebody else. If this is the case, it is difficult for an individual person to install an extensometer over a length of the slope. To avoid this problem, it is supposed here to embed an inclinometer at the bottom of the slope so that a precursor displacement is detected in a form of angular change. Note that lateral displacement of soil in the slope causes rotation or shear deformation of soil at the bottom of a slope.
4. Moisture content of soil at the bottom of a slope should be monitored as well. A warning should be issued if either inclination or moisture content exceeds critical levels. This is what is called "OR" logic.
5. The collected data of inclination and moisture content should be transmitted to a local disaster management center. After interpretation of those data, warning should be issued, if necessary, and this emergency information is automatically transmitted to local residents. The transmission of data and information is achieved at low costs by using the network of internet and mobile phones which are developing very fast in recent times.

With these points in mind, an equipment as shown in Fig. 16.13 is proposed. All the sensors are housed in this single equipment and are protected by the steel pipe from environmental impacts. Its embedment is made easy by this steel pipe and the cone tip. It is thus supposed that the steel pipe is embedded in a small hole which is made by a hand auger possibly together with hammer impacts

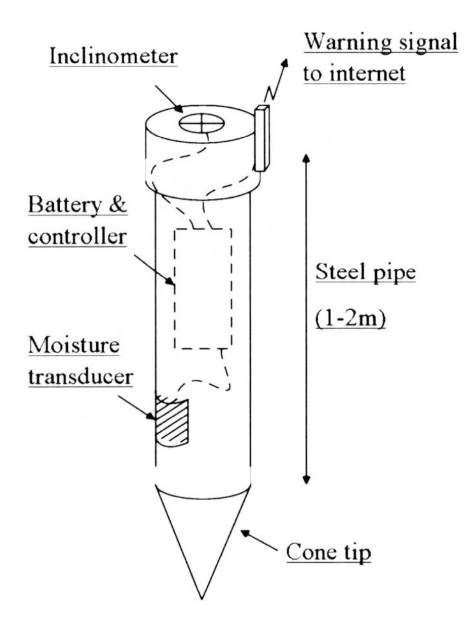

Fig. 16.13. Equipment which is embedded at bottom of unstable slope

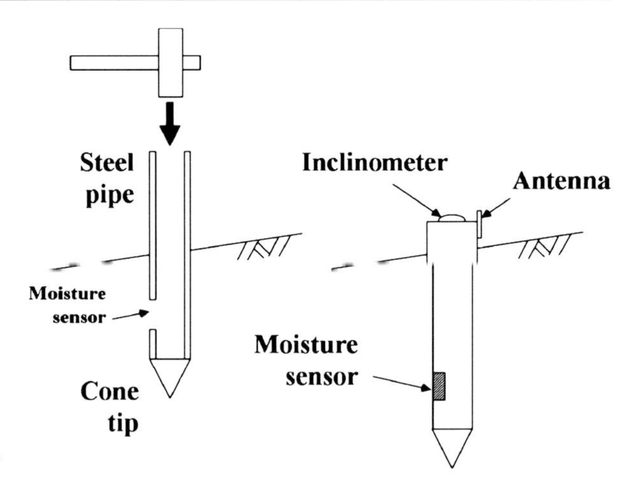

Fig. 16.14. Penetration and installation of transducers

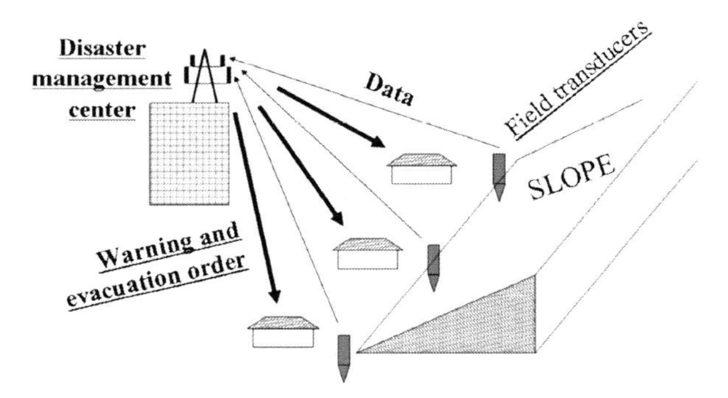

Fig. 16.15. Flow of collected data and warning information

at the head to some extent. This idea is illustrated in Fig. 16.14. Note that some sensible transducers are installed after hammer impacts. All the parts of this equipment are operated by a battery which can be easily renewed periodically. The flow of data and information is illustrated in Fig. 16.15.

In a more advanced stage of warning, it is desired to take into account the size of slope failure. In a more precise sense, it is important to take account of the distance of debris flow after failure. This is because a very small amount of failed soil mass would not affect any facility, and an evacuation order would be meaningless. On the contrary, a big slope failure which may influence a resident's house has to be detected in advance. To satisfy this requirement, several sets of transducers should be deployed along a slope. In the example of Fig. 16.16, three sets of transducers from the bottom detect imminent slope failure, while the top one does not. Thus, the range of instability is identified firstly, and by using an empirical knowledge of, for example, 30 degrees of angle between the top of failed soil mass and the end of the reach of mobilized debris, the range of influence is assessed. Warning should be issued if the assessed run-out of debris reaches facilities to be protected; otherwise, no warning is needed. It is apparently necessary now to assemble the described

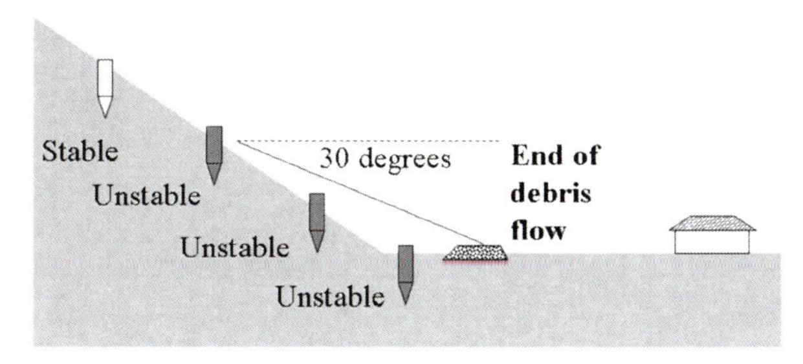

Fig. 16.16. Consideration of range of flow of debris in warning system

equipments, install them in a model slope, and run tests on slope failure by creating artificial rainfall so that the described logic of warning system may be validated.

16.5 Conclusion

Study has been carried out to develop a warning system which ordinary people can install on a personal basis and help themselves from a rainfall-induced slope failure. Based on former studies, it was decided to monitor displacement and water content near the bottom of a potentially unstable slope, and if the monitored data exceeds threshold values, a warning is issued. The whole system is made inexpensive so that people can afford it. Moreover, data communication is made easy by using mobile telephone system which is developing very fast in the recent times. It is now planned to run validation tests of the system by installing devices in a slope and causing landslides by artificial rainfall.

Acknowledgments

The early phase of the present study was carried out by Dr. R. Orense who used to be a member of the authors' group with a financial support from Japanese Railway East. A field excursion to landslide sites was helped by Dr. H. Toyota of Nagaoka University of Technology. Last but not least, the present research topic has been selected as one of the recommendable research topics by ICL. These supports are deeply appreciated by the authors.

References

Farooq K, Orense RP, Towhata I (2004) Response of unsaturated sandy soils under constant shear stress drained condition. Soils and Foundations 44(2):1–14

Orense RP, Towhata I, Farooq, K (2003a) Investigation of failure of sandy slopes caused by heavy rainfall. In: Proc. International Conference on Fast Slope Movements – Prediction and Prevention for Risk Mitigation (FSM2003), Sorrento

Orense RP, Shimoma, S, Honda T, Towhata I, Farooq K (2003b) Laboratory experiments on failure initiation in sandy slopes due to rainwater infiltration. In: Proc. Soils and Rocks of America (Panam. Conf. Soil Mech. Geotech. Engrg), Boston, vol. 2, pp 2465–2470

Orense RP, Farooq K, Towhata I (2004) Deformation behavior of sandy slopes during rainwater infiltration. Soils and Foundations 44(2):15–30

Shimoma S, Orense RP, Maeda K, Towhata I (2004) Experimental study on landslides caused by heavy rainfall. J Natural Disaster Sci 26(1):15–26

Part III Landslide Risk Analysis

Chapter 17

Geological and Geomorphologic Relationship of the Sub-Active Landslides of Cusco Valley, Peru

Raúl Carreño

Abstract. The valley of Cusco occupies a depression derived of the Andean folding of continental and marine redbeds series and of volcanic rocks of the Upper Palaeozoic until the recent Quaternary. Due to lithological as well as structural factors, numerous landslides exist (around 170 according to our inventory; most of them in sub-activity state) representing diverse levels of danger.

The landslides occur mainly in the Cretaceous evaporitic, calcareous and pelitics formations of the so-called Yuncaypata Group. In those formations the landslides respond mostly to the lithological genetic factor. The continental red beds of San Jerónimo Group (Lower and Middle Tertiary) show a smaller density than the Cretaceous formations and the landslides are conditioned by the structural factor (conform slopes). The volcanic rocks of the Permo-Triassic (Mitu Group) and Plio-Pleistocene (Rumicolca Formation) present a very low landslide occurrence, always conditioned by the structural genetic factor. At least two cases of deep Toppling becoming complex landslides have been identified in the San Jerónimo Group outcrops.

Keywords. Landslide, sub-active landslide, landslide inventory, Cusco, Peru

17.1 Introduction

The first detailed landslide inventory of the basin of Cusco includes around 170 identified large scale landslides (surfaces ≥ 1 hectare) (PROEPTI 1999; Carreño 1998/2005). Most of they belong to the sub-active landslide category and representing different levels of risk. Smaller landslides also exist, mainly related to unconsolidated Quaternary formations, especially in lacustrine and morainic sediments.

Some characteristics well define the genetic relationships among the different geological formations outcropping in this basin and the landslide style and density. Those relationships are fundamentally lithological for the Cretaceous rocks of Yuncaypata Group, and structural for the San Jerónimo Tertiary Group and other volcanic formations.

From the elevation point of view, there is a higher concentration of landslides between the 3 400 m and the 3 800 m, and on hillsides which slope varies between 25° and 40°.

17.2 Study Site

The basin of Cusco occupies an inter-Andean valley of the south-eastern Andes Range of Peru. The elevation of the bottom of the valley varies between 3 160 m and 3 450 m and is constituted by a series of lacustrine and fluvio-alluvial terraces, as well as by alluvial cones and Pleistocenic moraines.

In this basin the following geological units outcrop (Cabrera 1986; Córdova 1986; Mendívil and Dávila 1994):

- Mitu Group (Permian-Triassic): mainly volcanic series of the Pachatusan Formation (riolites, andesites, ignimbrites, breccias) and sedimentary rocks (sandstones, conglomerates, and lutites).
- Yuncaypata Group and Killke Formation (Middle and Upper Cretaceous): limestone, evaporites, lutites and sandstones of marine origin.
- San Jerónimo Group (Low and Middle Tertiary): continental red beds series (mainly sandstones and lutites including some levels of conglomerates).
- Rumicolca Formation: shoshonitic andesites of the Plio-Pleistocene.
- San Sebastian Formation (Pleistocene): not consolidated lacustrine sediments, including diatomite levels.

Structurally the basin is defined by a folding system following a SE-NW direction. Toward the western, another system which direction is N-S exists. The main faults systems follow SE-NW and N-S directions. There are also several active seismogenic faults; which of Tambumachay is the most active.

Morphologically we can define the following units:

- Mountain ranges: the so-called "Montañas de Cusco" (Pachatusan Mountain, 4 861 m a.s.l.) at northern part, and "Serranías de Vilcaconga", at south (Huanacaure Mountain, 4 451 m a.s.l.).
- Mesetas of Huacoto, Saqsayhuaman and Oqopata, developed on Yuncaypata outcrops.
- Glacis formed by juxtaposed alluvial cones (Larapa, Puquín, Tankarpata, Pillao…).

- Floodplain and fluvial terraces of Huatanay River.
- The landslides are situated mostly in the transition area between the mountains to glacises; their heads don't surpassing the 3 900 m a.s.l.
- Those landslides represent an important menace for the city of Cusco and several villages (which total population reaches around 350 000 inhabitants), as well as important communication and services infrastructure.

17.3 Methods

The location, boundaries and type of the sub-active landslides were determined by means of air photos and topographical maps analysis and field visits. Equally a GIS with AutoCAD Map was created using the 1 : 25 000 and 1 : 10 000 scale topographical maps. In some cases 1 : 2 500 topographical maps were available. For this kind of work those mapping scales has been relatively small, conditioning the calculation of areas and volumes of the involved material, and restricting the validity of some morpho-structural and genetic interpretations, which are only approximate.

The carried out work constitutes the first phase of the IPL project (M-123) "Cusco regional landslide hazard mapping and preliminary assessment", which has started with a general revision and correction of the inventory elaborated in previous years, using a new topographical base. Likewise, a revision of the geological map of the area was another task oriented to completing the existents not up-to-date maps, which showed many lacks and deficiencies.

17.4 Analysis

The analysis of the landslides inventory map put in evidence the geological and structural relationships for each lithological unit outcropping in this basin. The Yuncaypata and San Jerónimo Groups concentrate more than 90% of the large scale and sub-active landslides if the basin. The rest correspond to the volcanic formations, excepting the smallest short evolution phenomena, whose origin is mainly related to the erosion and the human action, and which are plentiful in the Quaternary formations.

17.4.1 Yuncaypata Group and Puquín Formation

Both geological units have lithological similarities allowing considering them like a single unit from a geodynamic point of view.

The Yuncaypata Group presents the highest landslides concentration of the basin: around 65%, mainly in the northern and western part. This relationship was already noticed by the first geological studies carried out in the Cusco by Gregory (1916), who denominated this area like a "morphological province of landslides".

The genetic relationship here is clearly lithological. The gypsum, anhydrites and lutites layers acting like rupture levels favoring the formation of the landslides which, in their majority, respond to rotational mechanism, evolving after to complex mechanisms.

The stratigraphic disposition of the rocks is chaotic; and it's not possible to recognize the original structural feature. This is due to the existence of those mentioned décollement levels (especially gypsum and intraformational breccias) that favoring the erosion and the irregular displacement of the rigid blocks.

Other geological factors have also an important role in the landslides genesis, like the high drainage density influencing the erosion regime as well as the high dislocation of the rocks thanks to the intense tectonic activity during the Upper Cretaceous and mainly during the Andean orogeny. Just as example: the formations situated in the plateau of Saqsayhuaman has suffered the effect of four tectonic phases in which happened folding processes, inverse faulting, horizontal faulting and finally, until the present time, normal faulting. All these factors contribute to the higher incidence of instability phenomena in the areas where these rocks outcropping.

In the plateaus where those units outcrop, especially in Saqsayhuaman, the Spread phenomenon is plentiful, affecting the calcareous blocks overlaying the lutites and gypsum levels. In some cases, the Spread has favored the formation of Toppling and small landslides.

On the southern slope of the Pachatusan Mountain many landslides respond to the conjunction of the lithological and structural conditioning, because of the presence of the evaporitic and pelitic rocks of the Yuncaypata Group which structural feature correspond to an conform slope on the southern flank of an Vilcanota anticline. In that case, the failure surfaces coincide with the contact levels of the Yuncaypata Group with older formations.

The analysis of the recent evolution of those landslides shows that they have a tendency to induce quick and vio-

Fig. 17.1. Partial view of a typical landslide developed on Cretaceous formations under lithological influence (Saphy, NW of Cusco)

lent slide episodes. The fluvial erosion evacuates the disintegrated materials pushed easily toward the riverbeds; that is due to the nature and fragile constitution of these marine redbeds (Fig. 17.1).

17.4.2 San Jerónimo Group and Chilca Formation

In this unit, the occurrence of landslides is quantitatively smaller, but its individual dimensions are bigger than others developed in the Cretaceous units. The four bigger instability phenomena of the basin corresponding to this unit: Huaynapicol, Saylla-Ch'akiqocha, Pumamarka and Huamancharpa (this last was responsible of the biggest geodynamic episode happened in this basin in recent times). Huaynapicol and Saylla-Ch'akiqocha are related to deep Toppling and complex deformation and rupture processes. Huaynapicol and Saylla-Ch'akiqocha are related to deep toppling and complex deformation and rupture processes (Carreño and López 1998).

The Huamancharpa landslide (that caused a violent episode that mobilized 40 million cubic meters in January 1982) is a typical Dip Slope landslide developed over a favorable structure where the sedimentary layers are parallel to the topographical slope. The involved rocks are sandstones, lutites and conglomerates. In the last two decades, several secondary blocks located toward the northeast (at foot and at north of the 1982 slided block) are showing strong movements and forming natural and ephemeral dikes strangling the Oqopata-Huancaro River. This process represents a serious danger for the neighborhoods located in the southern part of the city of Cusco. This activity could be announcing a general re-activation of the landslide in the next years.

The Saylla-Ch'akiqocha Landslide is more difficult to characterizing, due to its dimensions (approximately 6 km^2) and its complex structural feature. The failure surface has relationship with an unconformity surface, probably related to a deep complex Toppling accident. The intervention of neotectonic processes in its origin and evolution is not discarded. It is also probable that the foot of the landslide was buried by lacustrine and fluvio-alluvial sediments during the Holocene, that have stopped most of the deep movement; this would indicate that its origin goes back until the Pleistocene period (more than a hundred thousand years ago), when a lake occupied the whole tectonic origin basin. The active blocks are relatively superficial and their activity responds to infiltration coming from channels, to the seismic and also to erosion factors.

In this landslide, several secondary blocks located toward the north show signs of strong activity. In March 2001 one of these blocks suffered a catastrophic acceleration, damming the gulch of Jatunhuaycco; the derived debris flows reached the vicinities of the town of Saylla.

The landslide of Huaynapicol has a clear genetic relationship with a process of deep Toppling. An active fault passing near the foot of the instable slope probably also influenced its genesis. Strictly it is a landslide in Sagging phase. The development of the crown and lateral escarps has been spectacular in the last two decades; however, the failure plane is not still complete; but this situation doesn't impeaching the formation of continuous (but volumetrically not very important) debris flow and rock-falls in the gully formed in the middle part of the most active area.

The Pumamarka landslide (and some smaller neighboring landslides) is a peculiar case of structural genetic condition, because it is a landslide developed in a contrary slope. The rupture plane doesn't correspond to joints but to the mirrors of subsidiary faults, parallel to the main fault of Tambumachay. The intersection between those faults mirrors and the hillside allowed the detachment of a tabular mass of sedimentary rocks (without the intervention of the hydric or erosive factors), evolving to a translational landslide mechanism.

In conclusion, the genetic factor in the landslide developed in this geological unit is almost exclusively structural (Fig. 17.2): the slide planes have been developed along the hinges of complex Toppling (Huaynapicol and Ch'akiqocha), on the soft levels of lutites which strike is parallel to the slope (Huamancharpa) or in tabular masses defined by parallel faults (Pumamarka).

17.4.3 Mitu Group and Rumicolca Formation

In the Mitu Group only one relatively important instability phenomena has been identified: the Huacoto landslide; which is also the highest landslide in the basin; its head is situated at almost 4 000 m a.s.l. (Fig. 17.3). It has been developed in an opposed slope and its origin seems to responding to the fluvial erosion along an area of lineal fracture. Its importance resides in its capacity to cause debris flows that could affect the inferior basin of the river Huacotomayo and the road Cusco-Puno-Arequipa.

In the Rumicolca Formation only two landslides have been identified. The first one in the area of Choquepata, close of the archaeological group of Tipón, is a rocky landslide in formation; the second one, in Urpikancha, where the mass of volcanic disaggregated rocks evolved to very slow sliding mechanism, where the role of the water table is practically null because the volcanic mass has not a real retention capacity, due to the absence of fine sediments. In both cases the decisive factor is related to the jointing and disintegration taken place by the quick cooling of the shoshonitic lavas just after the Pleistocenic eruptions. For this same reason continuous rock falls occur in these rocky mass. The plastic behavior of the underlying material (evaporites of the Yuncaypata Group) also contributes to the genesis of landslides.

The main outcrop of the Rumicolca Formation, at southern part of the archaeological group of Pikillaqta blocked completely the valley of Huatanay during the early Pleis-

Fig. 17.2. The three types of landslides developed under influence of the structural feature (San Jerónimo Group): **a** Huaynapicol (deep Toppling); **b** Huamancharpa (dip slope); **c** Pumamarka (tabular body defined by parallel faults mirrors)

Fig. 17.3. Huacoto landslide (Pachatusan Formation)

tocene, deviating orthogonally the river. The lavas also buried the foot of the Pukaqhasa landslide (Yuncaypata Group), stabilizing part of the unstable mass.

17.4.4 Dioritic Stock of Saqsayhuaman

In this intrusive body only two small and well defined landslides have been identified. Their origin is strictly related to dense jointing systems, product of the tensional tectonic that allowed the Stock intrusion, especially in the gulch of Choquechaka. The cinematic style of these phenomena combines translational landslides with rock fall mechanisms.

17.5 Conclusions

The analysis of the giant (and mainly sub-active) landslides of the basin of Cusco shows that their origin responds fundamentally to lithological and structural factors. In the evaporitic rocks and marine red beds of the Cretaceous, the more determinant genetic factor is the lithological nature of the formations, with prevalence of the rotational mechanism. In the continental red beds of the San Jerónimo Group (Tertiary), and in the volcanic formations, the structural factor prevails; in some cases, deep and complex toppling processes also intervene, as well as discontinuities defined by parallel faulting planes of subsidiary faults and other joint systems.

There is a higher density of landslides in the Cretaceous formations, but the largest landslides belong to the San Jerónimo Group. The most of those phenomena is concentrated in an altitudinal strip going from 3 400 m to 3 800 m.

References

Cabrera J (1986) Néotectonique of the région of Cusco. 3rd cycle Dr. Thesis, University of Paris VI Orsay, France

Carreño R (1998/2005) Landslides inventory: Cusco Valley. GRUDEC AYAR investigation report

Carreño R, López R (1998) Two cases of landslides related to Toppling processes in the valley of Cusco. Proceedings of the Xth Peruvian Geological Congress (in Spanish)

Córdova E (1986) Les couches rouges continentales de la région of Cusco. 3rd cycle Dr. Thesis, University of Pau, France

Gregory W (1916) Geological reconnaissance of the Cuzco Valley, Peru. Am J Sci, vol. II, no. 241

Mendívil S, Dávila D (1994) Geología de los cuadrángulos de Cusco y Livitaca. INGEMMET Bull 52, Serie A, Lima

PROEPTI (1999) PROEPTI project final report. EPFL, Lausanne

Chapter 18

Measurement of Velocity Distribution Profile in Ring-Shear Apparatus with a Transparent Shear Box

Hiroshi Fukuoka* · Kyoji Sassa · Gonghui Wang · Ryo Sasaki

Abstract. Using a new ring-shear apparatus with a transparent shear-box and video image analysis system, drained speed-controlled test was conducted on coarse-grained silica sands to study the shear-zone formation process in granular materials. Velocity distribution profiles of grains under shear at various stages in the ring shear test were measured through processing the video image by Particle Image Velocimetry (PIV) program. In the initial stage of shearing, comparatively major part of the sample in the upper shear box showed a velocity distribution due to deformation and dilatancy behavior. Thereafter, velocity distribution profile changed into slide-like mode and thereafter showed almost no change. This study was conducted as a part of the International Programme on Landslides M101 "Areal prediction of earthquake and rain induced rapid and long-traveling flow phenomena (APERITIF)" of the International Consortium on Landslides (ICL). These results will contribute to understanding the mechanism of shear-zone development in granular materials as a basic knowledge for disaster risk mitigation of rapid long run-out landslides.

Keywords. Ring shear tests, speed-controlled test, drained condition, shear-zone development, transparent shear box, PIV, velocity distribution profile, rapid landslides

18.1 Introduction

The Nikawa landslide in Hyogo prefecture, Japan, which was triggered by the January 1995 Hyogoken-Nambu earthquake showed a high mobility and killed 34 people in a densely populated urban area. Sassa (1996), Sassa et al. (1996) conducted undrained cyclic loading ring shear tests, and proposed a concept of Sliding Surface Liquefaction (SSL) as a key mechanism of rapid and long travel landslides. Then, Sassa (2000) succeeded in simulating the triggering process by applying record of real seismic waveform, as well as simulating post-failure rapid movement process, with use of the undrained ring shear apparatuses.

Sassa (1996) found that SSL takes place along the sliding (shear) surface, and it does not need destruction of the soil structure. The main cause of this phenomenon is the grain crushing along the sliding surface which let the soil skeleton volume tend to shrink, and it leads to excess pore pressure generation to reduce effective stress, and shear resistance, as well. Thus, SSL is the key mechanism and assumed to be responsible for low apparent friction angle of rapid and long-travel landslides. The SSL is not like the conventional "liquefaction" concept defined by Castro (1969) and SSL can be assumed as a localized liquefaction within shear zone.

In order to reveal the mechanisms of rapid and long-travel landslides, it is essential to observe the shear behavior of soils during shear. However the shear-zone formation process in ring shear tests under undrained condition has not been observed because shear boxes of most ring shear apparatuses are metallic and invisible. For the purpose of study on mechanism of long-travel landslides, ring shear apparatus is most appropriate because there is no limitation for shear displacement. Bishop et al. developed low-speed ring shear apparatus and conducted tests on residual strength of clays (Bishop et al. 1971). Lupini et al., Bromhead and Dixon, and others followed ring shear tests for landslide mechanism studies (Lupini et al. 1981; Bromhead and Dixon 1986). Hungr and Morgenstern, Tika, and Fukuoka had conducted high speed ring shear tests on granular materials for study of rate effect on shear strength (Hungr and Morgenstern 1984; Tika 1989; Fukuoka 1991). Rapid shear faster than 1 m s^{-1} was possible by the apparatuses used in Hungr and Morgenstern (1984), and Fukuoka (1991), however there was no ring shear apparatus to simulate "extremely rapid landslides" (Varnes 1978) which is defined as faster than 3 m s^{-1}. Since 1984, Sassa has developed various high-speed ring shear apparatuses and undrained ring shear apparatuses (Sassa 1992, 2000; Sassa et al. 1996, 2003). However, their metallic shear boxes did not allow visual inspections during tests to examine the movement of grains in the samples.

Lang et al. (1991) developed transparent shear box for low-stress high-speed ring shear apparatus to observe movement of dry uncrushable glass beads during shear. The beads diameter was 2–6 mm, normal stress was 2.9 and 29 kPa, and shear speed was 1 cm s^{-1} and 10 cm s^{-1}. They found the shear-zone thickness decreased under higher normal stress in the test on beads of smaller diameter grains. They reported that shear-zone thickness was slightly thicker during faster shear. However, these tests used only glass beads and not used natural materials.

Wang and Sassa (2002) observed the development of a shear zone under undrained conditions at very large shear displacement. They inserted a column of different sand inside the sand sample to examine the thickness of shear zone when SSL takes place. They opened the shear box and observed the shear zone only after the test finished because the shear box is metallic and not visible from outside. Therefore, examining of the process of shear-zone development from outside was not possible.

Wafid et al. (2004) investigated the shear-zone development process by directly observing the shear zones by repeating tests under the same condition but different stages of shear displacement from the initiation of failure to the steady state by undrained ring-shear tests. Unique shear-zone sampler was developed and they succeeded to examine the structure of the shear zone and grain size distribution of many samples at various shear displacement. However, continuous monitoring from outside of shear box and detection of horizontal and vertical movement of grains were still impossible.

In this study, the main purpose is to observe the process of shear-zone development through measuring velocity distribution profiles of sand grains during shear by analysis of video images of shearing samples. To achieve this purpose, ring shear tests were performed by means of a ring shear apparatus DPRI Ver. 7 with a transparent shear box, which was developed in 2003 by K. Sassa et al. at Disaster Prevention Research Institute, Kyoto University (Sassa et al. 2004). Additionally, this apparatus is the first apparatus of which the maximum speed reached 3 m s^{-1}, i.e. the criterion of the "extremely rapid landslide".

Ishikawa (2004) conducted preliminary tests using the DPRI Ver. 7 with a transparent shear box. Video camera and CCD camera were used to take video images. PIV program was tested to analyze the velocity distribution of grains above shear surface only because no special video camera was prepared to take videos of the sample below the shear surface. Undrained and drained ring shear tests were conducted on samples of various grain sizes under the normal stress of 10 and 200 kPa, shear speed of 0.1, 3.0, and 100 cm s^{-1} were employed. Nevertheless, the velocity distribution was analyzed only when the sample was at the steady state. For better understanding of the initiation and post failure movement of landslide mass, it is of great importance to examine the shear-zone development process from the point of start of shearing.

This study is part of IPL M101 "Areal prediction of earthquake and rain induced rapid and long-traveling flow phenomena (APERITIF)" of ICL. The DPRI Ver. 7 apparatus was developed as a part of the APERITIF project. APERITIF is an extension study of the project "Aerial Prediction of Earthquake and Rain-Induced Flow Phenomena" (APERIF project, represented by K. Sassa), of the Special Coordinating Fund for Promoting Science and Technology of the Ministry of Education, Culture, Sports, Science and Technology of Japan (MEXT) for 3 years from 2001 to March 2004.

18.2 Ring Shear Apparatus with a Transparent Shear Box (DPRI Ver. 7)

Figure 18.1 is the schematic illustration of ring shear tests of this study. DPRI Ver. 7 which has a transparent sample box, was used. Speed-controlled ring shear test was conducted on coarse-grained silica sands with different shear speed under drained condition, and dynamic behavior was observed. Grains of the sands were observed by two video cameras and a high speed CCD camera to analyze the movement speed distribution inside the shear zone by the velocity measuring technique, Particle Image Velocimetry (PIV).

The overview of the ring shear apparatus DPRI Ver. 7 is shown in Fig. 18.2. The basic structure of this apparatus is the same as DPRI-5 and DPRI-6 which are introduced by Sassa et al. (2003, 2004). The size and the capacity of the DPRI Ver. 7 apparatus are shown in Table 18.1. Both of the outer-upper-shear-box and the outer-lower-shear-box are made of transparent acrylic resin and stainless steel. Detailed section of the shear boxes is shown in Fig. 18.3. Rubber edges are fixed on the top surfaces of the lower shear boxes. Constant contact force is applied to the rubber edge and its contact pressure is always kept automatically in a value greater than expected pore pressure to prevent the leakage of water and sample. A thin stainless steel plate of 2.0 mm thickness, is attached to

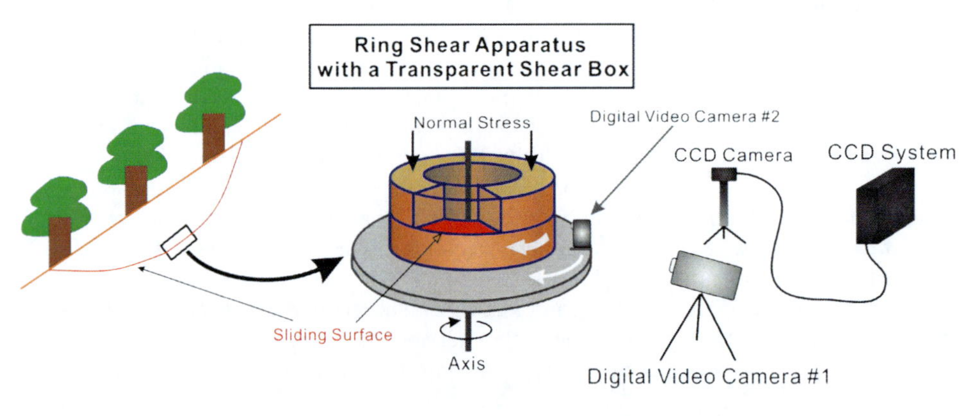

Fig. 18.1. Schematic illustration of ring shear tests of this study

the bottom of the upper half of the outer shear box contacting to the rubber edge. This plate is necessary to protect the acrylic shear box from damage due to heating during high speed shearing. This system could not observe the exact shear plane and its vicinity because the shear zone is shaded by invisible contacting parts consisting of the rubber edges and the metal plate as shown in Fig. 18.3. Teflon and grease is coated on the rubber edge to reduce friction during shearing. In addition, the laboratory temperature of a laboratory is kept almost constant by air-conditioner because acrylic is highly inflatable against room temperature rise.

Fig. 18.2. Overview of the ring shear apparatus with a transparent shear box (DPRI Ver. 7)

The actual normal stress on the sample through a loading plate can be precisely controlled by servo-system and measured (Sassa et al. 2003, 2004). The sample height after initial consolidation is about 11 cm and the predetermined shear plane is located at 4.0 cm above the bottom of the shear box. The lower half of the shear box (hereinafter referred to as "the lower shear box") can be rotated either in the torque control condition or in the speed control condition by a servo-controlled electric motor, while the upper half of the shear box (hereinafter referred to as the upper shear box) restrained by two load cells, which measure shear resistance.

Table 18.1. Properties of the Ring Shear Apparatus DPRI Ver. 7

Year of development		2003
Shear box	– inner diameter (mm)	270
	– outer diameter (mm)	350
Maximum height of sample (cm)		11.5
Maximum normal stress (kPa)		500
Maximum shear speed (cm s^{-1})		300
Cyclic loading (Hz)		0–5
Undrained capability (maximum pore pressure) (kPa)		300
Maximum data acquisition rate (data s^{-1})		1 000

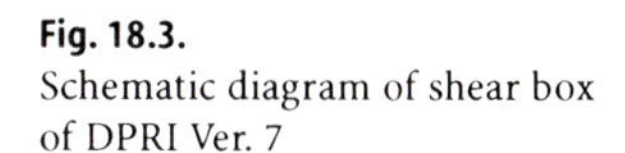

Fig. 18.3. Schematic diagram of shear box of DPRI Ver. 7

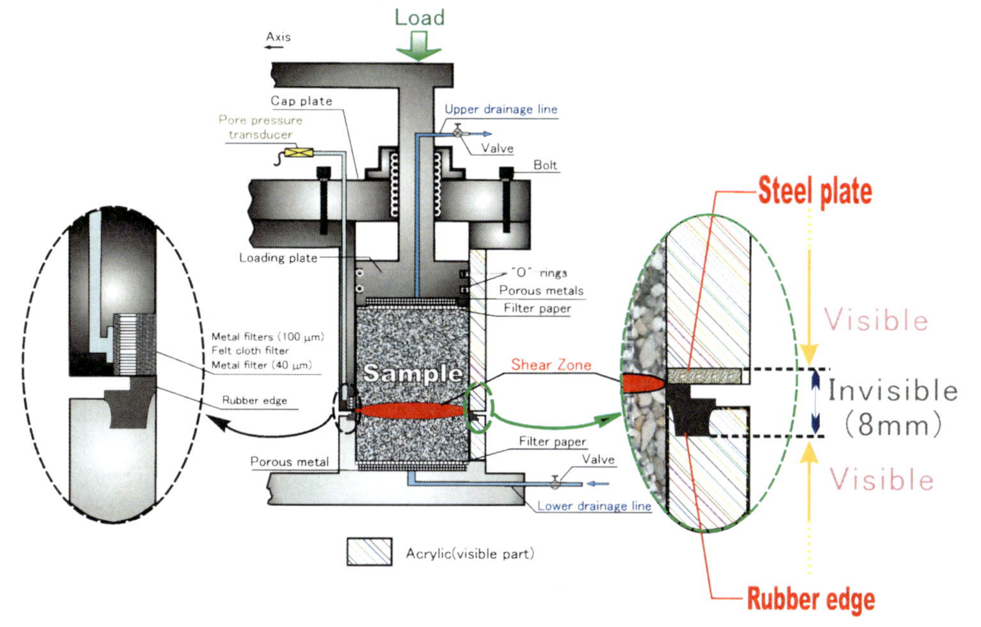

18.3 Video Image Acquisition System and PIV Analysis System to Obtain Velocity Distribution Profile

In order to acquire the images of the grains of the sample during shear, the authors used two digital video (DV) cameras and a CCD camera (Fig. 18.4). The CCD camera was installed with a high-speed image processing system, which is capable to acquire images of 640 × 480 pixels at every 1/60 s. One of the video cameras is located on the floor ("DV camera 1" in Fig. 18.4) and the another is fixed on the rotating table of the lower box ("DV camera 2" in Fig. 18.4). Those DV cameras take video images of 720 × 480 pixels (DV camera 1) and 320 × 240 pixels (DV camera 2) at every 1/30 s. The DV camera 1 takes the video images of the stable upper shear box and the DV camera 2 takes the video images of the rotating lower shear box. Use of the two video cameras enables to detect even small speed difference of sand grains near the shear plane in both of the upper and lower box. As for lighting of the upper and lower shear box for video shooting, a DC (direct current) light was used instead of AC (alternating current) light to avoid the effect of flickering of the AC light.

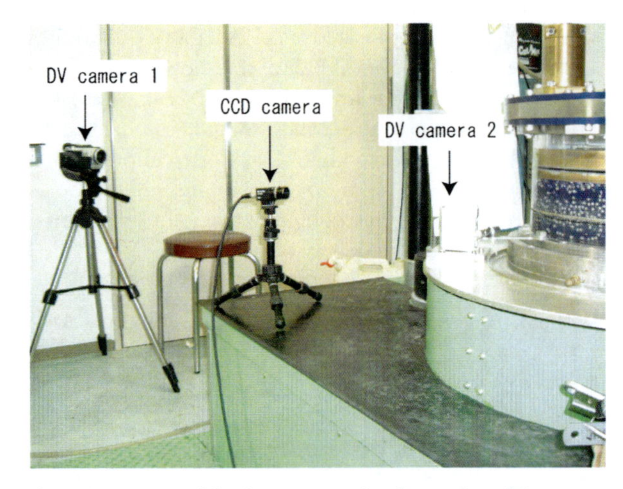

Fig. 18.4. Location of the three cameras for observation of shear zone during tests

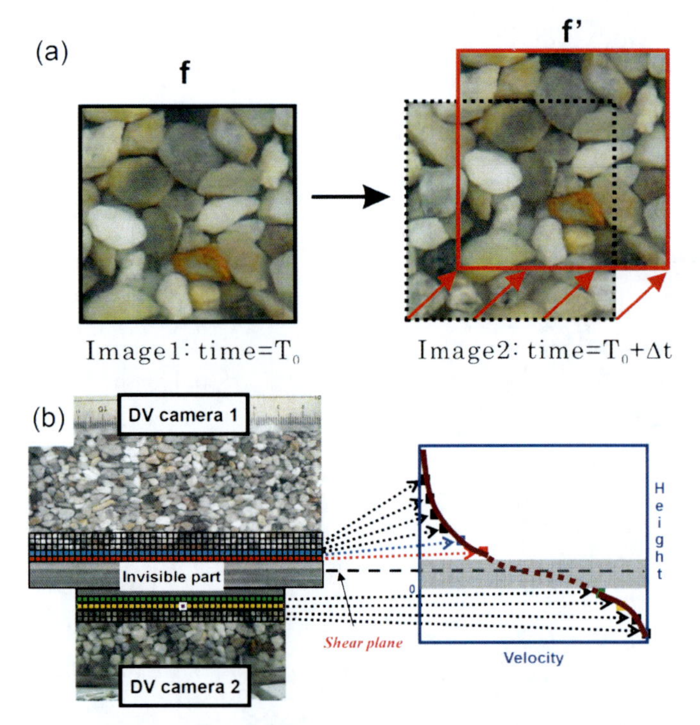

Fig. 18.5. Schematic illustration of pattern matching algorithm of PIV program and example of obtained velocity distribution

Particle Image Velocimetry (PIV) is a velocity measuring technique which was originally developed in the field of visualization of flows. PIV has two different analyzing algorithms of; (1) tracking marker particles, and (2) image pattern matching. Tracking marker particles algorithm was used for measuring velocity distribution in the flow of a fluid. The flow pattern was examined by mixing with marker particles and tracking the movement of the markers by image processing. Polystyrene balls or colored powder are often mixed with the fluid to provide identifiable texture on which the image processing can operate (Adrian 1991).

Image pattern matching was developed recently. The image pattern matching algorithm needs no marker particles. This algorithm is applicable for detecting movement of grains of sands because sands have their own texture in the form of different colors and brightness of grains.

The procedure of the image pattern matching algorithm is shown in Fig. 18.5. Whole image area is divided into smaller regions. This algorithm compares the pair of image segments (f) between the image 1 at time T_0 and the image 2 at time $(T_0 + \Delta T)$ by calculating the cross-correlation function and providing a displacement vector of a small segment (f) shown as red arrows. The PIV software repeats this procedure to extract distribution of vectors for selected image area in Fig. 18.5a.

Image processing was carried out in order to measure the velocity distribution within shear zone. Shear-zone video images, except invisible part, taken by DV cameras 1 and 2 are analyzed by PIV software to provide movement velocity of each smaller region. Then, the profile of particle velocity distribution $v(h)$ is represented by the brown curved line of Fig. 18.5 (b). Here, h is the height from the position of shear plane, at the top of rubber edges. At the same time, profile of velocity normalized by shear speed of the apparatus (V_s), $v(h) / V_s$ was also presented for each test of this study.

18.4 Physical Properties of Samples and Ring Shear Test Condition

In this study, Silica Sands no. 1 (S1) were used. These sands are industrial material for construction use. They are produced artificially by crushing of natural rocks and sieving. These sands consist of roughly round grains of which 92–98% are quartz and a little amount of feldspar. Basic properties of S1 sands are following: mean diameter of $D_{50} = 3.01$ mm, effective grain size of $D_{10} = 2.00$ mm, niformity coefficient of $U_c = 1.64$, and specific gravity of $G_s = 2.64$.

Test conditions are summarized in Table 18.2. A very low normal stress (22 kPa) was selected because visibility of grains was lost by muddy water formed by heavy grain crushing in the tests under high normal stress. The test was conducted under constant shear speed of 10 cm s^{-1} with acceleration of 2 cm s^{-2} and under drained condition.

Table 18.2. Test condition of the ring shear test

Normal stress (kPa)	Shear speed (cm s^{-1})	Acceleration (cm s^{-2})	Drainage	Initial void ratio
22	10	2	Drained	0.79

18.5 Velocity Distribution Profiles of Sand Grains

Figure 18.6 presents a series of photos showing the sample during the test. White and cloudy silty water was not obvious rather than the photos introduced in Sassa et al. (2004) of higher normal stress (200 kPa) test under undrained condition. Sassa et al. (2004) explained that when grain crushing created fine particles, pore water became cloudy and gradually diffused later. This test was conducted under 22 kPa and therefore did not cause grain crushing enough to make the sample cloudy.

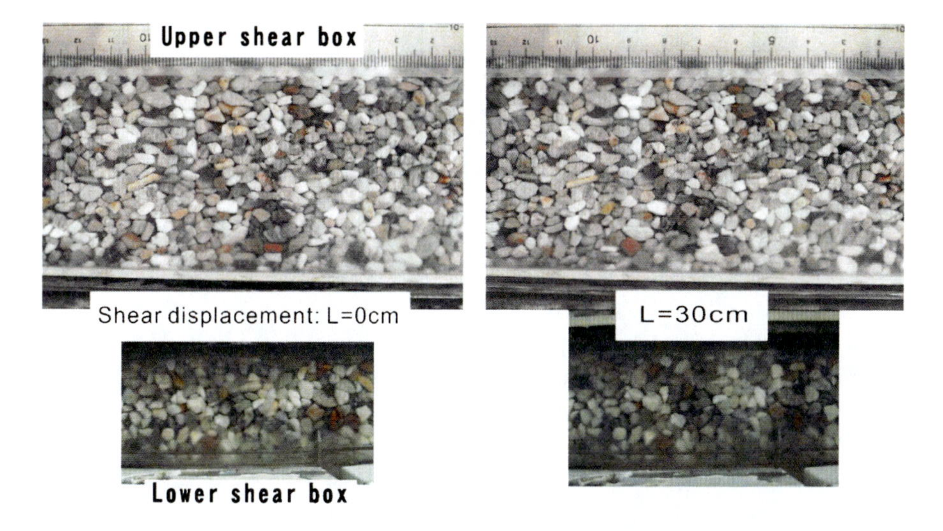

Fig. 18.6. A series of photos of the upper and lower shear box of the ring shear test

PIV analysis was carried out on video images taken during each test by 2 DV cameras and a CCD camera. In order to investigate the process of shear zone formation especially in accelerating stage, the development of velocity distribution profile within shear zone was observed and analyzed. Figure 18.7 shows the change of velocity distribution profile of the test. In this figure, horizontal axis is horizontal velocity of grains in cm s^{-1} and vertical axis is the height from top of rubber edge, i.e. shear plane at the boundary of upper and lower shear box. Velocity distribution profiles at seven stages of velocities of 1.0–10 cm s^{-1} are plotted in this figure. These colored lines shows the velocity distribution profile when shear speed reached at 1 cm s^{-1} (marked as A), 3 cm s^{-1} (B), 4 cm s^{-1} (C), 6 cm s^{-1} (D), 8 cm s^{-1} (E), 10 cm s^{-1} (F), and at the end of the test (at 10 cm s^{-1} and marked as G). A hatching and long bar that is notated as "invisible part" indicates the part invisible from outside between the heights between −0.6 to 0.2 cm. The (A)-(G) profiles are interpolated in the gray hatching bar and presented as broken lines.

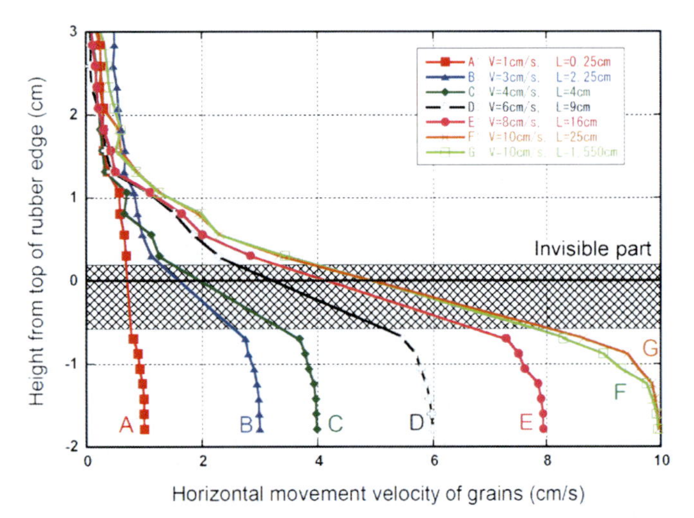

Fig. 18.7. Development process of the velocity distribution profiles in the ring shear test. *V:* Shear speed of the ring shear apparatus; *L:* shear displacement

Figure 18.8 shows the velocity distribution profile of the test normalized by shear speed of the apparatus (Vs). In Fig. 18.8, profiles of (C)-(G) are almost same and the normalized velocity changed from 0.4–0.8 in the thin invisible part. However, profiles of (A) and (B) are different from (C)-(G). Both profiles show relatively higher normalized velocity distribution in the upper shear box, and this means (A) and (B) profiles are relatively flow-like than (C)-(G), because normalized velocity indicates relative motion of grains inside sample.

Video movie around (A) and (B) showed grains of the upper shear box moved not horizontally, but small upheaving component is visible for a short moment. Velocity distribution of B could be interpreted as movement derived from dilatancy of sand grains in the initial stage of shearing. In order to examine distribution of dilatancy behavior around (B) in more detail, vertical upheaval movements of sand grains from shear displacement of 1.0 cm to (B) and to (C) were analyzed and the results are shown in Fig. 18.9. The shear displacement interval up to (B) is 1.0–2.25 cm and the profile is shown as blue plots and line. The shear displacement interval up to (C) is 1.0–4.0 cm and the profile is shown as red plots and line. In the two profiles, it is clear that the dilatancy at first takes place more intensively in and around shear surface and then this intense dilating zone diffuse to the upper zone. In other words, dilatancy began in the shear

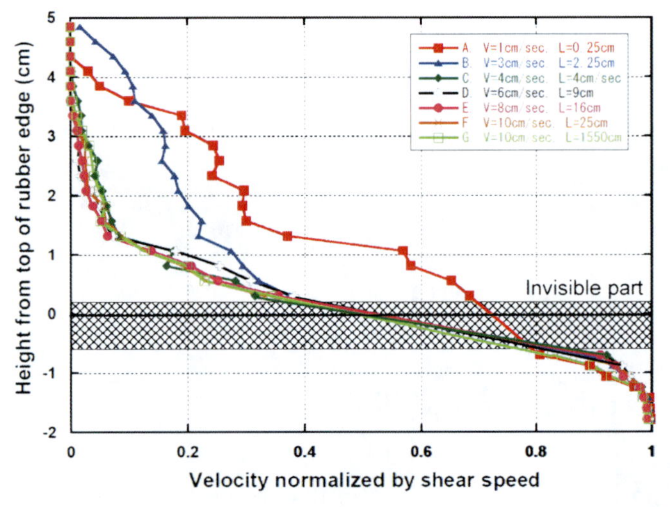

Fig. 18.8. Development process of the normalized-velocity distribution profiles in the ring shear test. *V:* Shear speed of the ring shear apparatus; *L:* shear displacement

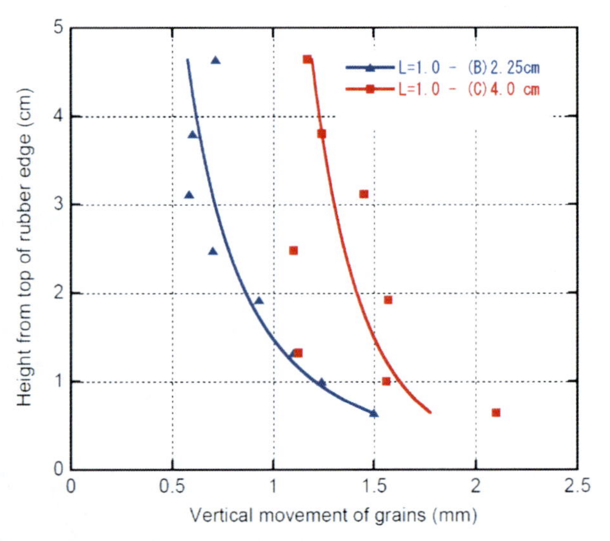

Fig. 18.9. Sample height versus vertical movement of grains obtained in the ring shear test. *L:* Shear displacement

zone and then to propagate upward. On the other hand, no such dilation was observed in the lower box because only sample in the upper box can dilate both positively and negatively as loading plate can move vertically, and also the sample in the lower box can not dilate downward.

18.6 Conclusions

By using a newly developed ring shear apparatus with a transparent shear box, a drained monotonic shear speed-controlled test was conducted on coarse-grained silica sands to study the process of shear-zone development in granular materials. With the help of CCD and DV cameras, the velocity distribution profiles of sample within the shear box during shearing was observed, and was analyzed by means of Particle Image Velocimetry (PIV) method. Obtained results are summarized as follows.

1. Authors succeeded to monitor the movement of sand grains within the upper and lower shear boxes simultaneously. The results revealed the existence of velocity distribution profiles even within the samples of the lower shear box.
2. The whole shear process could be divided into two significant steps in terms of velocity distribution: before and after the shear failure. Their characteristics were as follows:
   a In the initial stage of shearing, deformation due to dilatancy of sand structure within the major part of sand layer of upper shear box affected the velocity distribution profile. The reason might be that the sample is allowed to dilate upward or downward because loading plate can move vertically under drained condition.
   b After failure, velocity distribution changed from flow-like to slide-like mode. Flow-like mode is characterized by apparently thicker shear zone. Slide-like mode is characterized by localization of shear within a very thin shear zone, i.e., invisible zone of the apparatus.

Acknowledgments

Mr. Naohide Ishikawa, a past master course graduate student is acknowledged for conducting preliminary study for this research in our laboratory from 2002–2004. Authors appreciate deeply the members and students of the Research Centre on Landslides, Disaster Prevention Research Institute, Kyoto University for their assistance to this research.

This study is a part of the M101 Project, "Areal Prediction of Earthquake and Rain Induced Rapid and Long-travelling Flow Phenomen" (APERITIF, proposer: K. Sassa), of the International Programme on Landslides (IPL) operated by the International Consortium on Landslides (ICL) since 2002, and also is a part of the Japanese project "Aerial Prediction of Earthquake and Rain-Induced Flow Phenomena" (APERIF project, represented by K. Sassa), which was supported by the Special Coordinating Fund for Promoting Science and Technology of the Ministry of Education, Culture, Sports, Science and Technology (MEXT) of Japanese Government for 3 years from 2001–2004.

References

Adrian RJ (1991) Particle imaging techniques for experimental fluid mechanics. Annu Rev Fluid Mech 23:261–304

Bishop A, Green G, Garga V, Anderson A, Brown J (1971) A new ring shear apparatus and its application of the measurement of the residual strength. Geotechnique 21(4):273–328

Bromhead EN, Dixon N (1986) The field residual strength of London Clay and its correlation with laboratory measurements, especially ring shear tests. Geotechnique 36(3):449–452

Castro G (1969) Liquefaction of sands. Ph.D. dissertation, Harvard Soil Mechanics Series, no. 81. Harvard University, Cambridge, MA

Fukuoka H (1991) Variation of the friction angle of granular materials in the high-speed high-stress ring-shear apparatus, influence of re-orientation, alignment and crushing of grains during shear. Ann Disaster Prev Res I 41(4):243–279

Hungr O, Morgenstern NR (1984) High-velocity ring-shear tests on sand. Geotechnique 34(3):415–421

Ishikawa N (2004) An experimental study on the shear behavior of granular materials by means of a new ring shear apparatus with a transparent shear box and image processing. Master's thesis, Kyoto University

Lang Y, Ote K, Fukuoka H, Sassa K (1991) Image-processing the velocity distribution of particles in ring shear tests. In: Proceedings of the Conference of Japan Society of Erosion Control Engineering 1991, pp 302–305 (in Japanese)

Lupini JF, Skinner AE, Vaughan PR (1981) The drained residual strength of cohesive soils. Geotechnique 31(2):181–213

Sassa K (1992) Access to the dynamics of landslides during earthquakes by a new cyclic loading high-speed ring-shear apparatus (keynote paper). In: Proc. 6th International Symposium on Landslides, "Landslides," A.A. Balkema, Christchurch, 10–14 February, vol. 3, pp 1919–1937

Sassa K (1996) Prediction of earthquake induced landslides. Special lecture for 7th International symposium on landslides, Landslides, Balkema, vol. 1, pp 115–132

Sassa K (2000) Mechanism of flows in granular soils. In: Proc. GeoEng2000, vol. 1, pp 1671–1702

Sassa K, Fukuoka H, Scarascia-Mugnozza G, Evans SG (1996) Earthquake-induced-landslides: distribution, motion and mechanisms. Special Issue for the great Hanshin Earthquake Disasters, Soil and Foundations, pp 53–64

Sassa K, Wang G, Fukoka H (2003) Performing undrained shear tests on saturated sands in a new intelligent type of ring shear apparatus. Geotech Test J 26(3):257–265

Sassa K, Fukuoka H, Wang GH, Ishikawa N (2004) Undrained dynamic-loading ring-shear apparatus and its application to landslide dynamics. Landslides 1(1):7–19

Tika TM (1989) The effect of rate of shear on the residual strength of soil. Ph.D. Dissertation, University of London (Imperial college of Science and Technology)

Varnes DJ (1978) Slope movement types and processes. In: Schuster RL, Krizek RJ (eds) Landslides: analysis and control. Transportation Research Board National Academy of Sciences, Special Report 176, pp 11–33

Wafid MA, Sassa K, Fukuoka H, Wang GH (2004) Evolution of shear-zone structure in undrained ring-shear tests. Landslides 1(2):101–112

Wang GH, Sassa K (2002) Post mobility of saturated sands in undrained load-controlled ring shear tests. Can Geotech J 39(4):821–837

Seismic Behavior of Saturated Sandy Soils: Case Study for the May 2003 Tsukidate Landslide in Japan

Gonghui Wang* · Kyoji Sassa · Hiroshi Fukuoka

Abstract. During the 2003 Sanriku-Minami earthquake, a flowslide was triggered on a gentle slope of about 13 degrees in Japan. The displaced landslide mass deposited on a horizontal rice paddy after traveled approximately 130 m. Field investigation revealed that the landslide was originated from a fill slope, where a gully was buried for cultivation some decades ago, and shallow ground water exists. To investigate the trigger and movement mechanism of this landslide, a series of seismic simulating tests was performed on the sample taken from the source area, by using a newly developed ring shear apparatus. The seismic loadings were synthesized using the seismic records at the same earthquake. The undrained and partially drained tests revealed that shear failure and post-failure landsliding could be resulted due to the seismic loading, given the existence of saturated soil layer above the shear zone in certain thickness.

Keywords. Flowslide, earthquake, seismic loading, ring shear simulating test

19.1 Introduction

An earthquake with a moment magnitude of 7.0 occurred in northern Japan on 26 May 2003 (called as 2003 Sanriku-Minami earthquake in Japan), during which a massive landslide was trigged on Tsukidate area, northwest of Sendai, the capital city of Miyagi Prefecture (see Fig. 19.1, thereinafter we call this landslide as Tsukidate landslide). Tsukidate landslide partially destroyed two houses. Two people were buried almost to their necks by the displaced landslide mass. Fortunately, they had been able to flee from the mud easily on their own without any injury. This landslide was triggered on a gentle slope with the inclination of the sliding surface being approximately 13.5 degrees. The displaced landslide mass traveled a distance of about 130 m, spread and deposited on a horizontal rice paddy, showing some typical characteristics of rapid long runout flow phenomenon. Although landslides triggered by earthquake had been reported widely (Seed 1968, 1979; Ishihara et al. 1990; Sassa et al. 1996; among others), this landslide triggered by the earthquake without rainfall attracted great attention of the researchers of various fields. Understanding on the mechanism for this kind of rapid long runout landslides occurred on gentle slopes is of great importance to landslide hazard analysis and mitigation.

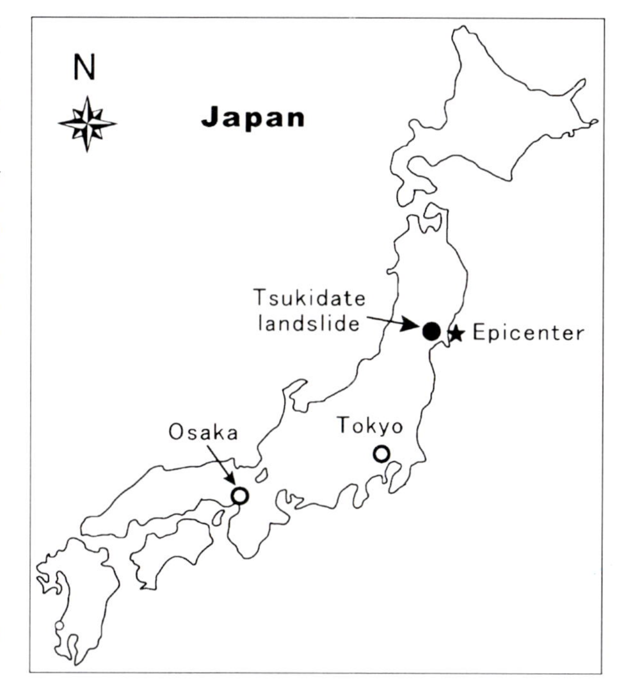

Fig. 19.1. Location of Tsukidate landslide as well as the earthquake epicenter

Field investigation revealed that the landside was originated from the poorly compacted filling of an old valley with abundant standing groundwater. By using a newly developed undrained ring shear apparatus and the real seismic waves recorded during the same earthquake, the seismic response of the soil from the landslide source area during earthquake was examined. Some preliminary results are presented in this paper.

19.2 Tsukidate Landslide

Figure 19.2 presents aerial view of the Tsukidate landslide (taken by Kokusai Kogyo Co., Ltd.). It is seen from this photo that the road on the right side is almost parallel to the landslide, indicating the nature of gentle slope. Figure 19.3 is the view from the toe of the landslide, which was taken four days after the event (at our survey time).

Fig. 19.2.
Oblique side view of the landslide (photo courtesy of Kokusai Kogyo Co., Ltd.)

Fig. 19.3.
View from the toe of the Tsukidate landslide

As shown in Figs. 19.2–19.3, the landslide mass slid out of the source area, traveled a long distance, and finally spread and deposited on the rice paddy below the slope. The source area was covered by blue sheet for the prevention of possible second disaster during rainfall. The bamboos originally on the source area were transported with the landslide mass; however, most of them were standing almost vertically on the rice paddy after a long traveling distance, with thin soil layer of landslide mass below them. From the deposition area on the horizontal rice paddy, it is inferred that the displaced landslide mass had great energy (great velocity) when slipped out of the gentle slop.

A detailed contour map of the Tsukidate landslide area after the landsliding is given in Fig. 19.4 (courtesy of Aero Asahi Corporation, Japan). This map with a contour interval of 50 cm was obtained by means of airborne laser scanner after the landslide event. The landslide area is outlined by a dashed line. A longitudinal section along the line A-A' in Fig. 19.4 is given in Fig. 19.5. The landslide descended 27 m over a horizontal distance of 180 m (from the source area to the toe of the landslide on the paddy). The original ground surface was inferred according to the shape of the neighboring ground surface. It is seen from Fig. 19.5 that this landslide originated from a gentle sliding surface with an inclination of about 13.5 degrees and moved with an average apparent friction angle (φ_a) of about 7.3 degrees (from the top to the toe). The displaced landslide mass was estimated as approximately

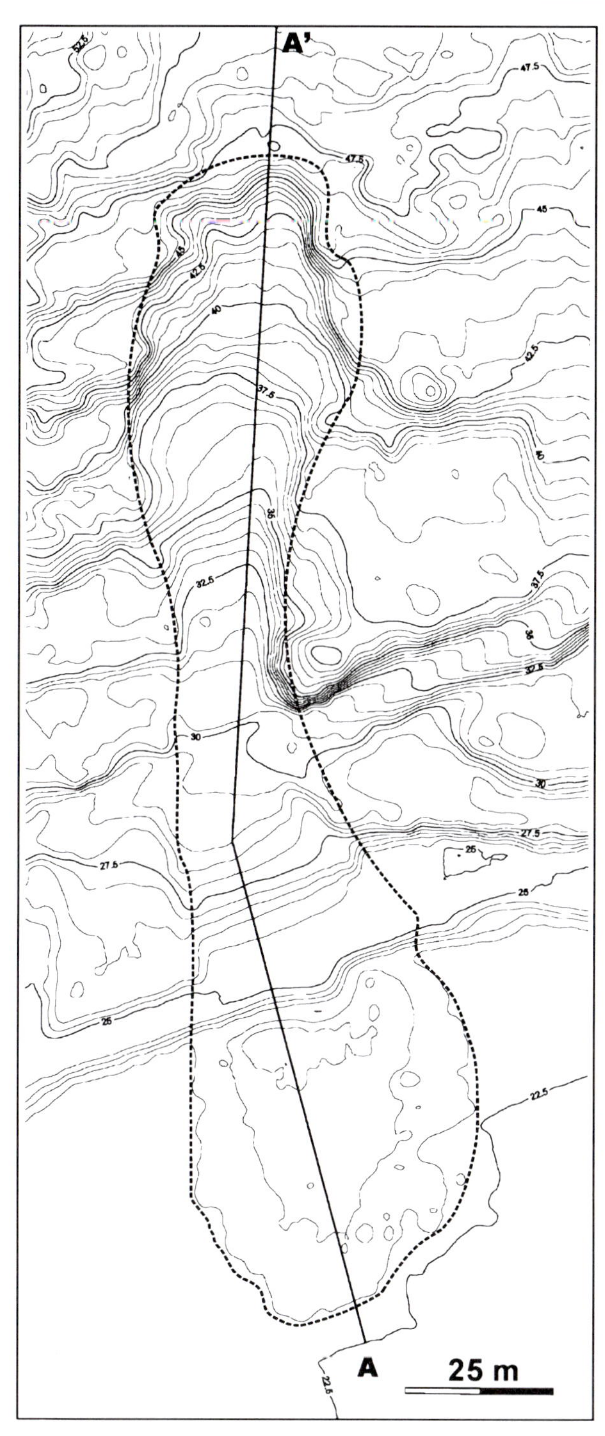

Fig. 19.4. Contour map of the Tsukidate landslide area after the earthquake obtained by airborne laser scanner The Contour interval is 50 cm, and the *red line* shows the landslide area (courtesy of Aero Asahi Corporation, Japan)

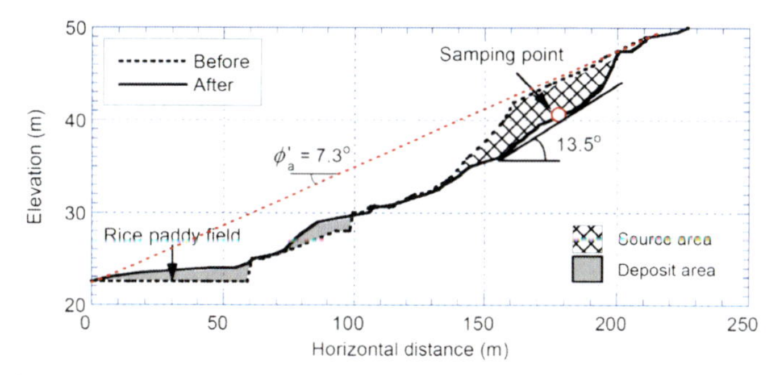

Fig. 19.5. Cross section of the Tsukidate landslide along line *A-A'* in Fig. 19.4. Dashed line shows the section before landslide, which was estimated from topographic map

Fig. 19.6. Sampling on the source area

5 000 m^3. It is worth noting here that φ_a is usually used as a parameter representing the mobility of a landslide (Hsü 1975; Okuda 1984; Sassa 2000; Legros 2002). Lower value of φ_a indicates higher mobility of landslide. Because this landslide was triggered in a very gentle slope with long travel distance, its triggering and traveling mechanisms attracted wide attention from both the geo-researchers and the disaster management organizations.

Comparing the present contour map of this area to that of about 40 years ago, it is made clear that the source area of this landslide was located on the fillings, where a gully was buried for cultivation (Fukuoka et al. 2004). To examine the triggering and traveling mechanism of the landslide, soil samples were taken from the filling on the source area, which was the main materials involving in movement, and ensured to have been liquefied. This sample was mainly composed of pyroclastic deposits. Figure 19.6 presents a photo showing the sampling site (on the left side of the landslide). As was revealed clearly inside the dashed line, the pyroclastic deposits changed its color from brown (due to oxidization) to blue-gray (due to deoxidization) (inside the dashed line). This means that the iron contained in the soil was not oxidized due to the long-term existence of ground water. Therefore, it was concluded that abundant ground water existed within the source area. In-situ penetration test results also revealed that the fillings composed of pyroclastic deposits were loosely compacted (Fukuoka et al. 2004). Because it is obvious that liquefaction phenomenon was the main reason for this long runout landslide, a series of ring-shear based seismic simulating tests was performed on the soils from the source area of the landslide, in attempt to examine the seismic response of the filling soils during the earthquake, and then clarify the possible condition leading to the catastrophic failure of the landslide. Recorded real seismic waves were used in the analysis of possible seismic loadings on the sliding surface during earthquake.

19.3 Properties of the Sample

Because the landslide originated from the loose filling, samples were taken from the filling left on the source. Field survey revealed that although this landslide involved in the movement of different type of soils, which distributed on different area of the landslide, the liquefaction failure was mainly triggered on those volcanic deposits. Therefore, the ring shear tests presented in this paper were mainly performed on those samples of volcanic deposits.

The grain size distribution of the sample of volcanic deposits is presented in Fig. 19.7. The mean grain (D_{50}) is 0.144 mm, effective grain size (D_{10}) is 0.0045 mm, and uniformity coefficient is (0.224 / 0.0045) 49.8. Due to a size limitation imposed by the shear box of the employed ring shear apparatus, gravels greater than 4.75 mm were removed by sieving before testing, assuming that the undrained shear behavior of these samples be mainly controlled by the matrix material when the gravel content is less than a certain percentage, say 40% (Kuenza et al. 2004). The sample had a specific gravity of 2.48 and a dry density of 10.2 kN m^{-3} in the field. Therefore, the void ratio for the sample in the field was calculated as 1.38. The permeability of the sample with the grains greater than 4.75 mm being moved is plotted against the void ratio in Fig. 19.7b.

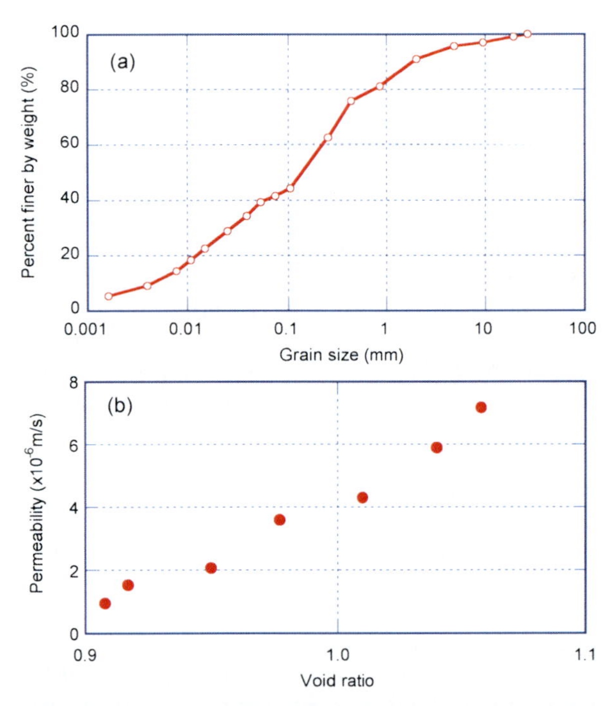

Fig. 19.7. Properties of samples: **a** grain-size distribution of sample from the source area; **b** variation of permeability of the tested sample against void ratio

19.4 Ring-Shear Apparatus and Test Method

In the present research, a newly developed ring-shear apparatus, DPRI-7 (Sassa et al. 2004), was employed. DPRI-7 poses a maximum shear speed of 3.0 m s^{-1}, and has a shear box sized 270 mm in inner diameter, and 350 mm in outer diameter. Detailed explanation on the design and construction principles of this ring-shear apparatus, as well as the testing method, could be obtained from Sassa (1998, 2000), Sassa et al. (2003, 2004), and Wang and Sassa (2002).

In all the tests except from that one presented in Fig. 19.13, the oven-dried sample was first placed in the shear box by means of dry deposition (Ishihara 1993), and then was saturated with the aid of carbon dioxide and de-aired water. After the sample was fully saturated and consolidated, seismic loadings were applied to the sample in undrained or partially drained conditions. To observe pore-pressure generation accompanying the shear displacement, samples were sheared to steady state, i.e., pore pressure and shear resistance remained constant throughout the test.

Because the tests aim to examine the seismic response of the soil on the source area, here a soil element on the sliding surface with the overlain soil layer being about 5 m thick was selected as the examining target. In calculating the acting stresses (normal and shear stresses) on the sliding surface, an average slope angle is measured approximately as 13.5 degrees from Fig. 19.5, and a saturated unit weight (γ_{sat}) for the soil mass in the field was calculated as 15.9 kN m^{-3}. Therefore, the initial normal stress (σ_i) is calculated approximately as 75.1 kPa, and initial shear stress (τ_i) 18.0 kPa. All the tests were originated from this initial stress state.

To examine whether shear failure could be triggered due to the introduction of seismic loadings during the earthquake, a series of seismic simulating shear tests was performed by using the recorded seismic waves during the earthquake. According to the records of High Sensitivity Seismograph Network Japan (Hi-net), which has been operated by the National Research Institute for Earth Science and Disaster Prevention (NIED), Japan, the maximum acceleration (recorded by K-NET MYG011 station) exceeded 1 000 gal (1 g) in the vicinity of the epicenter. Figure 19.8 presents the recorded seismic wave. There is also a seismic observatory station in Tsukidate area, which is very close to the Tsukidate landslide area. It is ideal and desirable to use the seismic records from this station. However, this station unfortunately failed to operate properly during the earthquake. Because Tsukidate area is very close to the epicenter (see Fig. 19.1), we used the seismic wave shown in Fig. 19.8 in the following simulating tests.

The main aim of this study is to examine the seismic response of the filling soils during the earthquake as well as the possible condition necessary for the long runout movement, an ideal 13.5 degrees infinite slope was assumed as shown in Fig. 19.9, and the soil layer is 5 m thick. The possible seismic loading acting on the sliding surface during the earthquake was synthesized. During the synthesis, considering that both the K-NET MYG011 station and the landslide site are very close to the epicenter, the possible attenuations in both the horizontal and vertical peak accelerations with the distance to the epicenter (Fukushima and Tanaka 1990; Ambraseys and Bommer 1991) were not taken into account. Meanwhile, considering that K-NET MYG011 station is located on the top of a slope, the possible enlargement of waves when the waves transfer from hard rock to weak soil layer (Fukushima and Tanaka 1990) was not accounted.

The seismic loadings on the potential sliding surface could be obtained following the steps shown in Fig. 19.10. The horizontal acceleration ($a_{HR(t)}$) during earthquake along the traverse line (Fig. 19.10a) is

$$a_{HR(t)} = a_{EW(t)} \times \cos\beta + a_{NS(t)} \times \sin\beta \tag{19.1}$$

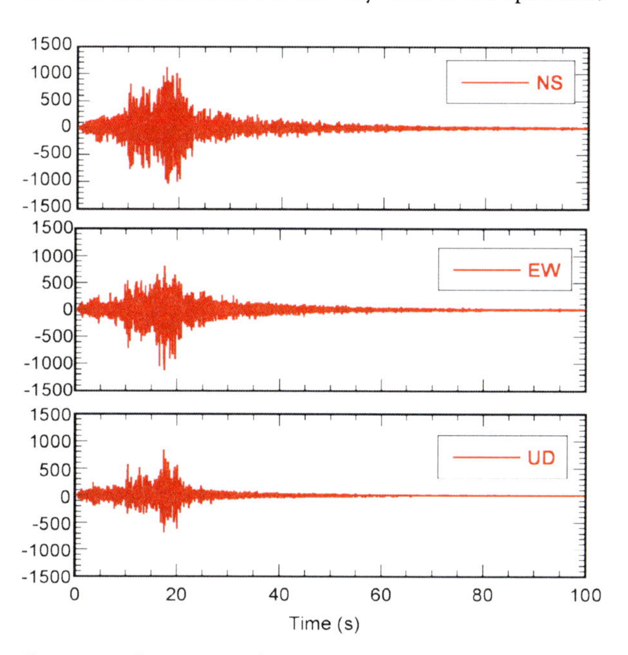

Fig. 19.8. Acceleration time history recorded during the earthquake (K-NET MYG011-1111gal)

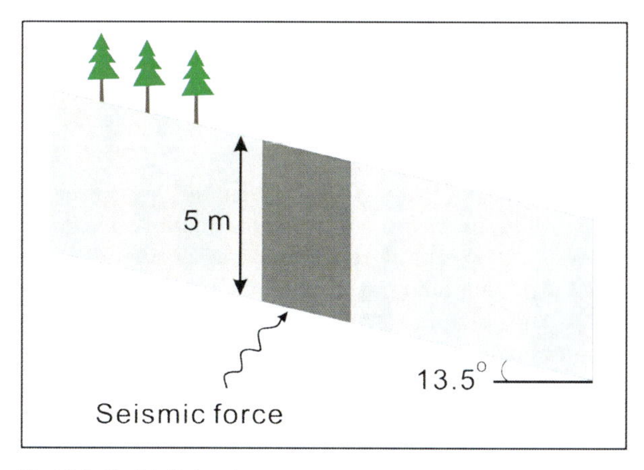

Fig. 19.9. Ideal infinite slope

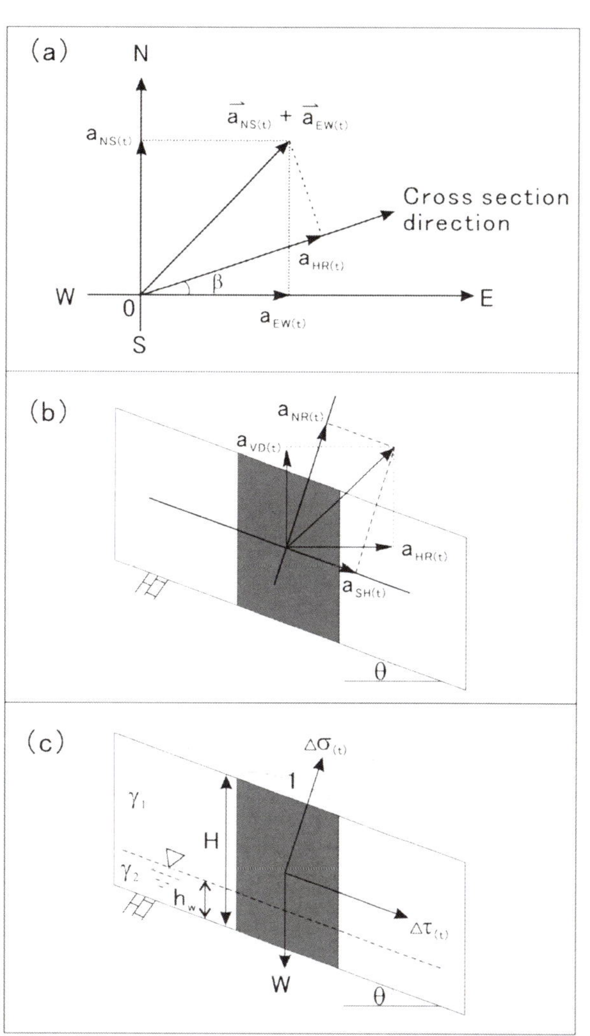

Fig. 19.10. Synthesis of the seismic loadings acting on the potential sliding surface. **a** Projection of the horizontal seismic accelerations along the cross section direction. **b** Projection of the horizontal and vertical seismic accelerations along the vertical and downslope directions. **c** Seismic loadings due to the seismic accelerations acting on the potential sliding surface

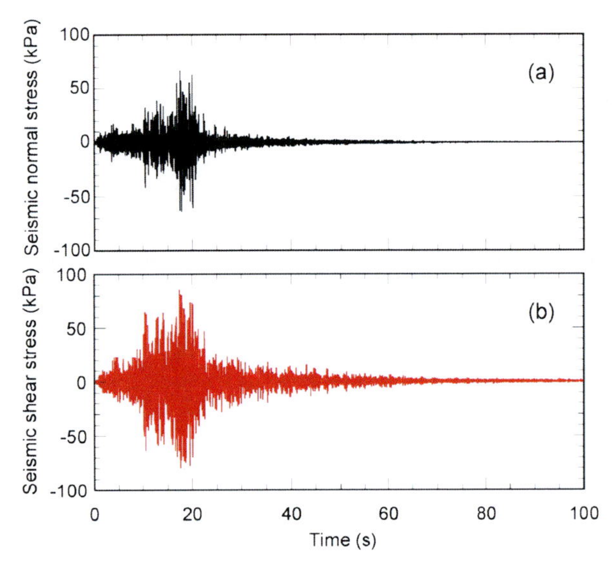

Fig. 19.11. Synthesized seismic loadings on the sliding surface

The accelerations parallel ($a_{SH(t)}$) and perpendicular ($a_{NR(t)}$) to the potential sliding surface (Fig. 19.8b) are

$$a_{SH(t)} = a_{HR(t)} \times \cos\theta - a_{VD(t)} \times \sin\theta \tag{19.2}$$

$$a_{NR(t)} = a_{HR(t)} \times \sin\theta + a_{VD(t)} \times \cos\theta \tag{19.3}$$

Therefore, the seismic normal stress ($\Delta\sigma = m \times a_{NR(t)}$) and shear stress ($\Delta\tau = m \times a_{SH(t)}$) on the potential sliding surface were computed. Figure 19.11 shows the calculated seismic normal stress and shear stress on the sliding surface during earthquake.

19.5 Test Results and Discussions

19.5.1 Undrained Seismic Simulating Test

Seismic simulating test was first performed on saturated specimen under undrained condition. After the sample was saturated and normally consolidated, the shear box was switched to undrained condition; thereafter the seismic loadings shown in Fig. 19.11 were applied. The test results are presented in Fig. 19.12. It is seen from Fig. 19.12 that the monitored normal stress is almost the same as the control signal shown in Fig. 19.11a. The monitored value for shear resistance shows the applied shear stress before failure; but after failure, it presents the mobilized shear resistance, because the target value for modified shear stress could not be applied to the sample completely. Hence, the shear strength in Fig. 19.12 behaved differently from the control signal in Fig. 19.11b. During the main shock, the excess pore-water pressure increased continuously, and the shear strength showed certain of reduction. Shear failure resulted, and shear displacement increased. After the seismic loadings ceased, shear displacement accelerated, while pore pressure elevated continuously and tended to a constant.

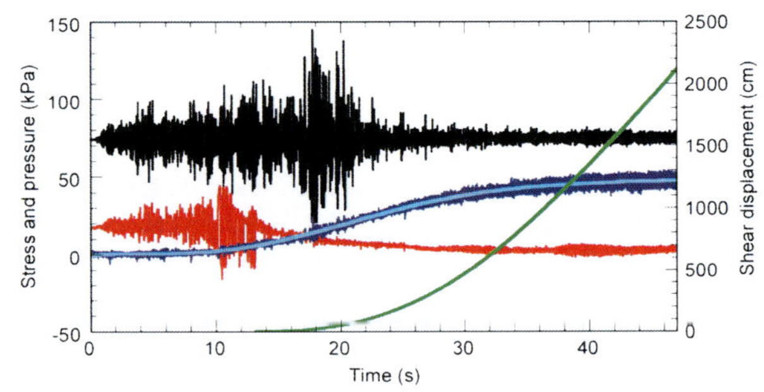

Fig. 19.12. Seismic simulating test on saturated sample under undrained condition ($B_D = 0.97$, $e = 1.10$)

19.5.2 Partially Undrained Seismic Simulating Test

In the analysis of liquefaction triggered by earthquake, rainfall, or some other factors, it is deemed that the loadings are applied in such a short time that the generated pore-water pressure could not dissipate immediately, such that almost all the laboratory works have been trying to mimic the soil behavior in undrained load cells. In fact, soils in a natural slope cannot be truly undrained, especially in the post-failure process of sliding (for landslides), where the pore-water pressure could be dissipated with progress of shear displacement. In this sense, partially drained shear tests may be more close to the practical situation. Therefore, partially drained seismic loading tests were also conducted. Considering that the overlaid soil layer above the sliding surface was 5 m thick for the natural slope, while the possible thickness of the soil layer above the shear surface in ring shear apparatus was less than 5 cm after consolidation, a fluid with its viscosity being 25 times that of water was used as the porous fluid in attempt to mimic the dissipation rate of generated pore-water pressure. During test, the specimen was made into slurry with this fluid and then set into the shear box. After normally consolidated, the specimen was seismically loaded, keeping the drainage valve in the top of the shear box open, while the drainage valve in the bottom of the shear box remained closed.

The results of the partially drained test are presented in Fig. 19.13, in the form of time series data. As indicated in Fig. 19.13, shear failure was initiated during the main shock. After the main shock, although the shear resistance did not show significant reduction, the shear displacement increased continuously, showing an accelerating failure process.

It is inferred that high pore-water pressure was built up within the shear zone as the combined result of the generation and dissipation. The dissipation of generated pore-

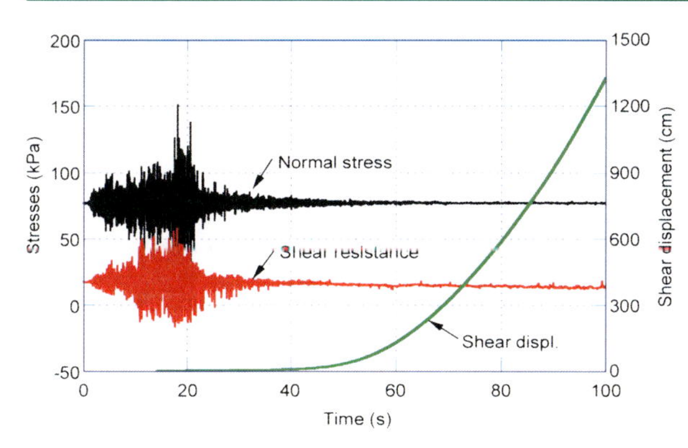

Fig. 19.13. Seismic simulating test on SM25-saturated sample under partially drained condition ($B_D = 0.96$, $e = 1.13$)

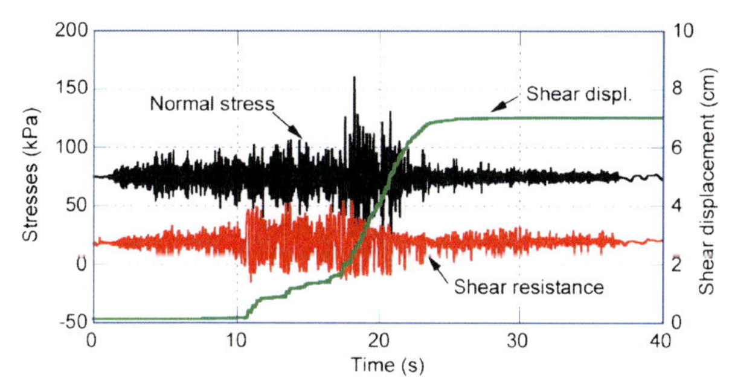

Fig. 19.14. Seismic simulating test on water-saturated sample under partially drained condition ($B_D = 0.97$, $e = 1.11$)

water pressure from the shear zone depends on the permeability of the soil itself and the drainage path, i.e., the thickness of the soil layer above the shear zone. When the generation rate is greater than the dissipation rate, excess pore-water pressure could be built up within the shear zone.

From Fig. 19.13, it is referred that dissipation of generated pore-water pressure from the shear zone is key to the after-main-shock movement of displaced landslide mass. To examine the role of possible dissipation of pore-water pressure, seismic simulating shear test was also performed on water-saturated sample under partially drained condition. The sample preparation procedures and test method were the same as those of the test shown in Fig. 19.12. The test results are presented in Fig. 19.14. As shown, shear failure was triggered due to introduction of seismic loadings. However, after the main shock, the shear movement ceased, indicting that no further shearing was continued, i.e., no long runout landslide was triggered. Therefore, from Fig. 19.12 and Fig. 19.14, it is inferred that long runout landsliding could be triggered in the case that the saturated soil layer above the sliding surface is thick enough, say 1 m. It can also be concluded that this kind of long runout landslides could be prevented by effective drainage works.

19.6 Summary and Conclusions

1. The Tsukidate landslide was originated on a gentle slope of about 13.5 degrees, and was a typical long runout flow phenomenon triggered by earthquake. The existence of the loosely compacted fills and abundant standing groundwater facilitated the liquefaction failure of landslide mass.
2. By using real seismic waves, the possible seismic loadings on the sliding surface due to the seismic excitation during earthquake were synthesized. Through a newly developed ring shear apparatus, the seismic loadings were applied successfully to the soils on the sliding surface to mimic the seismic response of soil during earthquake.
3. Undrained seismic simulating test revealed that due to the seismic loading, a certain amount of excess pore-water pressure was built up within the saturated sliding surface, and then led to the failure of the slope. After failure, high excess pore-water pressure was generated with increase of shear displacement. This finally resulted in great reduction in the shear resistance and rapid movement.
4. Dissipation of generated pore pressure from the shear zone plays key role on the movement of the displaced landslide mass. As revealed by those partially drained seismic simulating tests, the soil layer above the sliding surface having great thickness (long pore-water dissipation path) could suffer failure with long runout movement; whereas the soil layer with short dissipation path stopped its movement immediately after the seismic loadings ceased. Therefore, lowering the ground water table to shorten the dissipation path of generated pore-water pressure during earthquake, or elevating the drainage capacity will be a workable method to prevent this kind of flowslides like Tsukidate landslide.

References

Ambraseys NN, Bommer JJ (1991) The attenuation of ground accelerations in Europe. Earthquake Eng Struc 20:1179–1202

Fukuoka H, Wang G, Sassa K, Wang FW, Matsumoto T (2004) Earthquake-induced rapid long-traveling flow phenomenon: May 2003 Tsukidate landslide in Japan. Landslides: Journal of the International Consortium on Landslides 1(2):151–155

Fukushima Y, Tanaka T (1990) A new attenuation relation for peak horizontal acceleration of strong earthquake ground motion in Japan. Bull Seismol Soc Am 80(4):757–783

Hsü KJ (1975) Catastrophic debris streams (Sturzstroms) generated by rockfalls. Geol Soc Am Bull 86:129–140

Ishihara K (1993) Liquefaction and flow failure during earthquakes. Géotechnique 43(3):349–451

Ishihara K, Okusa S, Oyagi N, Ischuk A (1990) Liquefaction-induced flowslide in the collapsible loess deposit in Soviet Tajik. Soils and Foundations 30(4):73–89

Kuenza K, Towhata I, Orense RP, Wassan TH (2004) Undrained torsional shear tests on gravelly soils. Landslides 1(3):185–194

Legros F (2002) The mobility of long-runout landslides. Eng Geol 63:301–331

Okuda S (1984) Features of debris deposits of large slope failures investigated from historical records. Annual report of Disaster Prevention Research Institute, Kyoto University, no. 27 B-1, pp 353–368 (in Japanese with English abstract)

Sassa K (1998) Mechanisms of landslide triggered debris flows. Keynote Lecture for the IUFRO (International Union of Forestry Research Organization) Division 8 Conference, Environmental Forest Science, Forestry Science, Kluwer Academic Publishers, vol. 54, pp 499–518

Sassa K (2000) Mechanism of flows in granular soils. Invited paper. Proc. GeoEng2000, Melbourne, vol. 1, pp 1671–1702

Sassa K, Fukuoka H, Scarascia-Mugnozza G, Evans S (1996) Earthquake-induced landslides; distribution, motion and mechanisms. Special Issue of Soils and Foundations, pp 53–64

Sassa K, Wang G, Fukuoka H (2003) Performing undrained shear tests on saturated sands in a new intelligent type of ring shear apparatus. Geotech Test J 26(3):257–265

Sassa K, Fukuoka H, Wang G, Ishikawa N (2004) Undrained dynamic-loading ring-shear apparatus and application for landslide dynamics. Landslides 1(1):1–13

Seed HB (1968) Landslides during earthquake due to soil liquefaction. Journal Soil Mechanics Foundations Division, ASCE 94(5):1055–1122

Seed HB (1979) Soil liquefaction and cyclic mobility evaluation for level ground during earthquakes. J Geotech Eng-ASCE 105:201–255

Wang G, Sassa K (2002) Post-failure mobility of saturated sands in undrained load-controlled ring shear tests. Can Geotech J 39: 821–837

Chapter 20

Chemical Weathering and the Occurrence of Large-Scale Landslides in the Hime River Basin, Central Japan

Naoki Watanabe* · Naoshi Yonekura · Wataru Sagara · Ould Elemine Cheibany · Hideaki Marui · Gen Furuya

Abstract. The Hime River Basin is located in the northern part of Central Japan and is known as one of the areas where both erosional potential and sediment yield are extremely high in Japan. Landslides and debris flows triggered mainly by heavy rainfalls have frequently occurred in the basin. We have estimated the chemical weathering rates for nineteen watersheds in the Shirouma-Oike Volcano located in the western part of the basin. These rates have been simply estimated by the mass balance equation between solute fluxes of stream waters from each watershed and solute loss comparing fresh and weathered volcanic rocks and were calculated to be ranging from 0.15 to 3.24 mm yr^{-1}. A watershed showing the highest rate of chemical weathering and solute flux corresponded to the area where the large-scale landslide occurred in 1911 and debris flows and landslides have continually occurred until now. Unstable sediments yielded by chemical weathering are thought to be an important factor of sediment disaster occurrences in the research area. Solute fluxes of each stream could be useful for susceptibility mappings of landslides and debris flows in each watershed.

A cause of the high chemical weathering rate is the leaching of soluble elements from fresh bedrocks with sulfuric acid produced by the oxidation of pyrite in altered rocks by previous hydrothermal activities. Stream waters from the altered zones are characterized by high SO_4/Cl ratio. Such a simple hydrochemical signature could also be useful for detection of hydrothermally altered zones covered with vegetation and thick soil layers.

Keywords. Chemical weathering, sediment yield, landslide, debris flow, hydrochemistry, the Hime River Basin

20.1 Introduction

The combination of intensive rainfall and weathered materials on slope can lead to high landslide frequencies and the removal of sediments by fluvial erosion and debris flows. The removal of soils and sediments by landslides allows for the production of new weathered materials from exposed bedrock during physical and chemical weathering processes. Several investigators have recently reported that the chemical weathering rate is directly related to the rates of mechanical erosion in watersheds (e.g. Louvat and Allègre 1997; Gaillardet et al. 1999; Millot et al. 2002; Lyons et al. 2005). In similar climatic and geologic condition, high landslide frequencies or erosional potential are generally connected with high weathering rates and sediment yields.

Landslides and debris flows have frequently occurred in the Hime River Basin, Central Japan. Numerous studies on sediment disasters have been reported (e.g., Sato et al. 1991; Marui and Sato 1995; Marui and Watanabe 2001; Marui et al. 1997a,b). The basin is known as an area of the highest erosion yields in Japan. Although the basin is situated in the similar climatic, geomorphologic and geologic condition, the landslide frequencies or erosional potential are considerably different in each watershed.

Here we focus on chemical fluxes in stream waters from watersheds and attempt to estimate the chemical weathering rate for 19 watersheds in the basin. These estimated data from each watershed allow us to assess the potential of landslide or debris flow occurrences in each individual watershed. In addition, hydrochemical signatures allow us to detect the buried hydrothermally altered zones which are closely related to landslide occurrences.

20.2 Outline of Research Area

The Hime River Basin in the northern part of Central Japan (Fig. 20.1) is located along an N-S trend of the Itoigawa-Shizuoka Tectonic Line (ISTL). There is the Shirouma-Oike Volcano located in the west side of the basin. The volcano is composed of andesite lavas, pyroclastic and mud flow deposits, and their colluvial deposits. This volcanic activity is divided into two stages. The early stage volcanism occurred between 0.8 and 0.5 million years ago and the late stage volcanism is younger than at least 0.2 million years ago (Nakano et al. 2001). There are a few fumaroles in the western part of the volcano and several weakly altered zones during previous hydrothermal activities. The highest peak of the volcano is 2 489 m in altitude. The volume of the present volcanic products is estimated to be approximately 9.2 km^3. The restored volume including the loss by erosion is thought to be approximately 15 km^3. A volume of sediment has

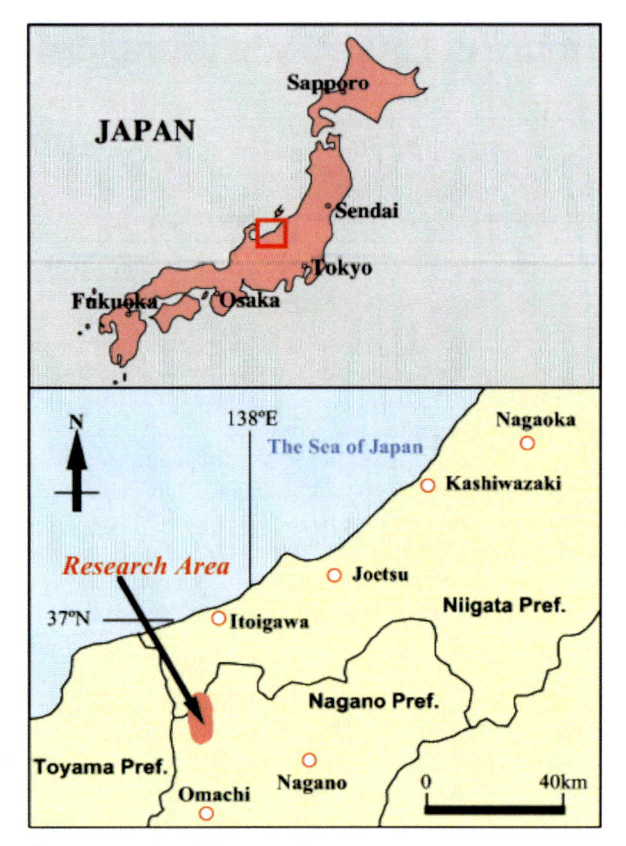

Fig. 20.1. Location of the research area

been yielded in watersheds of the volcano by weathering and erosion processes. The volcano and its surroundings, as our research area, have received extremely heavy snowfall with the maximum snow depth ranging from 150 cm around the riverbed to >400 cm in high elevation area. The annual precipitation varies from 2 000 to 3 000 mm. Such a large amount of precipitation accelerates the mechanical erosion and deepens the valley. Intensive rainfall and rapid snow melting are the main triggering factors of the frequent landslides and debris flows in the area.

In 1911–1912, three large-scale landslides associated with debris avalanches and subsequent debris flows occurred in the Hieda-yama area in the southeastern part of the volcano, and killed 23 persons, and damaged many houses along the Hime River. Landslides and debris flows have continually occurred in the Hieda-yama area since the above large-scale landslides. In July 1995, a seasonal rain front was activated and brought heavy rainfall of approximate 400 mm d^{-1} in the area. The rainfall induced serious sediment disasters in which houses, roads, and railway were damaged by landslides, debris flows and fluvial erosions along the Hime River. Debris flows particularly occurred in most watersheds. In December 1996, a debris flow occurred at the Gamahara-zawa watershed and killed 14 persons. It was a rare case of debris flow in snowy season in Japan.

20.3 Samples and Analytical Method

Water samples were collected from 19 main streams in each watershed once a month during base flow conditions for the period between June and November 2000, and from 48 streams as the tributaries of the main streams and 15 spring waters in September 2000. These samples were preserved in 250 ml polyethylene bottles for chemical analyses in laboratory. 15 rock samples of fresh and altered andesites were collected from lavas and breccias in pyroclastic flow deposits and debris flow deposits.

The pH, the electric conductivity (E.C.) and the temperature of the water samples were measured at the sampling locations. The alkalinity as HCO_3^- concentration was measured by titration with 0.02N-HCl. Other major ions (Na^+, Ca^{2+}, K^+, Mg^{2+}, Cl^-, SO_4^{2-}) in water samples were analyzed by ion chromatography using a Dionex DX-120 instrument. The mineral assemblage and chemical composition (SiO_2, TiO_2, Al_2O_3, FeO, MnO, MgO, CaO, Na_2O, K_2O, P_2O_5) of rock samples were determined by XRD using a Rigaku RAD-B SYSTEM instrument and by XRF using a Rigaku RIX3000 instrument, respectively.

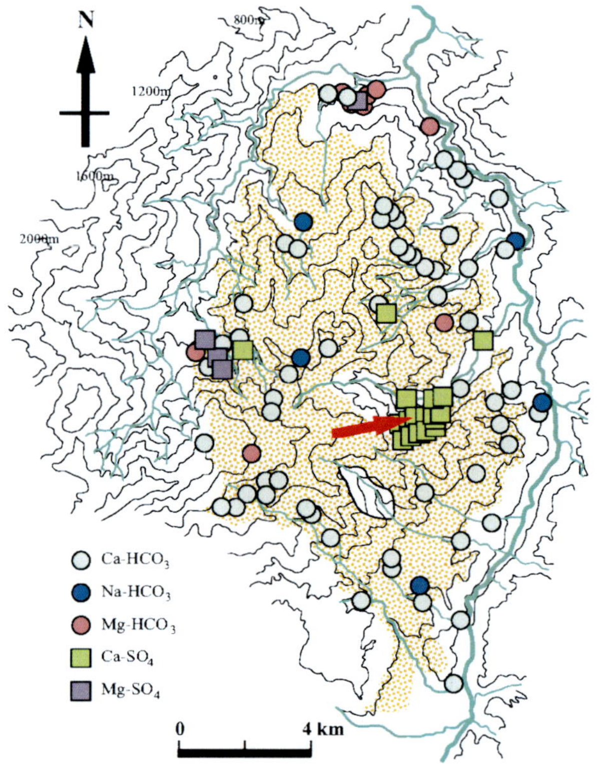

Fig. 20.2. Localities and chemical characteristics of water samples. The *yellow colored field* shows the distribution of the Shirouma-Oike volcanic rocks. An *arrow* shows the Hieda-yama landslide area (the Kanayama-zawa Watershed)

20.4 Results of Chemical Analyses

The mean of E.C. in stream waters ranges from 3.6 mS m^{-1} to 96.9 mS m^{-1} and is shown in Fig. 20.3. Ca-HCO_3 type waters showing low E.C. are extensively distributed in the area as shown in Fig. 20.2, and are formed by water-rock interaction consuming atmospheric or subsurface CO_2 gas during the chemical weathering process. Ca-SO_4 type waters are limited in several watersheds (Fig. 20.2) and often show high E.C. Ca^{2+} is a predominant cation in most of water samples. Figure 20.4 shows a good correlation between the E.C. and the total ion concentration (TIC) in meq l^{-1} of water samples. The E.C. reflects the rate of chemical weathering in each watershed.

The fresh volcanic rocks are composed mostly of clinopyroxene-orthopyroxene andesites and include a large number of plagioclase as both phenocrysts and groundmass minerals. The andesite rocks in the Hieda-yama landslide area partly underwent hydrothermal alteration. Secondary minerals such as pyrite, quartz and smectites in the altered rocks are distributed along joints and cracks in the rocks. Chemical compositions of fresh and weathered andesites are given in Table 20.1. Figures 20.5a

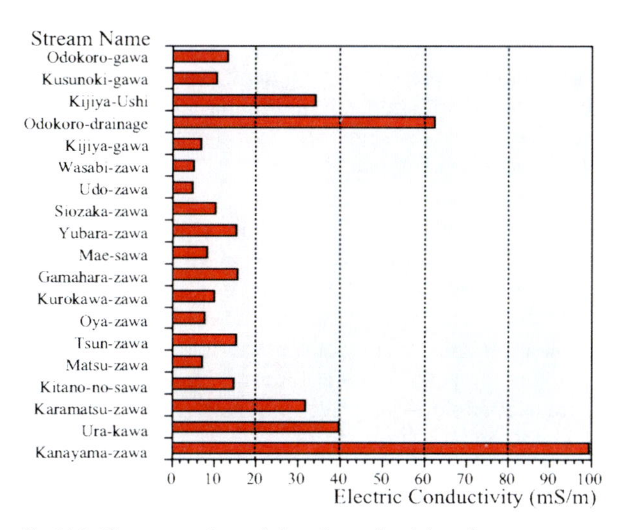

Fig. 20.3. The mean values of electric conductivity of stream waters

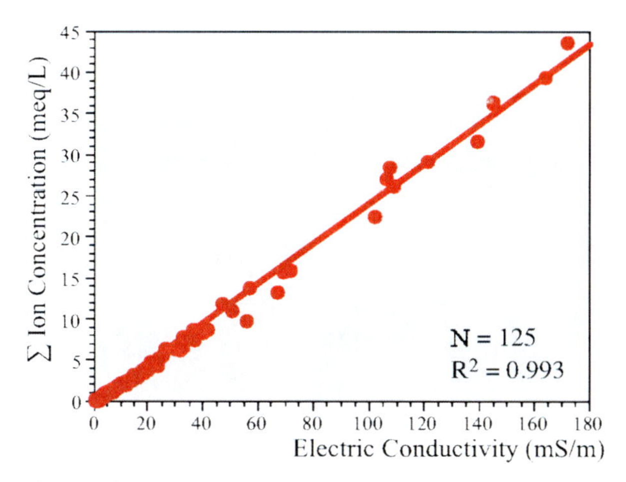

Fig. 20.4. The correlation between electric conductivity and total ion concentration of water samples

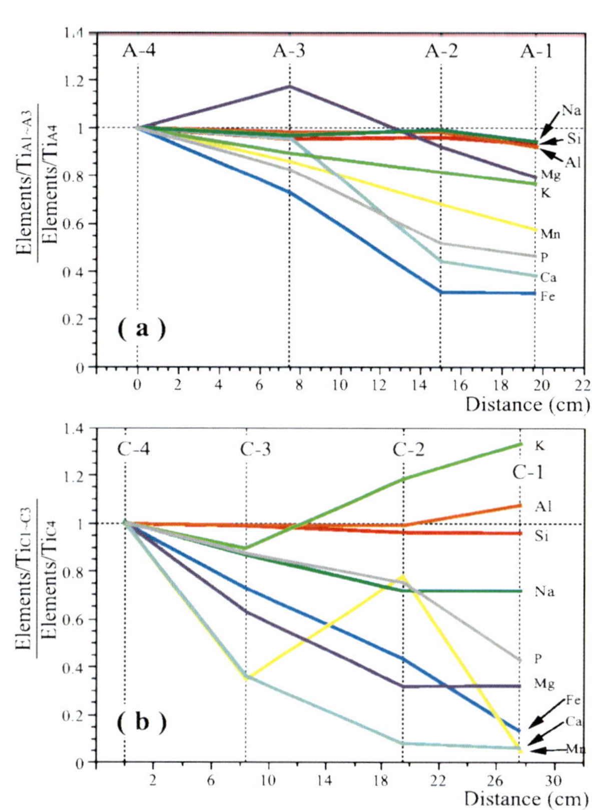

Fig. 20.5. Chemical behavior of andesite samples during chemical weathering. Soluble elements decreasing from the relatively fresh inner core to the weathered outer rim. **a** Sample A; **b** Sample C

Table 20.1. Chemical composition of fresh and weathered andesites

	Fresh andesite ave. (n = 10)	Weathered andesite ave. (n = 5)
SiO_2 (wt.%)	57.66	62.33
TiO_2 (wt.%)	0.90	1.15
Al_2O_3 (wt.%)	18.32	20.69
FeO* (wt.%)	7.64	4.68
MnO (wt.%)	0.18	0.27
MgO (wt.%)	3.15	3.18
CaO (wt.%)	6.54	3.61
Na_2O (wt.%)	3.28	2.59
K_2O (wt.%)	2.05	1.21
P_2O_5 (wt.%)	0.30	0.28

Data from this study, Yuhara et al. (1999) and Sagara et al. (1997).

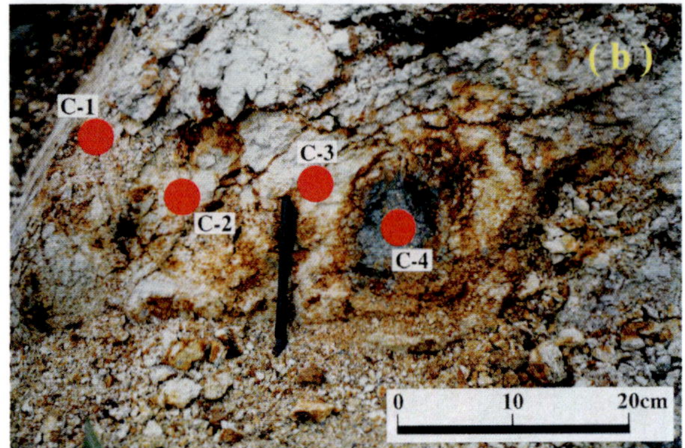

Fig. 20.6. Systematic sampling, in andesite A and C, from the relatively fresh inner core to the weathered outer rim. **a** Andesite sample A and **b** andesite sample C

and 20.5b show the chemical behaviors of major elements normalized with Ti content of the weakly altered andesites. Ti, Al and Si show nearly constant values between the relatively fresh inner core (A4 in Fig. 20.6, 1a and C4 in Fig. 20.6, 1b) and the strongly weathered outer rim (A1 in Fig. 20.6, 1a and C1 in Fig. 20.6, 1b), so they are immobile elements during chemical weathering processes. In contrast, Ca, Fe and Mn have been extremely removed during the chemical weathering process.

20.5 The Hieda-Yama Landslide Area (the Kanayama-Zawa Watershed)

Ca^{2+} and SO_4^{2-} are predominant ions in the stream water from the Hieda-yama landslide area. The E.C. and TIC of the water are the highest of all the watersheds in the research area. In addition, several spring waters showing pH < 3 were found in the landslide area. We recognized an interesting behavior of Fe depletion during the chemical weathering process in Figs. 20.5a and b. Fe is usually immobile and is retained in the weathered rock under the oxidative environment near the earth surface. The depletion of Fe is caused by leaching with the above acidic waters. Sulfuric acid (H_2SO_4) showing pH = 2–3 is easily produced by the oxidation of pyrite included in the altered andesite.

Soluble minerals, especially Ca-rich plagioclase and glass in fresh andesite are leached with H_2SO_4, and Ca^{2+} is released to stream water. Ca-SO_4 type waters of high concentration are formed through the unique chemical weathering resulted from the oxidation of pyrite.

We selected the Kanayama-zawa watershed (the Hieda-yama landslide area) as a model for estimating the annual rate of chemical weathering. The rate in the watershed was estimated using parameters such as the average annual precipitation, the catchment area, the Ca concentration in mg l^{-1} of stream water, and the CaO content in wt.% of fresh and weathered andesites.

20.6 Estimation of Chemical Weathering Rate

The rate of chemical weathering is represented as an integral function that consists of many parameters (e.g. climate, vegetation, topography, lithology, hydrology, etc.). However, several parameters can be excluded, because of the following reasons.

1. Andesitic rocks predominate in the research area. Such a monotonous lithology is useful for estimating the chemical weathering rate.
2. This area is narrow covering about 190 km^2 (12×16 km). Therefore, the average atmospheric temperature, the annual precipitation and evapotranspiration, and the vegetation in the area are similar condition.

Ca^{2+} is the predominant cation in most of the stream waters and is mainly derived from Ca-rich plagioclase in andesitic rocks. Considering a balance of Ca, we estimate the chemical weathering and the soil production rates at the Kanayama-zawa watershed as follows.

The Rate of Chemical Weathering:

$$\frac{(P - E) \times C}{L \times D_a} = 3.24\ \text{mm yr}^{-1} \tag{20.1}$$

The Rate of sediment yield (weight):

$$\frac{S \times (P - E) \times C}{L} = 10\,254\ \text{t yr}^{-1} \tag{20.2}$$

The Rate of sediment yield (volume):

$$\frac{S \times (P - E) \times C}{L \times D_s} = 9322\ \text{m}^3\ \text{yr}^{-1} \tag{20.3}$$

Table 20.2. Parameters for the estimation of the rate of chemical weathering at the Hieda-yama landslide area (the Kanayama-zawa watershed)

Area of watershed, S (km^2)	1.32
Annual preciptation, P ($mm\,yr^{-1}$)	2200
Evapotranspiration, E ($mm\,yr^{-1}$)	740
Ca concentration in water, C ($mg\,l^{-1}$)	126
Loss of Ca in andesite, L ($g\,kg^{-1}$)	23.6
Density of andesite, Da ($g\,cm^{-3}$)	2.40
Density of sediment, Ds ($g\,cm^{-3}$)	1.10

Table 20.3. Estimated results of the rate of chemical weathering in each watershed

Watershed	Weathering ($mm\,yr^{-1}$)
Kanayama-zawa	3.24
Ura-kawa	1.25
Karamatsu-zawa	0.99
Tsun-zawa	0.48
Gamahara-zawa	0.46
Yubara-zawa	0.50
Mae-sawa	0.27
Shiozaka-zawa	0.34
Kijiya-gawa	0.23
Kitano-no-sawa	0.46
Korokawa-zawa	0.32
Oya-zawa	0.23
Matsu-zawa	0.23
Kusunoki-gawa	0.32
Wasabi-zawa	0.16
Udo-zawa	0.15

The meaning of parameters represented as S, P, E, C, L, D_a and D_s is given in Table 20.2. The loss of Ca in weathered andesite is calculated by using data of Table 20.1. The Ca in the andesite is normalized with insoluble elements (e.g. Ti, and Al).

The rate of chemical weathering in each watershed is approximately estimated by the comparison of each E.C. instead of the Ca content, because the good correlation shown in Fig. 20.4 enables to use the E.C. as a key parameter to estimate the rate of chemical weathering instead of the solute concentration. Table 20.3 shows calculation results of the chemical weathering rate in each watershed and Fig. 20.7 shows that colored watersheds illustrate the degrees of the estimated chemical weathering rate.

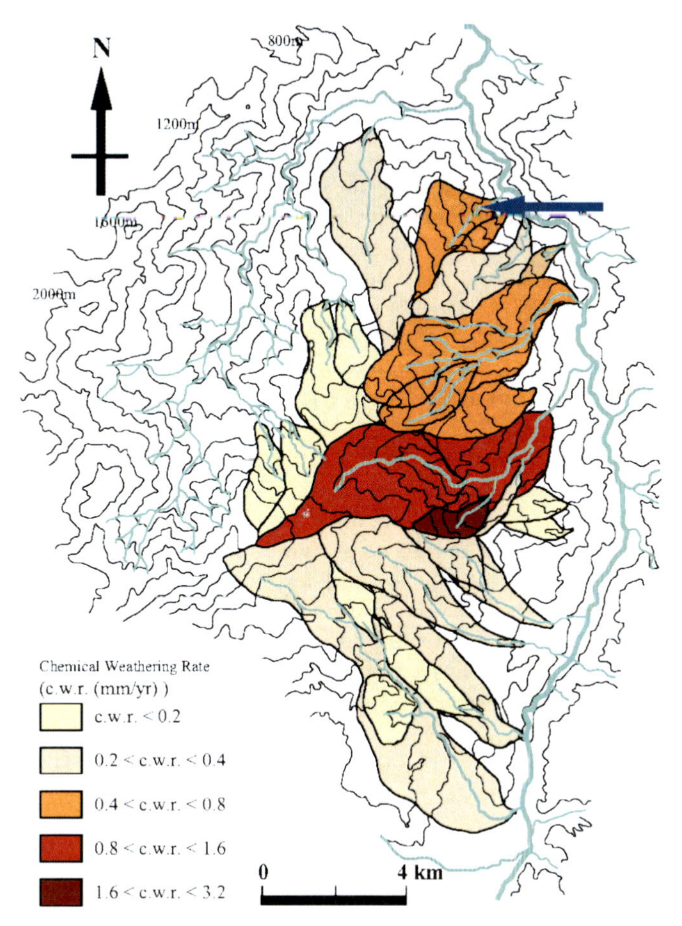

Fig. 20.7. Map illustrating the degree of the chemical weathering rate in each watershed by gradation of color. An *arrow* shows the Gamahara-zawa watershed

20.7 Connection between Chemical Weathering and Landslide Occurrence

Due to intensive rainfall unstable soils on the slope slide down and unstable sediments in the stream valley are removed by the fluvial erosion. The continual removal of weathered and unconsolidated materials forms new slope surfaces or exposes bedrocks that are easily affected by chemical weathering. The removal of soils and sediments from watersheds is a preparation for the production of new material from bedrock weathering.

The estimated rates of chemical weathering and sediment yield in the Hieda-yama landslide area were respectively 3.24 $mm\,yr^{-1}$ and ~5700 $m^3\,km^{-2}\,yr^{-1}$, and were extremely high values. The landslide area (the Kanayama-zawa Watershed), the reddish colored watershed in Fig. 20.7, shows actually the highest landslide frequencies of all watersheds in the research area. According to literatures and reports, the chemical weathering rate is likely to correspond to the disaster records of landslides and debris flows. For example, the Gamahara-zawa watershed where debris flow disaster occurred in 1996, one of the orange colored watersheds in Fig. 20.7, shows relatively high chemical weathering rate of 0.46 mm yr^{-1}. The Chemical weathering in this area is thought to be one of the most important factors for landslide and debris flow occurrences. Figure 20.7 represents a kind of susceptibility map for sediment disasters (e.g., landslides, slope failures and debris flows) in each watershed.

High SO_4/Cl waters are often found in several landslide areas (Fig. 20.8). The SO_4/Cl ratio reveals the existence of SO_4 rich waters that are usually hidden by the mixing with HCO_3 type waters or by dilution with fresh waters. The ratio provides the signatures of the altered zones inner mountain slopes or beneath thick soils. We also take particular notice of the inner altered zones in watersheds of the Shirouma-Oike Volcano. Such a simple investigation using hydrochemistry of stream and spring waters could be a useful substitute for geophysical explorations

20.8 Identification of the Buried Altered Zones

The hydrothermally altered zones including pyrite as a source material for sulfuric acid are thought to be another key factor in relation to the acceleration of the chemical weathering and the occurrence of destructive sector collapses or large-scale landslides. Most of the altered zones, however, have been covered with vegetation and thick soil layers. Therefore, it is difficult to directly recognize outcrops of the hydrothermally altered units by field investigation for surface geology.

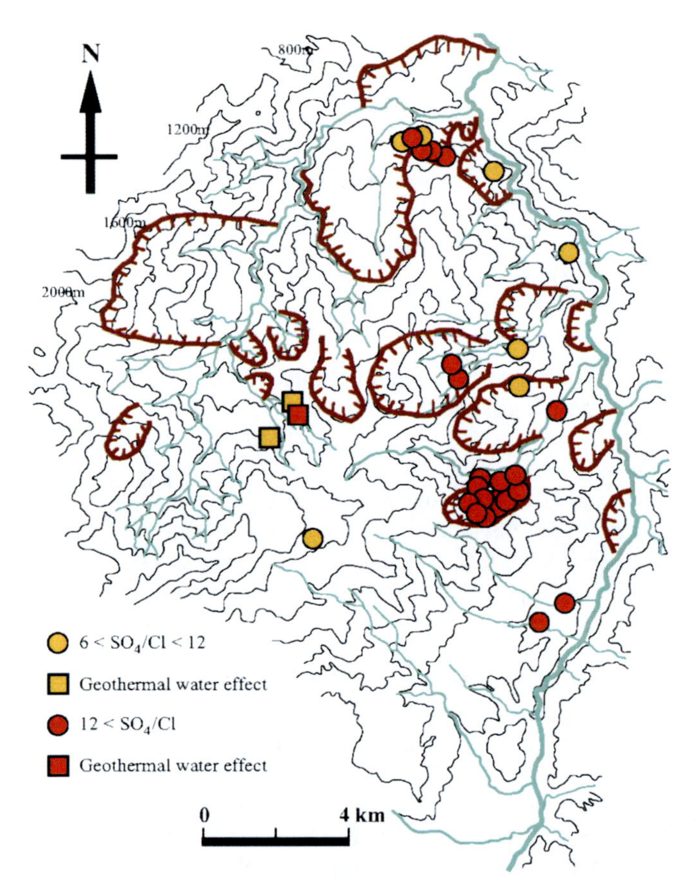

Fig. 20.8. Map showing the distribution of high SO_4/Cl waters. These waters emerging from several landslide areas

20.9 Summary

1. Ca^{2+} is a predominant cation in most water samples from each watershed.
2. There is a good correlation between the E.C. and the TIC in meq l^{-1} of water samples. This correlation enables to use the E.C. as an important parameter to estimate the rate of chemical weathering instead of the total solute concentration.
3. Stream waters from the Hieda-yama landslide area (the Kanayama-zawa watershed) are characterized by Ca-SO_4 type with the TIC of ~1 000–2 000 mg l^{-1}. The high concentration of total ions shows that the chemical weathering rate is the highest of all watersheds in the research area.
4. In particular, the high concentration of SO_4^{2-} and Ca^{2+} indicates the following two processes during the chemical weathering.
   I The oxidation of pyrite, a secondary mineral in hydrothermally altered andesite, produces sulfuric acid of pH < 3.
   II The acid leaches solutes out, especially Ca^{2+} from plagioclase, from fresh andesite.
5. The annual rates of chemical weathering and sediment yield are estimated by the mean of the annual precipitation and evapotranspiration, the catchment area, the Ca^{2+} concentration in mg l^{-1} in stream water, the density of soil and fresh andesite, and the Ca content in wt.% of fresh and weathered andesites. As a result, the rates of chemical weathering and sediment yield in the Hieda-yama landslide area are 3.2 mm yr^{-1} and ~9 300 m^3 yr^{-1}, respectively.
6. Comparing with the E.C. in each watershed, the chemical weathering rate in each watershed can be easily calculated. The calculation results suggest that several landslides were controlled by the rate of sediment yield related to the chemical weathering.
7. The high SO_4/Cl in stream waters might indicate the buried hydrothermally altered zones. Therefore, such a geochemical approach could be useful for hazard mappings, erosion controls, and the disaster prevention of landslides and debris flows

Acknowledgments

This research was supported by JSPS's Grant-in-Aid for scientific research no. 16310126 (Rep. H. Marui). We thank Dr. M. Yuhara of Fukuoka University and Dr. Y. Kakihara of Geological Survey of Hokkaido for their analytical expertise. We are also very grateful to former Professor Dr. O. Sato for his helpful comments and criticisms of the original manuscript.

References

Gaillardet J, Dupré B, Allègre, CJ (1999) Geochemistry of large river suspended sediments: silicate weathering or recycling tracer? Geochim Cosmochim Ac 63:4037–4051

Louvat P, Allègre CJ (1997) Present denudation rates in the island of Réunion determined by river geochemistry: basalt weathering and mass budget between chemical and mechanical erosions. Geochim Cosmochim Ac 61:3645–3669

Lyons WB, Carey AE, Hicks DM, Nezat CA (2005) Chemical weathering in high-sediment-yielding watersheds, New Zealand. J Geophys Res 110:F01008, doi:10.1029/2003JF000088

Marui H, Sato O (1995) Geotechnical and geochemical study on the debris flows in the Urakawa River Basin. Proceedings of XXth IUFRO World Congress, Finland, pp 1–12

Marui H, Watanabe N (2001) Process of slide to flow in the Gamahara torrent debris flow. Proceedings of Conference on Transition from Slide to Flow-Mechanisms and Remedial Measures, 25–26 August 2001, Trabzon, Turkey, pp 157–166

Marui M, Watanabe N, Sato O, Fukuoka H (1997a) Gamahara torrent debris flow on 6 December 1996, Japan. Landslide News 10:4–6

Marui M, Sato O, Watanabe N (1997b) Preliminary report on Gamahara torrent debris flow on 6 December 1996, Japan. Journal of Natural Disaster Science 18-2:89–98

Millot R, Gaillardet J, Dupré B, Allègre CJ (2002) The global control of silicate weathering rates and the coupling with physical erosion: new insights from rivers of the Canadian Shield. Earth Planet Sc Lett 196:83–98

Nakano S, Takeychi M, Yoshikawa T, Nagamori H, Kariya Y, Okumura K, Taguchi Y (2001) Geology of the Shiroumadake district. Geological Survey of Japan, AIST, 105 p (in Japanese with English abstract and 1:50 000 geological map)

Sagara W, Watanabe N, Sato O, Marui H, Kakihara Y (1997) Chemical weathering of andesitic rocks at the Hieda-yama landslide area, Nagano Prefecture. Annual Report of Research Institute for Hazards in Snowy Areas, Niigata University, no. 19, pp 83–96 (in Japanese with English abstract)

Sato O, Aoki S, Marui H (1991) Effect of ground water in initiation of debris flow in Urakawa River Basin. Proceedings of Japan – U.S. Workshop of Snow Avalanche, Landslide, Debris Flow Prediction and Control, pp 453–462

Yuhara M, Sagara W, Takahashi T, Watanabe N, Yamagishi H, Marui H (1999) Sr and Nd isotopic compositions of rocks and surface waters from the Hieda-yama landslide area, Nagano Prefecture. Annual Report of Research Institute for Hazards in Snowy Areas, Niigata University, no. 21, pp 73–82 (in Japanese with English abstract)

Chapter 21

Mechanism of Landslide Causing the December 2002 Tsunami at Stromboli Volcano (Italy)

Daniela Boldini · Fawu Wang* · Kyoji Sassa · Paolo Tommasi

Abstract. Between 29 and 30 December 2002 the NW flank of Stromboli Volcano (Sciara del Fuoco) was involved in a series of large-scale instability phenomena which culminated in submarine and subaerial destructive landslides provoking two tsunami waves with a maximum run-up of 10 m. In this paper, part of the results of a joint research between the National Research Council (Italy) and the Disaster Prevention Research Institute of Kyoto University (Japan) are presented. The activity has focused on the mechanical characterization of the volcanoclastic material forming the Sciara del Fuoco depression and the interpretation of landslide mechanisms on the basis of large-scale ring shear tests. Attention is given here to the initiation and propagation of the submarine landslide which caused the first tsunami. In order to investigate the material response to different displacement rates in terms of shear resistance, pore pressure generation and grain crushing, ring shear tests were conducted in both undrained and drained conditions. Experimental results indicate that a fully or partial liquefaction mechanism can be invoked to explain the failure of the submarine flank of the Sciara del Fuoco and the long run-out which followed, as it was suggested by comparing pre- and post-failure in situ observations.

Keywords. Landslide, tsunami, ring shear tests, liquefaction, Stromboli Volcano

21.1 Introduction

The island of Stromboli (Southern Italy) is the subaerial part of an active volcano whose products spread over the NW flank, called Sciara del Fuoco (fire stream).

Between 29 and 30 December 2002, the thrust exerted by the magma intruded during a major eruption, triggered a sequence of large-scale instability phenomena on the Sciara del Fuoco culminating in a submarine and two subaerial destructive landslides. Landslides involved about 20 millions m^3 of slope materials producing two series of tsunami waves which run up the inhabited coasts of the Stromboli Island for a maximum height of 10 m, damaged shoreline facilities of the facing Panarea Island (20 km far) and were felt on the northern coast of Sicily (60 km far).

After the first emergency phase, the Italian Department of Civil Defence promoted a research program on the eruption effects, including the reconstruction of instability mechanisms of landslides and subsequent deformation phenomena. In the framework of this program, a joint research between the National Research Council (Italy) and the Disaster Prevention Research Institute of Kyoto University (Japan) was conducted as a part of the UNITWIN Cooperation Programme by Kyoto University, UNESCO, and the International Consortium on Landslides "Landslide Risk Mitigation for Society and the Environment". The joint research is aiming at:

- characterizing the mechanical behavior of the volcanoclastic materials in the conditions that possibly developed during the landslide processes;
- refining and supporting with experimental evidences the mechanisms of landslide initiation and propagation that had been suggested on the basis of post-failure observations and surveys (Tommasi et al. 2005b).

This paper refers to tests performed with the large-scale ring shear apparatus at DPRI, Kyoto University in order to reproduce geotechnical conditions existing during the triggering phase of the rapid submarine failure that provoked the first tsunami. Different drainage conditions were adopted in order to investigate the influence of displacement rate on pore pressure increment and strength reduction during the shearing process.

21.2 Sciara del Fuoco Morphology and Instability Phenomena

The island of Stromboli is the subaerial portion of a 4 000 m high volcanic edifice that rises for about 900 m above the sea level. The Sciara del Fuoco, hereafter called SdF or simply Sciara, is a scree slope, which collects the products of the persistent volcanic activity and drives them to the sea (Fig. 21.1). Actually, the Sciara is a depression originated by the last lateral collapse occurred after 5.6 ±3.3 kyr B.P. (Tibaldi 2001), that is filled by a more than 200 m thick alternation of loose and cohesive beds resulting from the overlapping of thick volcanoclastic layers with minor layers of primary products (i.e. pyroclastites and lava flows).

Volcanoclastic materials result from continuous sliding of primary products and their incessant re-distribu-

Fig. 21.1.
View of the Sciara del Fuoco before the 2002 landslides from the sea

Fig. 21.2.
View of the Sciara foot after the 2002 landslides and detail of the volcanoclastic deposit (in the foreground the front of a major lava flow of the 1985 eruption)

tion over the slope by sheet slides and subsequent grain flows so as to produce a series of irregular, mostly reverse-graded grain-supported layer (Fig. 21.2).

21.3 The December 2002 Tsunamogenic Landslides

Detailed description of the instability events and their links with volcanic activity is reported in the paper by Tommasi et al. (2005b). A brief summary of the succession of instabilities is given below.

The eruption started in the early evening of 28 December when lava poured from the NE crater and rapidly reached the shoreline. However, photographs taken by helicopter in the early morning of 30 December indicate that a relatively deep-seated landslide had occurred in the preceding hours (Figs. 21.3 and 21.4). The movement, which is indicated as α landslide, had disarranged the slope, but did not evolve into a destructive landslide.

During the morning of 30 December, the slope was relentlessly deforming and the α slide body was fragmenting in "blocks". The larger block (which would have collapsed afterwards producing the second tsunamogenic landslide β) extended from 450–500 m a.s.l., down to the shoreline (Fig. 21.3).

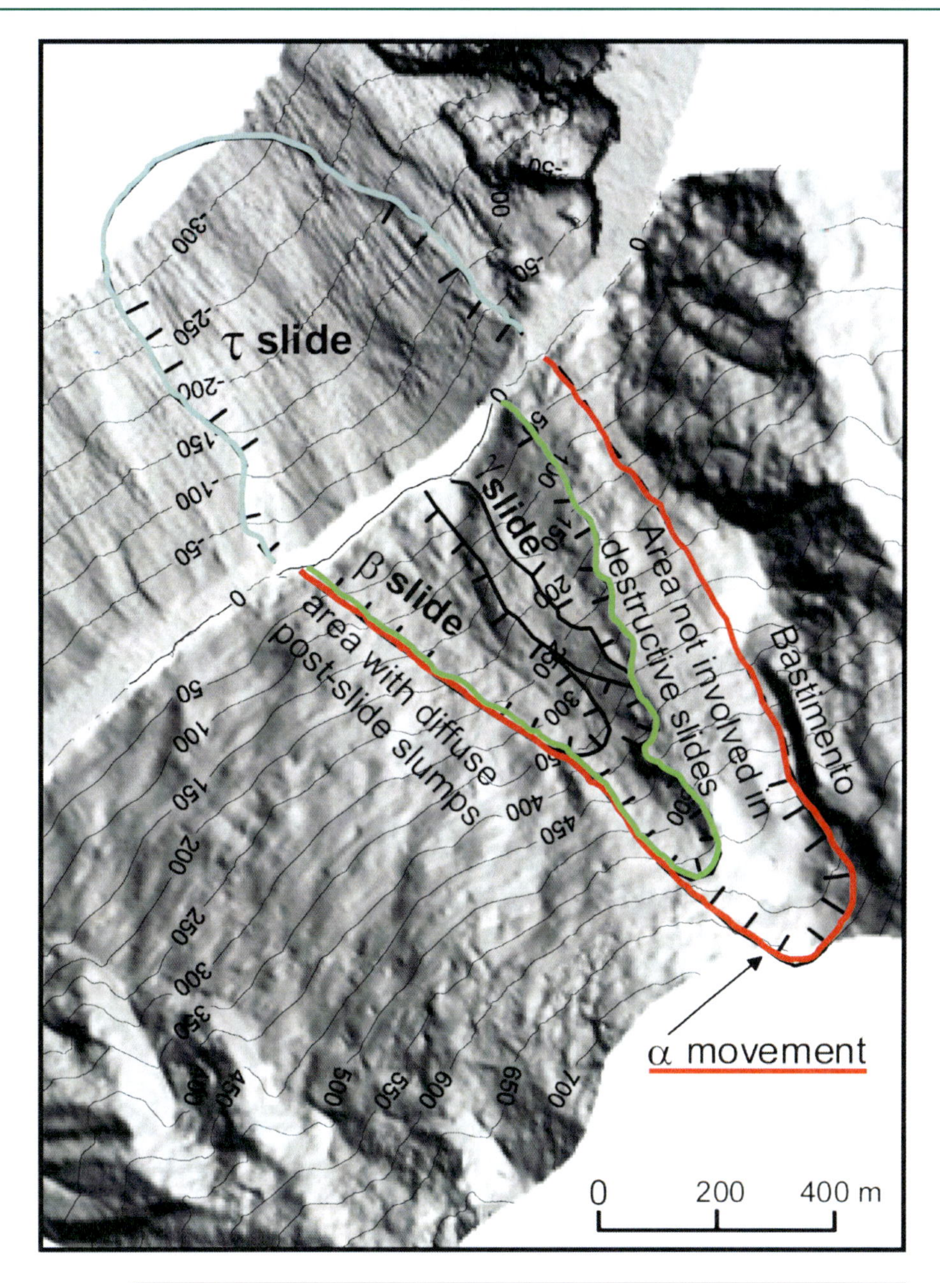

Fig. 21.3.
Limits of the major instability phenomena occurred on 30 December 2002 (from Tommasi et al. 2005b)

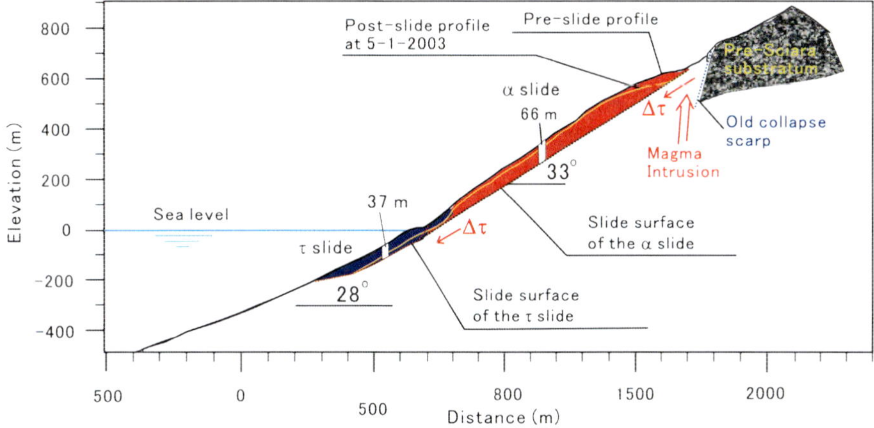

Fig. 21.4.
Profile of the Sciara del Fuoco slope. Slip surfaces reconstructed by comparison of pre- and post-slide morphology are shown

In the late morning two series of tsunami waves propagated from the SdF and hit the inhabited coastal areas of the island with a maximum run-up of more than 10 m (Tinti et al. 2003). The initial wave receding at Sciara del Fuoco suggests that the first tsunami wave was generated by a submarine slide, hereafter called τ slide (Figs. 21.3 and 21.4).

21.4 Lithological and Physical Properties of Volcanoclastic Material

Volcanoclastic layers represent the most abundant and weakest component of the Sciara deposit and extend with continuity over extremely large areas (up to some 10^5 m^2). Observations of the deposition process on the Sciara slope before and after the 2002–2003 events indicate that lithology, grain size and structure of volcanoclastic materials remain virtually unaltered in time. However, volcanoclastic layers deposited after the 2002 landslides possibly have higher extension and continuity. Therefore, these levels of relatively weak materials are likely to exist also at depth and can represent preferable initiation paths for slip surfaces.

Continuous falls of products ejected by the explosions inhibit the access to a large area around the craters. At lower elevations, incessant rock falls and small debris avalanches resulting from the shallow failures of the slope materials make large part of the Sciara slope not accessible excepting for few minutes.

In situ borehole investigations cannot therefore be carried out and geotechnical characterization of the volcanoclastic materials can only be conducted on samples taken at few accessible areas. However since the deposition process is repetitive, material sampled on outcrops is representative of the material at depth.

Sampling was carried out at the base of the slope, where transportation of large amounts of material is possible and access to the deposit is easier and less dangerous. Sampling was conducted in the early morning during the wet season, when increased soil suction due to a suitable water content in the shallow materials inhibits skin slides and subsequent spreading of debris downslope.

The volcanoclastic material is mainly composed by gravely and sandy layers (Fig. 21.2) that, in the sampling area, have angular to subangular clasts with high surface roughness and low sphericity. Rounding, smoothness and sphericity progressively increase proceeding downslope, due to grain sliding and rolling (Kokelaar and Romagnoli 1995).

Physical and mechanical properties of grains were determined by Tommasi et al. (2005a). Dry density from mercury pycnometer measurement on the different grain size classes resulted on average 2.34 Mg m^{-3}. Solid matrix density determined with a helium picnometer on powders of the whole is on average 2.90 Mg m^{-3} yielding a mean value of the total porosity of the grains equal to 19.2%, that is almost completely connected.

Grain strength estimated through point load tests on larger clasts revealed to be low and strongly dependent on grain porosity. For grain porosity of 20% and assuming that uniaxial compressive strength σ_c is related to the point load strength index $I_{s,50}$ by the relation $\sigma_c = 22 I_{s,50}$, a uniaxial compressive strength of 21 MPa can be estimated.

21.5 Stress Loading Ring Shear Test Results

Shear tests were performed with the ring shear apparatus DPRI-6 developed at the Disaster Prevention Research Institute at Kyoto University (Sassa 1997). The method of undrained shear tests using this new apparatus and the recent test results in simulating various landslide phenomena were reported by Sassa et al. (2003, 2004). This large-scale ring-shear apparatus is characterized by an inner diameter of 250 mm, an outer diameter of 350 mm and a maximum sample height of 150 mm and, in turn, is suitable for testing coarse-grained volcanoclastic material.

The DPRI-6 ring shear apparatus allows performing tests in both drained and undrained conditions thanks to a water-leakage tightness system constituted by O-rings on the upper loading platen and bonding rubber edges on the two confining rings of the lower rotary pair (Sassa et al. 2004). Pore pressure measurement is performed by two pore pressure transducers, the measuring points being located 2 mm above the shear surface.

Tests were performed on the fraction passing at the 8 mm sieve which assures a convenient ratio between the maximum grain size and the sample height in the ring shear apparatus.

Grain size distribution of the tested material, which can be classified as sand with gravel, is shown in Fig. 21.5. The sample was placed into the shear box by the dry pluviation method. In order to avoid segregation phenomena during the infilling, the whole sample was initially subdivided into eight smaller parts which were subse-

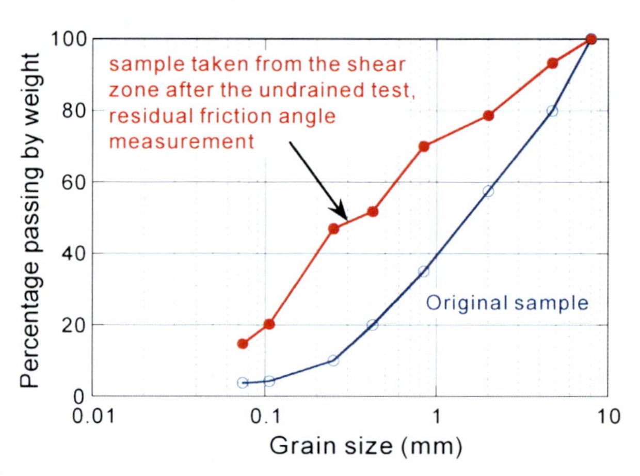

Fig. 21.5. Particle size distribution curves of the tested material before and after the tests

quently deposited in the apparatus in different layers without tamping the sample. In such a way, the in situ void ratio at the desired depth was obtained by applying the in situ state of stress during the consolidation phase. Furthermore, since after erosional phases the rapid deposition of new material re-establishes the original stress, lower values of void ratio deriving from overconsolidation are likely to be excluded.

The test program for investigating the initiation mechanism of the submarine landslide which caused the first tsunami wave was defined at DPRI in order to reproduce the following scenario, from magma intrusion to the rapid submarine landslide:

1. magma intruded into the slope and exerted a thrust which produced additional shear stress $\Delta\tau$ to α landslide;
2. α landslide moved downward and applied additional shear stress to τ landslide body under the sea level;
3. the movement of α landslide could have resulted in undrained loading or naturally (partially) drained loading of τ landslide body which was fully saturated under the sea. If high excess pore water pressures were generated in this undrained or naturally drained shear, the submarine landslide could have been a rapid landslide;
4. no excess pore water pressure was instead generated within the α landslide because the subaerial slope was likely to be dry.

Modeling of the undrained loading by the moving landslide mass from the upper slope and its reproduction by the undrained ring shear tests were presented for the Otari debris flow disaster by Sassa et al. (1997). The model for the Otari debris flow includes loading by both shear stress and normal stress. However, to simplify this case, we assumed that the angle of sliding surface of τ landslide body was the same with that of α landslide body. Then, only shear stress increments were applied in this test program to reproduce the case of Stromboli. The velocity of α landslide will not be so high like Otari debris flow, but the rate of effective stress change given by the movement of α land-slide is much greater than the rate of effective stress change during rainfalls. Then, the test program was made to give loading in the undrained condition, and also in the naturally drained condition. The comparison of shear behavior under the naturally drained condition in the ring shear test to that under the undrained condition was conducted for the risk evaluation of earthquake-induced-landslide in the upper slope of the Nikawa area (Sassa et al. 2004).

The first step in reproducing geotechnical conditions of this instability scenario was application of the initial stress state for τ landslide before the loading by α land-slide. Then, additional shear stress was applied to simulate loading by α landslide under two conditions: (1) an undrained torque-controlled ring-shear test; and (2) a torque-controlled ring-shear test with open drainages (possibly reproducing naturally drained conditions). At the end of each test the residual friction angle was also evaluated.

In both cases, for the slip surface of the landslide τ, an average depth and an average dip of 37 m and 28° were assumed, respectively. Therefore, the sample was first consolidated at an anisotropic state of stress applying a normal stress equal to 230 kPa and a shear stress equal to 122 kPa, which represent the lithostatic state of stress of a soil element lying on the sliding surface of landslide τ. In these tests, the dry unit weight at the end of the consolidation phase ranged between 16.6 and 17.2 kN m^{-3}.

Results of the undrained torque-controlled test are reported in Figs. 21.6 and 21.7. Figure 21.6 shows the time-histories of normal stress, shear resistance, pore-water pressure and shear displacement, whilst the effective and total stress paths are plotted in Fig. 21.7.

During the first stages of the undrained shearing, excess pore-water pressure generated as the shear stress increased, producing a decrease of the effective normal stress. This behavior inverted when pore pressure reached 35 kPa; because of material dilatancy, the pore pressure started to decrease down to negative values resulting in effective normal stresses higher than total normal stresses.

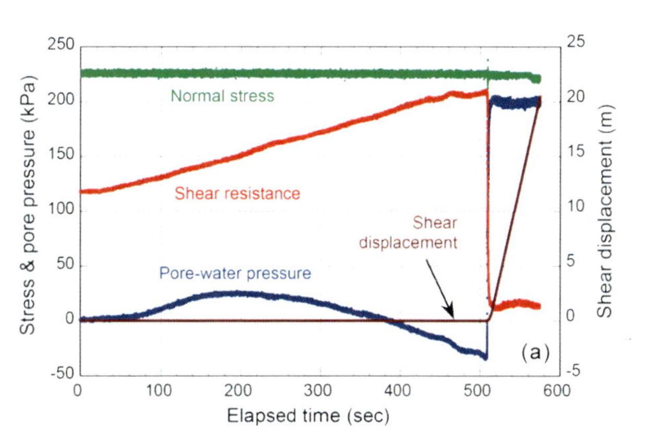

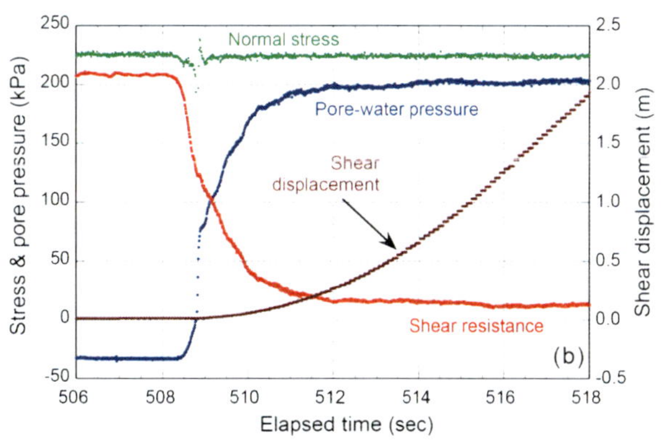

Fig. 21.6. Time-histories of normal stress, shear resistance, pore-water pressure and shear displacement recorded during the undrained ring shear test

The same trend is generally observed in the initial stage of triaxial tests on medium-dense sands that display a strain-hardening behavior after dilation takes place (e.g., Castro 1969). In contrast to this well-known material response, the Stromboli volcanoclastic material experienced liquefaction phenomena when negative pore pressure reached 45 kPa. As a consequence, shear resistance abruptly decreased down to few kPa and the shear displacement rate increased significantly (Fig. 21.6). Pore water pressure reached a value higher than 200 kPa. The apparent friction angle, which can be obtained by the initial normal stress and the mobilized shear resistance at the final stage, is about 3 degrees, showing a high mobility of the material under undrained shearing.

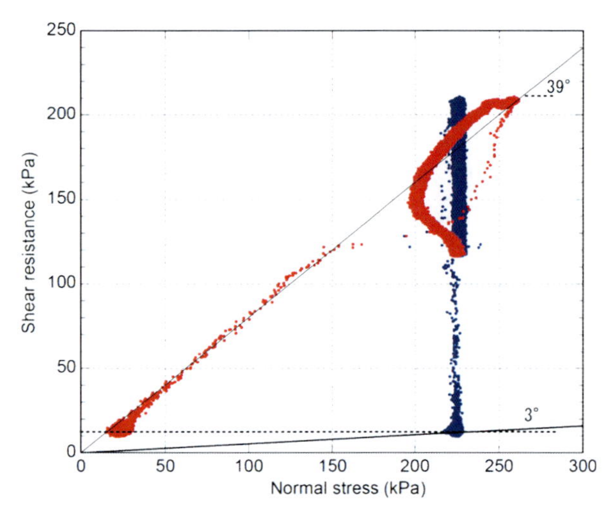

Fig. 21.7. Effective and total stress paths during the undrained ring shear test

Fig. 21.8. View of the shear zone after the test

Liquefaction phenomena for medium-dense to dense sands were already observed in ring shear tests performed in undrained conditions at DPRI by Sassa (1996) and by Wang and Sassa (2002). They demonstrated that, irrespective of the initial density of the soil, liquefaction can be triggered if the following two conditions are verified:

1. the shear stress has to be great enough to initiate the failure of the soil;
2. the grains of the soil should be crushable during shearing under the applied normal stress.

Stromboli grains demonstrated to be highly crushable under the applied testing conditions. Particle size distribution of the material taken from the shear zone after the shearing test is also shown in Fig. 21.5. A view of the shear zone after the test is given in Fig. 21.8, where grain crushing due to the shearing process is apparent.

The second test presented here was conducted by applying the shear at the same loading rate (55 Pa s^{-1}), but by letting the drainages open. Time-histories of the normal stress, shear resistance and shear displacement are plotted in Fig. 21.9.

The shear resistance increased gradually up to the maximum value corresponding to a peak friction angle of about 44°. Drained conditions can be reasonably assumed for this first part of the test. After peak, suddenly, the shear resistance dropped for some seconds before increasing again up to an equilibrium value that corresponds to the residual strength in conditions of high-speed shearing. As a consequence, shear displacement rate rapidly increased as well.

The drop of the shear resistance down to values lower than the residual ones is probably to be associated to lo-

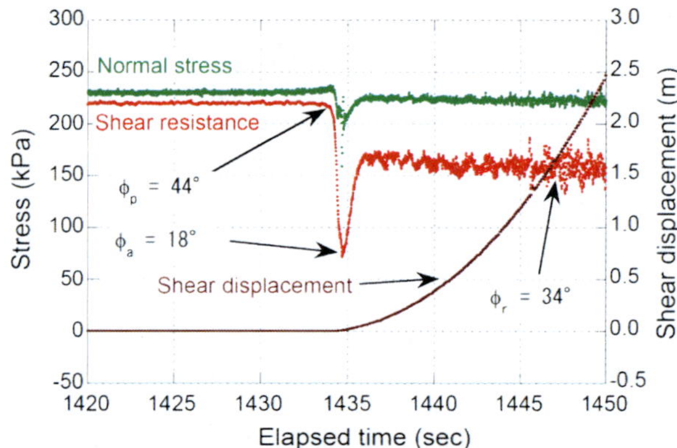

Fig. 21.9. Time-histories of normal stress, shear resistance and shear displacement recorded during the ring shear test with open drainages

cal generation of excess pore pressure within the shear zone, that can build-up due to the high crushability of the Stromboli volcanoclastic material. The amount of excess pore-water pressure depends on the combined effects of loading rate and the rate of pore pressure dissipation. Immediately after failure, high pore pressure was suddenly generated due to rapid grain crushing (the minimum apparent friction angle of 18° was reached). However, since the rate of grain crushing and, in turn, the rate of pore pressure generation decreased, excess pore pressure decayed soon, the shear resistance was recovered and the apparent friction angle reached about 34° (Fig. 21.9).

According to the process reproduced in the ring shear tests, the apparent friction angle mobilized in τ landslide ranged between 3° and 18°. However, since the drainage path in the natural slope (the distance from the shear zone to the free water is 37 m) was two order magnitude higher than that in the ring-shear apparatus (the distance from the shear zone to the free air is around 12 cm in this series of tests), the mobilized apparent friction angle should be actually closer to 3°.

Test results reported in Figs. 21.5–21.7 and in Fig. 21.9 experimentally support this instability scenario, which entails the passage from a dry and slow subaerial landslide to a fully-saturated rapid submarine landslide.

The residual friction angle in static conditions was determined after each test by shearing the sample to high displacement values and by applying shear through the shear – speed control (shearing rate = 0.103 mm s^{-1}) after having opened the drainages in the case of undrained test. After equilibrium conditions were reached, the normal stress was gradually reduced at a rate of 5 kPa min^{-1} and correspondingly the shear resistance decreased according to the residual strength envelope of the material. After the applied normal stress reached a value lower than about 35 kPa, it was again increased at the same loading rate, up to its original value of 230 kPa. Accordingly, also in this case, the shear resistance increased following almost the same limit strength envelope.

Evaluation of the residual friction angle with the procedure described above is shown in Fig. 21.10 for the first test. The residual friction angle resulted equal to 37° during both unloading and reloading of the normal stress.

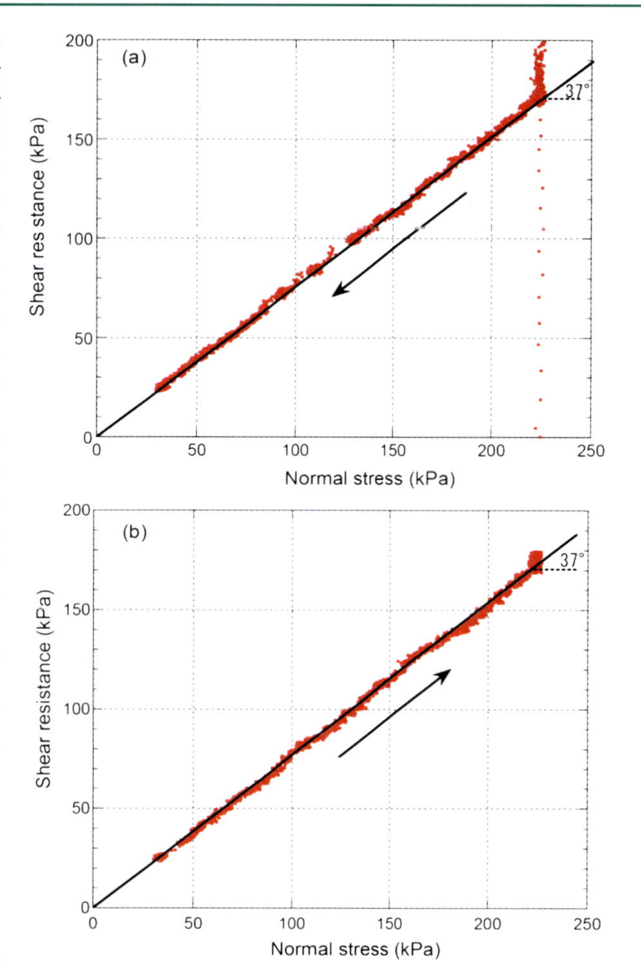

Fig. 21.10. Evaluation of the residual friction angle

21.6 Discussion and Conclusions

Results of the undrained ring shear test indicate that, if undrained conditions established in the submerged slope, liquefaction can be invoked to explain the submarine failure and the subsequent long run-out of the slide material (greater than 1 000 m).

Test performed with open drainages, however, provided experimental evidence that even during not fully undrained shearing, a significant drop-down of shear resistance was recorded that would have been sufficient to lead the slope to failure. Such a mechanical behavior is an effect of the high susceptibility to crushing exhibited by the grains, that in undrained or naturally drained conditions, results in rapid generation of excess water-pore pressure during shearing.

Undrained or naturally drained shearing was likely to occur during the slope deformation process preceding the submarine landslide (τ slide). Sudden increases in shear displacement velocity or in the shear stress were in fact continuously induced by local failures in the subaerial slope (Tommasi et al. 2005b) that produced the fragmentation of the initial slide body α.

Finally, it is to be noted that the normally consolidated Stromboli volcanoclastic material revealed to be highly dilatant in the initial shearing stage, displaying a high peak friction angle at failure, which may be attributed to the angularness and surface roughness of grains. The residual friction angle in static conditions was found to be also rather high (37°), even if lower values were obtained in condition of higher shear displacement rates.

Acknowledgments

Financial support by National Research Council within the research program "Short-term mobility" to visit the Research Centre on Landslides of Disaster Prevention Research Institute, Kyoto University was greatly appreciated by the first author. The UNITWIN Cooperation Programme by Kyoto University, UNESCO and the International Consortium on Landslides "Landslide Risk Mitigation for Society and the Environment" enabled this international joint research cooperation. Financial support to research activity in Italy by Italian Department for Civil Defence is greatly acknowledged. Mr. K. Kondo gave a valuable help in apparatus maintenance. Mr. R. D'Inverno and Eng. F. Cignitti prepared samples for ring shear tests. Mr. M. Zaia (Zazà) largely contributed to the access to sampling sites and sampling operations. Mr. O. Igwe helped to process the experiment data.

References

Castro G (1969) Liquefaction of sands. Ph.D. thesis, Harvard University, Cambridge, Massachusetts

Kokelaar P, Romagnoli C (1995) Sector collapse, sedimentation and clast population evolution at an active island-arc volcano: Stromboli, Italy. B Volcanol 57:240–262

Sassa K (1996) Prediction of earthquake induced landslides. In: Proc 7th Int Symp on Landslides. Trondheim, 17–21 June, vol. 1, A.A. Balkema, pp 115–132

Sassa K (1997) A new intelligent type dynamic loading ring shear apparatus. Landslide News 10:33

Sassa K, Fukuoka H, Wang FW (1997) Mechanism and risk assessment of landslide-triggered-debris flows: lesson from the 1996.12.6 Otari debris flow disaster, Nagano, Japan. In: Cruden DM, Fell R (eds) Landslide risk assessment. Proc Int workshop on landslide risk assessment, Honolulu, February 2003, A.A. Balkema, pp 347–356

Sassa K, Fukuoka H, Wang G, Ishikawa H (2004) Undrained dynamic-loading ring-shear apparatus and its application to landslide dynamics. Landslides 1(1):9–17

Sassa K, Wang G, Fukuoka H (2003) Performing undrained shear tests on saturated sands in a new intelligent type of ring-shear apparatus. Geotech Test J 26(3):257–265

Sassa K, Wang G, Fukuoka H, Wang FW, Ochiai T, Sugiyama M, Sekiguchi T (2004) Landslide risk evaluation and hazard mapping for rapid and long-travel landslides in urban development area. Landslides 1(3):221–235

Tibaldi A (2001) Multiple sector collapses at Stromboli Volcano, Italy: how they work. B Volcanol 63:112–125

Tinti S, Armigliato A, Manucci A, Pagnoni G, Zaniboni F (2003) Le frane e i maremoti del 30 dicembre 2002 a Stromboli: simulazioni numeriche. XXIInd Convegno GNGTS, Rome (in Italian)

Tommasi P, Boldini D, Rotonda T (2005a) Preliminary characterization of the volcanoclastic material involved in the 2002 landslides at Stromboli. In: Bilsel H, Nalbantoğlu Z (eds) Proceedings of the International Conference on Problematic Soils GEOPROB 2005, vol. 3, Famagusta, pp 1093–1101

Tommasi P, Baldi P, Chiocci FL, Coltelli M, Marsella M, Pompilio M, Romagnoli C (2005b) The landslide sequence induced by the 2002 eruption at Stromboli Volcano. In: Sassa K, Fukuoka H, Wang FW, Wang G (eds) Landslides: risk analysis and sustainable disaster management. Proceedings of the First General Assembly of the International Consortium on Landslides, Washington DC, 12–14 October 2005, this issue

Wang G, Sassa K (2002) Post-failure mobility of saturated sands in undrained load-controlled ring shear tests. Can Geotech J 39: 821–837

Chapter 22

Characteristics of the Recent Landslides in the Mid Niigata Region – Comparison between the Landslides by the Heavy Rainfall on 13 July 2004, and by the Intensive Earthquakes on 23 October 2004

Hiromitsu Yamagishi* · Lulseged Ayalew · Koji Kato

Abstract. Niigata Region, Japan, is known as that designated landslides are more than 2 000 sites which are the most abundant in Japan, and still now we have experienced with landslides in the mountainous areas in Niigata Region (Yamagishi and Ayalew 2004; Yamagishi et al. 2004). Most of the landslides are deep-seated and taking place on gentle slopes in the Neogene to Pliocene mudstone areas. These triggers are mostly by snow melting. However, on 13 July 2004, heavy rainfalls due to the intensive activities of rain front occurred in the Mid Niigata Region, Japan. They are as much as 400 mm in 24 hours, and brought about serious flooding by breaking the river banks, as well as landslides. The heavy rainfall-triggered landslides are recognized as 3 600 sites. Most of the landslides are shallow seated, but some of them are more or less deep-seated and associated with long-run mudflows. Followed by such heavy rainfalls, the southern region of Mid Niigata Region was attacked by intensive earthquake of *M* 6.8 on Richter scale on 23 October 2004. The main earthquake was followed by intensive and small after-shocks until December 2004. By these earthquakes, many landslides also occurred in the hilly and mountainous areas. These landslides are classified into three types; one is deep-seated slides, the second shallow landslides, the third is flowing slides. Namely, in 2004, Niigata Region has been experienced with different induced landslides which are also different from the used landslides characteristic of Niigata, Japan. Therefore, in this paper, we are describing the distribution and characteristics of the heavy rainfall-induced and the intensive earthquake-induced landslides, and then comparing in features and scales with the different-trigger landslides.

Keywords. Landslides patterns, heavy rainfalls, intensive earthquakes

22.1 Introduction

In 2004, ten typhoons landed and heavy rainfalls gave big damages throughout Japan. In particular, Mid Niigata Region (Fig. 22.1a) was damaged by not only 13 July heavy rainfalls, but also 13 October intensive earthquakes (Fig. 22.1b). The heavy rainfalls claimed the lives of 13 peoples by flooding of the two rivers such as Kariyatagawa and Ikarashigawa, both of which are branch tributaries of the Shinanogawa (the longest river in Japan). They also claimed only 2 human lives by the landslides, however, more than 3 600 landslides occurred in the region. Three months later, considerable intensive earthquakes attacked the southern area of the heavy-rainfall areas. Totally 40 persons were killed directly and indirectly by the earthquakes. The characteristics of the earthquakes were that strong shaking affected the hilly and mountainous areas. Therefore, many landslides gave damages to many infrastructures and residential houses. Both of the landslides by the heavy rainfalls and earthquakes attacked the similar geologic and geomorphologic conditions. In this paper, therefore, we are discussing the features of both of the landslides, and comparing in scales, types and other characteristics between the different-triggered landslides.

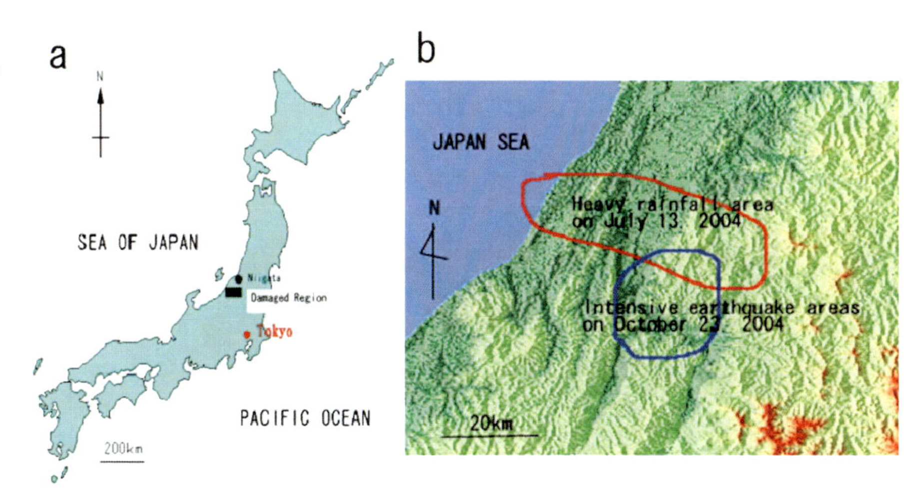

Fig. 22.1. Location maps of the Mid Niigata area affected by the heavy rainfalls on 13 July 2004 and by the intensive earthquake of 23 October 2004

22.2 Landslide Distribution by Heavy Rainfall on 13 July 2004

22.2.1 Precipitation Conditions of Heavy Rainfall

According to Ushiyama (2004), the rainfall began in the evening of 12 July 2004 and reached its peak in the early morning of 13 July in the lowlands and in the mountainous areas. Figure 22.1b presents the rainfall distribution on 13 July in the areas where flooding and landsliding mostly occurred. In the roughly 1 250 km^2 area considered, the Tochio and Kariyatagawa localities got the maximum rainfalls of the day. The 24 hour rainfall of AMEDAS station in Tochio City was recorded to be 421 mm.

Many landslides have been inventoried through aerial photographs of Asia Aerial Service. Particularly, Izumozaki map area (Fig. 22.2a) and Yoita map area (Fig. 22.2b) are recorded as high density of landslides. In particular, the Yoita map area is characterized by more than 500 landslides recorded.

22.2.2 Pattern of Landslides by the Heavy Rainfalls

Information on landslides occurring on 13 July 2004, was collected from field research data of 270 sites (on 31 May 2005). By the researching and aerial photographs, we have been recognizing slope surface.

22.2.3 Surface Failures

They were classified into planar type failures and spoon type failures. The planar type failures generated on steep slopes with thin engineering soils, and usually on anti-slope structures of hard mudstones as shown in Fig. 22.3a. Spoon type failures occurred at heads of valleys or concaved slopes showing gentle slopes with thick soft soils as shown in Fig. 22.3b.

22.2.4 Deep-Seated Landslides with Long-Run Mudflows

These types of slides are characteristic of that the collapsed materials spread downward long (Figs. 22.3c and 22.3d). The sliding scars are deep and wide and the sliding materials almost completely separate from the scars. Many of the large slides with wide and deep scars of tens of meters are associated with fine-grained debris (mudstone or clay) of several meters to tens of meters long. Some of the landslides took place as an earth-flow.

22.3 Landslides Triggered by the Intensive Chuetsu Earthquake on 23 October 2004

22.3.1 General Introduction to the 2004 Chuetsu Earthquake

The main shock attacked the Yamakoshi area at 17:56 on 23 October. This earthquake had the hypocenter located in 37°17.4' N in latitude, 138°52.2' E in longitude with a depth of 13 km, and had the magnitude is 6.8 on the Richter scale. In Kawaguchi Town, the earthquake marked seismic intensity of 7 of the Japanese Meteorological Agency scale, which is the maximum degree of this scale. The main shock was followed by a number of large aftershocks; in particular, three of them which attacked the area within two hours after the main shock, had magnitudes larger than 6. The maximum acceleration exceeded 1 000 gal, 7 km away from the epicenter. This Yamakoshi area is characterized by typical folding structures trending in NNE-SSW directions, some of which are associated with active faults and small folding. Many seismographic data such as Active Fault Research Center (2004), revealed that a NWW-SEE compression stress was forced, therefore, they presumed that the active reverse faulting took place in the same way in the underground of ca. 10 km deep. Recently, Active Fault Center, Japan, reported in the home page (2004) that a seismic fault appeared on the surface at Obiroo area.

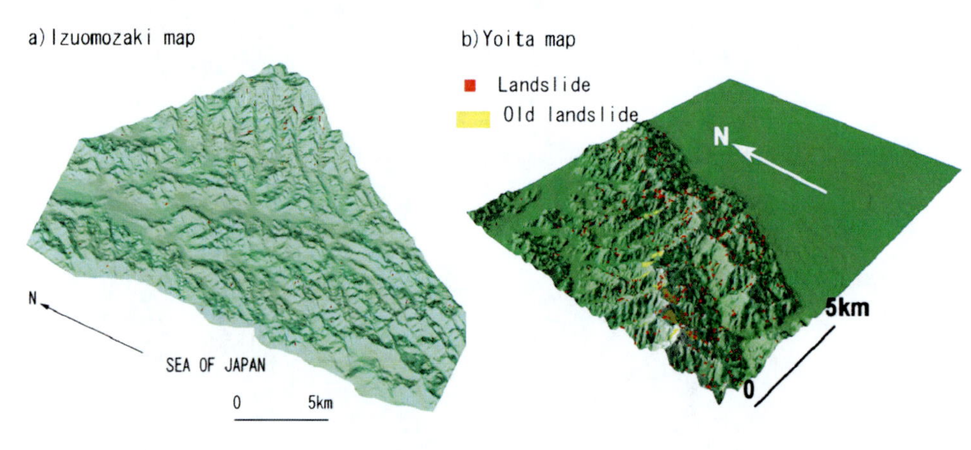

Fig. 22.2. 3D expression of landslide distribution maps of Izumozaki area (**a**) and Yoita area (**b**), using GIS. Notice that the large slide area in **b** indicate the old deep-seated landslides (by LDNIED 2005)

Fig. 22.3. Typical landslide types on 13 July 1994. **a** Planar type surface failure in Washima Beach. **b** Spoon-type failure in Aida, Izumozaki. **c** Deep slide with long-run flow at Tanokuchi, Tochio, **d** the simiar type at Ichinokura, Tochio. Courtesy of Nakanihon Air Service Co. Ltd.)

22.3.2 Distribution and Characteristics of the Landslides

By the earthquakes, many landslides were induced (Fig. 22.5a), because the earthquake area is the hills and mountains. Geology of the area is composed of Miocene to Pliocene sandstones and mudstones, some of which vary in hardness and weathering. Geological structure is characteristic of NNE-SSW trending anticlines and synclines, most of which are known as active. Some areas are recognized as questa topography due to dipping stratified sedimentary rocks.

Several hundreds of deep-seated landslides and incredible number of shallow disrupting landslides were generated by the earthquake, forming more than ten landslide dams, although the landslide hazard areas were limited only within the epicentral area of 15 by 30 km. According to Kawabe et al. (2005), total landslides are counted as more than 1 662, and the total sediments yields are estimated as 7×10^7 m^3.

The aerial photograph interpretation and field researching have been revealing that the landslides are classified into the following three types as shown Fig. 22.4.

1. Deep-seated slides (Fig. 22.4a): They are large-scale and took place mostly in the sandstone areas or sandy mudstone areas. They are regarded as primary or reactivation of old landslides. Most of them are gliding, along the dipping strata, but some are slumping type.
2. Flow type slides (Fig. 22.4b): They are long-run slides, and mixing with muddy materials and water probably derived from valleys or ponds for feeding carps.
3. Shallow-seated slides (Fig. 22.4c): They are small-scale and took place on steep slopes from the knick points.

Fig. 22.4. Typical pattern of the landslides by the earthquake (Coutesy of Nakanihon Air Service Co. Ltd.). **a** Deep-seated landslide at Higashitakezawa; **b** flow type landslide at Oguriyama; **c** shallow-seated landslides at Hanzogane

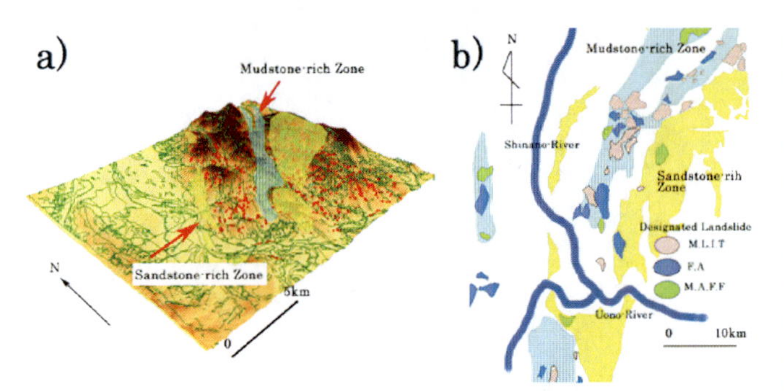

Fig. 22.5. a 3D map using GIS. It was combined of the earthquake-induced landslides, geologic maps (Takeuchi et al. 2004). **b** Landslide designated map (Niigata Prefecture 2004), related with brief geologic map. Notice that M.L.I.T, F.A and M.A.F.F indicate the landslides are governed by Ministry of Land, Infrastructure and Transport, Forest Agency and Ministry of Agriculture, Forest and Fisheries, respectively

In detailed, the (1) deep-seated slides took place, as shown in Fig. 22.5a, in the area of sandstone-rich zone with distinct stratification, and they occurred as dip-slipping. Therefore, geomorphologically, they occurred on more ore less gentle slopes. On the other hand, the (3) shallow-seated slides occurred on steep slopes which are in places anti-slope slipping. However, such anti-slope has large joints crossing the bedding, deep-seated slides occurred. Finally, the (2) flow type slides occurred mixing with abundant water from ponds for carp feeding, small rivers or channels and others.

22.4 Comparison between the Landslides by the Heavy Rainfalls and Those by Chuetsu Earthquake on 23 October 2004

Both of the landslide-devastated areas are hilly to mountainous area up to 600 m, named as Uonuma, Nishiyama and Higashiyama Hills located in the 80 km south of the Niigata City. These areas are composed mostly of young, weak sedimentary rocks from the Miocene to Pleistocene in age. Due to these geomorphic and geologic settings, many old landslides are distributed in this area. Therefore, they also characterize the geomorphic features in these hills.

The heavy rainfall-induced landslides took place mostly in the mudstone-rich areas and they are classified into two types of surface failures and deep landslides associated with long-running mudflows. The difference between the two depends mostly on thickness of the weathering materials rather than geologic structures of bedrock although some of them are controlled by jointing and dipping-strata, and the others took place as reactivation of the old landslides. While, the landslides triggered by the intensive earthquakes were strongly controlled by the geologic structures and rock facies. Because, deep-seated large scale landslides some of which blocked several rivers to form landslide dams, took place along the dipping of the strata and in the sandstone-rich zones rather than mudstone-rich zones (Fig. 22.5a). While, most of the designated landslide areas are concentrated into the mudstone-rich zone rather than sandstone-rich zones (Fig. 22.5b). It suggests that usual landslide movement has been triggered by hydrological causes, and that considerable prevention facilities were set already. In addition, the flow type slides and shallow-seated slides by the earthquakes are similar to the deep landslides with mudflows and the surface failures by the heavy rainfalls, respectively. Therefore, both of the landslides by the heavy rainfalls and intensive earthquakes have several common characteristics. However, shallow slides by the heavy rainfalls are more abundant in wider areas, but small in scale, because they occurred mostly along concave slopes. Therefore, we can evaluate each landslide site. On the other hand, the shallow slides by the intensive earthquake are also abundant, but in small areas, and are much more large in scale. In addition, the earthquake slides generated mostly from the knick points, and on planar or convex slopes.

22.5 Summary and Discussion

This study is being conducted on landslides due to the heavy rainfalls occurring on 13 July 2004, and landslides by the intensive earthquakes on 23 October 2004. Both of the landslides have been checked through aerial photographs and in the field. As the results, the patterns of landslides by the 7.13 heavy rainfalls, were classified into slope failures and deep-seated slides with mudflows. While, in terms of the landslides by the 10.23 earthquake we have identified the three types of landslides; shallow-seated slides, flow type slides and deep-seated slides or slumps.

Comparison in relationship to geology, has been revealing that the landslides by the earthquake were controlled by geologic structures and rock facies. However, the landslides by heavy rainfalls depend mostly on the weathering materials rather than bedrock geology.

Acknowledgments

We are grateful to Niigata Prefecture, Asia Aerial Service Co. Ltd., Aero Asahi Corporation Co., Ltd. and Shin Engineering Consultant Co. Ltd., for providing many data, and to Niigata University for providing funds for researching and getting much information.

References

Active Fault Research Center (2004) http://unit.aist.go.jp/actfault/activef.html

Kawabe H, Gonda Y, Marui H, Watanabe N, Tsuchiya S, Kitahara H, Osanai N, Sasahara K, Nakamura Y, Inoue K, Ogawa K, Onoda S (2005) The sediment related disasters caused by the mid Niigata Prefecture earthquake in 2004. J Jpn Soc Erosion Control Eng 57(5):39–46

LDNIED (Landslide Database of National Research Institute for Earth Science and Disaster Prevention) (2005) http://lsweb1.ess.bosai.go.jp/jisuberi/jisuberi_mini_Eng/jisuberi_top.html

Niigata Prefecture (2004) Landslide designated map of Niigata Prefecture

Takeuchi K, Yanagisawa Y, Miyazaki J, Ozaki M (2004) 1 : 50 000 digital geologic map of the Uonuma Region, Niigata Prefecture (Ver. 1). http://www.gsj.jp/jishin/chuetsu_1023/geomap.html

Ushiyama M (2004) Characteristics of heavy rainfall disasters in the Niigata Prefecture from 12 to 13 July 2004. Journal of Japan Society for Natural Disaster Science 23:293–302

Yamagishi H, Ayalew L (2004) Recent landslides in Niigata Region, Japan. Bull Geol Soc Hongkong 207–212

Yamagishi H, Marui H, Ayalew L, Sekiguchi T, Horimatsu T, Hatamoto M (2004) Estimation of the sequence and size of the Tozawagawa landslide, Niigata, Japan using aerial photographs. Landslides 1(4):299–303

Chapter 23

Slope Instability Conditions in the Archaeological Site of Tharros (Western Sardinia, Italy)

Paolo Canuti · Nicola Casagli · Riccardo Fanti*

Abstract. The archaeological site of Tharros in western Sardinia (Italy), of Phoenician-Punic origin, is exposed to different types of landslide hazard. This paper gives a description of the main geological and geomorphological features of the site with special reference to the impact of slope instability on the archaeological heritage.

Keywords. Archaeological sites, instability

23.1 Geologic, Geomorphic and Climatic Characteristics of the Area

The remains of the city of Tharros are located in the Capo San Marco Peninsula (western Sardinia, Fig. 23.1) which stretches out in a north-south direction and forms, together with the facing Capo Frasca, the large Oristano Gulf. The promontory, consisting of three low hills (Murru Mannu, Torre of San Giovanni Hill, Capo San Marco Hill) linked by a thin, flat isthmus, has sub-vertical walls on its western flank but gently slopes to the sea on its eastern margin. A stratigraphic sequence of relevant interest for the paleo-ecology of the Miocene and Pliocene outcrops in the area and is therefore object of numerous geology, stratigraphy and paleontology papers. (Ambrosetti 1972; Pecorini 1972; Cherchi 1973; Caloi et al. 1980; Lecca et al. 1983; Sanna 1983; Carboni and Lecca 1985; Sanna 1989).

Fig. 23.1. The peninsula of Capo San Marco with the ruins of Tharros

This stratigraphic sequence consists of (from the bottom):

- dark gray clays (Tortonian);
- green silts and white calcareous marls (Messinian, Capo San Marco Formation);
- breccias, arenites and silty clays (Lower Pliocene);
- dark gray basalts, closely fractured (Plio-Pleistocene);
- fossil-rich conglomerates, breccias and sandstones (Middle Pleistocene);
- massive, fossil-rich sandstones (Middle Pleistocene, San Giovanni Formation);
- fossil-rich conglomerate (Upper Pleistocene);
- fossil-rich conglomerates and sandstones (Upper Pleistocene, Santa Reparata Formation);
- layered and massive aeolic sandstones (Upper Pleistocene, San Giuseppe Formation);
- dune and beach sands (Holocene).

A geological survey of the Capo San Marco promontory, carried out to verify and integrate the information already available on the geological outlay of the area, led to the realization of the geologic map shown in Fig. 23.2. In the previously described sequence three principal sedimentary bodies of marine origin are recognizable. These are separated by the basalts produced during the Plio-Pleistocene volcanic phase, by continental deposits (aeolic sandstones of the San Giovanni and San Giuseppe Formations), by paleosoils and by at least one erosive surface. These are testimony of at least three transgressive phases starting from the Tortonian, the sediments of which have been ascribed, in the literature, to various Tyrrhenian episodes (Pecorini 1972; Carboni and Lecca 1985).

With the aim of identifying the types of mass movements present in the area a geomechanical and geotechnical characterization of the lithotypes was carried out. In particular, a point load index test was carried out on irregularly shaped samples collected on the surface, the results of which are shown in Fig. 23.3. The geologic outlay of the promontory strongly controls the morphological features: the structure of two of the three hills con-

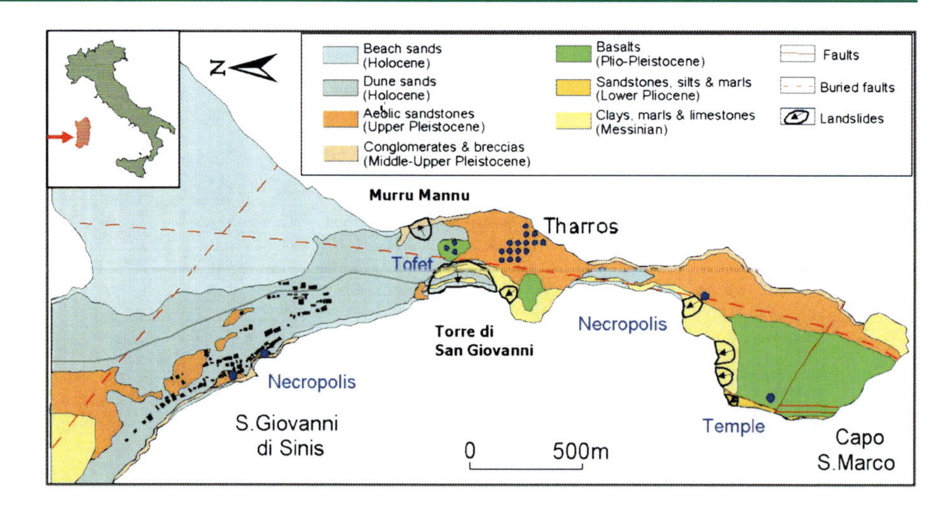

Fig. 23.2. Geological sketch of the San Marco Peninsula

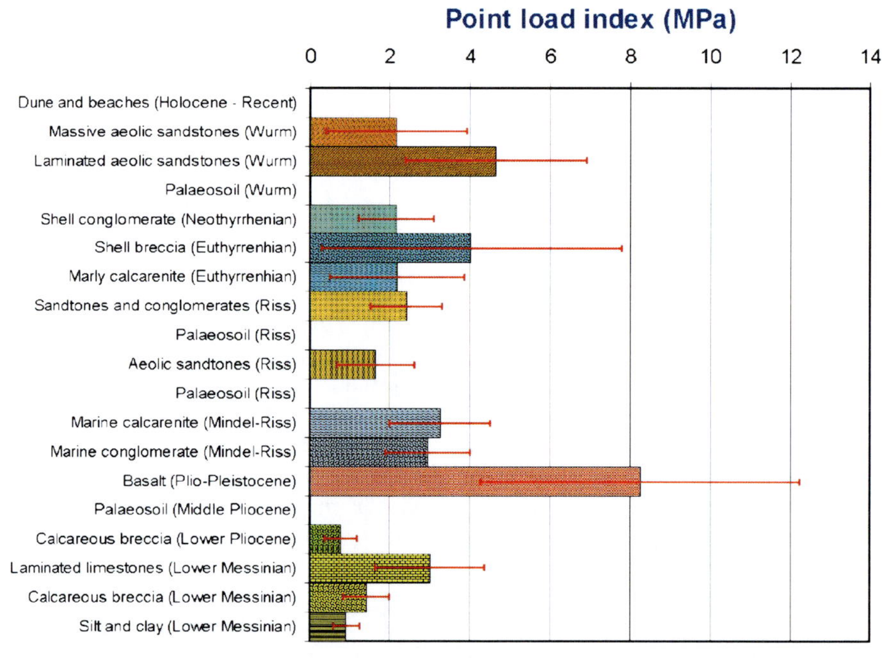

Fig. 23.3. Geomechanical stratigraphy based on the results of point load tests

sists of basalt outcrops, the margins of which, especially in the south-western sector, present sub-vertical walls. Here rock falls are frequent and concentrated in the zones where the basalt is more heavily fractured. Roto-translational slides are frequent in the clay outcrops and, in the particular case in which the clays are overlain by resistant rocks, rockfalls and toppling also take place. The largest earth slide occurs on the western slope of the Murru Mannu while a complex movement affects the northern slope of the Capo San Marco Hill, where the aeolic sandstones of the Upper Pleistocene overlay the Messinian clays. Another complex landslide, leading to a particular trench structure, is located on the western slope of the Cape, where basaltic flows overlay clays. Rockfalls are also visible along the San Giovanni di Sinis coast, where great part of the stratigraphic sequence outcrops: in this case the falls are caused by wave action on the sea cliff. This faces WSW, corresponding to the direction of the prevailing winds. The anemometric regime of the peninsula is, in fact, characterized by a definite predominance of the winds coming from the western quadrant and includes the dominant wind (ponente-maestrale, W-NW winds).

The average monthly rainfall in the area varies between 5 mm and 77 mm, for an annual total of 571 mm, the average monthly temperatures range from 10.7 °C in February to 24.2 °C in August for an annual mean temperature of 16.3 °C. Based upon these data, the climate of the Capo San Marco Peninsula can be defined as warm-temperate with characteristics of sub-aridity, due to the scarce precipitations in the lengthy summer period.

23.2 Historical and Archaeological Summary

The origins of the city of Tharros are, according to archaeological evidence, to be traced back to the colonization of Sardinia by the Phoenicians (770–730 B.C.). The morphological characteristics of the promontory, sporadically occupied since the Late Bronze Age, make it an ideal port. The city probably represented an important stop within western Sardinia's littoral transportation network: this route, however, was subordinate compared to the principal internal network, which passed through Forum Traiani (Fordongianus), leaving Tharros partially isolated, but not for this less important economically (Acquaro and Finzi 1986).

Other information from historical sources regarding Tharros does not come from direct citations but rather from indirect references to events and regional issues involving the whole of Sardinia. From these, however, emerges a picture of a vital center of the coastal fortification system during the lengthy administrative period of Phoenician and Punic rule. The city was conquered by the Romans shortly after the first Punic war (238 B.C.) and played an active role in the following, major anti-Roman uprising of the years 216–215 B.C. The next centuries testify alternate fortunes fall upon the city, first involved in the colonial events of the declining Roman Empire and later ruled by the Vandalic and Byzantine empires, until it was completely abandoned following the Saracen raids in the ninth century (Acquaro and Finzi 1986; Zucca 1993). The evidence of these sixteen centuries of history is numerous and of complex interpretation, due to the inevitable overlaying and intersection of the diverse urban philosophies. Of particular interest are the urban nucleus with its temples, the civil and thermal buildings, the Roman road network, the water works and the tofet zone, the two necropolises of Capo San Marco (southern necropolis) and of San Giovanni di Sinis (northern necropolis); the remains of the so-called "rural temple" of Phoenician-Punic age located on the western side of the Capo San Marco (Acquaro 1978, 1983; Ferrari 1984, Acquaro and Finzi 1986; Acquaro 1991).

23.3 The Murru Mannu Landslide (Tofet Area)

The Murru Mannu, on which the archaeological tofet area is located, represents a residual relief from the dismantling of the basaltic deposits, formed in a phase of fissural volcanic activity during the Plio-Pleistocene tectonic stage. In its upper portion, the relief is mantled by basalt blocks, used as construction material by the Romans, overlying the clay and marl deposits of the Miocene. On both the eastern and northern flanks the aeolic sandstone from the San Giuseppe Formation is encountered; the entire sequence is for the most part buried under Holocene sand dune and beach deposits.

The survey confirmed the existence, on the western slope, of a roto-translational landslide of approximately 20 × 100 m, which extends from sea level up to the base of the Roman walls around the tofet and includes the terminal part of the access road to the archaeological area (Figs. 23.4 and 23.5). The landslide crown cuts the Roman road near the ancient Cornensis Gate, causing a vertical displacement of 70–80 cm in the road pavement.

The dune sands, which cover great part of the area, partially hide the crown and the lateral limits of the landslide, therefore the slide surface geometry and the overall

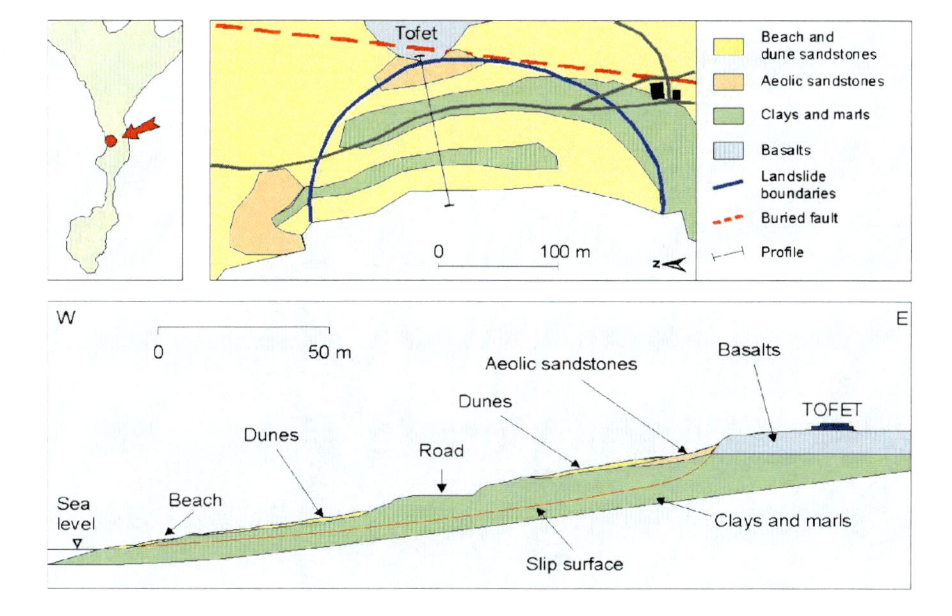

Fig. 23.4. Geological sketch of the Murru Mannu slide

volume of the landslide are not completely known. Several hypotheses are currently under investigation: *(a)* a deep sliding surface, presumably within the Miocene clays and marls; *(b)* coalescence of several shallow movements placed at the contact between the dune sands and the underlying clays. These two hypotheses, not necessarily in contrast with each other, are currently being investigated with the aim of both completely characterizing the substratum and of monitoring present landslide activity. Two boreholes were first instrumented with an inclinometer and an open pipe piezometer and successively with two fixed-depth inclinometric probes, a wire extensiometer and a pressure transducer. These sensors have been connected to a data logger powered by a solar panel, for continuous monitoring of the landslide parameters. Four cone penetration tests in the area (Fig. 23.6) identical with 3 topographic profiles that extend from the hillcrest down to the waterline were carried out.

It is interesting to note that these tests could help shed light on the dynamics of landslide movement in historic times and on how it influenced the evolution of the city. It is possible that the landslide damaged the Roman aqueduct and the main access road thus playing a crucial role in the demise and abandonment of Tharros in the ninth century. The spatial relationship between the present-day landslide and the above mentioned structures is shown in Fig. 23.7.

Fig. 23.5. General view of the Murru Mannu slide

23.4 Rockfalls in the San Giovanni Di Sinis Necropolis

The necropolis extends along the margin of the cliff in front of San Giovanni di Sinis and contains several rectangular trench-type tombs excavated in the aeolic sandstone of the San Giuseppe Formation. Several tombs near the vertical cliff wall are evidently exposed to the detachment and fall of various sized (from submetric to decametric) blocks so that the tomb chambers are visible in both the bedrock and in the blocks that have fallen on the narrow beach (Fig. 23.8).

The tombs are excavated in the upper massive portion of the aeolic sandstone over the layered basal level; these sandstones closeout a complex stratigraphic sequence, observable from its bottom member to the top along the San Giovanni beach. Below the sandstone, breccias are alternated with cemented conglomerates of medium strength; the sequence is then closed by a calcarenitic layer. The result is a "sandwich" structure in which a level of lower resistance (breccias and conglomerates) is enclosed between two resistant lithotypes (Fig. 23.9). In these conditions the wave action causes differential erosion with the formation of overhanging blocks: this is readily observable in the section of coast where the largest blocks have accumulated (Fig. 23.8).

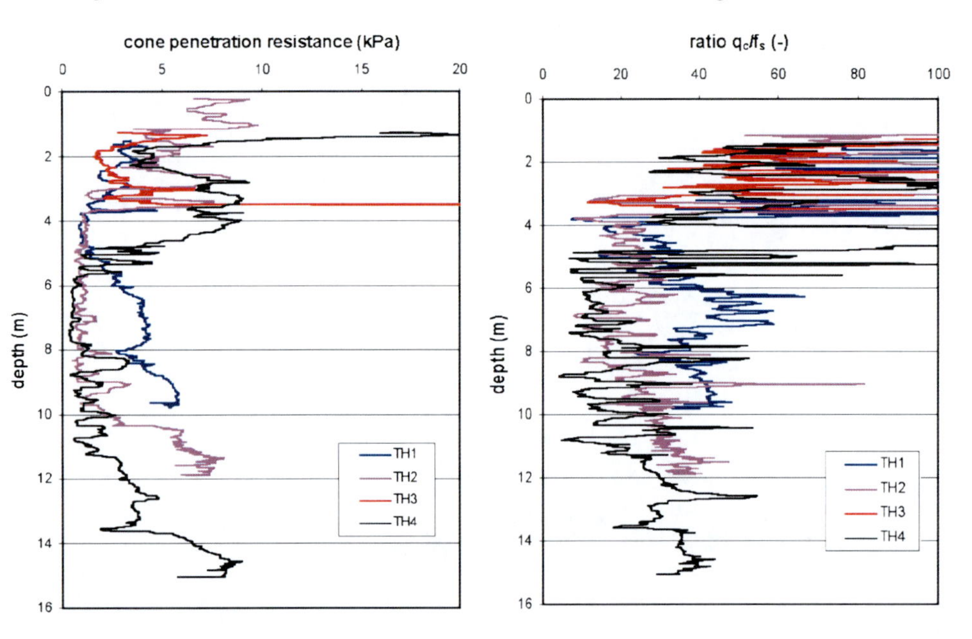

Fig. 23.6. Results of cone penetration tests in the Murru Mannu landslide

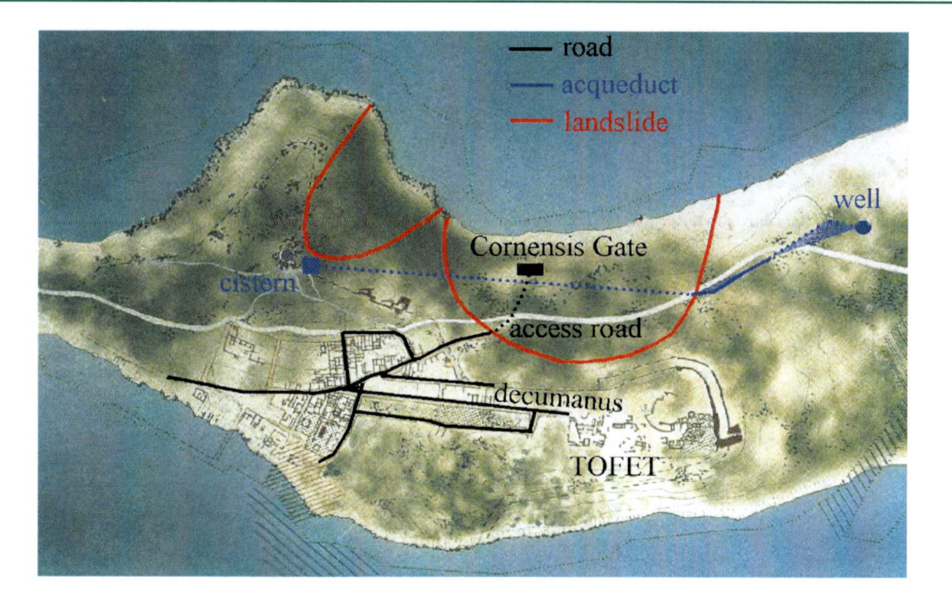

Fig. 23.7.
Relationships between landslides and Roman urban structures

Fig. 23.8.
Rock falls in the San Giovanni necropolis

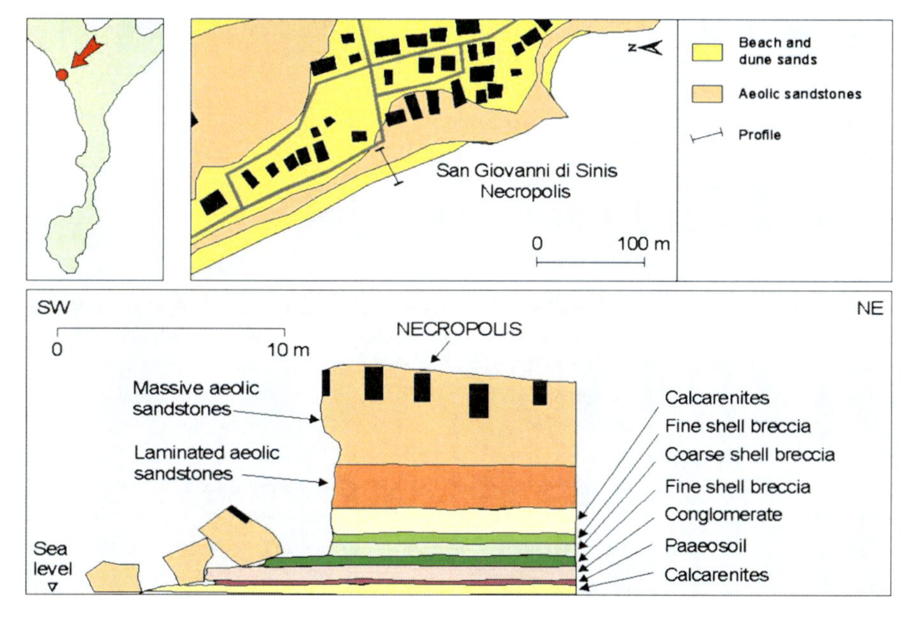

Fig. 23.9.
Geological sketch of the San Giovanni cliff

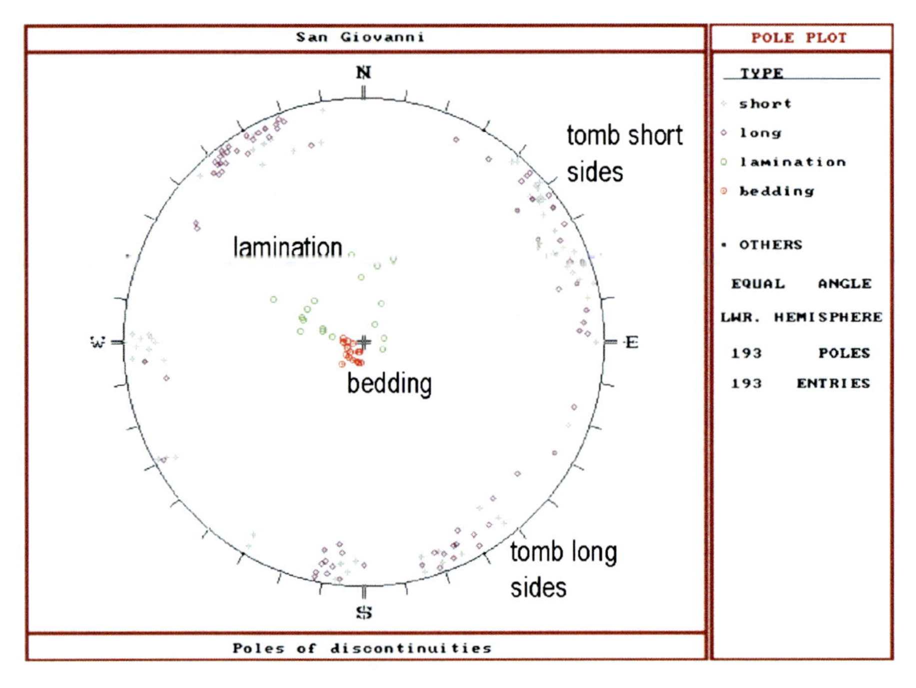

Fig. 23.10. Stereographic projection of the discontinuities in the rock mass

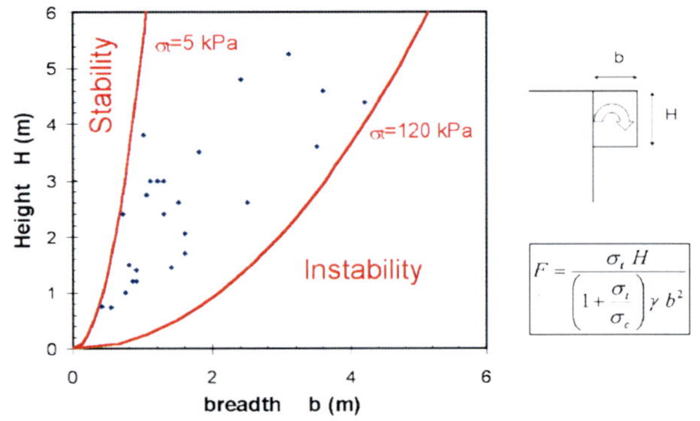

Fig. 23.11. Back-analysis of failures based on a cantilever mechanism

In these outcrops, however, the sandstone is layered and laminated but not jointed (Fig. 23.10), so that the tomb cuts are the principal factor controlling the conditions of blocks instability. These eventually fall when the portion of overhanging rock lengthens and becomes unstable following protracted wave action.

The size of fallen blocks over 10 cm was measured and blocks with at least one side delimited by a tomb cut were inventoried. These data were utilized for a back-analysis of the failure mechanism considering the cantilever model shown in Fig. 23.11: the resulting values of mobilized tensile strength are relatively low (5–120 kPa), especially if compared with the point load test results (2-5 MPa) carried out on the same sandstones (Fig. 23.3). This difference can be interpreted in two ways:

a weathering, mainly due to aloclastic processes and the wave impact, cause a marked decay in the tensile strength of the rock;

b the failure of blocks could have been artificially induced during the so-called "Tharros gold rush" that took place in the second half of the past century (Zucca 1993); during this indiscriminate archaeological campaign, explosives were utilized to knock down the dividing walls between the tomb chambers.

The rockfall mechanism, in any case, is strictly related to the dynamics of marine erosion, which in turn depends on the wind regime and is thus still to be considered active. The presence of large blocks at the foot of the cliff probably mitigates the effect of wave action, dissipating part of the marine energy. The inspection of the inscriptions present on the larger blocks has permitted to date the occurrence of the rockfalls to at least 20–30 years ago.

23.5 Mass Movements in the Capo San Marco Necropolis

The other necropolis of the Tharros area is located to the south of the isthmus that links the Capo San Marco basaltic promontory to the rest of the promontory and, as the tombs of San Giovanni di Sinis, is excavated in the aeolic sandstone (Fig. 23.12). Although it belongs to the same geologic formation, the structure of the sandstone is different in the two cases: at San Giovanni the sandstones overlay the breccias and conglomerates while on Capo San Marco they are in direct contact with the underlying clays and marls of the Miocene. The contrast in competence caused the development of a network of natural joints, that together with the artificial discontinuities of the tomb cuts, gave rise to complex block shapes (Fig. 23.13 and 23.14).

Fig. 23.12. View of the San Marco necropolis

To this already complex structural situation a landslide present in the pelitic deposits must also be added, the upper limit of which reaches the overlying aeolic sandstone. Consequently, the whole eastern sector of the necropolis is upset by several falls caused by the formation of numerous unstable blocks (geometrically dependent on the complex discontinuity network) and stemming from the mass movement (Fig. 23.13). On the NW side of the hill the fallen blocks, containing portions of the tomb chambers, accumulate on the slope beneath the cliff, while on the opposite side the tombs are dispersed on the flank that rises up to the basaltic plateau.

As with the San Giovanni cliff, an analysis of the joint network was carried out in which the natural joints were distinguished from the manmade tomb cuts. The results

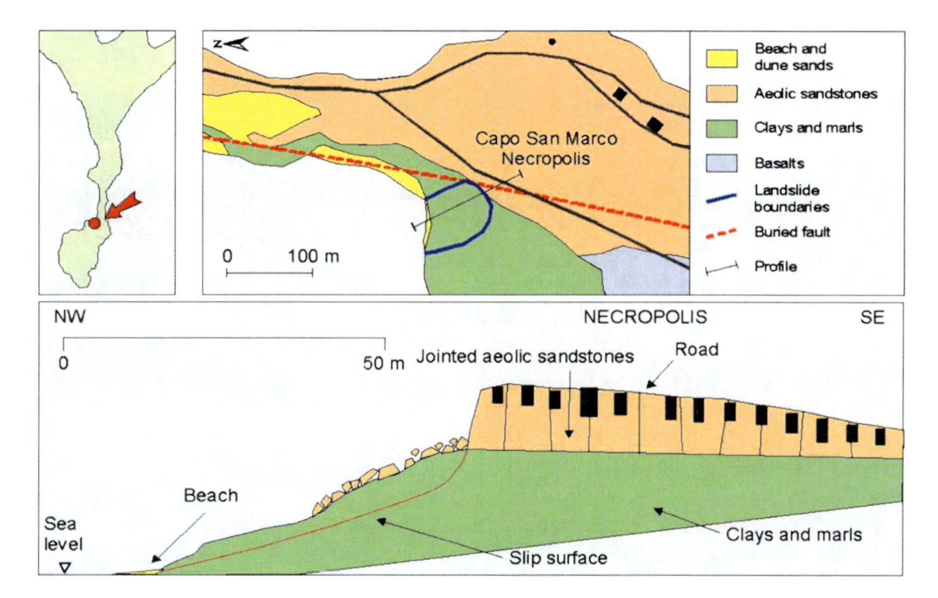

Fig. 23.13. Geological sketch of the San Marco cliff

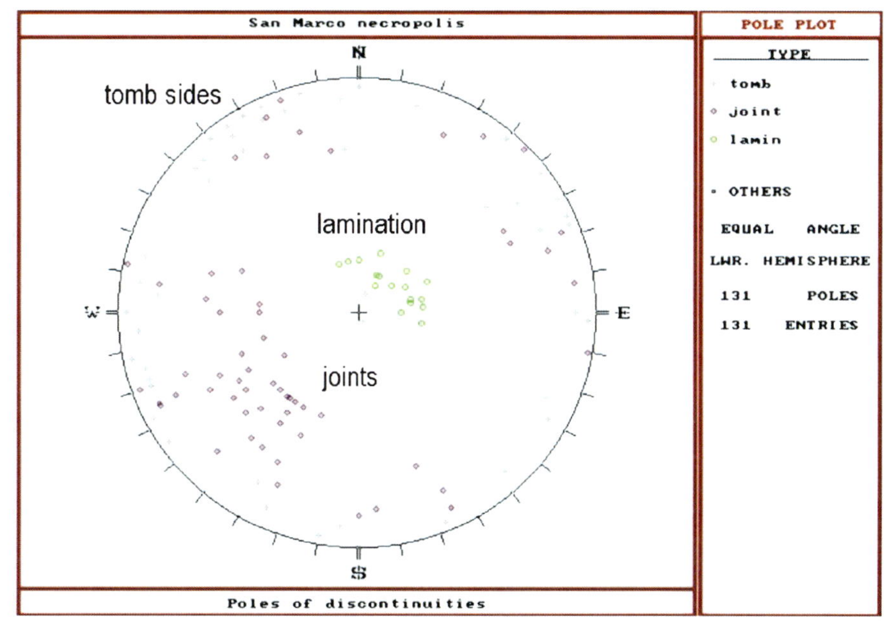

Fig. 23.14. Stereographic projection of the discontinuities in the rock mass

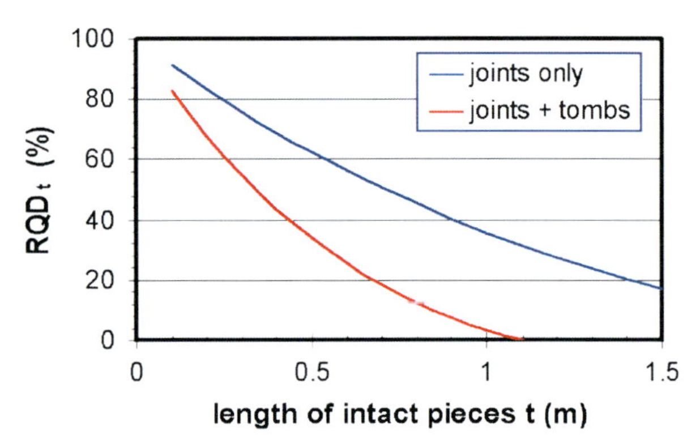

Fig. 23.15. Effect of the tomb cuts on the rock quality designation index distribution

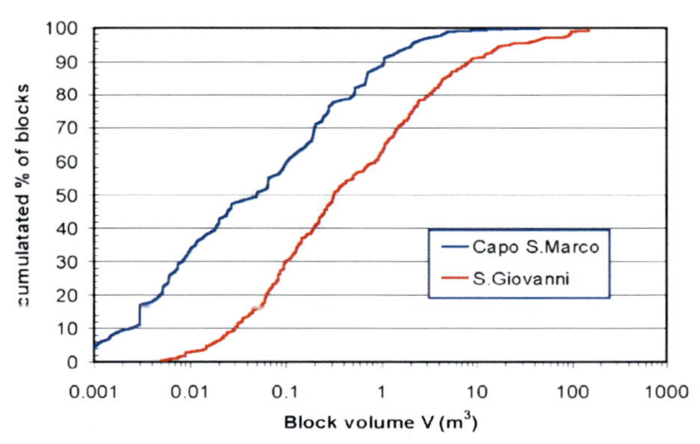

Fig. 23.16. Comparison of block volume distributions in the two necropolises

indicate that two different joint sets exist and are responsible for the formation of potentially unstable blocks of different sizes. The fallen blocks were measured as was the spacing of the discontinuities, both natural and manmade, that consented to determine the decay of the RQD index due to tomb excavation (Fig. 23.15). The effect of the interaction between the natural joint network and the artificial tomb cuts also results in smaller-sized blocks than the ones in the San Giovanni necropolis (Fig. 23.16).

The necropolis also extends beyond the maintenance road that leads to the Capo San Marco lighthouse and several tombs on the eastern portion of the promontory are located within fenced off private property that is partially shrouded by trees. The road almost appears to divide the area into two parts: to the west the tombs are in precarious conditions, being subject to rockfalls and to the retrogression of the landslide, while to the east the general layout of the terrain indicates that the tombs are intact.

23.6 The Complex Mass Movement of the Southern End of Capo San Marco

The eastern sector of Capo San Marco is a sub-horizontal basaltic plateau with a thickness of at least 35–40 m, overlying the sequence of clay and marl deposits. At the base of the basalt a paleosoil testifies of a period of emersion prior to the Plio-Pleistocene volcanism.

The basalt presents several joint networks, the spatial distribution of which is related to the cooling of the original flow. This determines falls of meter-sized blocks at the edge of the plateau along the western sea cliff; these accumulate at the foot and in the shallow water, where they are rounded by the wave action.

The remains of the Phoenician-Punic "rural temple" are located on the edge of the western cliff, in a precarious position, where they are threatened by future rockfalls of the underlying blocks. The area is also affected by a reactivated tectonic structure in which a complex network of faults has formed a series of parallel trenches (Fig. 23.17). These are

Fig. 23.17. Trenches in Capo San Marco

caused by uplifted blocks (horst) alternated with approximately 10 m deep trenches (graben) within which fallen rock debris is accumulated, thus covering the faults. This structure, although probably tectonically dormant, is complicated by a mass movement present within the underlying argillitic material that has caused the reactivation of the fault system (Fig. 23.18). The assessment of future evolution therefore also depends on the identification of the slide surface within the clay layers.

23.7 Conclusions

The geomorphic and geomechanical analysis presented in this note, permitted to identify the main sources of hazard which threaten the archaeological sites in the Tharros area. Marine erosion and the intrinsic weakness of the lithotypes have been recognized as the principal causes of destabilization. Countermeasures for risk mitigation should be aimed at completely halting marine undercutting in the unstable areas. Considering the peculiarities of the site, possible solutions could regard the realization of artificial gravel beaches to protect the cliff toes, associated with deep drainage measures within the pelitic formations.

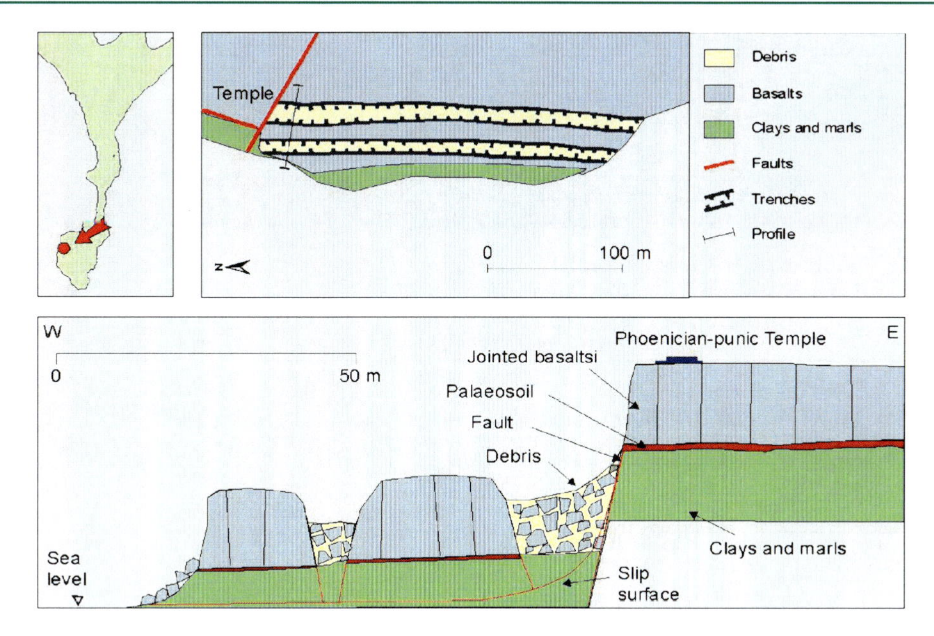

Fig. 23.18.
Geological sketch of the "rural temple" area

Acknowledgments

Research on geo-hydrological hazards affecting archaeological sites is carried out within the special Project "Safeguard of Cultural Heritage", of the Italian National Research Council, and the UNESCO-IUGS International Geological Correlation Program IGCP-425 "Landslide hazard in cultural heritage site". The authors are grateful to Dr. Giorgio Lollino (CNR-IRPI) for the cooperation during the monitoring activity and Dr. Giacomo Falorni for the revision of the text.

References

Acquaro E (1978) La necropoli meridionale di Tharros: appunti sulla simbologia funeraria punica in Sardegna. Orientis Antiqui Collectio, vol. XIII, pp 111–113

Acquaro E (1983) Nuove ricerche a Tharros. Atti del I Congresso Internazionale di Studi fenici e punici, Roma, pp 7–12

Acquaro E (1991) Tharros tra Fenicia e Cartagine. Atti del II Congresso Internazionale di Studi fenici e punici, Roma, pp 558–564

Acquaro E, Finzi C (1986) Tharros. Carlo Delfino Editore, Sassari

Ambrosetti P (1972) L'elefante fossile della Sardegna. Bollettino della Società Geologica Italiana 91:127–131

Caloi L, Kotsakis T, Palombo MR, Petronio C (1980) Il giacimento a vertebrati del Pleistocene superiore di San Giovanni di Sinis (Sardegna occidentale). Rendiconti di Scienze Fisiche, Matematiche e Naturali dell'Accademia Nazionale dei Lincei, vol. LXIX, pp 185–197

Carboni S, Lecca L (1985) Osservazioni sul Pleistocene medio-superiore della penisola del Sinis (Sardegna occidentale). Bollettino della Società Geologica Italiana 104:459–477

Cherchi A (1973) Appunti biostratigrafici sul Pliocene in Sardegna. Bollettino della Società Geologica Italiana 92:891–902

Ferrari D (1984) Per un recupero della necropoli punica di Tharros. Oriens Antiquus XXIII:97–106

Lecca L, Scarteddu R, Sechi F (1983) La piattaforma continentale sarda da Capo Mannu a Capo Marrargiu. Bollettino della Società Geologica Italiana 102:pp 57–86

Pecorini G (1972) La trasgressione pliocenica nel Capo S. Marco (Oristano, Sardegna occidentale). Bollettino della Società Geologica Italiana 91:365–372

Sanna MR (1983) Contributo alla conoscenza del Quaternario di San Giovanni di Sinis. Unpublished MSc thesis, Università degli Studi di Cagliari, Cagliari

Sanna MR (1989) Caratteri geografici e geomorfologici di Capo San Marco. In: Desogus P (ed) Tharros. La Poligrafica Solinas, Nuoro, pp 24–31

Zucca R (1993) Tharros. Edizioni Giovanni Corrias, Oristano

Chapter 24

'ROM' Scale for Forecasting Erosion Induced Landslide Risk on Hilly Terrain

Roslan Zainal Abidin* · Zulkifli Abu Hassan

Abstract. A study was initiated to classify and predict potential erosion induced landslide locations of occurrence at both well known resort areas of Malaysia namely Fraser Hill and Genting Highlands. The classification was done by determining the soil susceptibility for failure in terms of its soil erodibility index value with regards to the 'ROM' scale. Soil samples were taken on slopes at every 1 km stretch along the main road leading to both highlands. Concurrently, daily rainfall data of both areas were thoroughly examined to determine the erosive frequency.

From the soil samples analysis, Km 13-14 in Genting Highlands had been identified as the most susceptible location to landslide risk, while for Fraser Hill, Km 4-5 tops the ranking. The analyzed rainfall data however, had shown that the rainfall risk frequency is at the highest risk in the month of November and September for both Genting Highlands and Fraser Hill respectively.

Keywords. Highlands, erodibility, 'ROM' scale

24.1 Introduction

In Malaysia, erosion induced landslides pose substantial threats and over the years has caused severe damages. Factors like the infringement of the steep slope areas and accelerated rate of land development have contributed towards the escalation of the soil erosion related problems. Therefore, identification of potential erosion locations is crucial as it would lead to determination of landslide prone areas. To date, there are a number of places in Malaysia that have been identified as landslide prone areas. However, the problem still lies on when and where it will happen. A study had been carried out at two well known highland resort areas in Malaysia namely Genting Highlands (Fig. 24.1) and Fraser Hill (Fig. 24.2). This study was attempted to determine potential landslide hazard that would serve as a method to predict the possibility of landslides occurrence by the 'ROM' scale application (after the name of researchers *Ro*slan and *M*azidah).

In this study, locations of highly susceptible to landslide occurrence had been successfully identified, graded and ranked accordingly to the degree of soil erodibility risk potential. In other words, the locations were graded as Low, Moderate, High or Critical with respect to their degree of 'ROM' scale values.

Fig. 24.1.
Genting Highlands

Fig. 24.2.
Fraser Hill

24.2 Study Site

Fraser Hill which is located on the Main Range covering an area of 2 804 ha was developed as hill station for the British Administration which began back in 1919. This hill station is less developed compared to Genting Highlands where only 32 154 of tourist arrivals were recorded in 1999 and this is limited to Nature Tourism (WWF Malaysia 2001). At a mean altitude of 1 219 m a.s.l., the area experienced a temperature comparatively low from other parts of the country with daily temperature of 16.5–23.0 °C and receives annual rainfall between 2 000 and 2 700 mm. The access to the summit of Fraser Hill is via a place known as the Gap. The 9 km stretch road from the Gap to the summit is quite narrow and winding, allowing only a one-way traffic flow alternating between descending and ascending traffic time. The opening of the new road for public use which links Pine Resort to the Gap approximately 2 km longer than the old road is the new access road specifically used for ascending traffic due to its steep gradient while the old road for descending traffic. However this road was closed for public in the end of the year 2002 as it was found unsafe due to cases of landslide occurrences.

Genting Highlands is also located on the Main Range and it is Malaysian most modern and intensively developed hill resort. Daily temperature at Genting Highlands Range between 15–22 °C and receives annual rainfall between 2 000 and 2 600 mm. Most of the development is mainly concentrated at the altitudes of 1 000 m and 1 706 m. The Genting Highlands Resort is the main hub of activity with developments concentrated mostly on a 27 hectares plateau. This hill station, on the average attracts some 37 000 visitors, tourists and excursionists per day (Resort World Berhad 2001).

Access to the highland is from Genting Sempah, which is about 19 km. It is very steep as it rises from 550 m to 1 700 m in short distance of 19 km, giving an average slope of 6% with a well-maintained two-lane dual carriageway road.

24.3 Methods

In this study, the locations selected were localized at a specific area that can be described as the area where potential erosion may directly affect the road user to possible landslides occurrence. The study in Genting Highlands had been undertaken on slopes along the 20 km stretch of the leading main road from Genting Sempah – Genting Highlands resort (Fig. 24.3) and in the case of Fraser Hill, along a 9 km stretch of the leading main road from the Gap – Fraser Hill (Fig. 24.4). Soil samples were taken from the slopes at both main roads (Fig. 24.5 and 24.6) at every kilometer stretch.

As for the 'ROM' scale application, it requires the particle size distribution results from laboratory analysis (Roslan and Mazidah 2002). The results obtained from hydrometer analysis provide the input in terms of percentage of sand, silt and clay that were then substituted into the EI_{ROM} equation. The erodibility index value of the sample can be computed using Eq. 24.1. The applica-

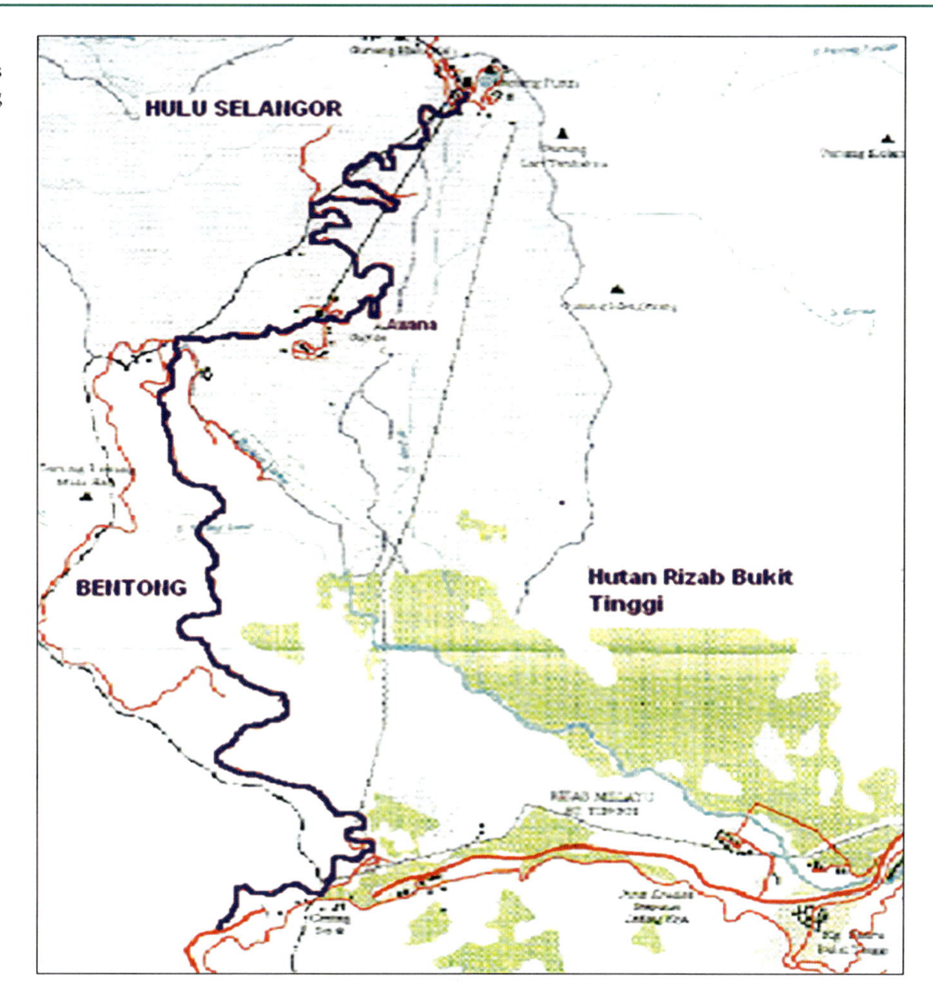

Fig. 24.3. Road map for Genting Highlands from Genting Sempah to Genting Highlands Resort (20 km)

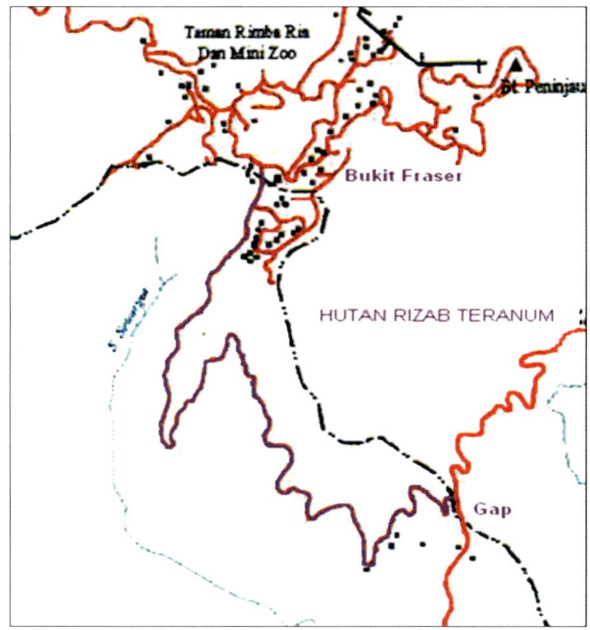

Fig. 24.4. Road map for Fraser Hill from Gap to Fraser Hill (9 km)

tion of this equation can be shown typically in the 'ROM' scale table.

$$\text{'ROM' equation} \quad EI_{\text{ROM}} = \frac{\%\,\text{sand} + \%\,\text{silt}}{2(\%\,\text{clay})} \qquad (24.1)$$

The 'ROM' scale index value was used as the ultimate indicator in determining the degree of soil erodibility. Although the grades are classified into Low, Moderate, High and Critical that correspond to its susceptibility towards erosion, the index value of each slope sample however is still significant as the value will be used to further rank the level of risk. Basically, the degree of risk were ranked based on their 'ROM' scale index value at every kilometer stretch of the road. The rank and the index value that remarks each slope sample subsequently could be used to interpret the spatial probability of potential erosion induced landslide occurrence.

While in the rainfall data analysis, the data that had been recorded over the period of five years, from 1997–2001 were analyzed. The frequencies of rainfall

Fig. 24.5.
Soil erosion feature at Genting Highlands

Fig. 24.6.
Soil erosion feature at Fraser Hill

on daily basis that exceeded 20 mm were taken as the amount of rainfall that is liable in creating soil erosion process in Malaysia (Roslan and Tew 1997). The numbers of erosive rainfall per month were comparatively analyzed for each and every year within the five-year period and the annual monthly patterns of the erosive rainfall were also observed.

Relative information on the time period of potential erosive rainfall at both areas in relation to the predicted spatial probability of erosion occurrence could therefore be converted into a temporal probability of landslide risk. In this case, the month by which the rainfall is classified as highly or critically potential to create erosion risk would lead to landslide occurrence.

24.4 Results

The ranking of soil erodibility index along the two main roads leading to both Fraser Hill and Genting Highlands are the results of the 'ROM' scale determination. The soil samples which is ranked on top is the one that has the highest possibility of being susceptible towards erosion or one which is potentially unstable. Table 24.1 and Table 24.2 show the sample ranking for both Fraser Hill and Genting Highlands respectively. For the erosive rainfall frequency analysis, Table 24.3 and Table 24.4 highlight the monthly erosion risk frequency recorded for the period of 1997–2001 where they are ranked accordingly.

Table 24.1. Degree of Erodibility Index based on 'ROM' Scale for Fraser Hill

Location (km)	'ROM' scale	Degree of 'ROM' scale	Ranking no.
4 – 5	5.75	High	1
6 – 7	5.35	High	2
3 – 4	5.33	High	3
7 – 8	5.31	High	4
5 – 6	5.10	High	5
2 – 3	4.96	High	6
1 – 2	3.92	Moderate	7
8 – 9	2.83	Moderate	8
0 – 1	2.37	Moderate	9

Table 24.2. Degree of Erodibility Index based on 'ROM' Scale for Genting Highlands

Location (km)	'ROM' scale	Degree of 'ROM' scale	Ranking no.
13 – 14	4.35	High	1
11 – 12	4.27	High	2
15 – 16	4.13	High	3
12 – 13	4.09	High	4
17 – 18	4.05	High	5
16 – 17	3.93	Moderate	6
14 – 15	3.80	Moderate	7
10 – 11	3.66	Moderate	8
6 – 7	3.50	Moderate	9
8 – 9	2.54	Moderate	10
9 – 10	2.49	Moderate	11
5 – 6	2.18	Moderate	12
7 – 8	1.96	Moderate	13
0 – 1	1.83	Moderate	14
1 – 2	1.73	Moderate	15
2 – 3	1.70	Moderate	16
3 – 4	1.32	Low	17
4 – 5	0.90	Low	18

Table 24.3. Ranking of erosive rainfall according to month for the period of 1997–2001 (Fraser Hill)

Ranking no.	Month	Erosive rainfall frequency
1	September	24
2	November	23
3	April	19
4	October	17
5	December	17
6	June	16
7	May	14
8	August	13
9	February	12
10	March	12
11	July	10
12	January	8

Table 24.4. Ranking of erosive rainfall according to month for the period of 1997–2001 (Genting Highlands)

Ranking no.	Month	Erosive rainfall frequency
1	November	26
2	October	23
3	April	22
4	August	20
5	December	19
6	May	18
7	June	16
8	July	16
9	March	15
10	September	15
11	February	11
12	January	3

24.5 Conclusions

For Fraser Hill, it was found that the general classification of the whole area on the 'ROM' scale is High with EI_{ROM} value of 4.55. It was also found that Km 4-5 as the location of highly susceptible to erosion which recorded 5.75 'ROM'. However, for Genting Highlands, the general classification of the whole area was found as Moderate on the 'ROM' scale with EI_{ROM} value of 2.91. The location of the highest potential to erosion had been identified to occur at Km 13-14 with EI_{ROM} value of 4.35 which is classified as High on the 'ROM' scale. This has shown that Fraser Hill area is more susceptible to erosion compared to Genting Highlands.

For the temporal assessment whereby rainfall is seen as a triggering or the impact factor in many events of landslide occurrences, the results of the erosive rainfall frequency analysis had revealed that September and November as the month where the landslide risk is the highest for Fraser Hill and Genting Highlands respectively.

The results of this study have therefore provided vital information on the susceptibility of the locations towards erosion induced landslides and the time of critical impact that would make way for landslide occurrence. Apart from that, the ranking list for soil samples in this study is also a form of hazard assessment that indicates the order in which landslide risk potential should be investigated.

References

Resort World Berhad (2001) Resort World Berhad annual report 2000, Kuala Lumpur

Roslan ZA, Mazidah M (2002) Establishment of soil erosion scale with regards to soil grading characteristic. Proc. 2nd World Engineering Congress, Sarawak, Malaysia, pp 235–239

Roslan ZA, Tew KH (1997) Rainfall analysis in relations to erosion risk frequency – case study (Cameron Highlands). Proc. 3rd International Conference on FRIEND, FRIEND 97, Postojna, Slovenia

WWF Malaysia (2001) Study on the development of hill stations, vol. 1. World Wildlife Fund Malaysia, Selangor

Chapter 25

Geotechnical Field Observations of Landslides in Fine-Grained Permafrost Soils in the Mackenzie Valley, Canada

Baolin Wang* · **Susan Nichol** · **Xueqing Su**

Abstract. Landslides in fine-grained permafrost soils have been paid much less attention compared to those in temperate regions. The lack of attention paid to those landslides can be attributed to their remote locations and relatively lower social and economic impacts. With recently increased interest and activities in the northern regions, especially from the energy sector, there is an increased need for better understanding of landslides in such regions. This paper describes some geotechnical field observations from a number of landslide sites recently visited in northern Canada. Evidence collected from the landslide sites provides valuable information for understanding the failure mechanisms and for further investigations. The information described includes: locations and orientations of the landslides, the slope and slide geometries, typical surface and subsurface material conditions, landslide flow phenomena, evidence of active layers (soils subject to annual freeze-thaw cycles), rate of head scarp surface ablation, and conditions of surface vegetation. Possible mechanisms triggering the landslides are discussed based on the evidence observed in the field. Landslide processes and stabilizing mechanisms are also discussed. Extreme weather conditions may have played a major role in one region with few or no trees, while forest fire may have been a dominant factor in another region with dense trees. The inherent characteristics of the slopes are also critical to the slope stability, for example, slope angle, thickness and strength of the active layer, soil moisture and ice contents, the insulating effect of the surface organic mat, and the reinforcing effect of roots. While changes in thermal regime are critical to slope stability in permafrost, the shear strength of the active layer can be a major contributor to slope stability under given thermal conditions. The paper discusses several aspects that are worth attention for further studies.

Keywords. Landslides, permafrost, active layer, transient layer, slope stability, Mackenzie Valley

25.1 Introduction

Recently renewed interest from the energy sector in oil and gas development in the Mackenzie Valley, Northwest Territories, Canada (Fig. 25.1) has emphasized the need for improved understanding of landslides in the region. While some landslide research has been carried out in the region (discussed below), in particular since the 1970s due to a pipeline initiative, various knowledge gaps about slope failure and movement mechanisms have been identified (Gartner Lee Ltd. 2003).

In response to industry need, a research project was initiated at the Geological Survey of Canada to carry out geotechnical studies on landslides in the region. As part of the research project, an initial field investigation was conducted in June 2005. The purpose of this field program was to visually inspect conditions at landslides and collect initial data required to plan future investigation programs.

During the June 2005 field program, a corridor approximately 750 km long by about 20 km wide was visually inspected from the air via helicopter. Several landslide sites were visited for visual inspections at a closer range. This paper summarizes the findings from the initial field program. It is anticipated that the data collected will be useful for continued studies of landslides in permafrost regions.

25.2 Overview of Previous Research in the Region

The following provides a general overview of some research carried out in the region.

Landslides in the Mackenzie Valley have been described and mapped since the 1970s (e.g., Code 1973; Aylsworth et al. 2000b; Dyke 2000, 2004). A number of common landslide types have been identified: for example, Code (1973) describes shallow active layer failures and larger scale multiple retrogressive slides of high banks; McRoberts and Morgenstern (1973, 1974a,b) describe solifluction, skin flows, bimodal flows, block slides and multiple retrogres-

Fig. 25.1. Location of study area. Inset area is shown in Fig. 25.2

sive slides; while Morgenstern (1981), Savigny and Morgenstern (1986a,b), and Dallimore et al. (1996) investigated the natural creep behavior of undisturbed clay permafrost and massive ground ice.

Isaacs and Code (1972), Hanna and McRoberts (1988), and Hanna et al. (1994) analyzed and monitored slope stability associated with pipeline design, construction, and operation. Quantitative analyses on slope stability were carried out by McRoberts and Morgenstern (1973, 1974a,b) which considered both the thaw-consolidation model and the ablation model. These analyses indicated the importance of excess pore pressure produced by thaw-consolidation on slope stability. Savigny et al. (1995) also noted that excess pore pressure may facilitate slope instability.

Pufahl and Morgenstern (1980) analyzed the energy balance that provided insight to understanding rate of ablation of landslide head scarps in permafrost. Lewkowicz (1988) discussed short-term rates of ablation and energy flux causing melting of exposed ground ice in the Mackenzie Delta.

Savigny et al. (1995), Hanna et al. (1998), and Dyke (2004) indicated the importance of forest fire on slope stability in ice-rich permafrost. The fundamental effect of forest fire is to reduce or eliminate the shielding vegetation canopy and the insulative quality of the organic mat. The surface albedo decreases and the ground surface temperature increases. As a result, long-term changes in the ground thermal regime may occur and annual surface thawing may deepen, which may lead to slope instability (e.g., Viereck 1982; Cleve and Viereck 1983; Harry and MacInnes 1988; Liang et al. 1991; Mackay 1995; Burn 1998; Tsuchiya et al. 2001; Lewkowicz and Harris 2005).

25.3 Surficial Geology and Site Conditions

The following summary is based on the detailed description of the surficial geology of the Mackenzie Valley given by Aylsworth et al. (2000a). Ground moraine (till) deposited by glacier ice covers much of the valley and also underlies many of the other surficial sediments. The portion of the valley shown in Fig. 25.2 is characterized by a cover of till that ranges from thin and discontinuous to thick and continuous (up to 60 m thick). Throughout the area there are local minor scattered glaciofluvial sand and gravel deposits, as well as fine-grained glacial lacustrine deposits occurring mainly along the Mackenzie River. Some coarse-grained glacial lacustrine deposits occur locally, mainly south of Fort Good Hope.

The till occurs as a variety of landforms. North of Little Chicago surficial sediments consist generally of hummocky, ridged or rolling till, although in some areas the till occurs as a flat to gently sloping plain. In this area the till consists predominantly of silt and clay with some coarser clasts. These fine-grained sediments commonly have 10–25% segregated ice occurring as thin, irregular

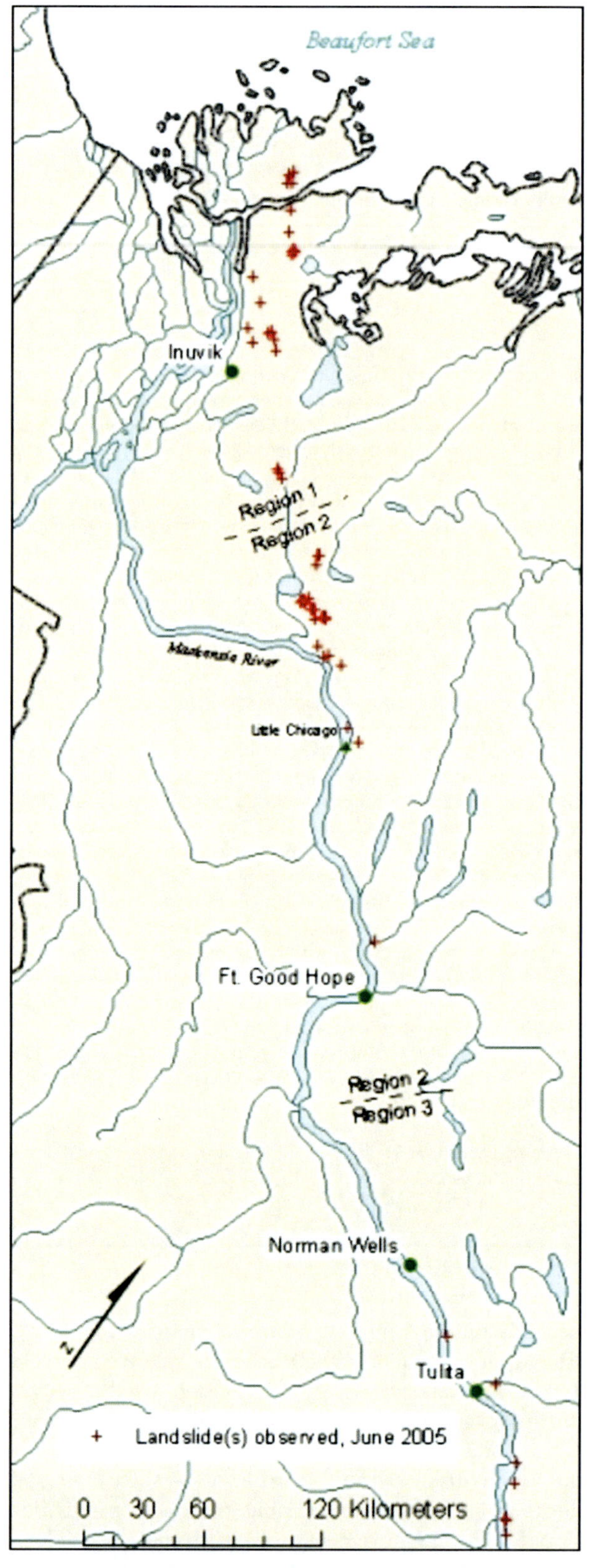

Fig. 25.2. Landslides observed in June 2005 along a 20 km wide corridor east of the Mackenzie River, with multiple landslides at some locations. See text for discussion of regions

discontinuous seams in the upper 2 to 3 m, with thicker lenses or irregularly distributed large masses of segregated ice at greater depth.

South of Little Chicago, the terrain is predominantly till up to 20 m thick in flat to gently sloping plains. The till in this area is generally coarse-grained, with 20–50% coarse gravel. South of Fort Good Hope the till occurs mainly as a veneer (<2 m thick) overlying bedrock, which consists of shale, sandstone and limestone. Bedrock generally occurs as prominent ridges, escarpments and hills, and surfaces are generally weathered or obscured by colluviums.

25.4 Geotechnical Field Observations

25.4.1 Landslide Distribution along Visited Corridor

The corridor visited in June 2005 was approximately 20 km wide and about 750 km long, from near the Beaufort Sea to south of Norman Wells (Fig. 25.2). The visited corridor can be classified into three distinct regions based on surface conditions and vegetation observed from the air:

- Region 1 From the Beaufort Sea south approximately 200 km: Gentle rolling terrain with few or no trees;
- Region 2 From 200 km to 500 km south of the Beaufort Sea: Gentle rolling terrain with slightly higher relief and with (often dense) trees; and
- Region 3 From 500 km to 750 km south of the Beaufort Sea: Hilly or rocky terrain, lowlands often with bedrock exposed, and often dense trees.

Landslides were noted at 97 locations along the narrow corridor visited in June 2005, as shown in Fig. 25.2. There are multiple landslides at several locations marked on Fig. 25.2. Most landslides with the potential to significantly impact pipelines are located in regions 1 and 2, where fine-grained permafrost soils were noted. Discussions in this paper are focused on the landslides found in regions 1 and 2. Landslides observed in region 3 are mostly related to rock slopes or riverbank failures, which are beyond the scope of the current paper, and are not discussed here.

25.4.2 Forms of Landslides Observed

Landslides observed in regions 1 and 2 are located mostly around lakes and sometimes on side slopes along creeks. Some slides appear to be relatively shallow in the form of skin flow, while others are relatively deep in the form of retrogressive failure.

The apparent shallow slides are often outlined by depressions of ground surface and distinct vegetation within the landslide zone (Fig. 25.3). Deep retrogressive failures often appear to be circular in shape as shown in Fig. 25.4. This type of landslide is sometimes called bi-modal flow (McRoberts and Morgenstern 1974a). It consists of a steep headscarp outlining the extent of ongoing failure and a tongue of flowing materials ablated from the headscarp. The headscarp is often nearly vertical. The height of the headscarps observed ranges from a few meters to over 20 m. The slope of the tongue is often very gentle and can be as

Fig. 25.3. A typical shallow flow

low as 1° to 3°. The active ablation of the exposed permafrost materials from the headscarp makes the entire landslide appear to be fresh although the flows have been active for years.

Some intermediate slides between shallow flows and deep retrogressive flows possess characteristics of the deep retrogressive flows. This type of flow appears to have a lower head scarp and sometimes a steeper tongue.

25.4.3 Soil Stratigraphy Observed at the Scarps of Deep Retrogressive Landslides

The headscarps of the deep retrogressive landslides are generally steep walls of exposed permafrost soils (Fig. 25.5). Continuous ablation of permafrost materials in summer makes it easy to observe the freshly exposed subsurface soils over the headscarp height, which can be as high as 15 m to 20 m.

Fig. 25.4.
A typical deep retrogressive failure

Fig. 25.5.
Head scarp of a retrogressive landslide

Fig. 25.6. Typical soil stratigraphy from a headscarp (note active layer marked by depth of roots exposed)

The most commonly observed soils in regions 1 and 2 discussed earlier are as follows (Figs. 25.5 and 25.6):

- surface 5 cm to 30 cm (locally thicker): organic mat consisting of roots or moss.
- under organic mat to about 100 cm to 150 cm depth: silty clay or clayey silt, with roots extending to the base of this layer, moist to wet with occasional ice wedges; and
- below 100 cm to 150 cm depth: silty clay or clayey silt with pervasive ice lenses throughout the exposed scarps.

Roots are exposed and hang on the scarp faces when the hosting soil is thawed and slowly ablated (Fig. 25.6). This feature was observed consistently at all landslide sites visited.

Tension cracks were often observed at the ground surface up to about 2 to 3 m from the crest of the headscarps. Some concentrated ice wedges were noted in the upper clay/silt layer, and are probably caused by freezing of surface water perched in the tension cracks. It was noted that thawing was only a few centimeters into the upper clay/silt layer at the time of the visit in June 2005. Some snow was still visible at some locations at that time.

Water content in the upper clay/silt layer appeared to be low compared to the lower clay/silt layer. Few or no ice lenses were visible in this upper layer except for the occasional concentrated ones discussed above. Photos of a soil sample taken from this layer are shown in Figs. 25.7a and b indicating frozen and thawed conditions of the sample. There is a clear contrast to that from the ice-rich layer below (discussed later).

Fig. 25.7. Soil sample taken from lower part of active layer. **a** Frozen state immediately after sample was taken from scarp. **b** Soil sample after being melted by hand

Fig. 25.8. Soil sample taken from upper part of permafrost. **a** Frozen state immediately after sample was taken from scarp. **b** Soil sample became liquid after being melted by hand

The ice content of soils below 100 cm to 150 cm depth is generally high as observed at all landslide sites visited in regions 1 and 2. Photos of a soil sample taken from this layer are shown in Figs. 25.8a (frozen) and 25.8b (thawed). The sample was taken from a scarp at a depth immediately below the exposed roots. The frozen sample changed to liquid when thawed, which is in clear contrast to the sample shown in Fig. 25.7.

25.4.4 Other Observations

In region 2, all landslides visited appeared to be associated with forest fire, as evidenced by burned trees. In region 1, on the other hand, no burning was evident around the landslides observed. In region 1, most slides seen were shallow and stabilized, while in region 2, more active deep slides were observed.

The slope appeared to be stable at one landslide site where a sand zone extended to depth on the scarp wall. This seems to have partially stopped or slowed down the sliding process in the area.

At the time of site visit in June 2005, the air temperature during day time ranged from 0 °C to about 20 °C. It was sunny and warm most of the time during site walks around the slides. It was observed that materials fell off the head scarps at a rate of a few cubic centimeters every few seconds here and there along the scarps. Larger pieces that can be measured in cubic meters (often less than 1 m^3) fell off the scarps occasionally from upper part of the walls.

25.5 Data Interpretation

Based on the information noted from the site visit, some factors contributing to landslides in the region are obvious, while others need to be investigated further. This section discusses the possible mechanisms of landslide processes based on the evidence collected.

25.5.1 Permafrost

The permafrost table in regions 1 and 2 is about 1 m to 1.5 m below ground surface. The upper 1 m to 1.5 m surface layer undergoes annual freeze-thaw cycles, and is conventionally called the active layer (discussed later). Most of the active layer was still frozen at the time of the visit in June 2005. However, the two zones (permafrost and active layer) were outlined on the observed scarps by two unique features: (1) ice content change; and (2) appearance of roots. Permafrost in fine-grained soils appeared to be ice rich. Visible stratified and occasionally massive ice is pervasive below about 1 m to 1.5 m depth in the scarps. Relatively little ice was observed in the active layer. The depth at which ice content changes coincides with that of the exposed roots on the headscarps (Figs. 25.5 and 25.6). It is obvious that the soil hosting roots has been undergoing freeze-thaw cycles.

A transient layer between the active layer and the permafrost must exist due to fluctuation of annual air temperatures. However, this layer was often not as obvious as might be expected. The transient layer was about 5 cm to 15 cm thick based on visual inspections at some locations. At other locations, ice-rich permafrost appeared almost immediately below the root-rich layer of soils. In any case, the transient layer may or may not hold a key for slope stability depending on the slope and temperature conditions as discussed later.

It is clear from Fig. 25.8b that the ice-rich fine-grained permafrost soil generates a significant amount of water when thawed. The soil changes from a solid to a liquid that cannot sustain shear stress. This material tends to flow if it thaws. If this happens in a slope and the excessive water does not have time to dissipate, it may trigger slope failure. Abnormal increase in ground temperature is therefore one of the most significant factors causing landslides.

25.5.2 The Active Layer

The active layer is subject to annual freeze-thaw cycles. Vegetation establishes roots in this layer. The relatively low water/ice content and the establishment of roots result in the active layer having much higher strength than the underlying ice-rich soil when thawed. The active layer is strong enough to support itself under normal conditions and also to provide some support when the underlying transient layer or permafrost is weakened by abnormal heat penetration. This supporting effect is limited to a certain extent depending on slope angle and the area or geometry supported. An imperfect active zone, such as might be expected near shorelines of lakes, may act as a weak zone and allow failure to initiate during the period when the permafrost zone has a tendency to flow. Pore water pressure may build up underneath the active layer if the underlying transient or permafrost layer thaws rapidly. This pore pressure could be due to thaw consolidation (Morgen-stern and Nixon 1971), or due to hydraulic gradient along the slope. In any case, the supporting effect of the active layer can be determined readily with conventional techniques for slope stability studies. Three dimensional slope stability analysis may be necessary to determine the critical size of the thawed zone for the active layer to fail.

25.5.3 External Heat

It is evident that all the landslides observed in region 2 are associated with forest fires. Monitoring results reported by Mackay (1995) and Viereck (1982) indicate that the depth of the active layer may increase by about 20 cm within a year following burning, and the increase may continue for a few years. As discussed earlier, the transient layer was observed to be about 5 cm to 15 cm thick at some of the visited sites. An increase in active layer depth of 20 cm within one year would extend thawing into those ice-rich permafrost soils and cause state change of the soils from solid to liquid. While quantitative studies are needed to confirm heat penetration rate and water seepage rate, this indicates that thawing-induced liquefaction of permafrost soils is very likely within one or two years following a forest fire. Depending on the duration and intensity of fire, and the time of burning when normal thawing may or may not be at the annual maximum, the impact of heat directly from the fire may or may not be significant. Viereck and Schandelmeier (1980) indicate that soil temperature change directly caused by fire is insignificant. This means that forest fire may not trigger landslide immediately when thawing of active layer is not at its maximum depth. Destruction of the surface insulative cover certainly results in more heat penetration into the ground in the following years.

In region 1, no evidence of fire was noted around the landslides observed. Viereck and Schandelmeier (1980) indicate that fire frequency is much less in Canadian tundra areas than in forested regions and fires are small, usually less than 1 km^2. Fire records in these areas are usually incomplete and evidence of fires disappears within a few years. Further investigation about fires at the landslide sites in this region is required. However it is probable that extreme weather conditions might have contributed to initiation of the slides by unusual heat penetration into permafrost.

It is interesting to note that all landslides observed in region 2 were associated with fire. In other words, no landslide observed in region 2 is suspected to have been caused by extreme weather alone. This is likely because the forest acts as a thermal shield between the atmosphere and the subsurface soil, and because roots from trees provide a reinforcing effect to the slopes, which are in addition to the insulative effect of the organic cover discussed above.

It is also interesting to note that large areas of burned trees were observed, but landslides occurred only at certain locations. Further investigation is needed to find out why landslides have not occurred at other locations that appear to be similar to the failed slopes. Also, compared to region 1 where there are few or no trees, the insulation conditions in the burned areas in region 2 in theory should be worse, as both the trees and the organic mat were destroyed partially or completely. This is another topic that is worth further study.

25.5.4 The Transient Layer

Depending on ice content, the transient layer may act as a cushion absorbing the impact of sudden abnormal heat. In other words, sudden thawing of this layer may not cause a serious threat to slope stability, as in theory it has at some point in the relatively recent past undergone at least one cycle of thawing without resulting in instability. However, the transient layer where observed is relatively thin (about 5 cm to 15 cm) and sometimes was not obvious. This implies that excessive heat may penetrate through the transient layer and cause thawing of ice-rich permafrost and hence slope failure. The thickness of the transient layer observed at the sites along the Mackenzie Valley is consistent with that summarized in a recent publication by Shur et al. (2005). Shur et al. indicate that the transient layer thicknesses at a number of sites vary mostly from 5% to 20% of the average active layer thickness. The reported sites are from other regions. Most of the sites have an average active layer thickness between 1 m and 2 m.

25.5.5 Landslide Process

It seems that the deep landslides have been active for years. Such landslides will not stop until certain conditions are met. These conditions include: (1) the natural slope diminishing to a point where no more ice-rich permafrost soils would be exposed further; (2) the headscarp encountering different materials; and (3) the slope of the flowing tongue becoming sufficiently steep that the exposure of the ice-rich permafrost soil on the scarp is eventually reduced.

Most deep landslides observed in regions 1 and 2 are circular in shape. The circular shape suggests that the slide likely started from a small area and expanded uniformly around the starting zone. As discussed earlier, the permafrost layer most likely thawed under abnormal conditions and a landslide was triggered. The permafrost layer was exposed in the headscarp when the surface active layer slid down. Ablation of the permafrost face starts from the exposed scarp and the scarp expands in radial directions around the starting area. The materials from the permafrost headscarp mostly become liquid and are very mobile when thawed (Fig. 25.8b).

When the headscarp is low, the materials from the active layer may become a significant component in the flowing mass, as the active layer would make up a greater proportion of the exposed surface. As discussed earlier, the water content of the active layer is relatively low. The ablated material mix is therefore less mobile if the headscarp is low. In this case the slope angle of the tongue is likely to be greater, which, in turn, may result in an even lower scarp and steeper tongue. Such slides may become stable sooner than the deeper slides. For example, one large landslide facing north was found near another large slide facing south. The south-facing landslide was more active, with a higher scarp and a gentle tongue slope, while the north-facing landslide appeared to be less active with lower scarp and steeper tongue slope.

Another possible factor for stabilization of a slide is the orientation of the slope. If a slope receives less heat, the ablation rate is lower, allowing slower movement of the tongue. The slower ablation allows longer time for the materials in the tongue area to lose moisture and become less mobile. This may result in a steeper tongue slope and a lower headscarp. Depending on the angle of the original slope, such slopes may become stable sooner. Literature has shown that slope orientation is not an important factor for the amount of energy the landslide scarps receives (Pufahl and Morgenstern 1980). This can be true as evidenced by the circular shape of the landslides and by other information provided in the literature. The difference of the two nearby landslides discussed above may be related to the difference in their orientations, although it could be due to other reasons as well.

A slower ablation rate may not always result in earlier stabilization of a slide. One large slide, also north-facing, was noted to have vegetation established in the tongue area, but the headscarp was high (about 15 m) and the tongue was relatively flat (about 1° to 3°) (Fig. 25.9). The vegetation, although young, indicates that the flow of the tongue has been slow. The newly ablated fine materials from the scarp were being carried towards the lake through channels across the vegetated areas.

25.6 Meaning to Pipeline Design and Recommendations for Further Investigations

Fine-grained ice-rich permafrost soil may become liquid when thawed. If the rate of thawing is faster than the pore water can dissipate, not only will the shear strength of the material be significantly reduced, but also pore water pressure will build up along the slope under a low-per-

Fig. 25.9.
Landslide with active scarp (background) and tongue with vegetation established (foreground)

meability active layer. A combination of thermal, seepage and slope stability analyses may help to determine whether a given heat source would cause detrimental impact on slope stability. Excessive thawing could be due to surface disturbance, removal of natural surface insulation layer, fire, difference between pipeline and ground temperatures and so on. The geometry or extent of thawing combined with other intrinsic characteristics of the slopes can be critical to slope stability and could be determined on case-by-case basis.

In addition to subjects discussed earlier, the following topics are suggested to be considered for further studies:

1. At what ratio of active layer thickness to height of head-scarp does a landslide stop without extensive retrogression?
2. At what slope angle and under what conditions does shallow flow occur without triggering extensive retrogressive failure?
3. Are the water/ice content and temperature conditions of material near shorelines of lakes critical to slope stability, and to what extent is the upper slope supported?
4. What are the effects of the transient layer between the active and permafrost layers?
5. Why do slopes fail in some areas but not in other areas within the same burned region?

Carrying out work in remote regions can mean significant costs due to logistics. More careful and thorough planning is required, which in turn requires data from the field. It is the intention of this paper to provide information that would be useful for further investigations.

References

Aylsworth JA, Burgess MM, Desrochers DT, Duk-Rodkin A, Robertson T, Traynor JA (2000a) Surficial geology, subsurface materials, and thaw sensitivity of sediments. In: Dyke LD, Brooks GR (eds) The physical environment of the Mackenzie Valley, Northwest Territories: a base line for the assessment of environmental change . Geological Survey of Canada, Bulletin 547:41–48

Aylsworth JA, Duk-Rodkin A, Robertson T, Traynor JA (2000b) Landslides of the Mackenzie Valley and adjacent mountainous and coastal regions. In: Dyke LD, Brooks GR (eds) The physical environment of the Mackenzie Valley, Northwest Territories: a base line for the assessment of environmental change. Geological Survey of Canada, Bulletin 547:167–176

Burn CR (1998) The response (1958–1997) of permafrost and near-surface ground temperatures to forest fire, Takhini River valley, southern Yukon Territory. Can J Earth Sci 35:184–199

Cleve KV, Viereck LA (1983) A comparison of successional sequences following fire on permafrost-dominated and permafrost-free sites in interior Alaska. In: Proceedings, 4th International Conference on Permafrost, Fairbanks, Alaska, pp 1286–1291

Code JA (1973) The stability of natural slopes in the Mackenzie Valley. Environmental-Social Committee, Task Force on Northern Pipelines, Task Force on Northern Oil Development, Report no. 73-9

Dallimore SR, Nixon FM, Egginton PA, Bisson JG (1996) Deep-seated creep of massive ground ice, Tuktoyaktuk, NWT, Canada. Permafrost Periglac 7:337–347

Dyke LD (2000) Stability of permafrost slopes in the Mackenzie Valley. In: Dyke LD, Brooks GR (eds) The physical environment of the Mackenzie Valley, Northwest Territories: a base line for the assessment of environmental change. Geological Survey of Canada, Bulletin 547:177–186

Dyke LD (2004) Stability of frozen and thawing slopes in the Mackenzie Valley, Northwest Territories. In: Proceedings, 57th Canadian Geotechnical Society Annual Conference, Quebec City

Gartner Lee Ltd. (2003) Identification of the biophysical information and research gaps associated with hydrocarbon exploration, development and transmission in the Mackenzie Valley Action Plan. Report prepared for DIAND, GNWT and ESRF. Ref. GLL 22-649, 94 pp

Hanna AJ, McRoberts EC (1988) Permafrost slope design for a buried oil pipeline. In: Proceedings, 5th International Conference on Permafrost 2:1247–1252

Hanna AJ, Oswell JM, McRoberts EC, Smith JD, Fridel TW (1994) Initial performance permafrost slopes: Norman Wells Pipeline Project, Canada. In: Proceedings, 7th International Cold Regions Engineering Specialty Conference, Edmonton, Alberta, pp 369–395

Hanna AJ, McNeill D, Tchekhovski A, Fridel T, Babkirk C (1998) The effects of the 1994 and 1995 forest fires on the slopes of the Norman Wells Pipeline. In: Proceedings, 7th International Permafrost Conference, Yellowknife, NWT, pp 421–426

Harry DG, MacInnes KL (1988) The effect of forest fires on permafrost terrain stability, Little Chicago-Travaillant Lake area, Mackenzie Valley, NWT. Geological Survey of Canada, Current Research, Part D, Paper 88-1D, pp 91–94

Isaacs RM, Code JA (1972) Permafrost, bank stability and pipeline construction, Mackenzie River, Northwest Territories, Canada. In: Proceedings, Canadian Northern Pipeline Research Conference, National Research Council, pp 229–241

Lewkowicz AG (1988) Ablation of massive ground ice, Mackenzie Delta. In: Proceedings, 5th International Conference on Permafrost 1:605–610

Lewkowicz AG, Harris C (2005) Frequency and magnitude of active-layer detachment failures in discontinuous and continuous permafrost, Northern Canada. Permafrost Periglac 16:115–130

Liang L, Zhou Y, Wang J (1991) Short communication – change to the permafrost environment after forest fire, Da Xi'an Ridge, Gu Lian mining area, China. Permafrost Periglac 2:253–257

Mackay JR (1995) Active layer changes (1968–1993) following the forest-tundra fire near Inuvik, NWT, Canada. Arctic Alpine Res 27(4):323–336

McRoberts EC, Morgenstern NR (1973) A study of landslides in the vicinity of the Mackenzie River mile 205 to 660. Environmental-Social Committee on Northern Pipelines, Task Force on Northern Oil Development, Report no. 73-35

McRoberts EC, Morgenstern NR (1974a) The stability of thawing slopes. Can Geotech J 11:447–469

McRoberts EC, Morgenstern NR (1974b) Stability of slopes in frozen soil, Mackenzie Valley, N.W.T. Can Geotech J 11:554–573

Morgenstern NR (1981) Geotechnical engineering and frontier resource development. Geotechnique 31(3):305–365

Morgenstern NR, Nixon JF (1971) One-dimensional consolidation of thawing soils. Can Geotech J 8:558–565

Pufahl DE, Morgenstern NR (1980) The energetics of an ablating head scarp in permafrost. Can Geotech J 17:487–497

Savigny KW, Morgenstern NR (1986a) In situ creep properties in ice-rich permafrost soil. Can Geotech J 23:504–514

Savigny KW, Morgenstern NR (1986b) Creep behavior of undisturbed clay permafrost. Can Geotech J 23:515–527

Savigny KW, Logue C, MacInnes K (1995) Forest fire effects on slopes formed in ice-rich permafrost soils, Mackenzie Valley, NWT. In: 48th Canadian Geotechnical Society Annual Conference, Toronto, Preprint 2:989–998

Shur Y, Hinkel KM, Nelson FE (2005) The transient layer: implications for geocryology and climate-change science. Permafrost Periglac 16:5–17

Tsuchiya F, Fukuda M, Tumbaatar D, Sharakuu N, Baatar R, Muneoka T (2001) Forest fire impacts to Mongolian permafrost (extended abstract). Tôhoku Geophysical Journal 36(2): 219–223

Viereck LA (1982) Effects of fire and firelines on active layer thickness and soil temperatures in interior Alaska. In: Proceedings, 4th Canadian Permafrost Conference, Calgary, pp 123–135

Viereck LA, Schandelmeier LA (1980) Effects of fire in Alaska and adjacent Canada - a literature review. U.S. Department of Interior, Bureau of Land Management Alaska Technical Report 6 (BLM/AK/TR-80/06), 124 p

Chapter 26

Slope-Structure Stability Modeling for the Rock Hewn Church of Bet Aba Libanos in Lalibela (Ethiopia): Preliminary Results

Giuseppe Delmonaco · Claudio Margottini* · Daniele Spizzichino

Abstract. Lalibela is located in the northern-central part of Ethiopia, approx. 600 km north of Addis Ababa in Northern Wollo, one of the most structural food deficit areas of the Amhara Region. The town, which has about 12000 inhabitants, is situated at an altitude of 2500 m (Fig. 26.1). In its center, a unique complex of 11 rock-hewn Christian Orthodox churches is loacted. The churches were cut out of the living rock some 800 years ago during the kingdom of King Lalibela (1167–1207) of the Zagwe dynasty. One of these churches, Biet Aba Libanos, is a monolithic church anchored to the rock from which it was carved. Two major damaging phenomena affect the church: weathering of volcanic tuff in the lower part of the edifice and sliding of the façade and lateral walls, as consequence of a prone discontinuity. A first destruction of the façade was already occurred in the past, as consequence of an old planar sliding, still in coincidence of the same joint. Presently, the walls prone to slide are the structures constructed to replace the original rock that collapsed during the slide as well as some of the original rock hewn lateral walls. Kinematical analysis and numerical modeling implemented, clearly evidence the hazardous conditions of the rock hewn church of Biet Aba Libanos and the need of a prompt and proper intervention.

Keywords. Planar sliding, tuff, Bet Aba Libanos, Lalibela, Ethiopia

26.1 Introduction

Lalibela is located in the northern-central part of Ethiopia, approx. 600 km north of Addis Ababa in Northern Wollo, one of the most structural food deficit areas of the Amhara Region (Fig. 26.1). The town, which has about 12 000 inhabitants, is situated at an altitude of 2 500 m. A unique complex of 11 rock-hewn churches is located in the center of the town, cut out of the living rock some 800 years ago during the kingdom of King Lalibela (1167–1207) of the Zagwe dynasty. The churches are still used daily for religious practices and ceremonies, whilst during major religious occasions large crowds of believers and pilgrims convey to the site. Since 1978, the churches and their surrounding areas have been included in UNESCO's World Heritage List.

Some of the churches are sculpted out of solid volcanic tuff (Fig. 26.2). Some are quarried enlargements of caves. They are almost all connected by long underground tun-

Fig. 26.1.
Locations of Lalibela in Northern Ethiopia

Fig. 26.2.
The church of Bet Gyorgis in Lalibela (Ethiopia)

Fig. 26.3.
The church of Bet Aba Libanos in Lalibela (Ethiopia)

nels and mazes. The maze of churches offers a wide variety of architectural styles: Grecian pillars, Arabesque windows, ancient swastika and Star of David carvings, arches, and Egyptian-like buildings. The main cluster of 11 churches is located in the village center: Bet Golgotha/ Bet Mia'el, Bet Maryam, Betk Meskel, Bet Danaghel, Bet Medhane Alem, Bet Amanuel, Bet Merkorios, Bet Aba Libanos, Bet Gabriel-Rufa'el, and Bet Gyorgis.

Over the years, the churches have been exposed to wind, rainfall, thermal changes and human activities. This has resulted in a severe degradation of the monuments, most of which are now considered to be in critical condition. The major causative factor for degradation is the large concentration of montmorillonite (largely expansive clay belonging to smectite family), and related weathering, in the poor silica volcanic rock.

However, because of the outstanding value and the nature of the monuments, the conservation of the Lalibela churches attracted long ago the interest of the international community. A first restoration attempt of the Lalibela rock-hewn churches was carried out in 1920.

Bet Aba Libanos is a monolithic church anchored to the rock out of which it was carved. By its excavation within the rock face (height of the circular tunnel: 7 m) and by its small dimension, it imitates a cave church (Bianchi Barriviera 1963) (Fig. 26.3).

Two major phenomena affect the church: weathering of volcanic tuff in the lower part of the edifice and sliding of the façade as consequence of a discontinuity that, in this case, is dipping to facilitate the moving down of the walls. It is important to observe that a first sliding, and destruction, of the façade as already occurred in the past.

26.2 Geological Setting of the Area

The region of Lalibela is characterized by the outcropping of tertiary volcanic rocks (Fig. 26.4), composed of basalt, trachyte and tuff (Kazmin 1973). The geological series of the area, from the oldest and geometrically lower, consist of (ACEL-SAVA 1997):

1. the Ashangi Formation (Oligocene-Miocene) which is mainly composed of olivine basalt alternating with agglomerates and tuff (Merla et al. 1979);
2. the Amba Aiba basalts (Oligocene-Miocene) which mainly consists of flood basalt with rare tuffs. The flows are always evident with columnar jointing and thickness up to tens of meters (Merla et al. 1979);
3. the Amba Alaji rhyolites (Miocene) cover generally the Aiba basalts. This formation consist of a succession of alkaline to perialkaline rhyolites, alkaline trachyalkaline and flood basalt. The acidic terms are large ignimbrites and tuffaceous levels, largely outcropping in the area and forming typical landscape elements such as steep walls and pyramids. Furthermore, the flat terminal surface of the Ethiopian Plateau is often constituted by the upper part of thick ignimbrites of the Amba Alaji rhyolites (Merla et al. 1979);
4. the Termaber basalt formation (Miocene to Pliocene) is at the top of local stratigraphic series. The Termaber unit is made of lenticular basalts with a large amount of tuffs, scoriaceous lava flows and typical red paleosoils (Merla et al. 1979).

Following is the general geological map of the area (Merla et al. 1979).

According to ACEL-SAVA (1997), the Churches are carved out within the Amba Aiba Formation, in two distinct units as the lower basalt and the basic tuff. On the top, it can be clearly seen the fill material (silt to stone), composed of silt to tuffaceous angular chips, that was deposited during the excavation of Churches and tunnels. Also a regolith, product of weathering processes on the basaltic rock, was identified by ACEL-SAVA (1997) in the northern side of Bet Medihanialem Church.

ICCROM (1978) details the local geological terms of Amba Aiba in (from the oldest and geometrically lower):

1. dark gray basalt, fine grained, very tough rock, igneous in origin. It contains mineral such as olivine, pyroxene and plagioclase, which in fault-brecciated zone disintegrate rapidly. Chloritization is prominent in the fault breccia;
2. scoria is an angular reddish (iron enriched) volcanic agglomerate. The particle size ranges from a fine ash to large lava bombs up to 2 m in diameter. Secondary mineralization is prevalent with calcite found along planes and a large amount of salt precipitation within the vesicles. The scoria is heavily differentially weathered with basalt pyroclasts standing out against the more easily weathered components. Secondary silica produced by metasomatic fluid had precipitated quartz in scoria vesicles.

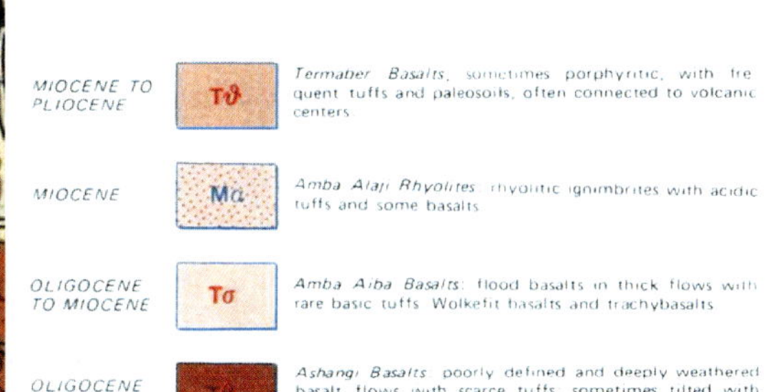

Fig. 26.4. Geological map of the Lalibela Region

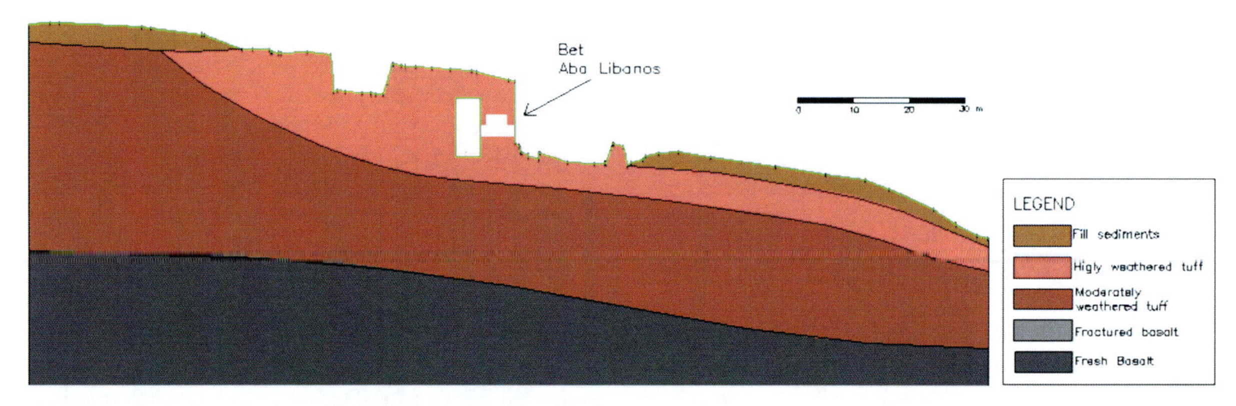

Fig. 26.5. Geological section thorough the church of Bet Aba Libanos

Table 26.1. Main properties of Lalibela weathered Tuff

Density (kN m^{-3})	Phi (°)	UCS (MPa)	Esec (GPa)	Etan (GPa)	Etan$_0$ (GPa)	V_p (km s^{-1})	V_s (km s^{-1})	Poisson ratio	G (GPa)	K (GPa)	E (GPa)
16.5–17.1	45 (s.d. 5.7)	5.53–7.93	1.72	1.54	1.89	1.85–2.03	1.10–1.14	0.214–0.266	2.04–2.21	3.02–3.99	4.99–5.60

26.3 Engineering Geology Investigation

The general geology of the area (ACEL-SAVA 1997; TEPRIN 2002) is characterized by the presence of a basalt bedrock, fresh in the lower geometrical part and fractured in the upper part of the formation. The basalt is covered by a red tuff, moderately weathered in the lower part and highly weathered on the upper. On the top a weathering regolith is described in a northern side of Bet Medihanialem Church and, finally, a large coverage of fill sediment was produced by the excavation of churches and trenches (Fig. 26.5).

The basalt is a good material with bulk density ranging 22.06–23.83 kNm^{-3}. Uniaxial compressive strength (UCS or sc) obtained in laboratory (TEPRIN 2002) is characterized by values of 7.8 MPa in the fractured unit and 21.4 MPa in the fresh material (data from the area of the 11 churches).

The tuff material outcropping in the Aba Libanos Site seems to belong only to the highly weathered unit of the basic tuff. The density was established in about 16.5–17.1 kNm^{-3}, while the bulk density was estimated by TEPRIN (2002) ranging from 20.00–21.18 kNm^{-3}. The laboratory investigation in samples from Aba Libanos tuff exhibits a σ_c, ranging from 5.53 to 7.93 MPa. The same σ_c obtained in the field with Schmidthammer (International Society for Rock Mechanics 1981), reveals higher values ranging from 27 till 42 MPa. Table 26.1 is synthesizing the laboratory tests performed in the samples of tuff from Aba Libanos.

Since the water was determined to play an important role on weathering of tuff, an in-situ slaking test was performed on a sample taken from Biet Aba-Libanos. The test has proved that the material is not affected by slaking phenomena.

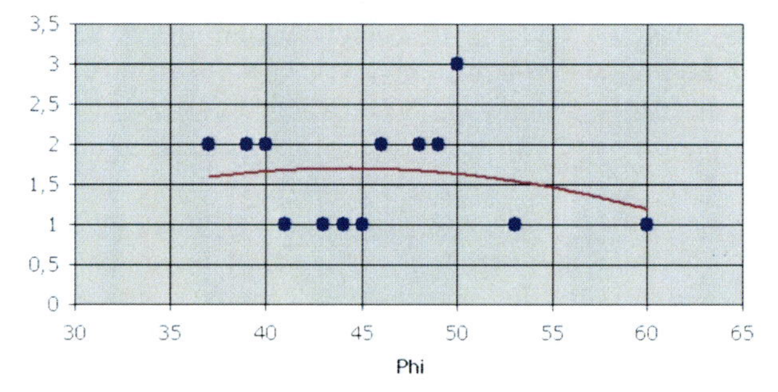

Fig. 26.6. Tilt test analysis on the Lalibela tuff of Amba Aiba Formation

Shear strength parameters were determined to investigate the potential resistance to sliding of rock-walls in Biet Aba Libanos. A tilt test (Fig. 26.6) was performed on site providing a friction angle of about 45°.

26.4 Structural Setting and Kinematic Analysis

The tuff formation shows sub-horizontal discontinuities of syngenetic origin, probably as consequence of lithostatic load of different pyroclastic flows. These discontinuities are spacing about 5–10 meters, with a limited dip likely following the original morphology at the time of the volcanic eruption. Nevertheless, this dipping is capable, like in Biet Aba Libanos, to produce the sliding of the church, if shear strength parameters are not providing an adequate resistance. In this church the frontal façade of Bet Aba Libanos partially collapsed in the past and has recently been reconstructed. The reason of collapse was the sliding along the detected discontinuities. Nevertheless, also after the reconstruction, according to the shear strength of the disconti-

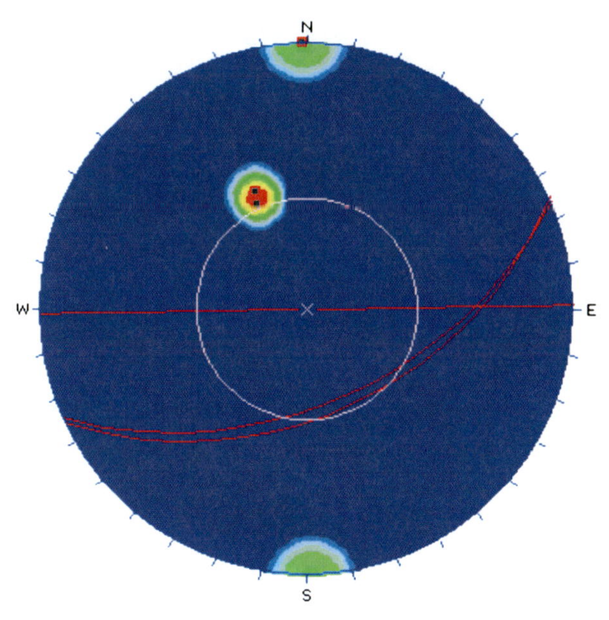

Fig. 26.7. Kinematic analysis of the church of Bet Aba Libanos

Fig. 26.8. The discontinuities detected on the lateral wall of Bet Aba Libanos; data refers to dip direction and dip

nuity based on tilt test, the kinematical analysis clearly shows that all the conditions leading to a new possible planar slide are satisfied (Fig. 26.7). In particular, the surface failure is kinematically viable if:

1. the dip direction of the planar discontinuity is within 20° of the dip direction of the slope face;
2. the dip of the planar discontinuity is lower than the dip of the slope face and thereby "daylights" in the slope face;
3. the dip of the planar discontinuity is higher than the friction angle.

All the above three conditions are satisfied in the case of Bet Aba Libanos.

The potential slide is affecting the two lateral walls where a clear discontinuity is dipping downwards, parallel to the wall sides; in addition, the discontinuity partially sustains the walls reconstructed after the previous fall (Fig. 26.8). To aggravate the situation there is the connection with an iron bar (tie rod) of the two lateral walls prone to slide. In such a way, the most unstable sides were tied together, when a correct intervention had to fix unstable walls with stable ground (direction of bar perpendicular to the present one). The joint roughness coefficient (JRC) of the discontinuity has been estimated with Burton equipment in about 14. In any case, likely, crack gauges installed by the restorer Angelini in 1965–1967 (International Fund for Monuments 1967), have not been recorded any displacement since that time.

26.5 Slope-Structure Stability Modeling

The church of Bet Aba Libanos is a particular engineering geology problem, since there is an edifice carved out from the rock, but with a structural behavior responding to the approach of rock mechanics. In particular, the church is now suffering for the stability of the two lateral walls, where an important discontinuity is causing the planar slide of the above rock and wall.

A similar planar sliding has occurred in the past and, after this, the entire façade was reconstructed with rock bricks. An attempt to estimate the condition of the church with the techniques of rock slope stability has been undertaken. Planar sliding was evaluated according to Hoek and Bray (1981) theory for strength of rock. The model is assuming the presence of a crack, as in the actual situation, with no cohesion and friction angle equal to 45°, obtained by tilt tests. The upper part of the cliff, above the church, was not considered in the first approach since the church is detached from the ceiling. In other calculation this was introduced as an overload (Fig. 26.9).

The results of the calculation show that the church is presently close, but below, the limit of equilibrium. When considering the overload the results are sensibly lower, confirming that the cliff is probably not affecting, with its load, the static load of Bet Aba Libanos. The reason is probably related to the lateral "arch effect", caused by the ex-

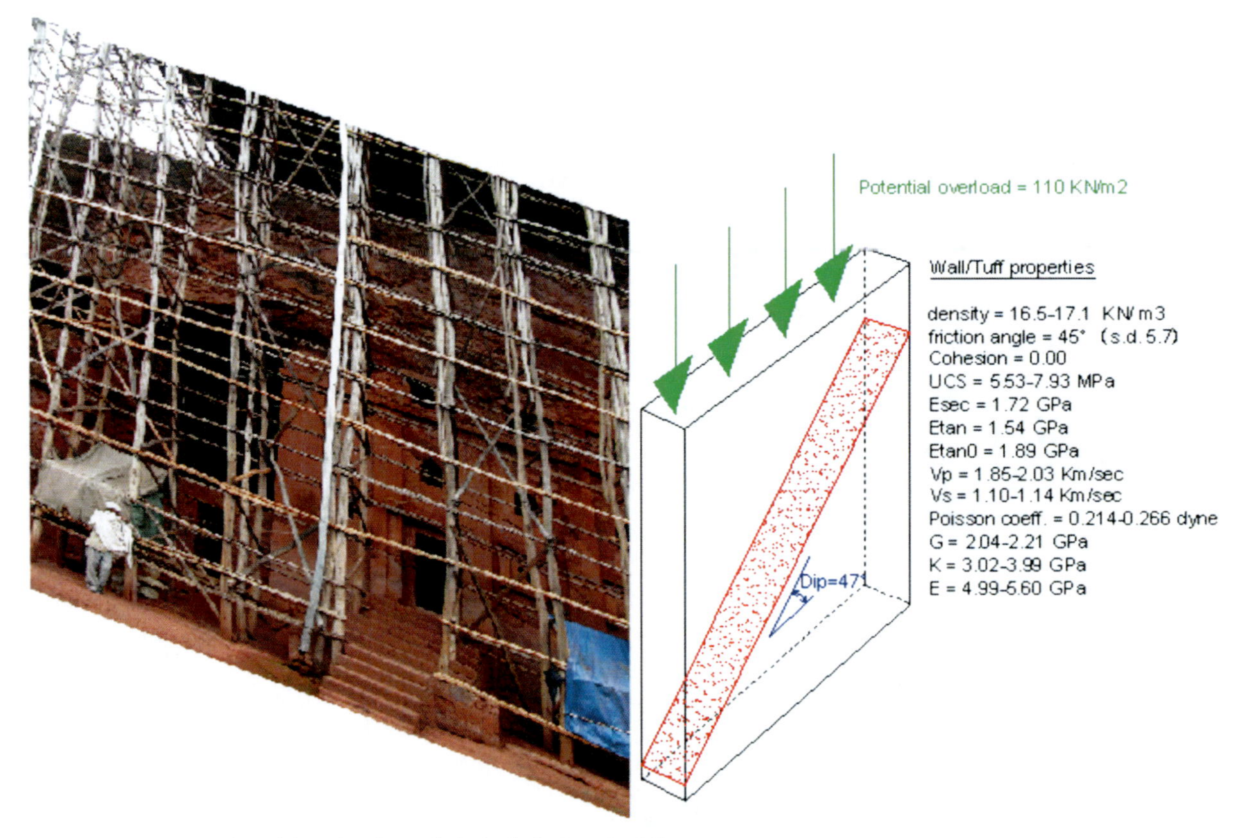

Fig. 26.9. The conceptual model used in the analytical calculation of stability

cavation of the tunnel surrounding the church. Therefore, the upper part of the cliff is statically sustained by the lateral part of the cliff and not from the church itself.

26.6 Conclusion

The Ethiopian town of Lalibela is included in the UNESCO World Heritage List for the 11 rock hewn churches. The churches were carved out in the Amba Aiba basalt and tuff formation, around twelfth century A.D. The churches are presently suffering for a severe weathering related with the diffuse concentration of montmorillonite in the volcanic rocks, a largely expansive clay mineral that may cause a decrease of shear strength of materials. In addition, in the case of Bet Aba Libanos, the orientation of the local system of joints is the main cause for the sliding of the façade and the lateral walls. A similar slide has occurred in the past and, after this, the entire façade was reconstructed with rock bricks.

The performed investigation has stressed the kinematical possibility of a planar sliding and a static equilibrium around the limit of stability, probably lower. The modeling was elaborated for the lateral wall (thickness about 1 m), as an element of unitary dimension in the model. The upper part of the cliff, inside which the church is excavated, seems not to play a major compressive stress. Nevertheless, the next step will be the 3D isostatic analysis of the church and cliff to better investigate the static condition of the whole area as well as a more sophisticated numerical modeling computation.

References

ACEL-SAVA (1997) Geology and engineering geology of Lalibela Churches. (restricted)

Bianchi Barriviera L (1963) Le chiese in roccia di Lalibela e di altri luoghi del Lasta. Rassegna di Studi Etiopici

Hoek and Bray (1981) Rock slope engineering, 3rd edition. Chapman & Hall, London

ICCROM (1978) Proceeding of the Symposia of International Experts on Stone Conservation. Lalibela, (restricted)

International Fund for Monuments (1967) Lalibela – phase 1. Adventure in restoration. New York

International Society for Rock Mechanics (1981) Rock characterization, testing and monitoring ISRM suggested methods. Pergamon, London

Kazmin V (1973) Geological map of Ethiopia. Imperial Ethiopian Government, Ministry of Mines, Geological Survey of Ethiopia

Merla G, Abbate E, Azzaroli A, Bruni P, Canuti P, Mazzuoli M, Sagri M, Tacconi P (1979) A geological map of Ethiopia and Somalia (1973) 1:2000000 and comment with a map of major landforms. Consiglio Nazionale delle Ricerche, Italy

TEPRIN (2002) Shelters for five churches in Lalibela project. Tender dossier, (restricted)

Chapter 27

Clay Minerals Contributing to Creeping Displacement of Fracture Zone Landslides in Japan

Netra P. Bhandary[*] **· Ryuichi Yatabe · Shuzo Takata**

Abstract. Various investigations on tectonically-induced landslides in Shikoku in west Japan have been carried out, most of which conclude at tectonic activities through the major fault lines and enhanced rock mineral decomposition as being mainly responsible for the landslide occurrence. Little work is found, however, on the mechanism of their creep activation and role played by the expansive clay minerals. This paper aims at looking into strength parameters of the landslide clays from mineralogical point of view. In addition, the strength behavior is analyzed from inclusion of weaker clay minerals such as smectites. It is found that the drop from peak to residual friction angles for the tested samples reaches as high as 20°. In addition, the residual strength of the landslide clays was found to decrease with higher amount of expansive clay minerals, which was estimated as being relative to chlorite mineral from the XRD patterns. Moreover, the presence of non-crystalline clay materials was found to considerably lower the friction angles of the landslide soils.

Keywords. Amorphous material, clay mineral, landslide, shear strength, smectite

27.1 Introduction

As a mountainous country, Japan has recognized a large number of active and potential landslide sites. For example, Shikoku in the southwest part of Japan (Fig. 27.1) has been designated as a highly landslide-prone area with a large number of active landslides induced primarily due to tectonic activities. As of March 1997, the number of large-scale creeping-type landslides under the direct jurisdiction of the Ministry of Land, Infrastructure, and Transport (abbreviated hereinafter as MLIT) in Shikoku alone is 670, which is nearly 21% of the number all over the country (MLIT 1997).

The mechanism of landslide occurrence in Shikoku is very often related with the region's peculiar geological conditions. A total of three major tectonic lines pass across the region in east-west direction. A main fault line called *chuokouzousen* (also known as median tectonic line and abbreviated as MTL) divides the Shikoku Island near the northern part into plains and mountains by making large scarps of Shikoku Mountains. Likewise, two equally important tectonic lines known as Mikabu Tectonic Line and Butsuzo Tectonic Line pass across the island and together with the MTL they divide it into four major geological belts (Fig. 27.1).

The geological belts recognized as being particularly prone to landslides are Izumi Group, Sambagawa Belt, Mikabu Belt, and Chichibu Belt. Each of these belts is separated by the tectonic lines. For example, the MTL separates the Izumi Group, which is a sedimentary deposit composed primarily of sandstone with the intrusion of thin layers of shale, and the Sambagawa Belt, which is a deposit of metamorphic rocks consisting mainly of green and black schist. Similarly, Mikabu tectonic line separates Sambagawa and Chichibu Belts and forms a narrow strip of Mikabu greenstone as a metamorphic deposit along the tectonic line itself. The rock type in the Chichibu Belt is sedimentary composed of green schist, mudstone, conglomerate, etc.

A highly favorable condition for the land sliding in Shikoku is supposed to be the fractured state of bedrocks, especially near the tectonic faults (MLIT 1997). It is widely believed that the fractured state of the bedrocks accelerates the rock mineral weathering by facilitating water ingress into the rock mass. Another concept that explains the formation of weak clay layer below certain depth in mountain slopes near the tectonic faults especially in Shikoku is hydrothermal alteration of rock minerals into

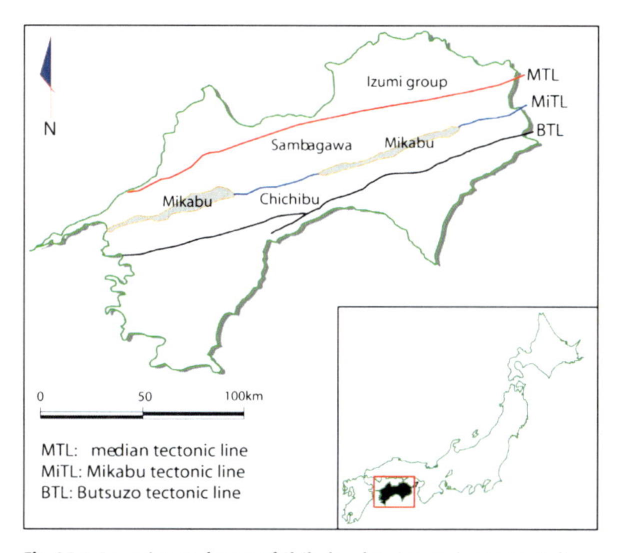

Fig. 27.1. Location and map of Shikoku showing major tectonic lines and geological belts

weaker clay minerals. In this process, high temperature deep undergroundwater under extremely high pressure is pushed up through fault fissures and spreads around through the rock joint planes to cause accelerated decomposition of the rock minerals. Such weathering or decomposition of the bedrock minerals on the planes of rock-mass fractures results in formation of clayey soil layers composed of weaker minerals such as expansive chlorite, smectite, and vermiculite, which in most cases form the slip layers for the landslides. Clays composed of such minerals possess significantly small angles of friction, which is primarily attributed to their capacity to absorb excessive water and swell significantly upon wetting.

Various research papers (e.g., Yagi et al. 1990; Yatabe et al. 1991a,b) have been published on the mechanism of landslide occurrence in fracture zone of Shikoku Region, but the explanations concerning the role of clay mineralogy still have unclear concepts, especially because slip layer clays have been found to contain expansive as well as non-expansive clay minerals. This paper therefore attempts to analyze the strength properties of landslide soils in Shikoku from clay mineralogical point of view. Emphasis is put on investigating the influence of expansive clay minerals and amorphous clay materials on the shear strength of slip surface soils.

27.2 Experiments and Results

27.2.1 Material and Method

The slip surface samples were obtained from boring cores of over 20 landslide sites in Shikoku, most of which were from Sambagawa and Mikabu geological belts (Fig. 27.1). The test program consisted of two parts: tests for strength properties involving triaxial tests and ring shear tests and tests for mineralogical composition involving X-ray diffraction analysis.

The triaxial tests were performed in isotropically consolidated-undrained conditions with pore-water pressure measurement (i.e., $\overline{\text{CU}}$ tests) at controlled rate of strain. The test specimens were prepared by one-dimensionally consolidating remolded pre-saturated samples passing through 425 µm sieve. Each prepared cylindrical test specimen measured 35 mm in diameter and 80 mm in height. The rate of compression was set at 0.044 mm min^{-1}, at which the development of pore water pressure throughout the specimen was confirmed to be uniform. To ascertain highest possible degree of saturation, a back pressure of 196.2 kPa was applied to each specimen, which raised the B-value (confining pressure-dependent coefficient of pore water pressure) up to 0.95 against a rise in confining pressure of 98.1 kPa.

Likewise, the ring shear tests were performed on remolded reconsolidated specimens under drained conditions. The annular specimen measured 120 mm in outer and 80 mm in inner diameter with a thickness of 10 to 12 mm. Before shearing, each specimen was consolidated until the completion of primary consolidation, which for the most samples took about 24 hours. Based on the method described by Garga (1970), the rate of shear until the peak strength value had been exhibited was set at 0.044° min^{-1} (equivalent to 0.038 mm min^{-1} of shear displacement), which was then increased 10 times (i.e., 0.44° min^{-1}) for the post-peak shear until steady states of shear strength and volume change were observed. Depending on the state of consolidation and sample type, the displacement required to achieve the peak strength value ranged from 2 to 7 mm. Likewise, the angular rate of shear of 0.44° min^{-1} in the ring shear apparatus employed could produce a displacement of 58 cm in 24 hours through the plane of shear. It was also ensured that the development of excess pore-water pressure during the pre-peak shear was zero.

27.2.2 Strength Characteristics

Figure 27.2 shows a plot between plasticity index and angles of shearing resistance obtained from the strength tests. The peak angles of friction were taken from the triaxial tests, whereas the residual angles of friction were taken from the ring shear tests. Theoretically, the peak friction angle measured in ring shear apparatus under drained conditions must be close (if not equal) to the effective angle of friction measured in triaxial tests. It is because the total stress during fully drained conditions is equivalent to the effective stress. Due to mechanical friction involved and unequal shear stress distribution across the width of the annular specimen in the ring shear apparatus, however, the values of peak friction angle obtained from the ring shear tests were found notably dif-

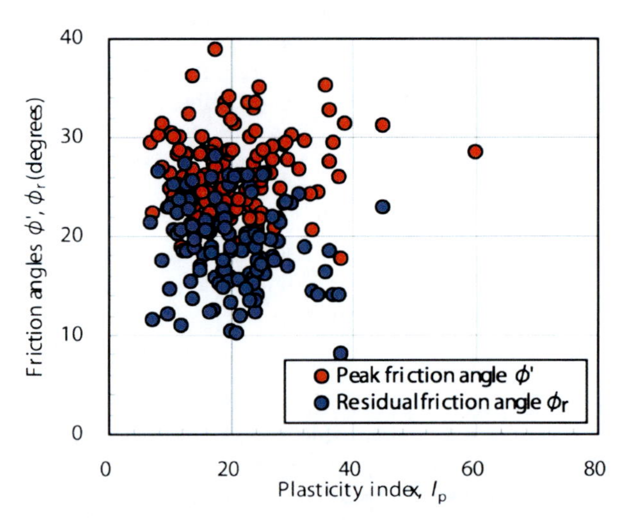

Fig. 27.2. Peak and residual friction angles plotted against the plasticity index

ferent from those obtained from the triaxial tests. So, the choice was made of effective friction angle to represent the actual value of peak angle of friction for the remolded reconsolidated test samples. Although it is true that the peak angle of friction for the slip layer clay in its undisturbed state must be different from that of remolded reconsolidated sample, the evaluation of strength can be made by considering the peak strength of samples prepared under the same initial conditions.

In Fig. 27.2, the effective angles of friction are seen to be ranging from 20° to 40°, whereas the residual angles of friction range from 10° to 25°. It is worth noting that the drop from peak to residual values of friction angle is high. Figure 27.3 shows that such drops for the landslide soils in Shikoku Range from a few degrees to as high as 20°. In general, the drop from peak to residual strength is considered higher in more plastic soils because higher drop means higher clay content and higher clay content means higher plasticity. However, it is interesting to note that the drops for the tested landslide soil samples do not definitely follow this trend. Except for some, most samples exhibit scattered trend.

As plotted in Fig. 27.4, the relation between clay fraction and residual angle of friction is not seen to be what is generally regarded (Lupini et al. 1981; Skempton 1985). Rather than having a fine reduction in residual friction angle with the increase in clay fraction, most data are seen to have scattered, which may be attributed to influence of proportional composition of different clay minerals. Different amounts of minerals composed might have given different shearing properties to the samples tested. Moreover, even at a clay fraction of as low as 10–30%, the average value of residual angle of friction is seen to be near 15°, which is generally considered a low value. To investigate the reason for the lower values of residual friction angle, it was therefore necessary to analyze the samples for clay mineral content.

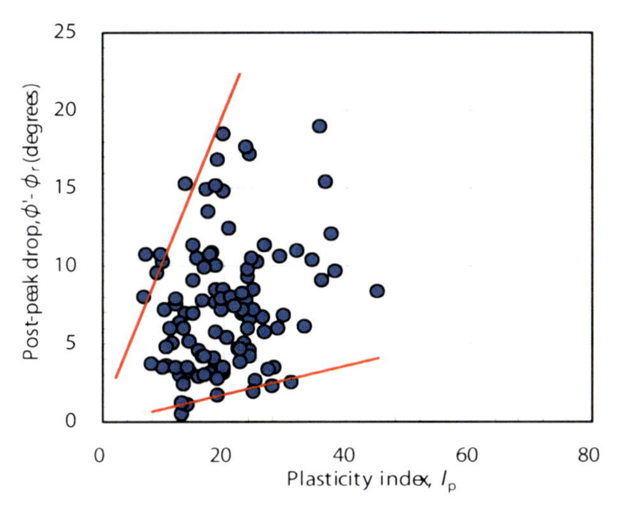

Fig. 27.3. Drop from peak angle of friction to residual angle of friction

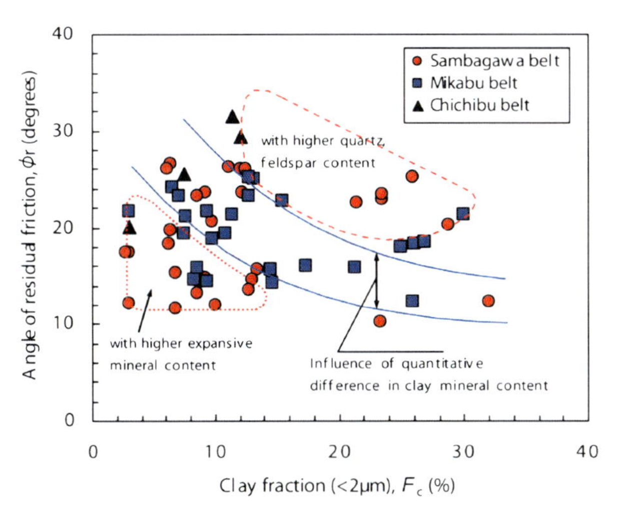

Fig. 27.4. Residual angle of friction plotted against the clay fraction

27.2.3 Clay Mineralogy

Samples Composed of Smectite Mineral

Figure 27.5 compares the angles of shearing resistance for some of the tested samples possessing expansive and non-expansive minerals. Although it is difficult to draw a clear relationship between expansive clay mineral content and shear strength properties, interpretations from this figure can be made as the angles of shearing resistance for the soils composed of expansive clay minerals are smaller than those for the soils not having them, especially in residual state. Therefore, it is considered that the presence of expansive clay minerals in slip surface soils greatly control the displacement behavior of creeping landslides.

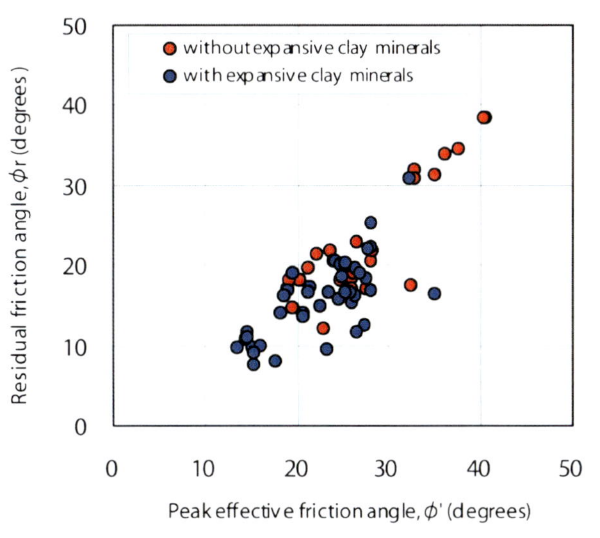

Fig. 27.5. Comparison of friction angles for samples with and without expansive minerals

Samples Composed of Expansive Chlorite

The term expansive chlorite here is used to represent an interstratified state of chlorite and smectite mineral. The method to estimate the relative amount of smectite in expansive chlorite is based on X-ray diffraction patterns. When an untreated sample containing expansive chlorite is placed on the X-ray diffractometer, the peak for chlorite mineral, which has a basal spacing of 14 Å, is exhibited at a diffraction angle between 6.1–6.4° and the peak for the expansive fraction of the mineral is exhibited near a diffraction angle of 6°. When treated with ethylene glycol, the position of the peak for the chlorite portion remains the same, whereas the position of the peak for the expansive fraction shifts to 5.2–5.5°, which is a range of diffraction angle for glycolated montmorillonite (smectite). When a ratio of peak intensity for montmorillonite to that of chlorite near the diffraction angle of 6° is calculated, as illustrated in Fig. 27.6, it should represent the relative amount of smectite mixed with chlorite in a test sample (i.e., A/B in the figure). The data obtained from such calculations are plotted against the angle of shearing resistance for the tested samples in Fig. 27.7. The relative amount of expansive clay mineral is termed as expansive mineral ratio and is represented by a ratio S/C, in which S stands for smectite and C stands for chlorite.

The figure shows that the angles of shearing resistance for the tested samples decrease with the increasing values of the expansive mineral ratio. The data are seen a little scattered but it can be attributed to experimental errors, especially with the ring shear tests. Moreover, the influence of non-clay minerals like mica, quartz, and tremolite is not considered, which might have also caused the scattering. So, if this effect is supposed to be minimal and the effect of expansive clay minerals is exaggerated, a drop of 10° in the angle of friction can be clearly seen against an increase of expansive mineral ratio by 0.5. Such a variation of angle of shearing resistance with the amount of smectite and similar minerals reveals that the degree of instability for a creeping landslide increases with increasing chances of other minerals turning into smectites.

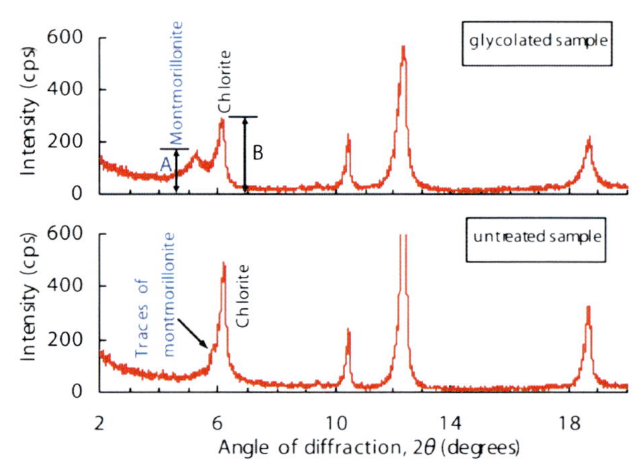

Fig. 27.6. X-ray diffraction patterns showing inclusion of montmorillonite mixed with chlorite

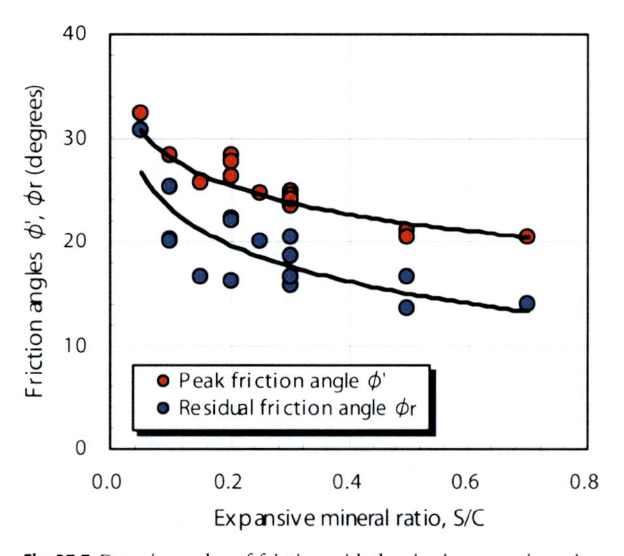

Fig. 27.7. Drop in angles of friction with the rise in expansive mineral ratio

Samples Composed of Non-Crystalline Clay Materials

It is generally understood that the amorphous or non-crystalline clay materials are formed by sudden cooling of magma or volcanic ash and decomposition of the clay minerals. In Japan, the former type of noncrystalline clay material can be found in the form of allophane composed in Kanto-loam and *shirasu* (volcanic ash). The amorphous clay materials resulted from decomposition of clay minerals, on the other hand, are oxides and hydroxides of aluminum, iron, and silicon. They may also occur as distinct crystalline units; for example, gibbsite, boehmite, hematite, etc. They often occur in clays as gels, precipitates, and coat mineral particles, or they often cement particles together.

The detection of amorphous clay materials in soil samples can be done by existing methods, but due to time constraints, a new method is proposed here and employed to determine the presence of the amorphous materials. For this, it is considered that the phase of minerals like mica, feldspar transforming into minerals like chlorite, smectite, vermiculite, etc. represent the noncrystalline minerals, and their proportional quantity can be obtained from following formula, which is based on X-ray diffraction pattern (Fig. 27.8).

$$Q_p = \frac{A - B}{B}$$

where Q_p is relative amount of the amorphous materials, A is average intensity for the angle of diffraction from 6° to 8°, which is a range for the phase transfer from miner-

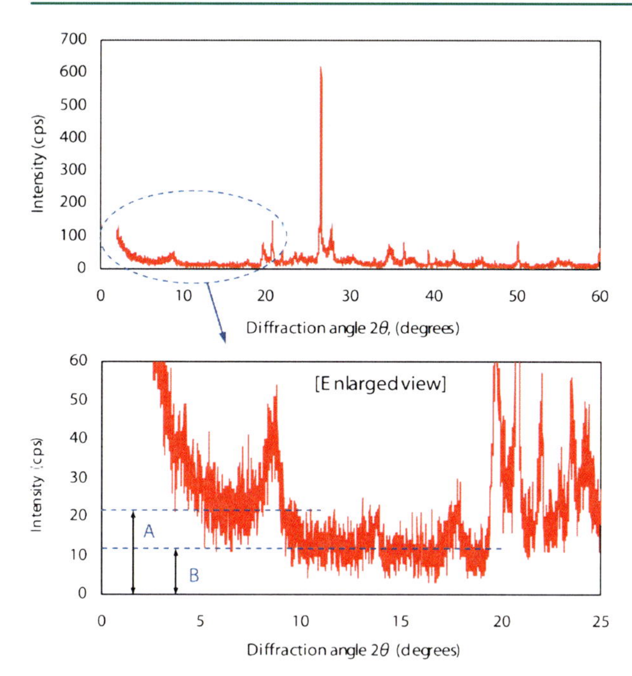

Fig. 27.8. Estimating relative amount of amorphous clay materials from X-ray diffraction patterns

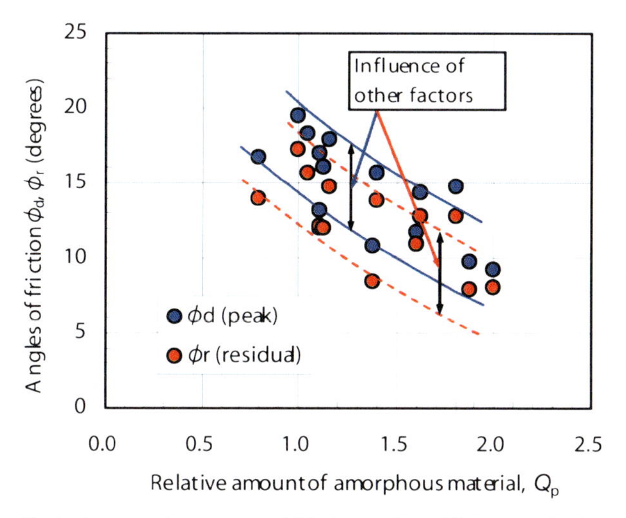

Fig. 27.9. Decreasing pattern of friction angles with increased relative amount of amorphous materials

als like mica and feldspar to chlorite, smectite, and vermiculite, as mentioned above, and B is the average intensity for the angle of diffraction from 14° to 16° (Fig. 27.8).

Figure 27.9 shows the results of the X-ray diffraction analysis for amorphous clay materials combined with the results of the strength tests on landslides soil samples from Sambagawa Geological Belt in Shikoku. It is clear from the figure that there is a decreasing pattern of friction angles but with a slight scatteredness. The scatteredness can be attributed to dissimilar composition of other minerals like quartz, feldspar, mica, etc.

27.3 Conclusions

This paper discussed the strength characteristics of slip layer soils from tectonically-induced landslides in Shikoku of Japan from clay mineralogical point of view. Attempts were made to evaluate the angle of friction for the landslide clays based on the influence of expansive clay minerals such as smectite and expansive chlorite and amorphous clay materials.

The peak effective angle of shearing resistance for the lab samples of landslide soils were measured to be 20° to 40°, and a significant drop of as high as 20° at the residual state was seen. Moreover, the drops from peak to residual values were not seen to have followed the increasing trend with the increase in plasticity index. Mineralogy of the tested samples, on the other hand, was found to have significant influence in the residual strength of the tested samples. Montmorillonite (smectite) was detected in most samples but with great variation in proportion. Samples containing montmorillonite were found to have lower angles of friction than those not containing the expansive mineral. An analysis of angles of shearing resistance and relative amount of smectite (i.e., relative to chlorite) in the tested samples showed a clear drop in the former with the increase in the latter. This indicates that the angle of residual friction decreases with the increase in the amount of smectite minerals, and even an insignificant variation in the inclusion of expansive clay minerals significantly changes the strength properties. Moreover, the presence of non-crystalline clay materials considerably lowers the friction angles of the landslide soils.

References

Garga VK (1970) Residual shear strength under large strains and the effect of sample size on the consolidation of fissured clay, PhD thesis, University of London

Lupini JF, Skinner AE, Vaughan PR (1981) The drained residual strength of cohesive soils. Geotechnique 31(2):181–213

MLIT (Ministry of Land, Infrastructure and Transport) (1997) Landslides in Japan. A booklet published by Sabo Publicity Center, Slope Conservation Division

Skempton AW (1985) Residual strength of clays in landslides, folded strata and the laboratory. Geotechnique 35(1):3–18

Yagi N, Enoki M, Yatabe R (1990) Consideration on mechanical characteristics of landslide clay. In: Bonnard C (ed) Landslides. Proc. the Fifth Int. Symp. on Landslides, 10–15 Jul. 1988, vol. 1, pp 361–364

Yatabe R, Yagi N, Enoki M, Nakamori K (1991a) Strength characteristics of landslide clay. Journal of Japan Landslide Society 28(1): 9–16 (in Japanese)

Yatabe R, Yagi N, Enoki M (1991b) Ring shear characteristics of clays in fractured-zone-landslide. Journal of the Japan Society of Civil Engineers 436/III-16:93–101 (in Japanese)

Chapter 28

Geotechnical Landslide Risk Analysis on Historical Monuments: Methodological Approach

Yasser ElShayeb* · Thierry Verdel

Abstract. During the second half of the twentieth century, and due to major advances in science and technology, industrial risk assessment studies became a must for any new/running industry. On the contrary, natural and environmental risk assessment studies lagged behind until the last two decades of the twentieth century. This chapter demonstrates a methodological approach for the assessment of natural hazards/risks (specifically landslides and block movements), with an application to the Tomb of Ramsis I at the Valley of the Kings, Egypt. The authors were able to develop this approach using different techniques of mathematical reasoning under certainty, and chose the fuzzy logic as the best applicable one. A comparison has been done with other "classical" methods of reasoning such as the probabilistic approach.

Keywords. Risk analysis, Valley of the Kings, fuzzy logic, multicriteria methods

28.1 Introduction

Landslides and ground movements are one of the main concerns for geotechnical engineers. In fact, landslides inherit many geotechnical aspects as well as geological, economic, and socio-economical aspects.

As land usage and urbanization sites approaches natural cliffs or undermined areas, they are sensible to natural landslides dangers. These sites might have economical values (the case of modern houses near a possible land slide location), environmental values (the case of natural or national reserves) or sometimes cultural or even world cultural values (the case of cultural heritage sites). These sites need great care and attention when dealing with, specifically in terms of danger and risk analysis.

The main objective of the present paper is to illustrate how the authors approached the establishment of a uniform strategy for the analysis and the evaluation of landslides and earth movements' risks for Historic Sites and Monuments.

The concerned methods have been developed and tested on a historical site (the tomb of Ramsis I at the Valley of the Kings, Egypt). In order to bring benefits to all other sites and configurations. It takes into consideration, the multidisciplinary character of the site parameters and we are proposing the generalization of these methods in order to get it out of the micro scale to the full macro scale of major historical sites. Figures 28.1 and 28.2 show

Fig. 28.1.
A general view of the temple of Deir EL-Bahari with the Theban Mountain behind it

Fig. 28.2. A general view of the valley showing the entrance of two tombs

two examples of world cultural heritage sites in Egypt. The first suffer from landslide risk and the second from underground rock-movements risk.

The final objective of the paper is to develop zone risk maps for major historical sites in Egypt and to be able, with the aid of the UNESCO, to define risk scales and further preservation and/or restoration projects on these sites.

28.2 Framework of the General Methodology

Our general methodology consists of two main phases; the first phase is a phase of investigation in the site in order to collect the necessary information needed for the analysis. These information could be in numbers "height of the slope" for example, or could be in words "degree of rock weathering" for example. This is where we apply extensively the fuzzy logic for the quantification of site's parameters.

The second phase, is the reasoning phase, where, with the help of multidisciplinary experts from all horizons, we set out sets of rules to be used to reason over the different classes of different parameters. A reasoning rule could be: "If the resistance is low and the slope inclination is big, then the risk is high".

The second phase, is more important and complicated than the first because it is the phase where we need more consultation, adaptation and back analysis in order to find out and settle upon a definite set of rules to be applied to all sites in all configuration.

This methodology has showed many advantages and was very interesting in comparing different parts of the same site, which allowed us to move to a further step to develop the same aspect using a multi-criteria method for decision aid.

The method of multi criteria consists of defining different criteria that judges the site evaluation "slope inclination, degree of fracturing, and humidity could be considered as criteria in an open surface site". Each criteria is assigned a weight that is given by a group of experts depending on the degree of importance of each criteria, and then for each site, these criteria are evaluated in quantitative "200, 300, etc." or qualitative "big, small, medium, etc." form. The output of this method is a classification of the sites with respect to each other, which could enable the comparison of different sites from different environments and also to produce a general risk scale for historical sites.

The development of the multi-criteria method is our main interest for the moment all together in accordance with the application of other methodologies or risk analysis, we are trying to develop an automatized codes for this purpose, and the use of the fuzzy logic is of big interest in the phases of site's evaluation and/or rules definition.

28.3 General Methodology of Natural Risk Analysis

In order to be able to define a general methodology for geotechnical risk analysis, we have combined many parameters into a general scheme presented in Fig. 28.3. The parameters used in the analysis are usually defined for each specific site. We were interested in generalizing the choice of parameters in order to produce a more general methodology applicable to all sites and all cases. This is why the parameters that are used in our general methodology are based upon the system of Rock Mass Rating

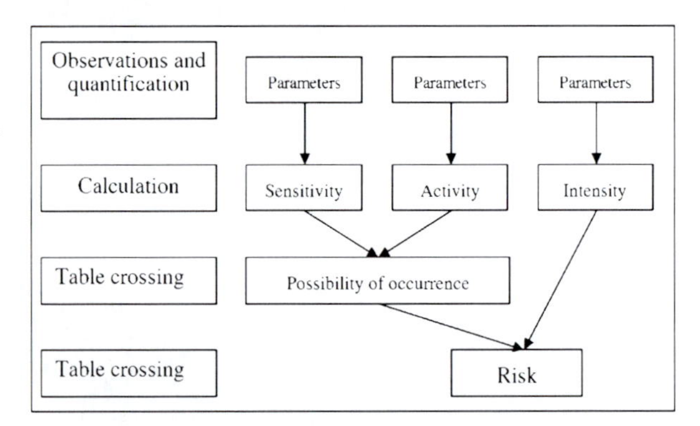

Fig. 28.3. General scheme for geotechnical risk analysis

Table 28.1. Table crossing between sensitivity and activity

Activity	Sensitivity (RMR)				
	Class I Very poor	**Class II Poor**	**Class III Fair**	**Class IV Good**	**Class V Very good**
Sleeping	Negligible	Negligible	Low	Low	Intermediate
Inactive	Negligible	Low	Low	Intermediate	Intermediate
Fresh	Low	Low	Intermediate	High	High
Active	Intermediate	Intermediate	High	High	High

RMR by Bieniawski (1989) classification for underground structures and the Slope Mass Rating SMR by Romana (1997) for slope and cliff structures.

The general scheme of a geotechnical risk analysis consists of four stages. The first stage is a stage of observation, where an intervention at the site is required in order to collect information about the site parameters such as discontinuity spacing, weathering of the rock mass, etc.

The second stage of the analysis consists of calculating (from the previously observed parameters) three main parameters namely sensitivity, activity and intensity of the site.

The sensitivity can be regarded as the potential capacity of the site to be in a situation that could generate an accident. In our case, the sensitivity is the result of the RMR or SMR evaluation.

The activity of a site indicates the capacity of site movements. It is defined from parameters like the opening of mechanical fractures and weathering of the rock mass.

In order to evaluate the severity of a possible accident, we observe also certain parameters that define the intensity of possible phenomena. Intensity is defined through the elementary size of possible falling blocks, or the total volumetric size of these blocks.

In a discrete approach, for example, each of observed parameters is quantified with an index and classified into a corresponding class and so on for the sensitivity, activity and intensity of the site.

The third and fourth phases of the analysis consist of table crossing between different classes of sensitivity and activity in order to produce the possibility of occurrence (third phase), and another table crossing between different classes of possibility of occurrence and intensity to produce the final risk (fourth phase). Table 28.1 shows an example of a table crossing between the sensitivity and the activity of a certain site.

28.4 Geotechnical Risk Analysis

For the application of the general methodology in the analysis of geotechnical risk in a specific site, care has to be taken in dealing with problems of uncertainty. For the discrete approach, three major problems are faced. First, the problem of variability that was mentioned earlier. Second, is the quantification of qualitative parameters such as rock mass weathering (Rock engineering experts usually tends to use linguistic variables such as highly weathered or moderately weathered for the description of the rock mass). The third problem is the class limits problem. This problem arises when the unique value of an observed phenomena falls near or at the limits of its corresponding class. This problem is illustrated in Fig. 28.4 that shows a typical problem of class limits for the RQD that is defined by Eq. 28.1:

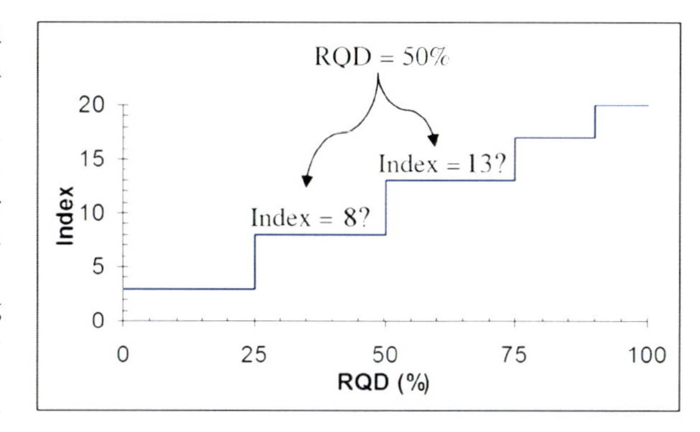

Fig. 28.4. The problem of class limits

$$\mathrm{RQD} = \frac{\sum \text{Individual rock core lengths} > 10\,\text{cm}}{\text{Total core length}} \tag{28.1}$$

Probabilistic approach can handle problems of variability of parameters' values providing that sufficient statistical observations are available and that statistical distributions are known. Although the problem of class limits could be considered as partially solved, the probabilistic approach is not capable of handling the qualitative parameters.

One of the disadvantages of the probabilistic approach is the necessity of making a large number of simulations in order to arrive at satisfactory results.

The fuzzy logic, on the other hand, is capable of handling all three basic problems of variability, class limits and representation of qualitative parameters. These three advantages in addition to the need of less computer power for the analysis encouraged us to adopt this approach for geotechnical risk analysis.

28.5 Uncertainty Analysis

At present, engineers and researchers usually use discrete approach when carrying out geotechnical risk analysis. This is a relatively old but easy approach. Whenever an uncertainty of a parameter's value is faced, an expert has to estimate a unique value for the parameter. This estimate is usually based on the experience or supported by few measurements. This estimation might be pessimistic (in case of a minimum value), optimistic (in case of a maximum value), or mean (in case of choosing a mean value). The analysis of the problem is carried on by estimating unique values for each parameter, supplying these values into the analysis, which leads to a unique final result for a specific problem.

This approach draws no attention to the nature of the parameter nor to its probabilistic distribution, and the parameters are arranged into classes that are assigned to indexes.

The probabilistic approach is based upon statistical distribution. For each parameter, a series of measurements is carried out. These measurements are fitted into a law of distribution (normal, log normal, uniform, etc.). The analysis is usually carried out using Monte-Carlo simulation where random values are drawn from distribution laws. The simulated values are introduced into the analysis as in the case of the discrete approach. Repeating this procedure for a large number of times makes it possible to overcome the randomness and the uncertainty of the phenomena. The results are collected in the form of histograms that represent the final result of the problem. The probabilistic approach is used massively in complex systems where parameters are random in nature and easy to be measured.

Monte-Carlo simulation can be only applied when sufficient statistical data are available for the estimation of distribution laws. In addition to the necessity of statistical data for this analysis, it cannot overcome the problem of non-random uncertainties such as qualitative parameters (weathering of the rock mass, roughness of joints, etc.).

The approach that is proposed by this study is based on the fuzzy theory introduced by Zadeh (1965). This theory underlines much of human ability to make decisions based on vague or imprecise information. It worth mentioning here that Many attempts has been performed using fuzzy/semi fuzzy techniques for the quantifications of qualitative parameters has been reported in the bibliography of Chowdhury (1986), Chowdhury et al. (1992), Kawakami and Saito (1984), Hudson et al. (1992), Nathanail et al. (1992), and Nguyen (1985).

Non-random uncertainties could be analyzed through fuzzy systems and fuzzy sub-sets. These subsets replace discrete numbers in the same way as statistical distributions replace parameters.

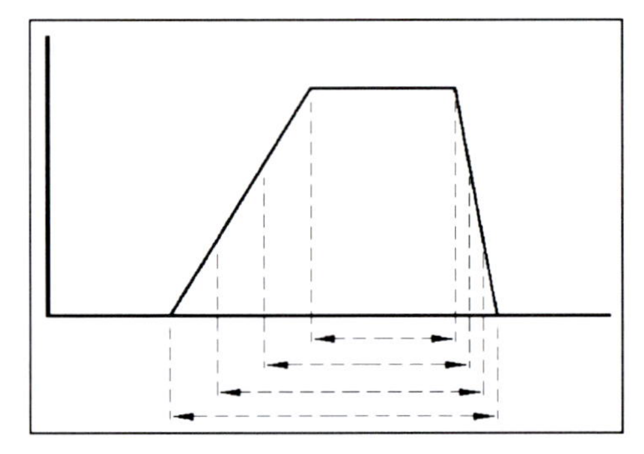

Fig. 28.5. Alpha cuts of a fuzzy number

Fuzzy numbers' calculations can be done in several ways. Juang et al. (1991) proposed a Monte-Carlo-like method of calculation that is similar to the classical Monte-Carlo simulation. This way of calculation that is really interesting could only be applied when fuzzy parameters are presented by Gaussian-like numbers.

The most widely used method of fuzzy calculation is the α-cuts method where all fuzzy numbers are transformed into a number of intervals at α level, and calculations are done on these intervals using the interval mathematics. Figure 28.5 illustrates the discretization of a fuzzy number into four α-cuts.

Fuzzy reasoning (sometimes referenced as approximate reasoning), is another point of research. Because of the nature and the properties of fuzzy numbers, we cannot perform a simple If-Then style of reasoning as presented by Eq. 28.2, because fuzzy reasoning consists of reasoning through statements like Eq. 28.3.

$$\text{If } X = 3, \text{Then } Y = 10 \tag{28.2}$$

$$\text{If Mechanical fractures are Evolving, Then Risk is High} \tag{28.3}$$

where *Mechanical fractures* and *Risk* are fuzzy parameters, and *Evolving* and *High* are fuzzy subsets.

Fuzzy reasoning is performed through the use of tables of rules that are defined by experts. Among the methods of fuzzy reasoning is the min-max procedure defined by Cox (1994) and illustrated in Fig. 28.6.

28.6 Risk Analysis inside the Tomb of Ramsis I at the Valley of the Kings

We have applied the general methodology of geotechnical risk analysis to a case of underground historical tomb at the Valley of the Kings, Egypt. The tomb is facing typical problems of unstability showing possible risks of block movements, and requires a global risk analysis study. Figure 28.7 shows a plan of the tomb altogether with the main discontinuities that were observed at the site.

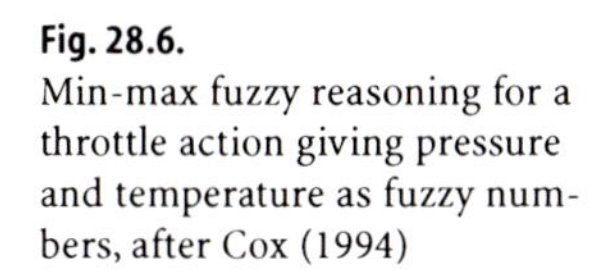

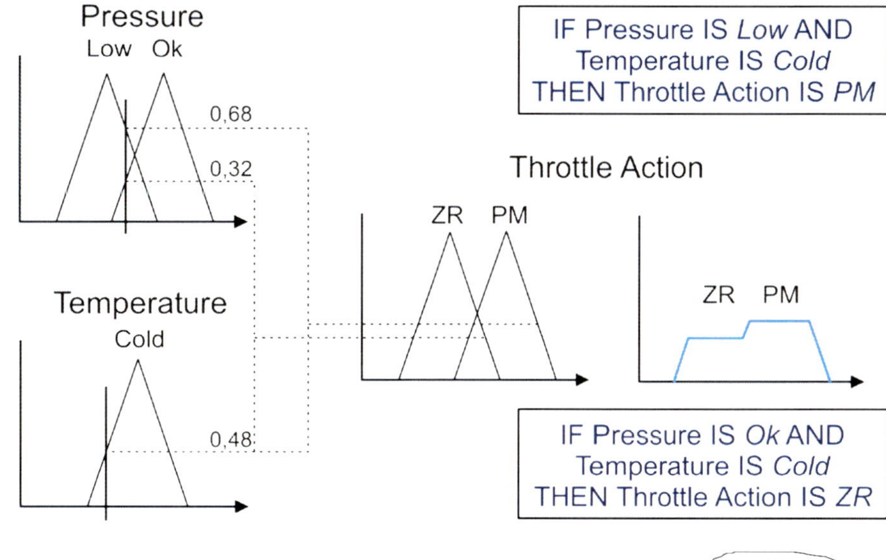

Fig. 28.6. Min-max fuzzy reasoning for a throttle action giving pressure and temperature as fuzzy numbers, after Cox (1994)

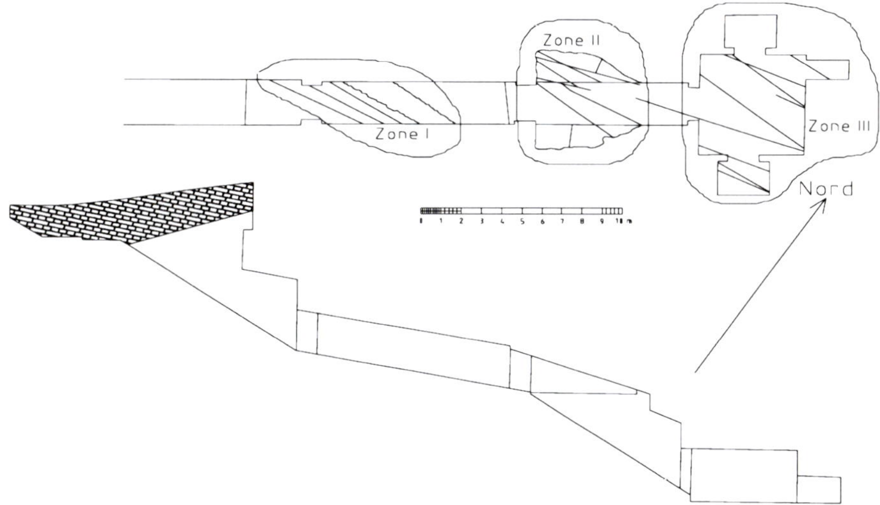

Fig. 28.7. Plan of the tomb of Ramses I in the Valley of the Kings, Egypt

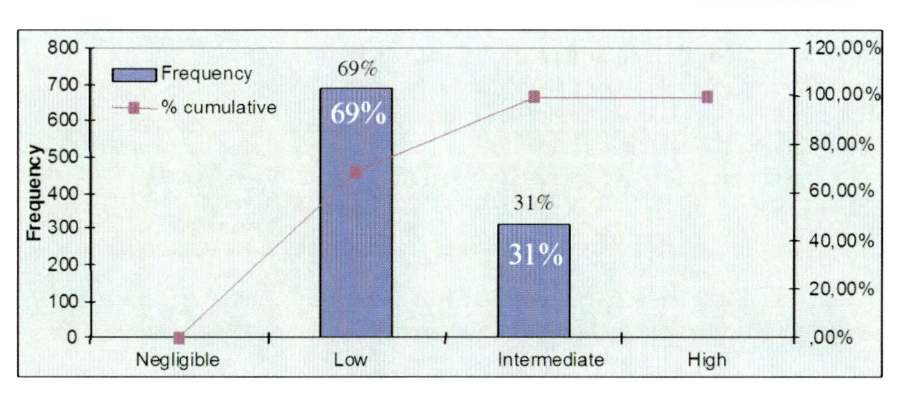

Fig. 28.8. Histogram of the risk at zone III (probabilistic approach)

We have applied the rock mass classification system for the definition of sensitivity altogether with other parameters defining the activity and the intensity of block falling phenomena. A discrete approach analysis indicated low risk at zone III. Figures 28.8 and 28.9 show the results of the analysis at zone III using the probabilistic and the fuzzy logic approach.

We can see from these results that the probabilistic and fuzzy approaches gives more information about the risk state at the site. For example, the fuzzy analysis shows 71%, 24%, and 5% potentiality to get low, intermediate and high risk respectively. The probabilistic approach gives no indication for a high risk and the discrete approach indicates only low risk.

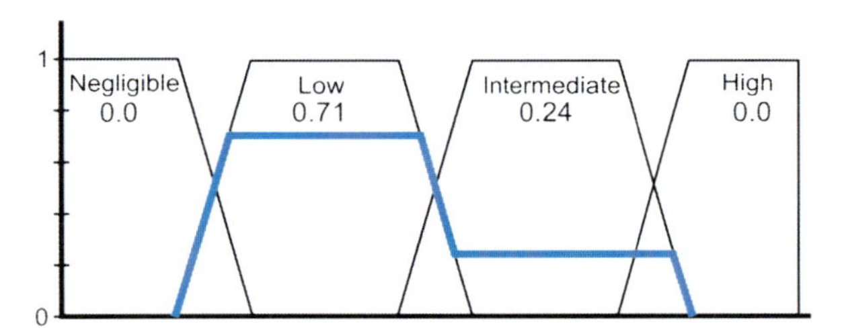

Fig. 28.9. Presentation of the risk at zone III by fuzzy logic

28.7 Conclusions

The analysis of geotechnical risks is today a difficult work and must not be confused with deterministic stability analysis. A risk analysis study has to take into account the uncertainty or the lack of information on the site under consideration in a way or another. We have proposed a general methodology that is based partly on the rock mass classification systems and that is better in considering parameters' uncertainties. The methodology is run using fuzzy parameters and fuzzy reasoning, and we have demonstrated a typical example for the use of this method in an underground opening giving more information than the classical discrete approach or the probabilistic approach.

28.8 Past Experiences

In our laboratory at Nancy School of Mines, France, and since more than ten years, we have started working with cooperation with Cairo University, Faculty of Engineering, on geotechnical problems facing some historical sites. This work has been conducted through different aspects "Numerical modeling of some underground tombs in Saqqara, study of the stability of slopes at El-Deir El-Bahari Temples in Luxor, the implementation of a project of geotechnical monitoring for some underground historical tombs in Saqqarah, etc."

In recent years, we have started a cooperation with Aleppo University in Syria about similar projects concerning numerical and geotechnical modeling of the citadel and the great mosque of Aleppo.

Lately, and since few years, we have started concluding these studies by the introduction of the notions of hazard and risk analysis. Many publications has been done through the years for these various studies "Please find enclosed a list of some publications on that subject".

These studies for hazard analysis and risk evaluation are done extensively through the usage of artificial intelligence techniques such as expert and fuzzy systems, and we are developing a general methodology for the evaluation of ground motion risks around geotechnical sites. We were also able to develop, some computer tools and programs based on fuzzy logic for the analysis and the evaluation of ground motion risks.

Landslide risk analysis studies are very rare in or around historical sites, and this is why this work is considered genuine and important. The utility of these studies is to compare between different sites at different configuration in order to be able to define risk and hazard scales that could be a starting point for any justifications for preservation and/or conservation projects.

References

Bieniawski Z (1989) Engineering rock mass classifications. John Wiley & Sons

Chowdhury R (1986) Geomechanics risk model for multiple failures along rock discontinuities. Int J Rock Mech Min Abstr, 23(5):337–346

Chowdhury R, Zhang S, Li J (1992) Geotechnical risk and the use of grey extrapolation technique. Proc. 6th Australian New-Zealand Conf. on Geomechanics, pp 432–435

Cox E (1994) The fuzzy systems handbook. Academic Press professional

Hudson J, Sheng J, Arnold P (1992) Rock engineering risk assessment through critical mechanism and parameter evaluation. Proc. 6th Australian New-Zealand Conf. on Geomechanics, pp 442–447

Juang C, Huang X, Elton D (1991) Fuzzy simulation processing by the Monte Carlo simulation technique. Civil Eng Syst 8:19–25

Kawakami H, Saito Y (1984) Landslide risk mapping by a quantification method. Proc. 4th Int. Symp. Landslides, Toronto, Canada, pp 535–540

Nathanail C, Earle D, Hudson J (1992) Stability hazard indicator system for slope failure in hetrogeneous strata. EUROCK'92, pp 111–116

Nguyen V (1985) Overall evaluation of geotechnical hazard based on fuzzy set theory. Soils and Foundations 25(4):88–18, Japanese society of soil mechanics and foundation engineering

Romana E (1997) The geomechanical classification SMR for slope correction. Proc. Tunneling under Difficult Conditions and Rock Mass Classification Basel, Switzerland, pp 1–16

Zadeh L (1965) Fuzzy sets, information and control, vol. 8, pp 109–141

Chapter 29

Collection of Data on Historical Landslides in Nicaragua

Graziella Devoli

Abstract. Systematic studies of landslides in Nicaragua started only after the disastrous impact of Hurricane Mitch at the end of October 1998, which caused widespread and devastating slope failures. An attempt to collect, integrate and analyze historical data is made in order to improve the current information on landslides. In the period between 1570 and 1988, 135 historical landslides were found through the review of catalogues, newspapers, monographs, technical reports, bulletins and scientific papers. The type and quality of information collected, and the methodologies and techniques used to analyze the data are described. The analysis has allowed verifying that debris flows and rock falls have been the most common types of movement. Historical landslides have been triggered by high-intensity or long-duration rainfalls, earthquakes, hurricanes and volcanic eruptions confirming that landslides are not isolated phenomena but usually a consequence of other events. Besides hurricane Mitch, also other hurricanes have triggered landslides in the past. Data on damage and casualties are collected and integrated in a database. The spatial distribution of historical landslides shows that they have occurred mainly along the Pacific Volcanic Chain and few events were found within the Interior Highlands, the hilly relief of the Pacific Coastal Plain and the Atlantic Coastal Plain. The historical data help in improved understanding of landslide processes, their spatial and temporal distribution, as well as the economical and human losses caused by them, all of which are necessary for future landslide hazard and risk assessment in Nicaragua.

Keywords. Landslides, historical data, Nicaragua

29.1 Introduction

In recent years an attempt to collect landslide data in Nicaragua and to establish a national landslide database has been made. However, because of limited investigation of past landslides, especially before 1990, the information is still deficient. The importance of including historical data in proper landslide hazard assessment has been pointed out by several investigators who have attempted to reconstruct historical records for individual landslides or landslide-prone areas (Ibsen and Brunsden 1996; Wieczorek and Jäger 1996; Guzzetti 2000; Glade 2001; Dominguez Cuesta et al. 1999; Calcaterra and Parise 2001; Guzzetti and Tonelli 2004). The work presented herein attempts to provide a better understanding of landslide processes in Nicaragua and shows the usefulness of historical archives, especially newspapers, in providing data on spatial and temporal distribution, number of victims and types of damages, and to establish the relationship between landslides and their triggering factors.

29.2 Study Site

Nicaragua is located in the central part of the Central American isthmus with both Caribbean and Pacific coasts. With a total surface of 130 370 km^2, Nicaragua is the largest country in Central America having a population of about 5.6 million (INEC 2004). Administratively the country is divided in 2 autonomous regions and 15 departments. 63% of the country is flat with small and isolated hills, 34% has an elevation between 200 and 1 500 m a.s.l. and 3% has elevation higher than 1 500 m a.s.l., up to 2 107 m a.s.l. (INETER 1995). Its climate is classified as tropical with a national average temperature around 25.4 °C (temperature is strongly dependent on altitude, especially above 800 m a.s.l.) and rainfall varying greatly, between 800 mm yr^{-1} and 5 000 mm yr^{-1} (INETER 2004). Nicaragua forms part of the Caribbean Plate which is composed of various blocks. Among them, Chortis and Chorotega blocks have been recognized in Nicaragua. Five physiographic provinces that closely correspond to geological provinces have been distinguished (Fig. 29.1). Low-grade metasedimentary Paleozoic rocks intruded by granitoid rocks and overlain by continental and marine Mesozoic sedimentary rocks are exposed in the northern part of the country. Tertiary volcanic rocks are superimposed on it and cover most of the Interior Highlands forming extensive plateaus. Mesozoic and Tertiary sedimentary rocks prevail in the western part in the Pacific Coastal Plain. The Pacific Volcanic Chain, a result of the Cocos Plate subduction beneath the Caribbean Plate, is made of active volcanoes as well as historically active and dormant ones. Volcanic rocks are represented by intercalation of tephra deposits, individual andesitic and basalts lava flows, tuffs, pumice, volcanoclastics sediments, and laharic debris. The Atlantic Coastal Plain and the Nicaraguan Depression are full of recent sediments (McBirney and Williams 1965; Parson Corporation et al. 1972; Weyl 1980; Dengo 1985; Fenzl 1989; van Wyk de Vries 1993).

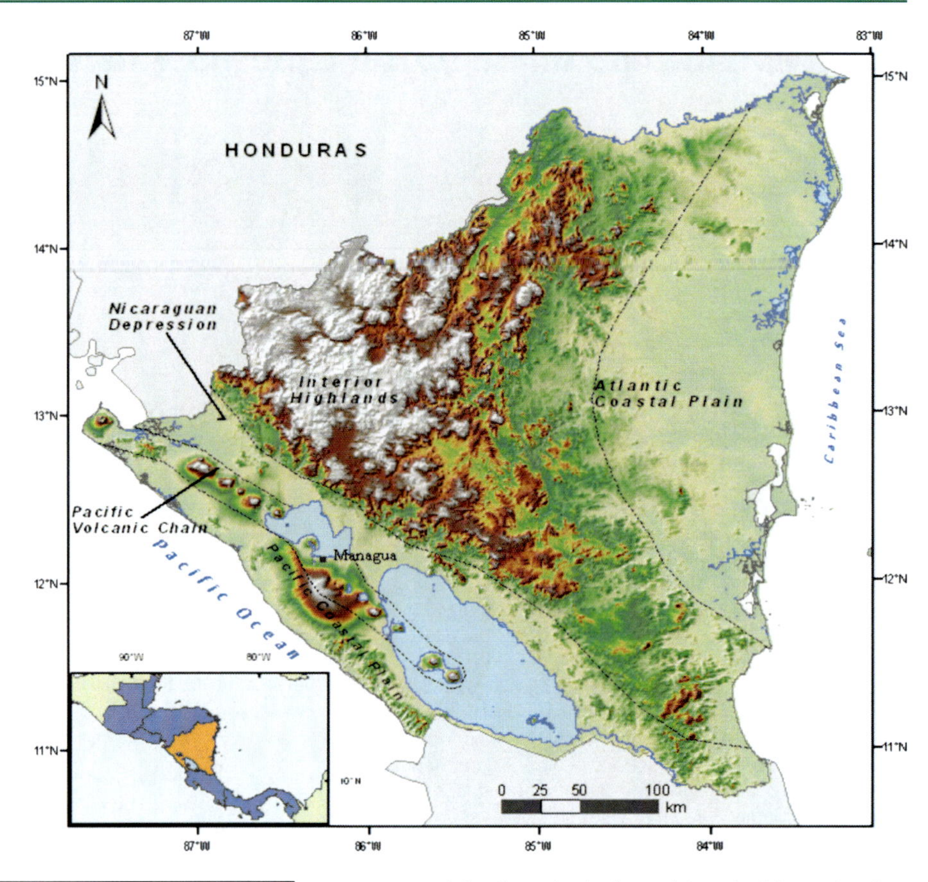

Fig. 29.1. Map of relief and Nicaraguan geomorphological units (based on Fenzl 1989)

29.3 Methodology

The methodology developed for this study includes research and collection of the available information on landslide processes and their impacts in Nicaragua; analysis of the status of knowledge of landslide processes; collection of historical landslide events through catalogues, newspapers, technical reports, scientific papers, bulletins and monographs; storage of the data in the national landslide database (Devoli et al. 2005), and the analysis of landslide data comprised in the period from 1570 to 1988 using interactive queries with SQL server. Only landslide events prior to 1990 are collected and analyzed because up until that time very little information exists. Sporadic scientific reports are available, but landslide data is incidental as the studies were focused on geological mapping and volcanic activity. The data that were looked for consisted of date of the event, location and magnitude of the event, slope movement typology, damage caused by the landslide, area affected, triggering factors and secondary hazards. Problems in preparing the historical catalogue of landslide events and difficulties in the extraction of historical information have been encountered, as also documented by others (Wieczoreck and Jager 1996; Ibsen and Brunsden 1996; Dominguez Cuesta et al. 1999; Guzzetti 2000; Calcaterra and Parise 2001). The problems encountered, both typical of any historical investigation and specific to this work, were complicated by the complexity of Nicaraguan history and its social, political, scientific and natural environment.

29.4 Results

The historical research permitted the identification of 135 major landslide events in the period from 1570 to 1988. 65% of data were obtained from newspapers, 25% from technical reports and 10% from other sources. 115 articles were found describing landslide processes in 9 different types of national newspapers. La Prensa newspaper provided the largest number of landslide events. Figure 29.2 shows the temporal distribution of landslide events in the period examined. Major landslide activity was observed during 1951, 1954, 1955, 1956, 1960, 1972 and 1988. There seems to be a tendency of increasing landslide activity in time, but this is most likely due to better reporting of more recent events. The oldest landslide recorded is the Acahualinca mud flow, dated by Bice (1980) to have occurred 7 500 years ago. The landslide's deposit is exposed in the western sector of the capital Managua. Another pre-historic event is the debris avalanche that occurred in the north-eastern flank of the Mombacho Volcano (van Wyk de Vries et al. 2005). 118 landslide events have been re-

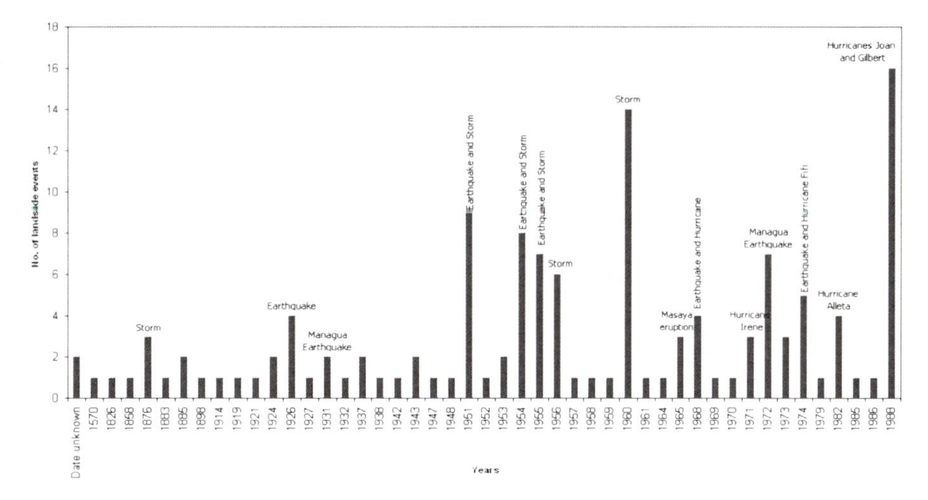

Fig. 29.2. Temporal distribution of historical landslides in the period from 1570 to 1988

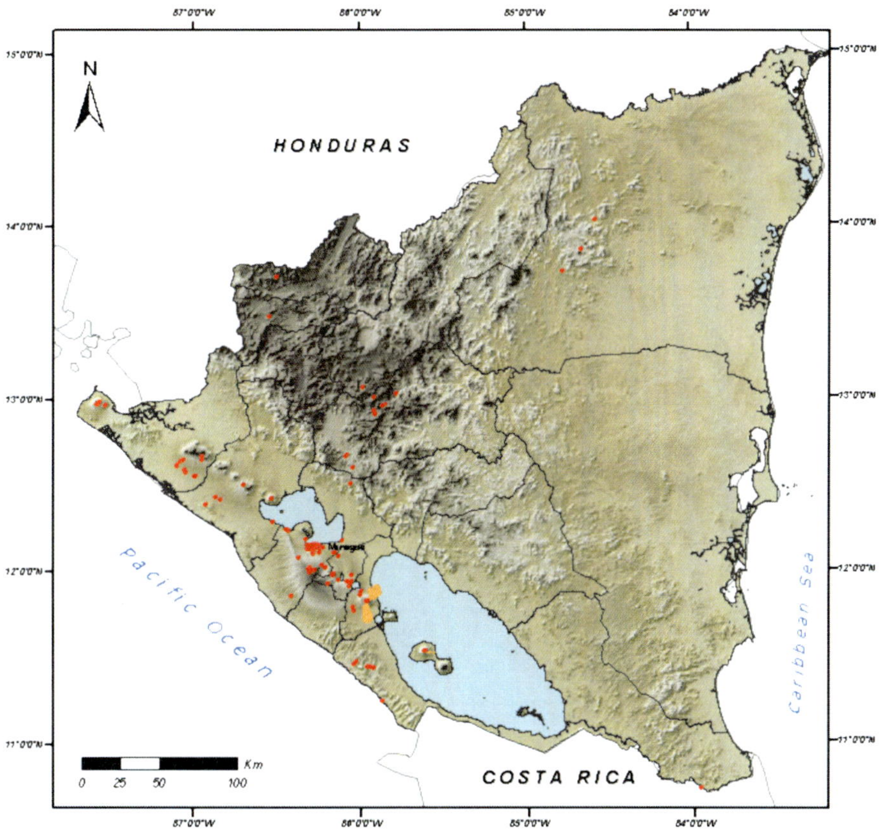

Fig. 29.3. Geographical distribution of the sites affected by landslides in the period from 1570 to 1988. Red dots indicate the locations of landslides for which detailed spatial information were not available. *Orange polygons* represent deposit or both deposit and scarp of historical landslides

ported to have occurred in the Pacific area within the departments of Managua, Masaya, Chinandega and Rivas. The capital Managua shows the highest records of landslides, but also Nindirí, Altagracia, Chichigalpa and Granada municipalities have frequently been affected by landslides. In the Interior Highlands, 12 landslides were recorded, with half of them in the municipality of Matagalpa (Fig. 29.3). Debris flows (a term denoting also mud flows and lahars in this paper) and rock falls have been the most common types of mass movement in the past. The biggest landslide events recorded are the debris avalanches at Mombacho Volcano for which Ui (1972) estimated a volume of 1 km^3. Sixty-six geohazard events were found to have triggered landslides, with high-intensity or long-duration rainfalls, earthquakes, hurricanes and volcanic eruptions as the main triggering factors. The few eruptions that have triggered landslides in the past have been related to the Masaya, Concepción and Cerro Negro Volcanoes. Besides Hurricane Mitch (1998), also other hurricanes have triggered landslides in the past. The Hurricane Joan (1988) triggered the highest number of landslides followed by Hurricanes Alleta (1982) and Irene (1971).

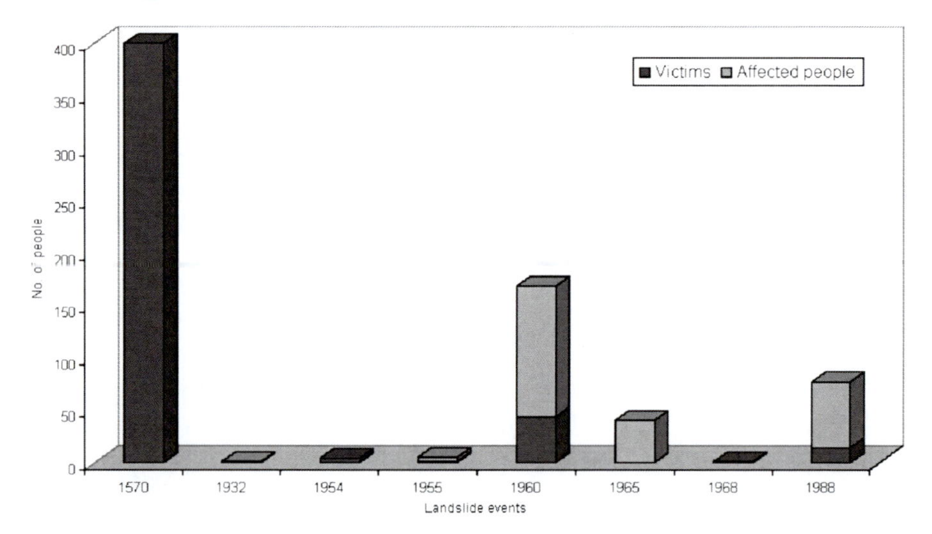

Fig. 29.4.
Temporal distribution of landslide events that have produced casualties and affected people in the period from 1570 to 1988

Twenty-one landslide events in the past have caused fatalities and seriously affected people. 463 people were killed and 233 people were affected by landslides in the period examined. Since 1998 much more disastrous landslides have occurred in Nicaragua. The debris avalanche in 1570 on the southern flank of the Mombacho Volcano, which killed 400 people, was the most catastrophic one (Fig. 29.4). Most frequently debris flows have been the most catastrophic slope failures. 14 debris flow events have caused 61 fatalities and affected 185 people. Roads and the old railway line were the most affected infrastructure in the country, followed by damaged crops, houses and bridges. Limitations and problems were encountered related to the type of sources analyzed and related to the type of information extracted from them. Despite the limitations, the inventory of historical landslides presented here is the most comprehensive source of information on landslides in the past (1570–1988) in Nicaragua.

References

Bice DC (1980) Tephra stratigraphy and physical aspects of recent volcanism near Managua, Nicaragua (Ph.D. thesis). Berkeley, California University of California Berkeley, 422 pp

Calcaterra D, Parise M (2001) The contribution of historical information in the assessment of the landslide hazards. In: Glade T, Albini P, Frances F (eds) The use of historical data in natural hazards assessment. Kluwer Academic Publishers, pp 201–217

Dengo G (1985) Mid America: tectonic setting for the Pacific margin from southern Mexico to northwestern Colombia, vol. 4. In: Nairn HEM, Stehli FG, Uyeda S (eds) The Pacific Ocean, the ocean basins and margins. Plenum Press New York, pp 123–180

Devoli G, Chavez G, Blanco R (2005) Landslide database for Nicaragua and its use. Geophys Res Abstr, vol. 7, 04157, European Geosciences Union

Dominguez Cuesta MJ, Jiménez Sánchez M, Rodríguez Garcia A (1999) Press archives as temporal records of landslides in the North of Spain: relationships between rainfall and instability slope events. Geomorphology 30, 125–132

Fenzl N (1989) Nicaragua: Geografia, Clima, Geologia y Hidrogeología. Belém, UFPA/INETER/INAN

Glade T (2001) Landslide hazard assessment and historical landslide data – an inseparable couple? In: Glade T, Albini P, Frances F (eds) The use of historical data in natural hazard assessments. Advances of Technological and Natural Hazard Research, Kluwer, pp 153–169

Guzzetti F (2000) Landslide fatalities and the evaluation of landslide risk in Italy. Eng Geol 58:89–107

Guzzetti F, Tonelli G (2004) Information system on hydrological and geomorphological catastrophes in Italy (SICI): a tool for managing landslide and flood hazards. European Geosciences Union, Natural Hazards and Earth System Sciences 4:213–232

Ibsen M-L, Brunsden D (1996) The nature, use and problems of historical archives for the temporal occurrence of landslides, with specific reference to the south coast of Britain, Ventnor, Isle of Wight. Geomorphology 15:241–258

INEC (2004) Población total por área de residencia y sexo según departamento. Proyección de la Población al 2004. Instituto Nacional de Estadísticas y Censos. Managua, Nicaragua, http://www.inec.gob.ni/estadisticas/proyeccion2004.htm

INETER (1995) Atlas escolar de Nicaragua. Segunda Edición. Instituto Nicaragüense de Estudios Territoriales, Managua, Nicaragua

INETER (2004) Atlas Climático de Nicaragua. Instituto Nicaragüense de Estudios Territoriales, Managua, Nicaragua

Mc Birney AR, Williams H (1965) Volcanic history of Nicaragua. Univ Calif Publ Geol Sci 55:1–65

Parsons Corporation, Marshall & Stevens Incorporated and International Aero Service Corporation (1972) The geology of Western Nicaragua. Final Technical Report, vol. IV. 221 p, Tax Improvements and Natural Resources Inventory Project, Nicaragua

Ui T (1972) Recent Volcanism in Masaya-Granada Area, Nicaragua. Bull. Volcanol., 36, 174–190

van Wyk de Vries B (1993) Tectonic and magma evolution of Nicaraguan volcanic systems. (Ph.D. thesis) Milton Keyenes, United Kingdom, Department of Earth Sciences, Open University, 328 p

van Wyk de Vries B, Shea T, Pilato M (2005) Mombacho: a volcano falling to bits. European Geosciences Union, Geophys Res Abstr, vol. 7, 03033

Weyl R (1980) Geology of Central America. Second, completely revised edition, 371 pp, Gebrüder Bornträger, Berlin Stuttgart

Wieczorek GF, Jäger S (1996) Triggering mechanisms and deposition rates of postglacial slope movement processes in the Yosemite Valley, California. Geomorphology 15:17–31

Chapter 30

World Heritage "Monasteries of Mount Athos" at Rock Slide Risk, in Greece

Basile Christaras* · Anastasios Dimitriou · George Dimopoulos · Spyros Pavlides · Paul Marinos

Abstract. Four Monasteries have been investigated as pilot monuments in the area of Mount Athos. In the present research the geotechnical conditions of the Monasteries foundation area are studied, regarding the description and the classification of the rock mass and the analysis of the rock slope stability. The area is very fractured and traversed by joints of various directions (mainly NNW, NNE and SSW). These discontinuities can generate unstable geotechnical conditions in the rock mass especially at the slopes of the construction areas.

Keywords. Religious monuments, UNESCO World Heritage

30.1 Introduction

Mount Athos (Holy Mountain) is located in Macedonia (Northern Greece) and it is a place of great historical and religious interest, where only Monasteries for men were built.

The complex geological structure of this area causes significant engineering geological problems at the foundation sites of the Monasteries. Furthermore, the most historical buildings (with a long history of about 1 000 years) were naturally built or restored without the consideration of the geomechanical particularities of the construction area.

In the frame of this research we studied the neotectonic and geotechnical conditions of a particular area of Mount Athos at the sites where many important monasteries are built. Four Monasteries were selected for investigation as pilot monuments in the area to provide the influence of the existing tectonic conditions to the stability of the buildings. Object of the present study is the investigation and documentation of the existing stability conditions of the geological formations at the foundation sites of the above four Monasteries of Mount Athos. The purpose of this research is to point out the existing instability phenomena of the geological formations at the foundation sites of the Monasteries and to make an attempt to interpret the mechanisms which create these problems in order to accomplish the best solution possible.

30.2 Study Case

The foundation stability conditions of the historical monuments are of crucial interest, especially in regions like the Mediterranean Basin and particularly the Greece, where the seismotectonic regime is active and the geological structure is complex. North Aegean and especially Chalkidiki/Mount Athos Peninsula are extremely seismic active areas.

The geological and geotechnical conditions of the foundation area are an important factor that influences the conservation of the historical buildings. Phenomena like settlement and slope movements as well as earthquakes and tectonic activity contribute to the damage of the historical buildings. The ground water activity is also an important factor, especially in cases where monuments are buried in the soil or they are founded on steep slopes. The environmental factor is also necessary to be taken in mind, when different protection measures are decided to apply.

The methods used for the stabilization of the slopes can be classified in the following categories: *(a)* change of the slope geometry to decrease the driving forces or to increase the resisting forces, *(b)* control of surface water infiltration to reduce seepage forces, *(c)* control internal seepage to reduce the driving forces and to increase material strength and *(d)* provide retention to increase the resisting forces.

The more common used drainage methods are related to the construction of deep wells, vertical drains, sub horizontal drains, drainage galleries, interceptor trench drains and relief trenches.

The various methods of retaining hard rocks slopes are the following: *(a)* concrete pedestal, *(b)* rock bolt for jointed masses, *(c)* bolts and concrete straps for intensely jointed masses, *(d)* cable anchors to increase support depth, *(e)* wire mesh to constrain falls, *(f)* impact walls to deflect or contain rolling blocks, *(g)* concrete to reinforce loose rock, with bolts and drains, *(h)* concrete to retard weathering and slaking of shales.

In soil slopes, the various types of retaining walls are classified into *(a)* gravity walls, *(b)* non gravity walls (basement walls, bridge abutment and anchored concrete curtain walls), *(c)* rigid walls (concrete walls) and *(d)* flexible walls (gabion walls).

Mount Athos is located in Macedonia (N. Greece, Fig. 30.1) and is administratively connected directly to the Patriarchate of Constantinople. It is a place of great his-

Fig. 30.1. Location map of Mount Athos

torical and religious interest, where only Monasteries for men were built. The Monasteries are historical buildings of the tenth to fourteenth century. During the centuries they were destroyed and burned down several times, while they were rebuilt, enlarged and expanded including newly constructed buildings.

The geotechnical instabilities observed at the sites of the Monasteries of Simonos Petra, Dionisiou, St. Gregory, and St. Paul, in Mount Athos are mainly due to the presence of active faults and the geometry of the tectonic discontinuities, in relation to the active seismotectonic regime. These discontinuities can create rock wedge sliding or planar failure in relation to the direction of the slopes. The groundwater activity also decreases the shear strength of the rock mass at the foundation area.

Geologically, the broader study area is mainly part of the Serbomacedonian Zone, an old massif. It consists of Palaeozoic or older two mica gneisses, biotite gneisses, plagioclase-microcline gneisses, marbles, amphibolites, peridotites and dunites, as well as Mesozoic intrusions (biotite granite with transitions into biotite-hornblende granite to granodiorite). Sills and dykes of leucocratic aplitic muscovite granite are frequent. The southern edge of the peninsula of Mount Athos is part of the Circum Rhodope Zone, which consists of greenschists, recrystallized limestones and marbles, plagioclase-microcline gneisses and ultramafic rocks.

The distribution of earthquake foci during historical times and their epicenters during the present century show that the present tectonic activity in the area is very high (Papazachos and Papazachou 1989). It is clearly distributed along two distinct seismic zones: *(a)* the NNW-SSE trending Serbomacedonian Zone, and *(b)* the ENE-WSW striking North Aegean Trough (Pavlides et al. 1990). South Mt. Athos is the angular-cross of these two seismic zones.

The morphotectonic criteria which used to study the neotectonic pattern of the area are landscape contrasts, e.g. steep sea shores, high gradient relief, very steep scarps, gravitational slides associated mainly with normal faulting, which constitute a typical environment of active tectonics and high rates of uplifting. Normal faults (N-S and E-W), typical extension joints and open fractures, which post-date the ductile deformation, characterize the neotectonic pattern of the area. Polished steep fault surfaces in the granite, mainly uneroded, striated mirrors, rock falls, open cracks establish the neotectonic, if not active, tectonic regime.

Under these circumstances morphotectonic and brittle tectonic data are of special importance in defining certain seismic fault parameters. On the other hand striae, which represent the slip vector on the fault plane, can be used in calculating tectonic stress (strain) ellipsoid.

The stress field is extensional with the σ_3 axis trending more or less N-S or as a more precise analysis shows it can be distinguished in ENE-WSW and NNW-SSE directions (σ_3). It is known from the broader area (Pavlides and Kilias 1987; Mercier et al. 1987; Pavlides et al. 1990) that NE-SW trending extension is an older neotectonic phase, while N-S to NNW-SSE is the active one.

The *Monastery of Simonos Petra* is located on the SW coast of Mount Athos Peninsula. It was built around A.D. 1257 by the Blessed Simon. It was burned down several times and consequently only the lower parts of the construction, close to the rock base, are of that age. The western part of the present building was built in A.D. 1590 while the eastern part was built after the fire of A.D. 1891.

The Monastery is built on an isolated and uplifted rock (altitude: 305 m), at the S/SW side of the mountain. The construction presents a particularity caused to the morphology of the rock-hill. The slopes of this rock are steep and the difference of altitude between the lower and higher points is more than 90 m. The rock mass consists of a typical, coarse grain, dark color granite that belongs to the Serbomacedonian mass.

The area is very fractured and traversed by joints of various directions. Many important faults cut the studied area in E-W and N-S general directions. These discontinuities can cause unstable geotechnical conditions, especially at the slopes of the construction area. These instability phenomena are related to the neotectonic conditions of the broader area. The presence of an important neotectonic fault or large scale joint of SW dip direction (216°/80°) distinguishes two sections in the rock mass at the western side of the Monastery, decreasing the stability of the Monastery (Fig. 30.2).

A slope stability analysis was performed with the determination of important unstable wedge and plane failures and the calculation of their factors of safety, using both field measurements and laboratory tests results (Christaras et al. 1994). The rock mass quality was estimated at several representative sites, and a geomechanical classification was performed. According to the results of

Fig. 30.2. The Simonos Petra Monastery. Important faults of E-W direction are occurred

the data elaboration, the rock mass quality in the southern (Fig. 30.3) and western slopes of the foundation area is very low and of limited stability, causing damage to the monument (Christaras et al. 1995).

In order to protect this rockmass a net bolt is necessary to be applied, in the sites where the approach is possible. Grouting could be used only in the cases where the material is very broken and the discontinuities open. All the protection techniques have to respect the environment.

The *St. Gregory Monastery* is an historical building of the fourteenth century lying at the southwestern coast of Athos Peninsula (Fig. 30.4). The area where the Monastery is built, consists of a typical, coarse grain, dark color granite (Gregory type), that is part of the Serbomacedonian massif (Kockel and Mollat 1977). Laboratory tests were carried out in order to determine the physical and the mechanical properties of the granite. The material is very competent, durable with high compressive strength (about 150 MPa). The weathering of the rock is high on the surface but it decreases with depth. The chemical weathering, the mechanical erosion from the waves which undercut the coastal slopes and the temperature fluctuation contribute to the decomposition of the granite causing a significant decrease of its physico-mechanical characteristics. Important neotectonic faults and joints, mainly of E-W and N-S directions, exist in the area, affecting the stability of the foundation rock mass. Joints are generally medium to closely spaced. Their aperture is narrow and they are usually unfilled. Some individual fractures however are filled with breccia of granite or rather with soil. The surfaces of the joints are planar to slightly rough; according to the chart of Barton and Choubey (1977), their roughness coefficient (I_f) is estimated at approximately 6–8. Field measurements were statistically interpreted and the results were plotted in stereographic projections, determining sites (mainly at the western part) where safety factors are lower than 1 (Dimitriou et al. 1997).

Fig. 30.3. Details from the southern slope

Fig. 30.4. Southern part of St. Gregory Monastery

The *Monastery of Dionisiou* is dated back to the second half of the fourteenth century (A.D. 1370–1374). The enlargement of the Monastery and generally its present shape, the tower and the aqueduct were built at the end of the fifteenth century. A big part of the Monastery was burnt down in A.D. 1535, but some years later the eastern part and the six-storey wing were rebuilt (Kadas 1989).

The Monastery is built on a very impressive isolated greenschist rock. The slopes of this rock are steep and the difference of altitude between the lower and higher points is 60–80 m. The area consists of a dark green and brownish greenschist, fine to medium grained, which is part of the Chortiatis magmatic suite of the Circum Rhodope Zone (Kockel and Mollat 1977).

The area is very fractured and traversed by megajoints of various directions (Fig. 30.5). The unfavorable orientations of the joints, which daylight on the slopes, create unstable rock wedges of important size. There are also important discontinuities cutting through the rock mass, which could cause unstable geotechnical conditions to the rock mass, especially at the slopes of the construction area (Dimitriou et al. 1997).

The *Monastery of St. Paul* is located at the western toe of mountain Athos (altitude 2 013 m) and its establishment is dated at the second half of the tenth century (Fig. 30.6). During the centuries the Monastery was destroyed several times by various causes like the fire of 1902 and the flood of 1911, so it was rebuilt almost from the beginning. The northern part was built at the fifteenth century, the defending tower at the beginning of the sixteenth century and all the other buildings during the last two centuries (Kadas 1989).

The largest part of the building has been founded on the soil-weathering mantle and only the north part has been founded directly on the bedrock. The bedrock in the area is described as schist with well-developed schistosity planes of SE to SSW dip direction.

Field measurements were statistically interpreted by using Schmidt diagrams (Dimitriou et al. 1997). The slope planes as well as the envelopes of joints that daylight on the corresponding slope face are presented in the stereonets in order to determine the probable surfaces of sliding. The intersections of joint sets that determine probable wedge and planar failure conditions are also given for representative sites, using the tests proposed by Markland (1972), Hocking (1976) and Hoek and Bray (1981). The interpretation of these stereonets determined wedge and plane failures and the factors of safety (SF) were calculated for each one of them providing unstable areas.

The statistical elaboration of the collected tectonic data show that the northern part of the studied rock mass, presents a fair quality and correspond to the most unstable part of the Monastery. At the western part, the poles of the calculated joint sets do not fall inside the daylight envelope (DE). Furthermore, the intersections of the joint sets do not correspond to unstable wedge failure conditions.

Fig. 30.5. Photo of Dionisiou Monastery

Fig. 30.6. Photo of St. Paul Monastery

30.3 Conclusions

In Mount Athos, the field observations and the data analysis have shown that the geomechanical stability problems of the Monasteries are mainly due to the geometry of the discontinuities (large-scale normal faults, open fractures and closely spaced joints). These discontinuities can create rock wedge or planar failures in relation to the direction of the slopes.

The origin of the geotechnical problems, mentioned in this paper, differ from site to site related to the foundation rock and the tectonic features that prevailed in each area.

Nevertheless most of these failures are related to the neotectonic frame of the broader area showing creep movements of various velocity.

A final classification of the instabilities in categories related to local or general conditions will be utilized both as a recording of the current conditions of our historical and religious monuments and as a guide for the suggested protecting measures which have to be applied.

For the retention of the slopes, rock bolts are placed at suitable sites. In cases where a thick weathered mantle covers the bedrock, the created circular failures are retained using retaining wall and piles founded at layers lower than the slip planes.

References

Barton N, Choubey V (1977) The shear strength of rock joints in theory and practice. Rock Mech 6:1–54

Christaras B, Pavlidis S, Dimitriou A (1994) Slope stability investigation in relation to the neotectonic conditions along the south-western coast of Mount Athos (N. Greece). The case of Symonos Petra Monastery. Proceedings of the 7th Int. Congr. I.A.E.G., pp 1577–1583

Christaras B, Dimitriou A, Mazzini E (1995) Rockmass quality at the foundation area of the Simonos Petra Monastery, in Mount Athos, Greece. La Citta Fragile in Italy, 1° Convegno del Gruppo Nazionale di Geologia Applicata con la participatione della IAEG. Giardini Naxos (ME), pp 191–196

Dimitriou A, Christaras B, Dimopoulos G, Pavlides Sp (1997) Geotechnical aspects at four Monasteries in Mount Athos (N. Greece). A first approach of the stability conditions of the geological formations at their foundation sites. Int. Symp. of IAEG, Engin. Geol. and Env., Athens, Balkema Publ., pp 3113–3122

Hocking G (1976) A method for distinguishing between single and double plane sliding of tetrahedral wedges. Int J Rock Mech Min 13:225–226

Hoek E, Bray JW (1981) Rock slope engineering. Inst. Mining and Metal., London, pp 1–358

Kadas S (1989) Agion Oros. Ekdotiki Athinon, Athens

Kockel F, Mollat H (1977) Erläuterungen zur geologischen Karte der Chalkidiki und angrenzender Gebiete 1 : 100 000 (Nord-Griechenland). Bundesanstalt für Geowissenschaften und Rohstoffe, Hannover

Markland JT (1972) A useful technique for estimating the stability of rock slopes when the rigid wedge sliding type of failure is expected. Imperial College Rock Mech Research Report 19:1–10

Mercier JL, Sorel D, Simeakis K (1987) Changes in the state of stress in the overriding plate of a subduction zone. The Aegean Arc from Pliocene to Present. Ann Tectonicae 1:20–39

Papazachos B, Papazachou K (1989) Earthquakes of Greece. Ziti, Thessaloniki, pp 1–356

Pavlides S, Kilias A (1987) Neotectonic and active faults along the Serbomacedonian Zone (SE Chalkidiki, Northern Greece). Ann Tectonicae 1:97–104

Pavlides S, Mountrakis D, Kilias A, Tranos M (1990) The role of strike-slip movements in the extensional area of Northern Aegean (Greece). A case of transtensional tectonics. In: Boccaletti M, Nur A (eds) Active and recent strike-slip tectonics. Ann Tectonicae 4(2):196–211

Chapter 31

The Archaeological Site of Delphi, Greece: a Site Vulnerable to Earthquakes, Rockfalls and Landslides

Paul Marinos* · Theodora Rondoyanni

Abstract. Seismotectonic activity and slope instability are a permanent threat to the archaeological site of Delphi and the nearby Arachova Center of winter sports in central Greece. In this paper the geological conditions as well as the major active faults of the broader area are presented and discussed, with emphasis on the stability of the limestone cliffs above the Delphi archaeological site. The archaeological site is located in a complicated geological environment, while the steep rock slopes overhanging the Delphi monuments are intensively fractured. These unfavorable conditions cause deformation, displacements, rotations and partial destruction of the monuments, triggered mainly by seismic events. Protection measures are proposed, taking in to account that any construction must not disturb the view of the site. Any intervention must be preceded by detailed mapping of the rock joints and discontinuity planes as well as study of free and underground water drainage.

Keywords. Monuments, rockfalls, earthquake, Delphi, Greece

Fig. 31.1. View of Delphi site. The mountain Parnassos, overlooking the Pleistos Valley and the Corinthian Gulf in the distance

31.1 Introduction

The archaeological site of Delphi is located about 150 km to the north-west of Athens, on the southern slopes of Parnassos Mountain, at elevation of 700 m, near the northern coast of the Corinthian Gulf. Delphi, the "oracle of the Earth" for the ancient Greeks, is a site of exceptional beauty. This superb cultural center is located in a wild landscape of high mountains overlooking the sea (Fig. 31.1 and 31.2).

At this site, a human history of three thousand years has been closely related to the geodynamic development of an active natural environment. The Sanctuary of Apollo lies at the center of Delphi built up on 700 B.C. (Fig. 31.3). The Temple is surrounded by monuments (the treasures) donated by many Ancient States commemorating great events. Delphi was the home of the Oracle Pythia. The theater, of the fourth century B.C., is built to the NW, while further up on the top, is the stadium (Fig. 31.4).

Many important geotechnical problems caused by rock falls and toppling slides, mainly triggered by seismic events, but also foundation subsidence and ground creep, are present in the Delphi area.

Fig. 31.2. Part of the archaeological site of Delphi

Fig. 31.3. The temple of Apollo and the adjacent Phedriades limestone cliff

Fig. 31.4. Aerial view of the site, where the stadium (1), the theatre (2) and the temple of Apollo (3) are marked

The site is characterized by the presence of:

- intensely fractured steep limestone slopes
- weak flysch formations
- at least three different generations of scree
- four sub parallel important normal faults

The stability of the site has been the subject of previous short studies resulting in publications or reports (Mario-lakos et al. 1991; Constantinidis et al. 1988; Koroniotis et al. 1988). Kolaiti completed a postgraduate report at the National Technical University of Athens in 1992. The imposing and magnificent geomorphological environment was thought to be an excellent setting for the selection of the site of the Oracle (Parke and Wormell 1956; de Boer and Hale 2000; Piccardi 2000).

31.2 Geological and Tectonic Setting

The broader region of Delphi consists mainly of the alpine formations of the Parnassos geotectonic zone, covered in places by Quaternary terrestrial loose deposits. A limited area, near the coastal zone is covered by the Pindos Zone Formations (Fig. 31.5).

The geological basement is represented by neritic, medium to thick bedded limestones of Triassic to Cretaceous age, followed by the sequence of flysch formations. These are alternations of brown-red siltstones, mudstones and sandstones with substantial amount of calcareous material. The formations are intensively deformed and fractured by the alpine compressional tectonism, resulting in the formation of folds and overthrusts of upper Eocene age.

This tectonism has reversed the normal sequence of the geological formations by a huge reverse fold, which resulted in places in an overthrust of the older limestones onto the younger flysch formations, as observed in the northern part of the archaeological site. The whole site rests on flysch with the thrusted limestone overhanging along its northern boundary. The front of this thrust in the area of Delphi has an E-W direction and dips towards the N, at an angle of 40–60° (Fig. 31.6).

The limestones are karstified and supply large springs to the broader area. The famous Castallia Spring of ancient Delphi is however a small spring, from a hanging local aquifer at the contact of the thrusted limestones over the flysch. Post alpine extensional tectonism has also affected these formations through a series of normal faults forming an E-W graben (Fig. 31.7).

The recent geological formations consist of scree and talus cones which cover the major part of the archaeological site and have a thickness varying from 2 m in the northern (uphill) part to 12–15 m in the southern part of the site. The deposition and the thickness of this material is closely related to the steep morphology of the mountain slopes in conjunction with the intense faulting and the systems of discontinuities creating rock falls and toppling slides, mainly during earthquakes.

Observations in the broader region of Delphi-Arachova provides the basis for a detailed section, showing the succession of the Quaternary formations based on lithostratigraphic correlations. The older deposits outcrop near Delphi, in the Amfissa Corridor, from a height of 1 200 m stepwards down to the sea level. About 500 m west of

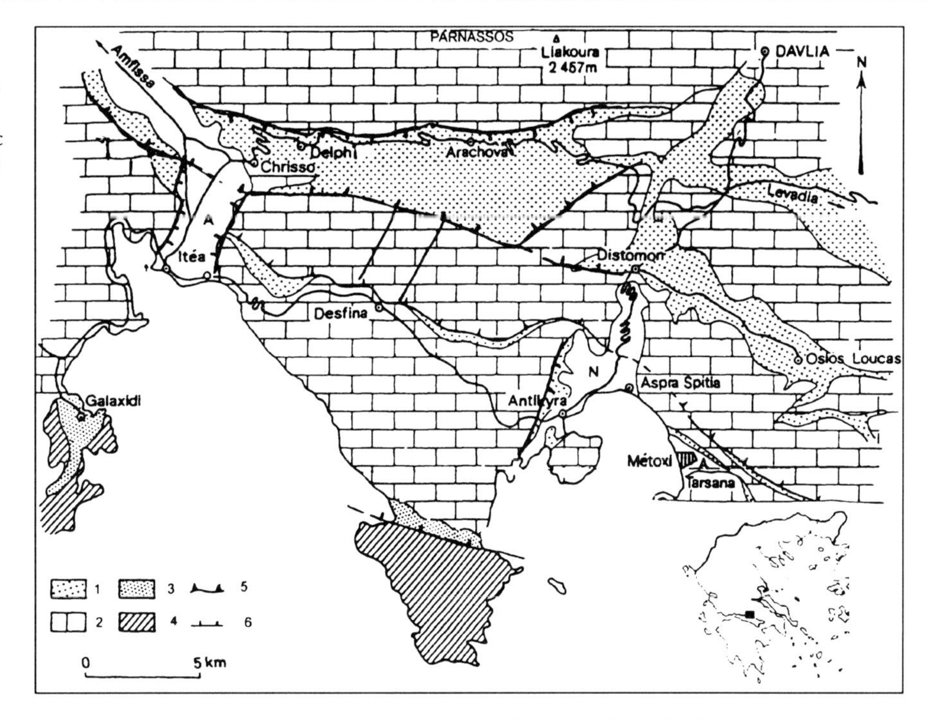

Fig. 31.5.
Simplified geological map of the broader Delphi area. Parnassos geotectonic zone: *1:* flysch; *2:* limestones. Pindos geotectonic zone: *3:* flysch; *4:* limestones and cherts; *5:* overthrust; *6:* fault (modified from Celet 1971)

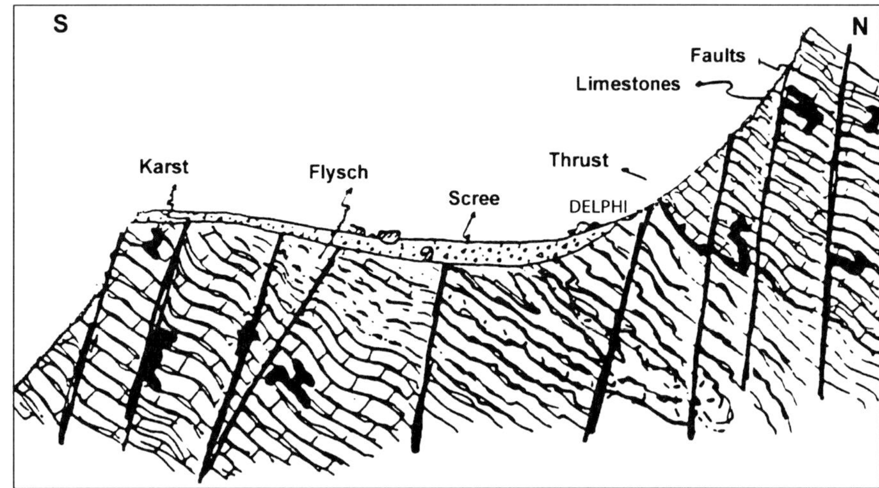

Fig. 31.6.
Schematic geological cross section, not in scale (from Mariolakos et al. 1991)

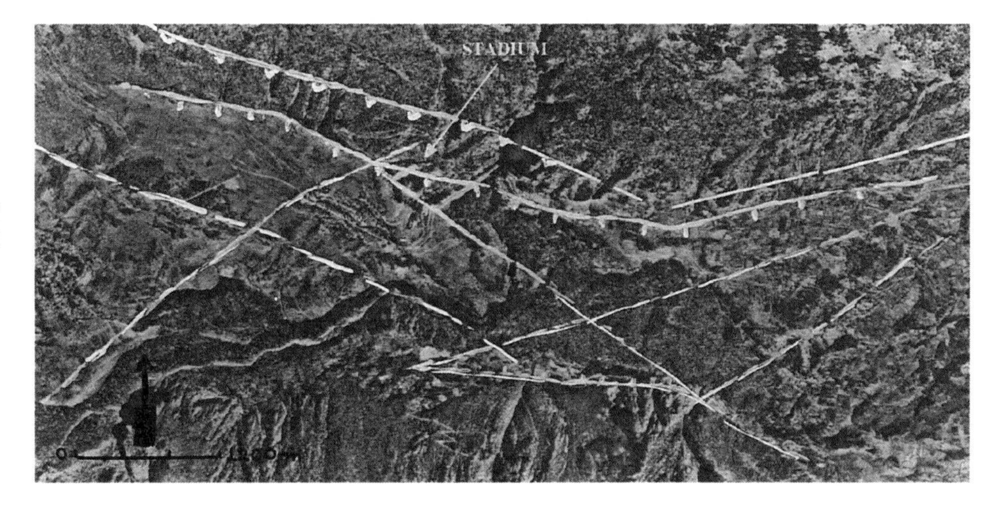

Fig. 31.7.
Air photograph of the Delphi area showing the normal faults. The main fault zone is represented with small teeth towards the hanging wall. The line with the small triangles shows the thrust structure. The location of the ancient Stadium is also noted (detailed view on Fig. 31.11)

Delphi, a 30 m thick conglomerate formation containing limestone angular blocs, is pervasive. The conglomerate is well cemented, insensitively karstified and disconnected from its roots, facts that show a Lower Quaternary age according to Sebrier (1977). It is covered by a younger breccia, 15 m thick, attributed by Pechoux (1977) to the morphogenetic cryoclastic process of the Upper Pleistocene. Both Middle and Upper Pleistocene fans and scree are related to the rhexistasy crisis, which interfered with tectonic activity.

The Plio-Quaternary extension caused the progressive fracturing of the region and the subsidence to the Corinthian Gulf, one of the most important active tectonic structures of the Greek territory (Fig. 31.8). This process is still active with the extension of the Corinthian Gulf reaching the value of 0.6 cm per year, according to a recent geodetic survey (Billiris et al. 1991).

The Delphi graben, trending E-W, underwent repeated activity during the Plio-Pleistocene, as shown by the faulted Middle and Upper Pleistocene breccia. This breccia includes some fragments of older polished fault surfaces. The cliff of Phedriades, to the north of Delphi, is rapidly uplifted north of the Gulf of Corinth. Higgins and Higgins (1996) estimated the combined movement of horst upward and graben floor downward at 3 km in 10 million years, creating a number of large impact earthquakes.

The major normal fault zone of Delphi-Arachova separates the tectonic horst of the Parnassos Mountain to the North from the tectonic graben of Itea to the South and crosses the northern boundary of the archaeological site. This is the main active fault of the area. The base of this active fault zone to the north of the archaeological site (and mainly in the area of Phedriades) is covered by four generations of scree material and talus cones, evidently the result of successive rockfalls, toppling slides

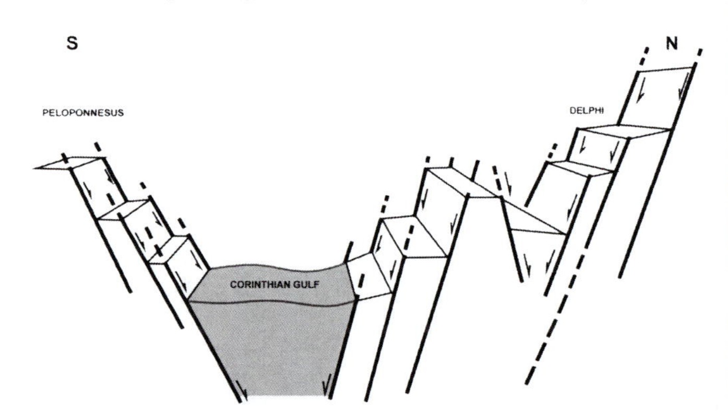

Fig. 31.8. Schematic cross section of Corinthian Gulf and Delphi graben structures

Fig. 31.10. The striated surface (mirror) of Delphi-Arachova active normal fault. Brownish breccia is located to the right of the photo

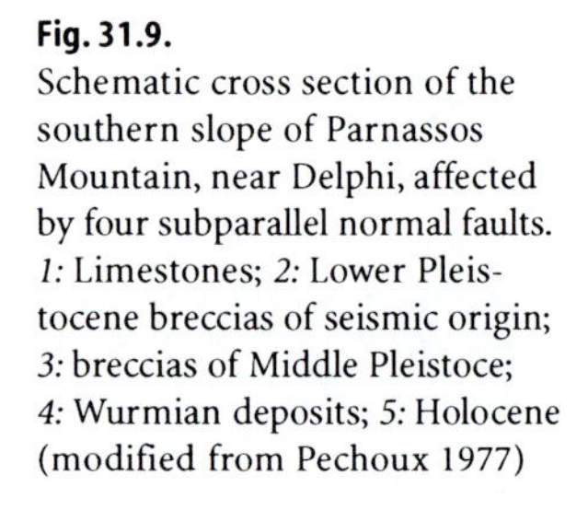

Fig. 31.9. Schematic cross section of the southern slope of Parnassos Mountain, near Delphi, affected by four subparallel normal faults. *1:* Limestones; *2:* Lower Pleistocene breccias of seismic origin; *3:* breccias of Middle Pleistoce; *4:* Wurmian deposits; *5:* Holocene (modified from Pechoux 1977)

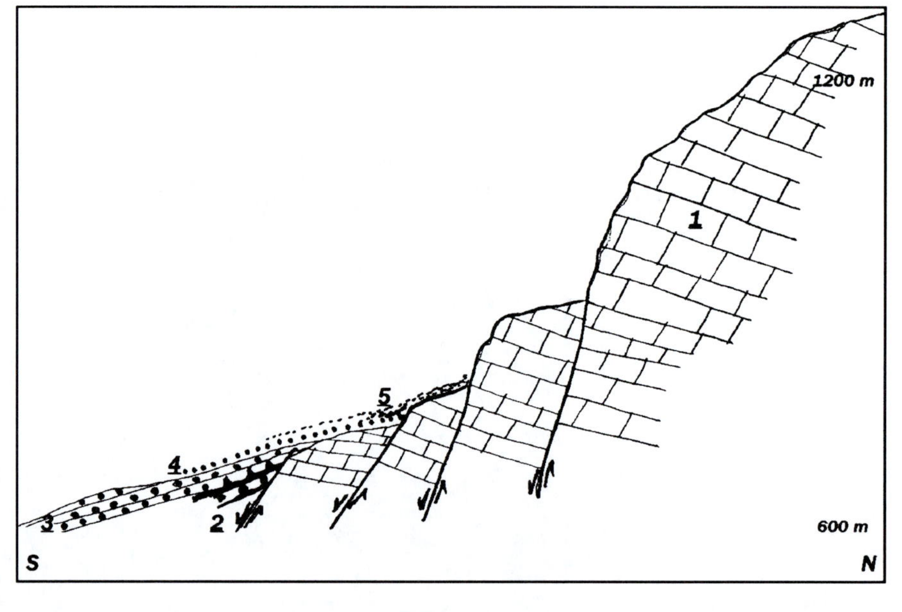

and surface weathering (Fig. 31.9). On both sides of the Delphi archaeological site, the big fresh fault surface (mirror) with subvertical slickensides is observed. Along this fault the direction varies from 65–100° and the dip is 60–70° (Fig. 31.10).

A large number of secondary faults also affect the alpine and post-alpine formations in the major area. Some of these are considered to be active and are thought to be the source of seismic events during historical time. The surfaces of the faults contain a zone of cataclastic rock and mylonitized fine material.

Fig. 31.11. View of the stadium crossed by the trace of the normal fault shown with the *white line* (details of Fig. 31.7)

31.3 Seismotectonic Activity of the Area

The great number of historic earthquakes that have affected the area, accompanied by rockfalls, landslides and other secondary phenomena show the continuation of the effects of the extensional stress field during recent times. The southern wall of the sanctuary of Apollo was built on a big fallen rock of limestone while similar fallen rocks are also frequent in the foundation of the temple of Athena Pronaia.

The historical and recent seismic activity of the major area is intense, with a large number of seismic epicenters in the eastern part of the Corinthian Gulf graben (Fig. 31.12). An analysis of the mechanisms causing the shallow depth earthquakes in the area shows that the tensile stress axis is horizontal in the N-S direction and perpendicular to the compressional axis (Papazachos and Papazachou 1997). Furthermore, the evolution of this tectonic graben is active, with the following average return periods of the expected seismic events: for events exceeding $M = 7.0$, $T = 70$ yr, for $M = 6.0$, $T = 8$ yr and for events exceeding $M = 5.0$, $T = 1$ yr. The expected ground acceleration having a probability of non-exceedence of 63% in 50 yr is 250–275 cm s^{-1} (Makropoulos et al. 1986).

The major recorded seismic events (exceeding a magnitude of $M = 6.0$) in historic time were occurred at 600 B.C., 373 B.C., 348 B.C., 279 B.C., A.D. 551, A.D. 996, 1402, 1580,

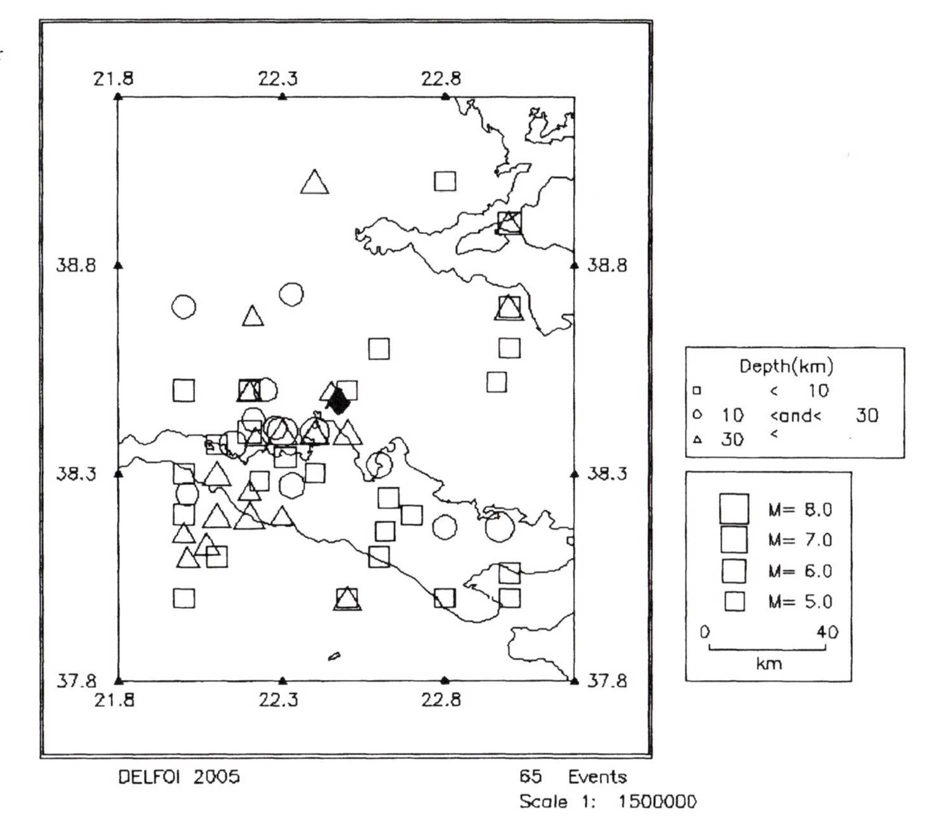

Fig. 31.12. Seismic epicenters of the broader Delphi area with magnitude $M = 5$. *Black little rhomb* shows the Delphi site

1870, 1887, 1938, 1965 and 1970. The most strong earthquake was this of the year 551, with an estimated magnitude of $M = 7.2$. The above seismic events have significantly affected the antiquities of the Delphi sanctuary and their results are evident today in the form of foundation subsidence, destruction of extensive deformation of retaining walls, damage caused by rockfalls and toppling slides as well as ground creep.

The most destructive of these events in the narrow Delphi area have been the following:

- The earthquake of 600 B.C. which caused the complete destruction of the Sanctuary
- The event of 361 B.C. Oribasius reports that "it is no longer true that Apollo has a house and a laurel plant for his oracle and a talking spring, the talking water too, has drained away".
- The earthquake of 373 B.C. which devastated the flourishing town of Helice, across the Corinthian Gulf and caused extensive damage to the archaic temple of Apollo in Delphi, mainly by rockfalls from the Phedriades Cliff.
- The earthquake of 1 August 1870, which occurred during a reactivation of the major Delphi-Arachova fault zone and caused significant damage to the monuments at Delphi as well as in the nearby towns of Chrisso, Itea and Arachova. In the epicentral area ground cracking and differential subsidence were abundant. From the existing descriptions, most of these features seem to be associated with slumping and rapid earthquake induced movement of the alluvium in the valleys and along the coast, as well as with development of tension cracks in the rock behind steep faces and cliffs. The 5 to 6 km long series of surface breaks terminating with a 2 m throw is the only evidence in the historical record. (Ambraseys and Pantelopoulos 1992).

Considering the potential for future seismic events, a magnitude of approximately $M = 6.5$ is expected, taking into account that the trace of the Delphi-Arachova fault can be identified for at least 20 km in surface, from air photos and field observations. Such conditions may result in a possible co-seismic displacement of about 100 cm, according to the existing empirical relationships between earthquake magnitude, fault length and surface displacement from the ground failure data of historical and recent earthquakes. Such relationships, where data from Greek earthquakes are also included, are showing in Fig. 31.13.

31.4 Stability Conditions of the Monuments

The monuments in the archaeological site of Delphi, during their long-time history, have suffered from rockfalls, toppling, foundation subsidence and ground creep, mainly triggered by seismic phenomena or intense rainfalls. The results of these activities are evident on the present appearance of the monuments in the form of displacements, rotations, toppling or even complete demolition of the stone masonry, deformation and subsidence of the foundations and the retaining walls built to support the monuments on the slopes.

31.4.1 Rockfalls

Rockfalls represent the main threat in the area and occur long before the development of the sanctuary, as shown by the observation that the foundations of several ancient structures have been constructed on large fallen blocks of rock, and given the abundance of such blocks in the scree material. The rockfalls are mainly due to the release of the blocks from the joints of the rockmass and bedding planes and are usually triggered by seismic events. Characteristic examples of incipient rockfall areas are the almost vertical rock cliff over the Kastalia Spring (where the oracle used to drink water before uttering its usually ambiguous prophecies) and in the rock cliff above the parking area next to the national road.

The presence of groundwater at the interface between the overlying thrusted limestone and the underlying im-

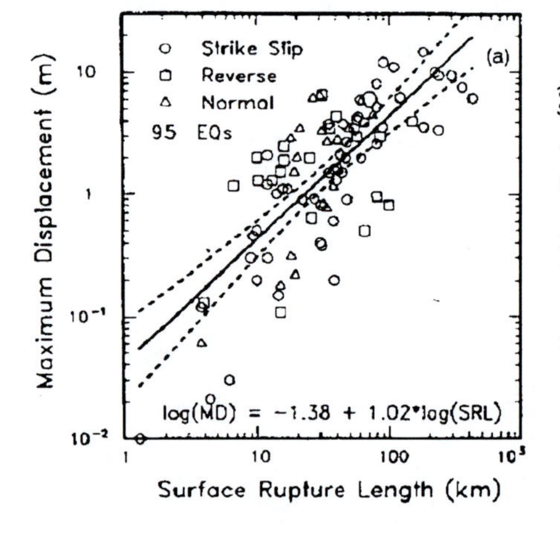

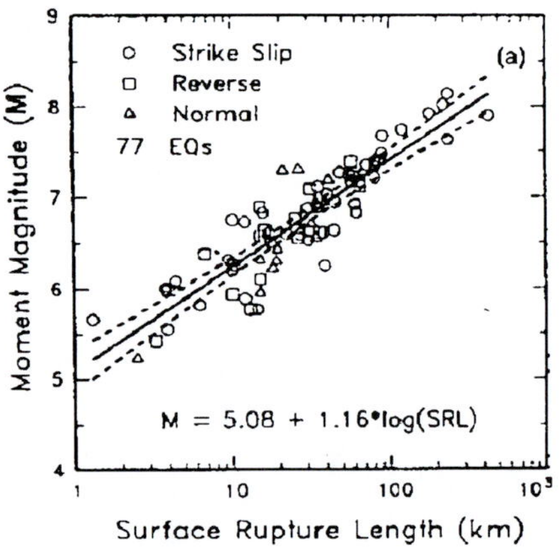

Fig. 31.13. Relationship between surface rupture length and **a** maximum displacement, **b** seismic event magnitude (from Wells and Coppersmith 1994)

permeable flysch, softens the flysch, inducing tensile forces from below, resulting in the fracture of the overhanging limestone and facilitating the rockfall phenomena.

31.4.2 Ground Creep

Creep phenomena are observed to the west of Kastalia Spring, to the west of the Theater and in the region of the Stadium. The slowly progressing creep deformation is evident in the curvature of the tree trunks. The creep is mainly observed in the areas of flysch covered by colluvia or scree from the overhanging limestones. Groundwater contributes by infiltrating into the scree and lubricating the underlying impermeable flysch. Displacements may also be induced by the earthquake activity in the area. Characteristic examples of creeping ground are in the area of the Gymnasium, the temple of Athena Pronaia, the south retaining wall of the Apollo sanctuary, the seats of the Stadium, etc.

31.4.3 Foundation Subsidence and Deformation

Such phenomena are very common in the area of the archaeological site and are usually associated with evolving ground creep movement of the foundations on soils with variable properties. Characteristic cases of differential foundation movements are found on the north side of the foundation of the Apollo sanctuary, the base of the Treasure of Corfu, etc.

31.5 Geotechnical Modeling of the Limestone Cliffs

The stability of the limestone cliffs above the Delphi archaeological site was modeled with discrete elements using the UDEC Computer Program based on the already discussed geological assessment. The analysis was performed by the Laboratoire de Mechanique des Terrains de l'Ecole des Mines de Nancy in collaboration with the Geotechnical Department of the National Technical University of Athens in 1996 (Fig. 31.14 and 31.15).

The stability of the slope was analyzed for seismic excitation using:

- A sinusoidal excitation having a frequency of 5 Hz, a time duration of 3 s and a peak bedrock acceleration of 0.25 g.
- The El Centro 1940 earthquake record (the initial ten seconds) scaled to a peak ground acceleration of 0.35 g.

The resulting displacements due to the dynamic loading are significant and the whole analysis implies that a critical equilibrium is reached incrementally, along the steep rock slope as a result of the previous incidents (earthquakes, rockfalls). Thus, the actual stability of the

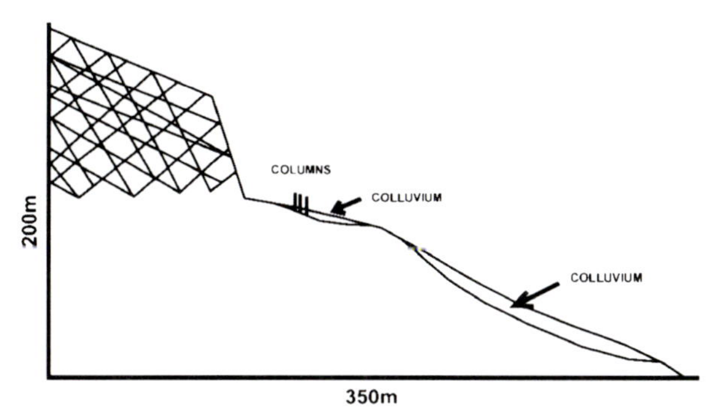

Fig. 31.14. Discrete element model of the slope above the archaeological site (from El-Shabrawi et al. 1996)

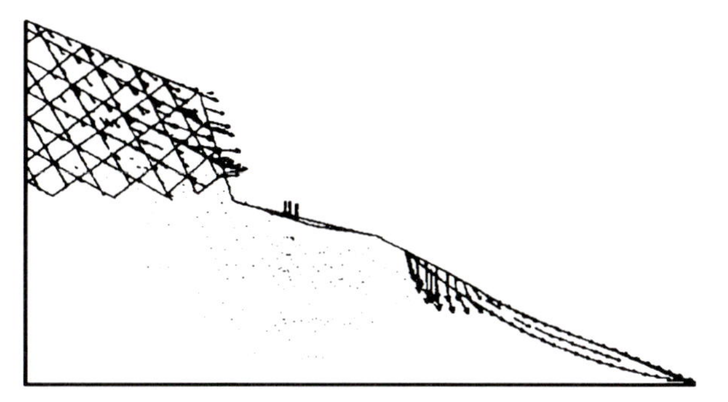

Fig. 31.15. Maximum displacement profiles during the seismic excitation – 3 mm of displacement are shown by the longest *arrows* (from El-Shabrawi et al. 1996)

Delphi site seems to be very sensitive to any modification of its geotechnical conditions (El-Shabrawi et al. 1996).

31.6 Rockfall Protection Measures

31.6.1 Identification of Potential Rockfall Problems

When dealing with boulders on the top of slopes, or steep slopes in a fractured rock mass, the rockfall hazards are obvious. However, the most dangerous types of rock failure occur when a block is suddenly released from an apparently sound face by relatively small deformations in the surrounding rock mass. This can occur when the forces acting across discontinuity planes, which isolate a block from its neighbors, change as a result of water pressures in the discontinuities, development of earthquake acceleration forces, or a reduction of the shear strength of these planes because of long term deterioration due to weathering. This release of "keyblocks" can sometimes precipitate rockfalls of significant size or, in extreme cases, large scale slope failures.

Usually it is suggested that all rock faces should be carefully inspected for potential rockfall problems but it should not be assumed that all rockfall hazards will be detected by such inspections. Consequently some judgment is needed in order to decide where rockfall protection is required.

31.6.2 Active Protection

Active protection is the most appropriate measure but it can be used only in limited areas, immediately affecting any adjacent structure, road or visiting facility. Any intervention must be preceded by thorough removal of any plants, roots and soil material inside the fissures or joints. The next step would involve careful hard scaling to remove all unstable blocks from the cliff face. Mapping of the face would then be carried out to determine the location of all remaining open joints. Light, dental concrete would then be used to fill these joints, taking care that it is not visible at the cliff face. The joints behind the filling would be grouted. Free water drainage must be maintained throughout this process. Finally, non corrodible steel anchors may be applied to retain any potentially unstable blocks.

31.6.3 Physical Restraint of Rockfall

If it is accepted that it is not possible to prevent all rockfalls, then methods for restraining those rockfalls, which do occur, must be considered. Some of these methods are illustrated in Fig. 31.16. Berms are a very effective means of catching rockfalls and are frequently used on permanent slopes. The effectiveness of these berms can be checked by means of a rockfall analysis in which the trajectories of falling rocks are simulated mathematically. In the case of Delphi this cannot be accepted, as construction of berms will detract from the appearance of the imposing cliffs at the site.

One of the most effective permanent rockfall protective systems is the construction of a catch ditch at the toe of the slope. The base of this ditch should be covered by a layer of gravel to absorb the energy of falling rocks and a sturdy barrier fence should be placed between the ditch and the site for protection.

The location of the barrier fence can be estimated by means of a rockfall analysis. The criterion for the minimum distance between the toe of the slope and the rock fence is that no rocks can be allowed to strike the fence before their kinetic energy has been diminished by the first impact on the gravel layer in the rock trap.

Fig. 31.16. Possible measures to reduce the damage due to rockfalls. Selected after Spang and Rautenstrauch (1988)

Rock traps work well in catching rockfalls provided that there is sufficient room at the toe of the slope to accommodate these rock traps. In Delphi this may be a problem in some stretches, where there is not sufficient room to accommodate rock traps.

Catch fences or barrier fences in common use are estimated to have an energy absorption capacity of 100 kNm (The kinetic energy of a falling body is given by $0.5 \times \text{mass} \times \text{velocity}^2$). This is equivalent to a 250 kg rock moving at about 20 m s^{-1}. More robust barrier fences, such as those used more frequently in the European mountainous regions (wire mesh fence which incorporates energy absorbing slipping joints) have an energy absorbing capacity of up to 2 500 kNm which means that they could stop a 6 250 kg boulder moving at approximately 20 m s^{-1}.

Another restraint system is the use of mesh draped over the face. This type of restraint is commonly used for permanent slopes. The mesh is draped over the rock face and attached at several locations along the slope. The purpose of the mesh is not to stop rockfalls but to trap the falling rock between the mesh and the rock face thereby reducing the horizontal velocity component which causes the rock to bounce onto the site below.

The problem with this system is it results in a significant change in the appearance of the slope.

Note that, in all of these cases, it is necessary to maintain the rockfall protection system. Rock traps that are full of fallen rock can be very dangerous since they can allow falling rocks to bounce to lower elevations. Draped mesh that has caught large numbers of falling rocks can be torn down by the weight of these rocks. Hence, regular inspection and maintenance of the rockfall protection systems is necessary in order to keep them fully operational.

31.7 Conclusions

The geological observations on the broader Delphi area, the evaluation of seismotectonic data and the geotechnical study focused on the important stability problems of the archaeological site, leading to the following main conclusions:

The archaeological site is located in a complicated geological environment, characterized by flysch formations, thrusted limestone and different talus scree slopes. The steep rock slopes overhanging the Delphi monuments are intensively fractured by a series of subparallel normal faults that form the Delphi graben, a secondary structure of the Corinthian Gulf, a major active neotectonic structure. These structures are active and related to the occurrence of strong earthquakes, one of the causes of instability and rockfalls phenomena occurring in the area.

The results of these phenomena are marked on the existed monuments as deformation, displacements, rotations, subsidence of the foundations and partial destruction of the stone masonry as well as of the retaining walls built to support the monuments on the slopes.

Rockfalls happened already before the development of the sanctuary in ancient times, as shown by the observation that several foundations have been constructed on large fallen blocks of rock. These rockfalls are usually triggered by the seismic events. The presence of groundwater at the interface between the overlying thrusted limestone and the underlying impermeable flysch, softens the flysch, inducing tensile forces from below, resulting in the fracture of the overhanging limestone and facilitating the rockfall phenomena. Creep phenomena, observed mainly in the areas of flysch, are evident, causing subsidence and differential foundation deformation.

The stability of the limestone cliffs above the Delphi archaeological site was modeled with discrete elements and the resulting analysis implies that a critical equilibrium is reached incrementally, along the steep rock slope, as a result of the previous incidents. The actual stability of this site seems to be very sensitive to any modification of its geotechnical conditions.

The discussion on protective measures is based on the principle that no construction should be visible at the cliff face and disturb the imposing view of this unique site.

References

Ambraseys N, Pantelopoulos P (1992) Long-term seismicity of central Greece. Mineral Wealth 76:23–32

Billiris H, Paradissis D, Veis G, England P, Featherstone W, Parsons B, Cross P, Rands P, Rayson M, Sellers P, Ashkenazi V, Davison M, Jackson J, Ambraseys N (1991) Geodetic determination of tectonic deformation in central Greece from 1900 to 1988. Nature 350

Celet P (1971) Quelques aspects de l'hydrogeology des regions calcaires meridionales du parnasse-Helicon (Grece). Ann Scient Pays Helleniques 3(15):13–16

Constantinidis CV, Christodoulias J, Sofianos AL (1988) Weathering processes leading to rockfalls at Delphi archaeological site. In: Marinos G, Koukis K (eds) Engineering geology of ancient works, monuments and historical sites. pp 201–206

de Boer JZ, Hale JR (2000) The geological origins of the oracle at Delphi, Greece. In: McGuire et al. (eds) The archaeology of geological catastrophes. Geological Society London, Special publication 171, pp 399–412

El-Shabrawi A, Verdel T, Piguet J-P, Kavvadas M, Marinos P (1996) Etude et modélisation geotéchnique du site de Delphes, Grèce. Rapport "Platon" Univ. Nat. Tech. d'Athènes et Ecole des Mines de Nancy, 25 p

Higgings M, Higgings MR (1996) A geological companion to Greece and the Aegean. Duckworth, London

Koroniotis KS, Collios AA, Basdekis AP (1988) Stabilization of the rock slopes at the region of Kastallia spring at Delphi. In: Marinos G, Koukis K (eds) Engineering geology of ancient works, monuments and historical sites, pp 207–211

Makropoulos C, Staurakakis G, Latousakis S, Drakopoulos J (1986) A comparative seismic hasard study for the area of Greece. Proc. Int. Semin. on Earth. Progn., 24–27 June, Berlin

Mariolakos E, Logos E, Lozios S, Nassopoulou S (1991) Technicogeological observations in the ancient Delphi area, Greece. In: Almeida-Teixeira ME, Fantechi R, OliveiraR, Gomes Coelho A (eds) Prevention and control of landslides and other mass movements. Report EUR 12918 EN, pp 273–283

Papazachos B, Papazachou C (1997) The earthquakes of Greece. Ziti ed. Thessaloniki

Parke HW, Wormell DEW (1956) The Delphic oracle: the oracular reponses. Oxford, U.K., Blackwell, 436 p

Pechoux PV (1977) Nouvelles remarques sur les versants quater-naires du secteur de Delphes. Rev Geogr Phys Geol, XIX(1):83–92

Piccardi L (2000) Active faulting at Delphi, Greece: seismotectonic remarks and a hypothesis for the geologic environment of a myth. Geology 28:651–654

Sebrier M (1977) Tectonique récente d'une tranversale à l'arc Egéen. Le golfe de Corinthe et ses régions périphériques. Thèse 3eme cycle, Univ, Paris-XI

Spang RM, Rautenstrauch RW (1988) Empirical and mathematical approaches to rockfall prediction and their practical applications. Proc. 5th Int. Symposium on Landslides, Lausanne, vol. 2, pp 1237–1243

Wells D, Coppersmith (1994) New empirical relationships among magnitude, rupture length, rupture area and surface displacement. Bull Seismol Soc Am 84(4):974–1002

Chapter 32

The Landslide Sequence Induced by the 2002 Eruption at Stromboli Volcano

Paolo Tommasi* · Paolo Baldi · Francesco Latino Chiocci · Mauro Coltelli · Maria Marsella · Massimo Pompilio · Claudia Romagnoli

Abstract. The complex sequence of large-scale tsunamogenic instability phenomena occurred on the subaerial and submarine NW flank of the Stromboli Volcano soon after the beginning of the December 2002 eruption is reconstructed and its relationship with volcanic activity is evidenced. After a brief description of slope morphology and stratigraphy, geometry and kinematics of the landslides are described. Finally, instability mechanisms that controlled the subaerial and submarine slope failures are proposed with reference to the different geotechnical, hydraulic, and loading/strain conditions that characterized the different stages of the slope evolution.

Keywords. Landslide, tsunami, volcanoclastic materials, Stromboli Volcano

32.1 Introduction

On 30 December 2002, two days after the beginning of a major eruption, large destructive, subaerial and submarine landslides on the NW flank of the Stromboli Island (Sciara del Fuoco) occurred. They produced tsunami waves that hit the inhabited coastal areas of the island, damaged buildings and infrastructures and threatened the population. Analyses conducted during the following days revealed that failures were preceded by a larger deep-seated movement.

Before December 2002, on the Sciara del Fuoco slope only frequent but shallow slides, rock falls and small debris avalanches had been documented, often induced by the eruptive activity. Large-scale instabilities have been only inferred from geological evidences in the form of huge sector collapses occurred in the recent geological past. Tsunami events (at least 6) were also reported by eyewitnesses during major eruptions occurred in the last two centuries.

Therefore, the instability phenomena triggered by the December 2002 eruption provide new key elements for understanding slope evolution at Stromboli Volcano and for formulating risk scenarios for civil defence decisions.

In the following sections, a reconstruction of geometry, sequence and mechanisms of the landslides, is proposed in the framework of the activities supported by the Italian Civil Defence Department. This reconstruction, that was essential to immediately re-assess slope stability conditions and to predict possible new slope failures, is currently being reviewed on the basis of more detailed analyses of geological/morphological data of the Sciara slope, a refined geotechnical modeling of the instability phenomena and extensive investigations on the mechanical behavior of volcanoclastic materials.

32.2 Sciara del Fuoco Morphology

The island of Stromboli is the subaerial portion of a large volcanic edifice that rises for a total height of some 3 200 m above the basin floor (about 930 m above the sea level). The broadly conical subaerial morphology is broken by several semicircular escarpments. They were originated by vertical and lateral volcano-tectonic collapses that alternated to the constructive volcanic activity so as to form the structure of the present Stromboli (Hornig-Kjarsgaard et al. 1993; Pasquaré et al. 1993). The most prominent evidence of the youngest collapses is the Sciara del Fuoco depression (Tibaldi 2001) (hereafter indicated as Sciara or SdF). It forms a large part of the NW flank of the volcano (Fig. 32.1). The SdF is bounded by lateral escarpments up to 140 m high (Fig. 32.1) and extends offshore down to 700 m b.s.l. (Kokelaar and Romagnoli 1995) with a strict continuity between the submarine and subaerial morphology (Fig. 32.1).

A sharp conical morphology characterizes only the area below the crater (between elevations 600 and 800 m). From 280 m a.s.l. down to hundreds of m b.s.l., the SdF slope is only slightly curved with an average dip direction of 320°. The dip is homogeneous over most of the subaerial part of the SdF (about 35°) except for two flat zones symmetrically located underneath the craters. The dip of the submarine slope instead decreases proceeding offshore, especially below 300 m b.s.l.

32.3 The Sciara del Fuoco Deposit

The Stromboli Volcano is characterized by a persistent eruptive activity in the form of frequent, mild, explosive events and episodic, large explosive and/or lava flow

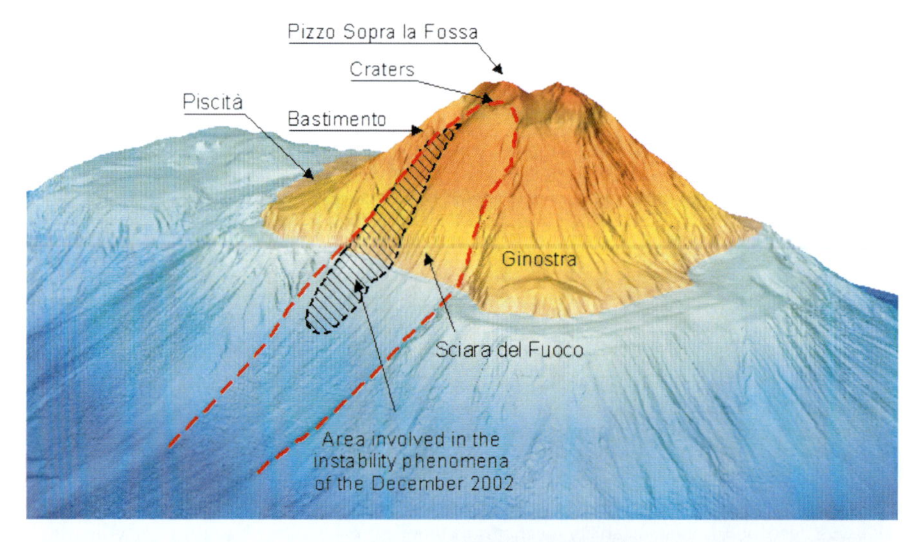

Fig. 32.1.
Shaded relief of the Stromboli Volcanic edifice down to a depth of 1 000 m b.s.l. drawn from the pre-slide DTMM (in blue the submerged part of the volcanic edifice)

Fig. 32.2.
a View of the NE Sciara foot after the 2002 landslides with a continuous volcanoclastic layer in the foreground. **b** Detail of a typical sequence in the background of Fig. 32.2a, formed by a thin lava-flow with a breccia sheet at the top, that is superimposed to a volcanoclastic layers

events that spread their products on the SdF slope. Pyroclastites and lava flows produced by eruptive activity are overlapped with epiclastic sediments thus forming vertical and lateral alternations of more or less competent beds (Fig. 32.2a) that fill the SdF depression.

Primary pyroclastic products are mainly thin layers of relatively coherent fallout spatters varying in thickness from a few centimeters, in the lower part of SdF, up to several decimeters close to the vents. Depending on the effusion rate, lava flow eruptions can form: *(a)* fan shape alternations of thin lava sheets and layers of small loose blocks (typical of steep slopes); *(b)* thick piles of small lava flow units in the two less steep zones located in the upper part of the SdF.

Epiclastic sediments consist of irregular, reverse-graded grain-supported packs of volcanoclastic layers, resulting from fragmentation and rounding of pyroclastic materials and small blocks of scoriaceous lava transported downslope by grain flows and small debris avalanches (Fig. 32.2b). They represent the most abundant component of the SdF deposit and cover large areas of the SdF slope, which in turn looks like a scree slope. Coarser layers consist of gravel and sands with interspersed cobbles and small blocks, whilst finer layers are coarse-to-medium sand with gravel. Grains have high porosity and are characterized by low strength and high crushability even in the finer fraction (Tommasi et al. 2005).

The scree slope represents the medium-lower part of SdF where grain and debris flows dominate, whilst primary fallout products accumulate in the upper part of the slope where they form a typical cinder cone (above 600 m a.s.l.).

Down to a minimum depth of 300 m the submarine SdF deposit is an apron with its typical morphology made up of coalescing fan-shaped bodies of volcanoclastic materials (Fig. 32.3). Debris fan and chutes extending with continuity from the lower subaerial slope down to the submarine slope indicate that the deposit mostly results from underwater accumulation of debris re-mobilized on the subaerial slope by continuous sliding. Successive side scan sonar images of the scarps of the 30 December slide clearly show that a stratification parallel to the present seafloor characterizes the apron.

Materials forming the submarine deposit were found to be extremely variable in size (gravel to sand) and rounding even within very short distances. In fact the seafloor deposit is formed by volcanoclastic materials coming from various zones of the slope, which are characterized by different grain rounding and sphericity (Kokelaar and Romagnoli 1995).

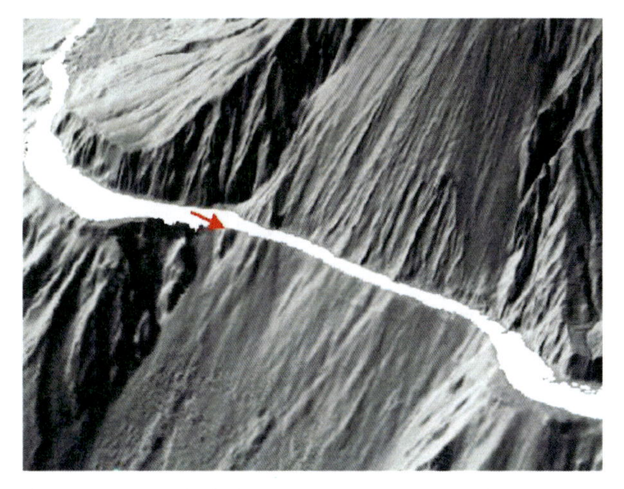

Fig. 32.3. Shaded relief of the coastal portion of SdF from the pre-slide DTMM, showing the interfingering of debris fans in continuity between subaerial and submarine slope. The 1985 lava flow front on the shoreline is indicated

Mechanical and physical characterization of the SdF deposit is extremely difficult due to:

- continuous volcanic activity and instability along the entire SdF slope, that avoid in situ testing and endanger in situ sampling operations;
- variability in structure and lithology of both the lithic and volcanoclastic layers.

The latter complexity can be overcome considering that volcanoclastic layers are extremely continuous and extend over large areas so as to control the shear behavior of the deposit. Testing therefore was focused on volcanoclastic materials. Due to the grain size of the coarser layers large testing devices were used (i.e. large direct shear boxes and ring shear test devices). The results of the geotechnical characterization of the volcanoclastic materials are described in detail by Migliazza et al. (2003), Tommasi et al. (2005), Boldini et al. (2005) and Cignitti (2005).

32.4 Investigations

A prompt qualitative analysis of the instability phenomena and of their interaction with volcanic activity was only based on the interpretation of few air-photo stereo pairs and photographs taken from the helicopter. The dynamics of the events described in the following section results from a close integration of on-shore and off-shore investigations, including aerophotogrammetric and bathymetric surveys as well as interpretation of stereo air photographs and oblique photographs. Quantitative reconstruction of geometry and kinematics of instability phenomena was only possible when the first on-shore and off-shore survey were performed. In this respect, adverse wind conditions (i.e. vapors spread from the craters over the slope) and continuous weather disturbances, maintained the visibility poor until the end of January (27 January). However on 5 January a slight increase in visibility allowed the first flight to be performed which largely lacks of stereoscopic coverage due to vapors and difficult flight conditions above 700 m a.s.l. (Baldi et al. 2005).

Adverse meteo-marine conditions also influenced off-shore investigations which started on 9 January, when a first multibeam bathymetric survey of the submerged NW flank of the island between 20 m and 1 000 m of depth was organized.

The availability of pre-slide aerophotogrammetric and bathymetric surveys, carried out in the framework of the 2000–2003 research program of the National Group for

Volcanology, allowed to compare pre- and post-slide detailed submarine and subaerial morphology of the NW flank. Pre- and post-slide morphology reported in the following sections result from merging photogrammetric with bathymetric surveys dated May 2001/February 2002 and 5/9 January 2003, respectively. In both cases bathymetric data were acquired up to a depth of 20 m.

Since no field activity could be conducted until the end of January, remote sensing techniques were, over a long time, the only quantitative source of data for reconstructing geometric and kinematic features of the instability phenomena and, in turn, for assessing slope stability conditions and formulating scenarios of slope evolution.

32.5 The Sequence of Landslide Events Triggered by the December 2002 Eruption

The eruption started at about 18:30 local time of 28 December when lava poured from the NE crater and rapidly reached the shoreline. First evidence of slope instabilities can be found only in oblique air photographs taken in the early morning of 30 December, that indicated a relatively deep-seated landslide had occurred in the preceding hours (possibly during the night between 29 and 30 December). No sign of deformation or failure was visible on the lower two thirds of the slope whilst a thick cloud blanket covered the upper third of the slope for a whole day after the eruption.

The landslide body was delimited by a high main scarp at the top of the slope, extending from the base of the NE flank of SdF up to the base of the NE cone (in continuity with a large breach that had opened on the NE side of the cone) and by a sharp lateral scarp along the SW limit of the lava field (Figs. 32.4 and 32.5). The movement, that will be indicated as α slide, did not evolve into a destructive landslide but intensely disarranged the slope as it is demonstrated by steps and terraces which were as sharp as to drive the lava paths.

The upper third of the slope suffered significant deformations testified by steps representing seaward-dipping shear discontinuities. Steps were clear in the first photographs taken after the slide (Fig. 32.5), but they were likely to be present before 30 December as diversion of lava paths indicated.

During the morning of 30 December, the slope was relentlessly deforming and the α slide body was fragmenting into large blocks separated by intermediate slide scarps

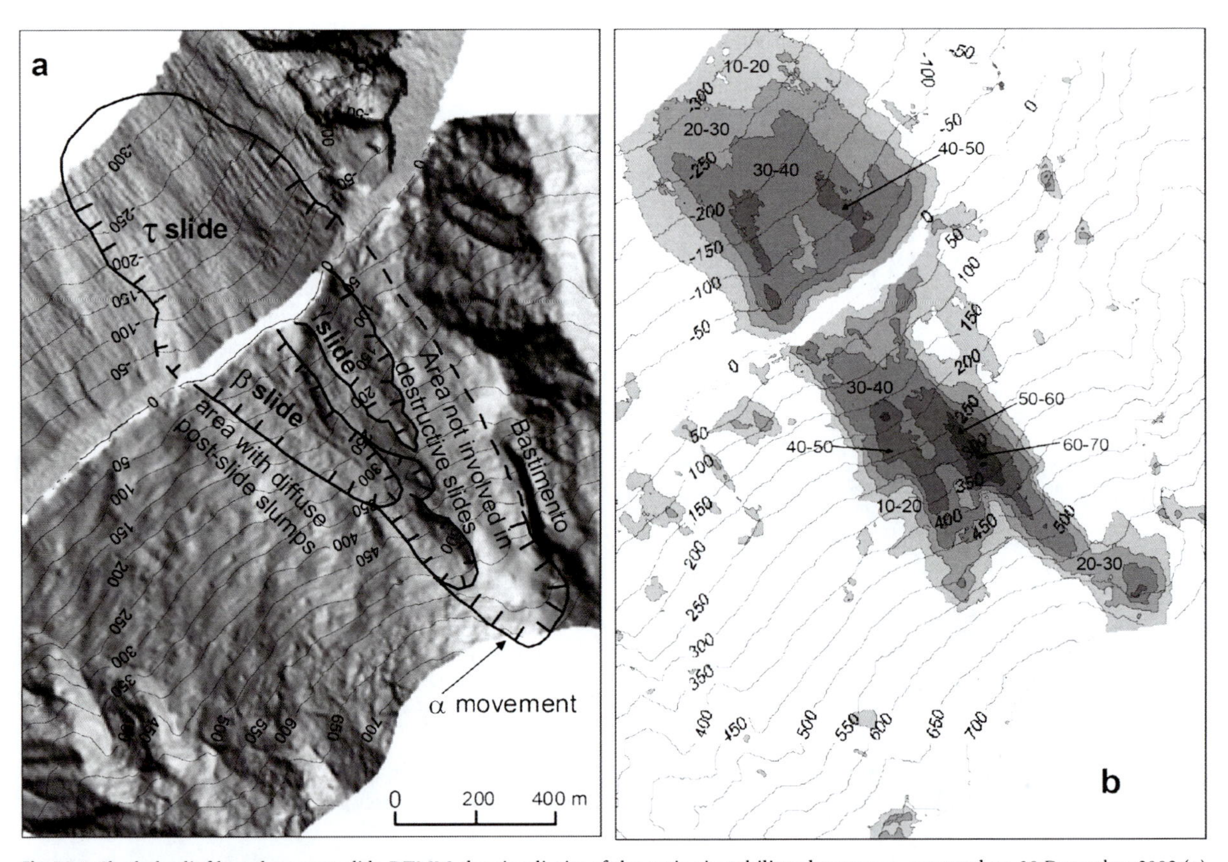

Fig. 32.4. Shaded relief based on post-slide DTMM showing limits of the major instability phenomena occurred on 30 December 2002 (**a**) and elevation differences after the slides, computed from pre- and post-slide DTMMs (**b**)

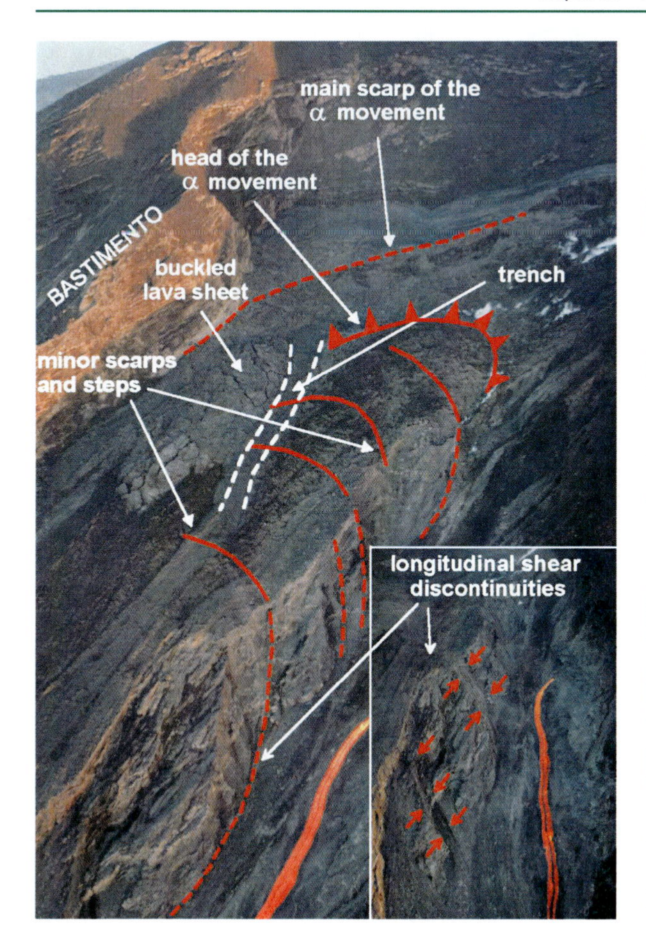

Fig. 32.5. Oblique photograph of the upper part of the slope, taken on 5 January with morphological elements of the subaerial instability α. The head of the impressive scar left by the β slide is apparent (photograph by M. Coltelli, INGV)

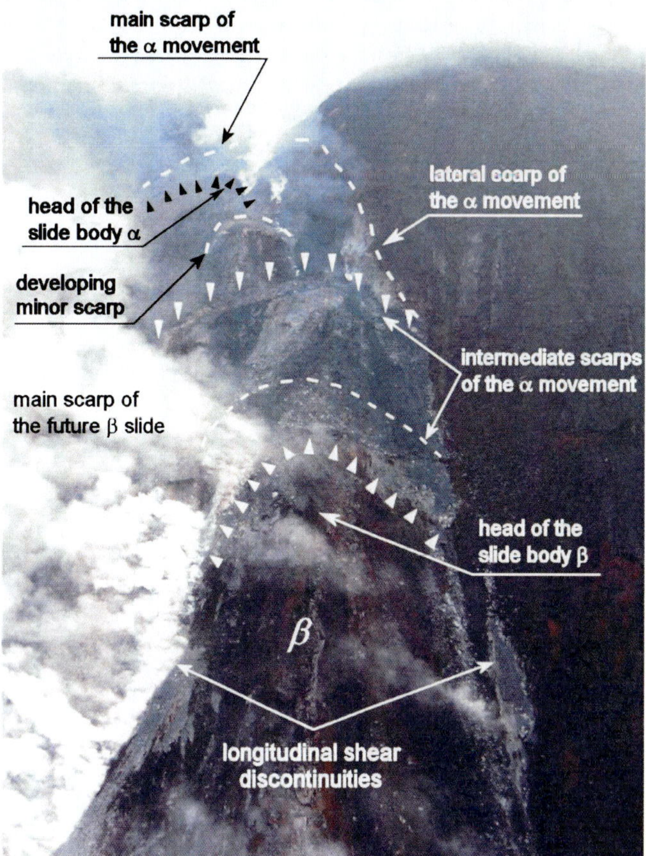

Fig. 32.6. Frontal view of the central and lower part of the slope before the following destructive events. The body of the imminent β slide is completely isolated by a major intermediate scarp of the deep seated α movement and by two along-slope discontinuities (photograph by S. Calvari, INGV)

(Fig. 32.6). The largest block (which would have collapsed afterwards producing the second tsunamogenic landslide β) extended over the southernmost part of the area involved in the deep seated movement α, from the shoreline up to 450–500 m a.s.l., where the highest secondary scarp developed (Figs. 32.4 and 32.6). Ongoing disruption of the slide body is demonstrated by an avalanche detached at high elevation, that reached the shoreline at about 13:15.

Few minutes later two series of tsunami waves propagated from the SdF and hit the inhabited coastal area of the island with a maximum run-up of more than 10 m (Tinti et al. 2003). The tsunami waves affected also the coast of the facing Panarea Island (20 km far from Stromboli) and reached the Milazzo Harbor, on the northern coast of Sicily (60 km far).

The initial retreat of the sea water at SdF suggests that the first tsunami wave was generated by a submarine slide (see e.g. Watts 2000), hereafter named τ slide. Observations made after the tsunami, when visibility improved, revealed that the submarine slide had been followed by a large slide on the subaerial slope, corresponding to the already detached β slide body (Fig. 32.7), accompanied to a minor shallow planar slide (γ) separated from the β scar by a thin pillar.

32.6 Geometry of the Slid Masses

The geometry of the slide body α could be directly observed only in the upper and southern part of the slope, where large vertical displacements (at least 20 m measured on the 3D model) produced a high main scarp and a sharp southern lateral scarp (Fig. 32.5). The lateral extension of the movement in the lower part of the subaerial slope and in the submarine slope is uncertain.

Deformations at the base of the NE escarpment, evidenced by buckling of a lava sheet (Fig. 32.5), indicate that the slip surface should have extended laterally up to the northernmost part of the SdF slope. Furthermore the location of the largest elevation residuals underneath the shoreline and the relevant depth of the slip surface suggest that the α movement involved also the submarine slope, at least in the near-shore area (down to 100 m b.s.l.). Under these hypotheses the volume displaced during the α movement can be estimated in some 33×10^6 m^3. Almost half of it corresponds to the NE part of the subaerial portion that was not involved in the successive destructive slides.

Fig. 32.7. View of the slope taken after the 30 December. The scars of the β (right) and γ (left) slides are apparent (photo by M. Coltelli, INGV)

In the subaerial slope the depth of the α movement was equal at least to that of the successive β slide (i.e. 75 m), calculated comparing post- and pre-slide DTMs (Figs. 32.4 and 32.8).

The scar left by the submarine τ slide was evident down to 350 m b.s.l. (Figs. 32.4 and 32.8). It was up to 45 m deep, regular and relatively smooth, thus suggesting the slide material was completely removed after failure. The submarine scar is wider than the subaerial one (especially to the north) and is characterized by a bottle-neck at about 100–150 m below the sea level (Fig. 32.4b) that could be the effect of a superimposition of two scars (Chiocci et al. 2004): a larger one with a sub-spherical slip-surface, which extended below 100 m b.s.l., and a shallower one that could represent the submarine foot of the mass displaced before the tsunami (α movement). In the submarine SdF slope, 10.5 millions of m^3 correspond to the slide scar (from 0 to 350 m b.s.l.) but at least 20.5 millions were wasted down to 1 000 m b.s.l.

The largest subaerial scar observed after the tsunami (Fig. 32.7) results from the coalescence of the β slide with retrogressive slumps (Figs. 32.4 and 32.5). The failure involved a volume of 5×10^6 m^3 and exposed impressive subvertical lateral scarps up to 40 m high (Figs. 32.5 and 32.7).

The smaller northern γ failure involved a tabular sector about 20 m thick, with a volume of $0.5–1 \times 10^6$ m^3. The pillar separating the two scars was dismantled in few days.

32.7 Mechanisms of Instability

The sequence of instabilities started with the α movement is believed to have been induced by the back-thrust exerted by a nearly N23° E striking dyke intruded in the uppermost NE part of the SdF slope. Uplifting forces due to magma intrusion along the slip surface were likely to

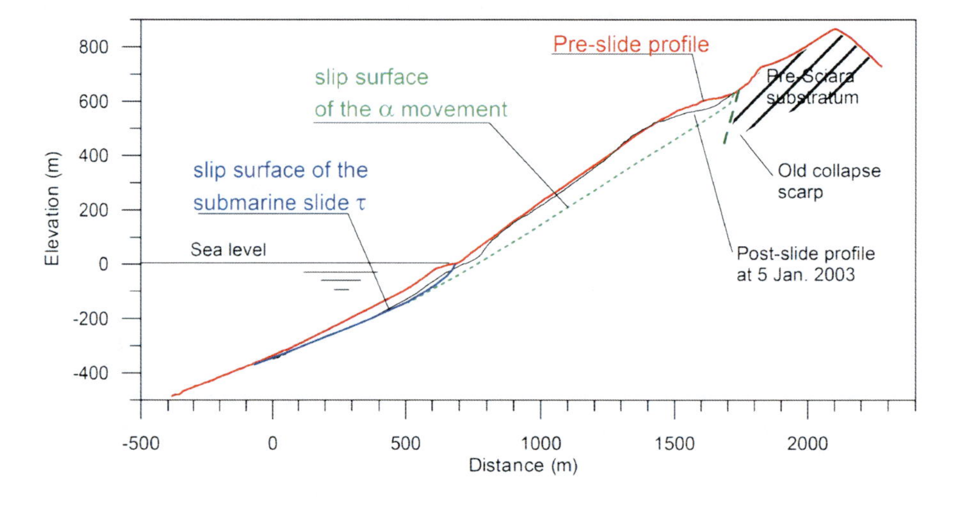

Fig. 32.8. Profile of the SdF slope with reconstructed slip surfaces of subaerial and submarine slides

add to the back-thrust only after a basal shear surface had developed.

The magma probably intruded at the contact between the SdF deposit and escarpment of the youngest lateral collapse, and did not penetrate the stiffer SdF flank, thus indicating a relatively low magma pressure. This latter element and the absence of huge gas venues suggest that the dyke was fed from the upper part of the conduit system, where pressure was controlled by the weight of the magmatic column. Three- and two-dimensional limit equilibrium analyses, described in detail by Tommasi et al. (2003) support the hypothesis that dyke intrusion had a fundamental role in initiating instability of the whole SdF slope. Further insight in the failure and successive deformation processes are being provided by analysis of state of stress and deformation, which are currently being performed.

The initial α movement, which can be considered as a "first time slide", mobilized the maximum shear strength which could be provided by the SdF deposit in dry conditions, at least within the weakest layers. However continuous displacements of the sliding mass rapidly produced a shear surface formed by highly comminuted material (Boldini et al. 2005) where shear strength reduced down to residual values. During the continuous deformation that had started soon after the α failure, excess pore pressure should have generated at the submerged base of the sliding mass, especially during sudden increases in shear displacement velocity and applied shear stress, induced by local failures in the subaerial slope (Tommasi et al. 2003). Increase of pore pressure reduced the effective stresses, possibly down to values compatible with static liquefaction of the material (Tommasi et al. 2003). Boldini et al. (2005) conducted undrained and partially drained ring shear tests at Disaster Prevention Research Institute of Kyoto University under the co-ordination of Professor K. Sassa in order to analyze the mechanism of excess pore pressure generation in the submerged slope. They evidenced that excess pore pressures are not uniquely produced by a sudden increase in the rate of shearing (necessary to determine undrained conditions) and could occur also along coarser horizons.

The regular and smooth morphology of the slip surface of the τ slide and the absence of remnants of the slide body support the hypothesis that the submerged sliding mass rapidly reached liquefaction. Further investigations have to be carried out in order to asses if the rapid loading induced by the avalanche deposit reported few minutes before the tsunami could have influenced stability conditions.

The large submarine slide deprived the foot of the subaerial slope of any support producing a sudden major increment of shear stresses. Under these conditions, the β and γ slide definitely failed, possibly with the collapse of the slide bodies, that were consequently dispersed towards the deep sea zone when they entered the sea.

The subaerial failures involved only the southernmost part of the NE sector of the SdF. After a general deformation, demonstrated by the presence of steps over the whole area in the early morning of 30 December (Fig. 32.5), the northernmost part of the NE sector (i.e. a slab elongated in the NW-SE direction) did not experience intense displacements and fragmentation as the southernmost part did. The boundary between these two zones characterized by different deformation rates was the persistent along-slope shear zone at the center of the NE sector of the SdF slope (Fig. 32.6), which also represented the NE lateral limit of the β slide. The reduced displacements of the northernmost slab can be interpreted as the effect of a constraining action exerted by a more rigid and resistant foot which actually prevented the uphill soil mass from large sliding even after the slope failures.

Acknowledgments

The research has been funded by the Italian Department for Civil Defence that, together with the Scientific Committee for the Stromboli Emergency, also encouraged and supported the technical activities. The staff of the CIGA (Italian Air Force) performed the aerial photogrammetric surveys. The first bathymetric survey was carried out on board of the Coast Guard unit CP-875 whose captains and crew are greatly acknowledged. Volcanologic guides provided assistance during field surveys; special thanks are due to Mario Zaia for his unfailing support. Two of the authors (P. Tommasi and M. Marsella) wish to thank Prof. Barry Voight of Penn. State University, who encouraged the study and gave suggestions for investigations and data interpretation. The reconstruction presented in this paper takes into account also data and considerations expressed by scientists involved in Stromboli Emergency during briefings and meetings held in Stromboli and Catania.

References

Baldi P, Fabris M, Marsella M, Monticelli R (2005) Monitoring the morphological evolution of the Sciara del Fuoco during the 2002–2003 Stromboli eruption using multi-temporal photogrammetry. ISPRS J Photogramm 59:199–211

Boldini D, Wang F, Sassa K, Tommasi P (2005) Mechanism of landslide causing the December 2002 tsunami at Stromboli Volcano (Italy)

Chiocci FL, Romagnoli C, Tommasi P, Bosman A (2004) The December 2002 submarne landslide at Stromboli Volcano: morphologic definition, tsunamogenic potential and scar evolution. Abstract at 32th Geological Congress, Florence 20–28 August 2004

Cignitti F (2005) Geotechnical characterisation of the volcanoclastic materials of Sciara del Fuoco and analysis of the stability of the slope. Thesis in rock mechanics, University of Rome La Sapienza (in Italian)

Hornig-Kjarsgaard I, Keller J, Koberski U, Stadlbauer E, Francalanci L, Lenhart R (1993) Geology, stratigraphy and volcanological evolution of the island of Stromboli, Aeolian Arc, Italy. Acta Vulcanologica 3:21–68

Kokelaar P, Romagnoli C (1995) Sector collapse, sedimentation and clast population evolution at an active island-arc volcano: Stromboli, Italy. B Volcanol 57:240–262

Migliazza R, Segalini A, Tommasi P (2003) Experimental studies on the mechanical behavior of pyroclastic material. Soil-Rock America 2003, Cambridge (MA), pp 501–506

Pasquaré G, Francalanci L, Garduño VH, Tibaldi A (1993) Structure and geologic evolution of the Stromboli Volcano, Aeolian Islands, Italy. Acta Vulcanologica 3:79–89

Tibaldi A (2001) Multiple sector collapses at Stromboli Volcano, Italy: how they work. B Volcanol 63(2–3):112–125

Tinti S, Armigliato A, Manucci A, Pagnoni G, Zaniboni F (2003) Le frane e i maremoti del 30 dicembre 2002 a Stromboli: simulazioni numeriche. XXII Convegno GNGTS, Roma

Tommasi P, Chiocci F, Marsella M, Coltelli M, Pompilio M (2003) Preliminary analysis of the December 2002 instability phenomena at Stromboli Volcano. In: Picarelli L (ed) International Workshop on Occurrence and Mechanisms of Flows in Natural Slopes and Earthfills. Sorrento, pp 297–306

Tommasi P, Boldini D, Rotonda T (2005) Preliminary characterization of the volcanoclastic material involved in the 2002 landslides at Stromboli. In: Bilsel H, Nalbantoğlu Z (eds) Proceedings of the International Conference on Problematic Soils GEOPROB 2005, Famagusta, vol. 3, pp 1093–1101

Watts P (2000) Tsunami features of solid block underwater landslides. J Waterw Port C Div 126(3):144–151

Chapter 33

Slope Phenomena in the Region of the Historical Monument "The Horseman of Madara" in NE Bulgaria

Margarita Matova* · Georgi Frangov

Abstract. The monument "The Horseman of Madara" (ninth century A.D.) includes a bas-relief and petroglyphs-historical chronology. The both of them are with importance for the ancient Bulgarian history. The monument was carved in the Upper Cretaceous calcareous sandstone of the abrupt Madara Plateau slope. The sandstone is fractured and eroded. The sandstone and the investigated monument lay over the Lower Cretaceous marls and clays. The slope processes take place in the Madara Plateau slope. They represent a significant danger for the monument, subject of study.

The contemporary movements along the fractures in the monuments are estimated by the extensometric measurement's monitoring. The extensometric measurements shows significant displacement as result of the 1999 Izmit (Turkey) earthquake ($M = 7.4$).

Keywords. Reserve area, seismotectonics, slope phenomena, rock characteristics, extensometer measurements

33.1 Introduction

The Madara Reserve area (Neolithic to twenty-first century A.D.) includes numerous historical monuments. The Horseman of Madara (ninth century A.D.) is the most important monument of the reserve. It proposes information for the role, the authority and the great importance for the Bulgarian State in the Balkan Peninsula. The monument is among the protected by UNESCO sites of the world cultural heritage.

The monument is situated near the first Bulgarian capital and among other significant historical constructions of the reserve. The reserve is in the region of the strategic Madara Plateau. Now the occurrence of slope processes provokes slowly increased danger for the destruction of the historical monument.

33.2 Madara Reserve Area

The reserve area is situated in NE Bulgaria, the territory of the first 3 capitals of the Bulgarian Kingdom, the towns of Pliska, Preslav and Veliko Tarnovo. The reserve is in the eastern periphery of the village Madara, 17 km to the east of the town of Shoumen, 70 km to the west of the town of Varna and 10 km to the SSW of the first Bulgarian capital – the town of Pliska.

The reserve area is situated at the foot, on the slope and on the top of the Madara Plateau. It includes several archaeological and historical monuments of world and national importance. Numerous traces of the long development of the eastern part of the Balkan Peninsula take place in the reserve area.

Two caves in the basement of the Madara Plateau slope represent very old monuments. They were populated from the Neolithic time. The Big Cave was considered sacred and used mainly for the spiritual rituals. The Little Cave has traces of long-term use for life and works of ancient people. The Big Cave has a 70 m high entrance. It is also known as the Cave of Nymphs, the Nymphs of the nature, water and forest. The Thracian, Roman, Slav and Bulgarian people had inhabited the Little Cave. The Caves of Madara were formed under a complex influence of the natural phenomena (karst, gravity phenomena, tectonic processes) and of the human activities (Matova 2003). The people made certain corrections in the natural forms.

Traces of Thracian tumuli and settlement (centuries B.C.) were discovered on the top and at the foot of the Madara Plateau. They testify that the studied territory was well-populated.

The relicts of a large Roman villa (second to fourth centuries A.D.), including 45 rooms and covering an area of 5 000 m^2, took place on the territory between the village Madara and the slope of the Madara Plateau. The villa testifies the good economic development of the locality.

The Madara castle (sixth to fourteenth centuries A.D.) indicates the important strategic position of the Madara Plateau in a close vicinity with the first Bulgarian capitals, the towns of Pliska and Preslav, and not far away from the third one – Veliko Tarnovo. Its position permits visual contacts between the capital Pliska and the castle. The castle has long-term development. It was used from the Byzantine period of Justinian the First (527–565) up to the end of the Bulgarian Middle Age (1386).

The well-known monument The Horseman of Madara (ninth century A.D.) was carved on the almost vertical slope of the Madara Plateau. It is situated at a height of 23 m above the terrain. The monument was in close vicinity with the first Bulgarian capital, the town of Pliska (Fig. 33.1).

Fig. 33.1.
The monument
"The Horseman of Madara"

The monument involves a bas-relief and petroglyphs. The petroglyphs are only partially preserved up now.

The bas-relief represents the Bulgarian Khan Tervel as a horseman, who is piercing a lion with a spear and who is following by a dog. That is an allegoric illustration of the glorious appearance of the Bulgarian State in the Balkan Peninsula and in Europe. The sizes of the bas-relief are the following: 6.5 m in height and 7.2 m in width.

The petroglyphs are placed on the left, the right and the bottom sides of the bas-relief. They represent the first short chronology for the Bulgarian State foundation and development.

The text interpretation shows relations with three historical periods. The first period is of the Khan Tervel with his significant military help to the Byzantine Emperor Justinian the Second in 705. The next two periods notes the successful State development during Khan Kormisiy (Kormisosh) and during Khan Omurtag in the eighth to ninth centuries A.D.

The bas-relief and the text are of great importance for the Bulgarian culture and history. The Horseman of Madara is the first document where the name Bulgarian is used and where significant historical events with the Bulgarian participation are noted. The monument testifies the triumphant recognition of the Bulgarian Kingdom by the Byzantine Emperor. Furthermore, it was create to celebrate the appearance of new European state called Bulgaria by his neighbor – the powerful Byzantine. The rock monument is the only of its kind in Europe. It is among the sites included in the List of the World Cultural Heritage Sites of UNESCO.

The relicts of Proto-Bulgarian Pagan sanctuary (ninth century A.D.) and of Orthodox monastery (ninth to tenth centuries A.D.) were situated at the foot of the Plateau. Another rock monastery (twelfth to fourteenth centuries A.D.) was placed on the western Madara Plateau slope. Certain small karst niches, fractures and man-made works had permitted the creation of the rock monastery. The monastery relicts are marked now only by the holes in the high part of the western slope of the Madara Plateau. The historical relicts propose very useful information for the development during the First (ninth to eleventh centuries A.D.) and the Second (twelfth to fourteenth centuries A.D.) Bulgarian Kingdom.

The Bulgarian geologist Raphail Popov was the initiator for the establishment of the Madara Museum and the Madara Reserve area in the first half of the twentieth century. His works were of great importance for the region and for the state.

Now the national archaeological and historical Madara Reserve area with the famous The Horseman of Madara is among the most attractive tourist sites in Bulgaria.

33.3 Geological and Tectonic Preconditions for the Development of Slope Processes

The most attractive monument of the reserve area, The Horseman of Madara, is carved in the northern slope of the Madara Plateau in the rocks of the *Madara sandy-calcareous Formation*. The formation is of Cenomanian-Middle Turonian age (Upper Cretaceous). The formation includes calcareous sandstone, sandy limestone and limestone (Fig. 33.2).

The lower part of the Madara Formation, respectively the lower part of the investigated slope, is represented by Cenomanian calcareous sandstone and sandy limestone (Venkov and Kossev 1974; Matova 1999). The cal-

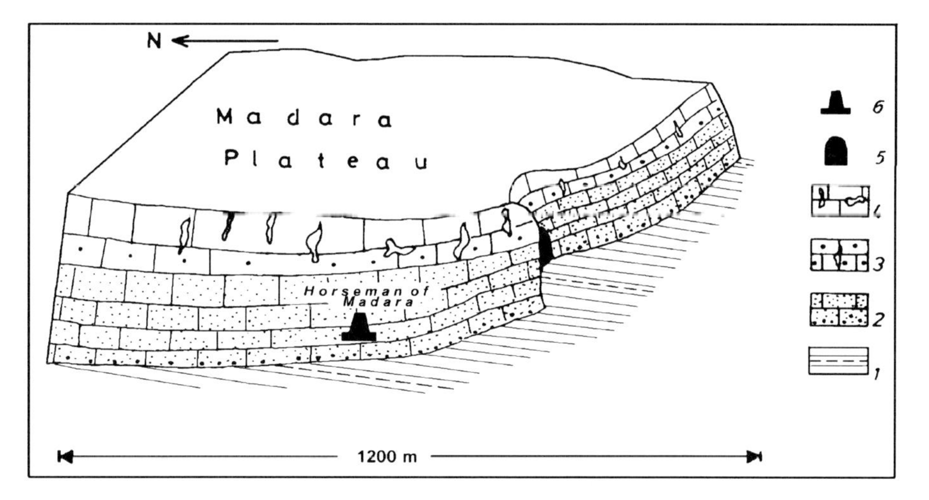

Fig. 33.2.
The monument and the formation's contact. *1:* Gorna Oryahovitsa marl formation; *2–4:* Madara sandy-calcareous formation; *2:* calcareous sandstone; *3:* sandy limestone; *4:* karstified limestone; *5:* Cave of Madara; *6:* The Horseman of Madara monument

careous sandstone from the bottom of the formation includes terrigenous components of different sizes – from the gravel up to the fine grained sands. There are cross-bedding manifestations at the bottom of the formation. The Caves of Madara are situated in the calcareous sandstone. These sandstone are composed by more homogeneous middle and fine grained terrigenous and biogenic components in the higher levels. The Horseman of Madara was carved in relatively homogeneous calcareous sandstone.

The organogenic and terrigenous calcareous material becomes the most considerable in the upper part of the formation. The sandy limestone and the limestone are well represented there. The mediaeval rock monastery was placed in the limestone. The Madara Formation has thickness more than hundred meters.

The sediments from the Madara Formation are moderate or strongly weathered. Locally they are eroded very intensively. Considerable karst manifestations take place in the limestone of the higher Madara Plateau slope.

The Horseman of Madara was carved in slope with moderate weathering and the erosion after good selection of one relatively compact and stable block. These characteristics of the local conditions are of great importance for the natural protection of the rock monument in the future. The long-term manifestations of various geological processes in the calcareous sandstone are the main cause of the recent degree in the efface of the bas-relief and of petroglyphs in the monument (Fig. 33.1).

The weathered, eroded and karstified rocks of the Madara Plateau slope permit the development of rockfall phenomena. The rockfalls in the middle, the lower and mainly in the upper parts of the slope create considerable danger for the investigated monument. The major part of these manifested rockfalls are of gravitational origin.

The Madara calcareous-sandstone formation is placed over the Gorna Oryahovitsa marl formation. The contact is transgressive.

The Gorna Oryahovitsa marl formation is of the Hotrivian-Aptian age (Lower Cretaceous). It includes very well represented marls (Fig. 33.2). The layers of the clay, the sandstone, the aleurite and the limestone are rarely established among the formation's sediments. The Gorna Oryahovitsa Formation has thickness of several hundreds meters. The tick marls, covered by thin clay layers are represented at the basement of the Madara Plateau slope.

The formation has very low hydraulic permeability. The upper surface of the formation is marked by several considerable springs. One of them is placed in the vicinity of the Big Cave of Madara.

The Horseman of Madara was carved not far away from the contact of the Gorna Oryahovitsa and the Madara Formations (Fig. 33.2). The contact permits the accumulation of big quantities of groundwater and superficial water. The contact of the clay, the marl and the water destabilizes the Madara Plateau slope. The contact creates preconditions for the manifestation of landslides and rockfalls. Now a lot of the landslides are covered by rockfalls.

The monument The Horseman of Madara is situated on the boundary of two structural units, the Moesian Platform and the Fore-Balkan. It is the boundary between two different Alpine structural units, namely the relatively stable Moesian Platform and the mobile Fore-Balkan unit. The Madara Reserve area is placed on the Fore-Balkan fault zone. The zone is long and wide. It has a general subequatorial direction.

Faults and fractures, also photolineaments with subequatorial, submeridional and oblique directions are well developed in the investigated territory (Gocev and Matova 1977; Matova 1999). These structures cause the block fragmentation and the destabilization of the studied area.

The NW slope of the Madara Plateau, the slope with the monument, is deformed mainly by local faults and fractures with NNE and WNW direction. The faults predetermined the general orientations of the NW slope surface of the Plateau. The Madara Plateau, including its NW slope, is cut by numerous subequatorial faults.

33.4 Seismotectonic Conditions for Activation of Slope Processes

Some of the above mentioned faults and photolineaments are active during the earthquakes. The seismotectonic data aid the interpretation for the rock stability and for the occurrence of the Madara Plateau slope processes.

An analysis for the influence of the strong earthquakes with magnitude $M = 7.0–7.5$ and for the moderate and weak earthquakes with magnitude $M < 7.0$ to the fault and photolineament mobility of the region will be performed.

The epicenters of the strong earthquakes are concentrated generally in the NE Bulgarian coastal and shelf zone, but not in close vicinity of the monument. The seismic information for the coast (Ranguelov and Gospodinov 1994) is very representative for a long period of time (first century B.C. to twentieth century A.D.).

A considerable number of faults and photolineaments with various directions are represented in the investigated territory. A lot of them are favorable for the distribution of the seismic influence in the studied area of the monument and for the Plateau slope destabilization.

The faults and photolineaments with subequatorial direction (the Devnya and the Kamchiya ones) and with submeridional direction (the Balchik and the Tyulenovo ones) are of great importance for the seismic destabilization of the region. The subequatorial faults and photolineaments contribute the distribution of several earthquake effects from the Black Sea Region to the Madara one.

The epicenters of the strong earthquakes are not very far away from the reserve area. The epicenters of the first century B.C. Bizone earthquake ($M = 7.0$), of the 543 Black Sea ($M = 7.5$) and the 1444 Varna earthquakes ($M = 7.5$), also of the 1901 Shabla earthquake ($M = 7.0$) are located in distance of 75–130 km to the E and ENE from the Madara Reserve area (Fig. 33.3).

Mainly the last two earthquakes (Matova 2002b) could have considerable influence to the stability of the slope with the monument. There are not written documents for seismic damages. But the cited strong seismic events are capable to provoke certain limited destabilization of the rock massif and respectively slope processes.

The submeridional photolineaments to the N-NNE of the reserve area could facilitate the distribution of the effects of the 1892 Dulovo intermediate earthquake ($M = 7.3$). The Dulovo epicenter is 55 km to the N of the Madara Reserve area (Fig. 33.3). It is logic to suppose that this earthquake could provoke certain activation of the slope processes in the territory of the monument.

The faults and photolineaments with subequatorial direction (the Zlataritsa, the Devnya and the Kamchiya ones) are also capable to provoke very limited effect in the studied territory related to the 1913 Gorna Oryahovitsa earthquake ($M = 7.0$). The G. Oryahovitsa epicenter is 110 km to the W of the investigated historical monument (Fig. 33.3). Its influence to the slope processes is not documented.

The moderate and weak earthquakes ($M < 7$) in the surrounding of the Madara Reserve area (Fig. 33.4) provoke very limited effects on the slope processes.

There are strong earthquakes in neighboring countries with possible influence on the slope processes in the Madara Reserve area. They are mainly the strong seismic events in Romania and Turkey.

The Vrancea (Romania) intermediate earthquakes with magnitude $M = 7.0–7.7$ and with the epicenters 210–240 km

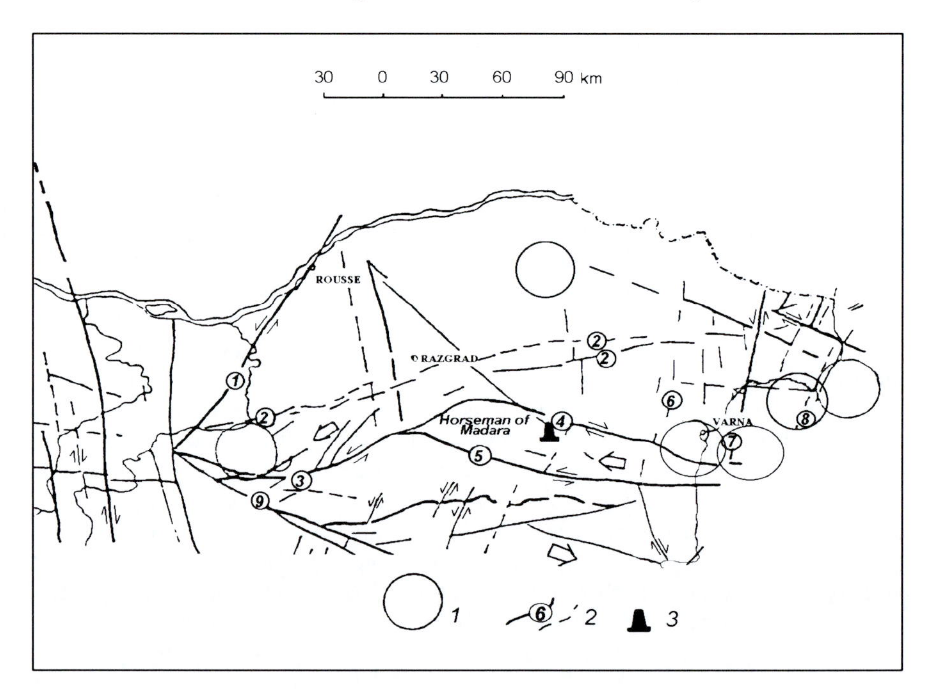

Fig. 33.3. Seismotectonic scheme for NE Bulgaria (according Gocev and Matova 1977, with author's compliment). *1:* Epicenters of earthquakes with magnitude $M \geq 7.0$; *2:* faults and photolineaments of different significance (*1:* Gorno Ablanovo; *2:* Draganovo-Doplya; *3:* Zlataritsa; *4:* Devnya; *5:* Kamchiya; *6:* Dobrich; *7:* Balchik; *8:* Tyulenovo; *9:* Kilifarevo); *3:* The Horseman of Madara monument in the Madara Reserve area

to the N of the reserve area provoke the increase of the dangerous situation in the studied territory. Some more, the strong Vrancea earthquakes have generally short recurrence periods of several tens of years. The seismic effects could be generally weak to moderate in the Madara Reserve area. The summarized influence of the strong and frequently repeated Vrancea earthquakes (Matova 2002a) could instabilize the Madara Plateau slope.

The 1999 Izmit ($M = 7.4$) and the 1999 Duzce ($M = 7.2$) crust earthquakes in the neighboring territory of NW Turkey have epicenters 360–420 km to the SSE of the reserve area. The strong Izmit earthquake caused limited destructive effects mainly in SE Bulgaria. The faults with submeridional and NW-SE direction could be of significance for the distribution of the influence of the strong Izmit earthquake in the eastern part of Bulgaria, including the Madara Reserve area. The earthquake could contribute for the activation of slope processes.

Certain destructive effects in the Madara Reserve area and Madara Plateau slope could be related to combined influence of the Bulgarian paleo- and archaeological earthquakes (Matova 1999), as well as the Romanian and the Turkish strong seismic events. The presence of a lot of faults and photolineaments in the investigated region is an important factor for the specific spatial distribution of the seismic influence. The submeridional and the subequatorial faults and photolineaments are very often seismically activated.

The earthquakes could deform the studied NW slope of the Madara Plateau where the monument is situated. The complicated geoenvironment of the region could provoke the development of fractures, faults, the block fragments and the activation of the slope processes.

In several cases the slope processes of different scales are partially or totally activated by seismic events. There are archaeological, historical and recent instrumental data for the seismic deformation in the investigated territory of the monument and for the destabilization of the Madara Plateau slope. The deformation is weak or moderate.

33.5 Mechanism of Slope Processes

The geological conditions, the physical and mechanical characteristics of the rocks, the geomorphology and the hydrogeology determine the appearance and activation of different types of slope processes.

The general physical and mechanical characteristics of the rocks depend generally on their lithology, compactness, density, porosity, weathering, erosion, fracturing and faulting. Marls and clays of the Gorna Oryahovitsa Formation (Fig. 33.2) represent the rocks in the basement of the Madara Plateau. They are compact and dense sediments with low porosity and very low hydraulic permeability, but they have high plasticity and low elastic deformation modules. The overlaying rocks of the western slope of the Madara Plateau are represented by the calcareous sandstone, sandy limestone and limestone of the Madara Formation (Fig. 33.2). They are compact and dense sediments with considerable porosity and high hydraulic permeability. They have high compressive strength and they are characterized with brittle failure. The rocks of the Madara Plateau slope are also fractured and faulted.

The investigated slope with the carved The Horseman of Madara is intensively weathered and eroded. The weathering and the erosion effects are more representative in

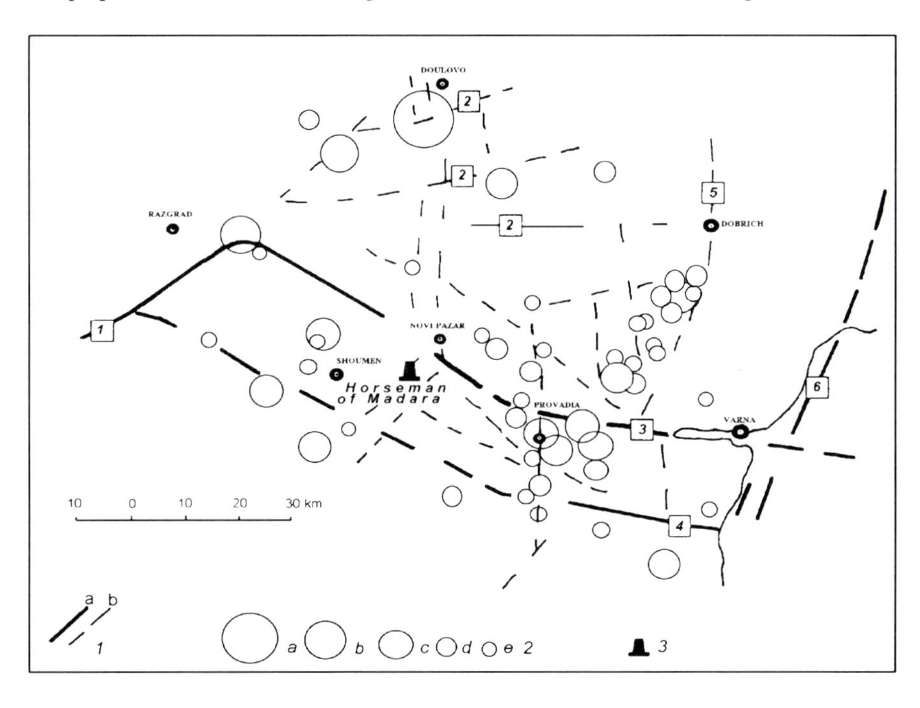

Fig. 33.4. Moderate and weak earthquakes in the region of the Madara Reserve area. *1:* Faults and photolineaments with different significance: *A:* certain (*1–6* the same as in Fig. 33.2); *B:* uncertain; *2:* epicenters of moderate and weak earthquakes: *A:* $M > 6.1$; *B:* $M = 5.1–6.0$; *C:* $M = 4.1–5.0$; *D:* $M = 3.1–4.0$; *E:* $M = 2.1–3.0$; *3:* The Horseman of Madara monument

the porous and permeable calcareous sandstone, sandy limestone and limestone of the Madara Formation than in the compact and impermeable marls of the Gorna Oryahovitsa Formation.

The slope processes are of complex type: toppling (in the upper part of the slope) and lateral spreading (in its lower part). The preparation of toppling is continuos process with relatively rapid occurrence. The lateral spreading is realized generally in the basement of the slope and provoked by the creeping of the Gorna Oryahovitsa Formation's clays and marls. The landslides are slow and permanent.

The main factors which influence on the rate of the slope processes have endo- and exogenic origin. The endogenic factors include the active faults and the lineaments, the temporary tectonic movements, the local and the regional earthquakes. The exogenic factors are the temperature fluctuations, rainfalls, weathering, erosion etc. The degree of significance of the factors could be determined generally by field investigations and monitoring measurements. Sometimes it is possible to apply the reinterpretation of the old documentation (Fig. 33.5).

For example, the rockfall effects were observed on a photo taken by St. Boyadjiev in connection with a study of the Madara castle and its surrounding during 1892 (Ovcharov 1982). The photo shows a considerable rockfall at the foot of the Madara Plateau near The Horseman of Madara. It could be related to the 1892 Dulovo intermediate earthquake ($M = 7.3$) (Fig. 33.5). The authors could only suppose that the same quake had caused also a landslide activity in the basement of the Madara Plateau where is the contact of the Madara and the Gorna Oryahovitsa Formations.

Some new formed and spatially very limited earthquake-induced rockfalls and landslides could be related to the 1986 Strazhitsa earthquake ($M = 5.9$). That was a moderate earthquake, but with the most significant influence in the studied region at the end of the twentieth century.

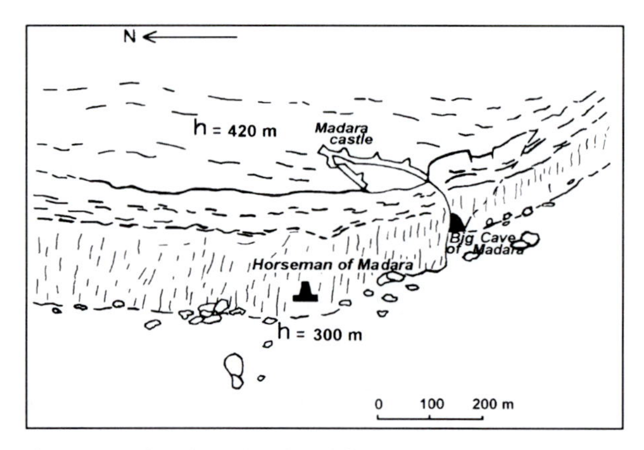

Fig. 33.5. Earthquake-induced rockfall related to the 1892 Dulovo earthquake ($M = 7.3$) (according the photo of St. Boyadjiev, cited by Ovcharov in 1982)

The summary effects of the gravitational and earthquake-induced landslides are very largely distributed in the studied region. We should underline that the earthquake-induced rockfalls and landslides are more rare phenomena than the gravitational ones. The earthquakes cause also temporary acceleration of the gravitational processes. This acceleration could have different values. Sometimes it could significant.

The gravitational (Frangov et al. 1998) and earthquake-induced landslides are related to the impressive deformation of the stairs at the foot of The Horseman of Madara monument. The stairs were built in the first half of the twentieth century. They are situated between the monument and the Museum of the Madara Reserve area. Now the Madara Museum is also partially destroyed by the above-cited types of landslides.

The delimitation of the gravitational and the seismically-induced landslides is very important and complicated task in the Madara Reserve area. The represented analysis is only at the beginning of the research.

The formation and the activation of the gravitational and earthquake-induced rockfalls and landslides are very dangerous phenomena for the rocks of the Madara slope. They create considerable hazard for The Horseman of Madara monument (Iliev-Bruchev and Avramova-Tacheva 1988) that is carved in the highly eroded, considerably fractured and relatively destabilized slope of the Madara Plateau.

33.6 Monitoring of Contemporary Movements in the Marginal Zone of the Plateau

The monitoring research for the contemporary movements is based on instrumental measurements. The measurement network is established within a sector of the slope where the historical monument The Horseman of Madara is carved.

A new man-made correction is provided in the investigated slope the last years. A construction with a concrete roof is built above the main fractures of The Horseman of Madara monument. The roof limits the infiltration of the atmospheric water into the fractures.

Several types of sediments of the Lower and the Upper Cretaceous age are represented in the slope of the Madara Plateau. They are described in the text above. The contact of the Upper Cretaceous and the Lower Cretaceous sediments is marked by springs. The springs are in small distance from the investigated monument.

The geological and hydrogeological conditions are favorable for the development of failures of the Madara Plateau slope.

The fracturing, the faulting and the block fragmentation of the marginal zone of the Madara Plateau are considerable. The fracturing, the faulting and the block frag-

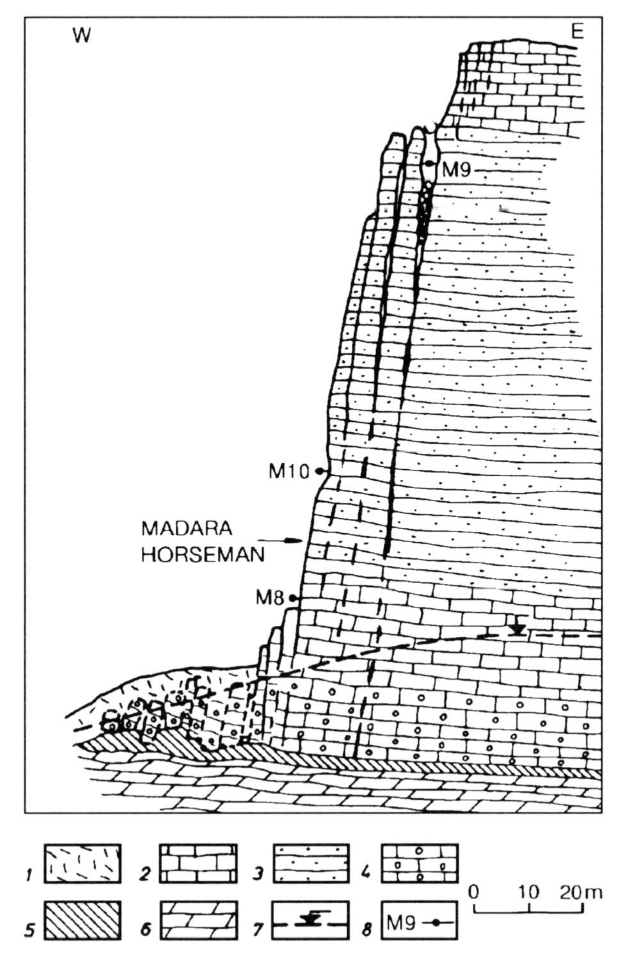

Fig. 33.6. Engineering-geological profile of the Madara Plateau (according Frangov et al. 1992; Kostak et al. 1998). *1:* Disintegrated blocks with clayey-sandy material of the Quaternary; *2:* whitish, well compacted limestone of the Upper Cretaceous; *3:* yellowish calcareous sandstone of the Upper Cretaceous; *4:* sandy-gravel limestone of the Upper Cretaceous; *5:* clay of the Lower Cretaceous; *6:* marl of the Lower Cretaceous; *7:* ground water table; *8:* point of monitoring measurement

mentation of the marginal zone of the Madara Plateau are significant. They are results generally of the creeping on the contact between the clay layers from the Gorna Oryahovitsa marl Formation and the calcareous sandstone, sandy limestone and limestone from the Madara sandy-calcareous Formation (Fig. 33.6).

The frontal part of the slope, especially with the sector of The Horseman of Madara, is an object of the monitoring measurements (Fig. 33.6, 33.7). The slope part with the monument and the other its parts are affected by the general tendencies of the creeping and the block disintegration of the Madara Plateau.

The limestone and the calcareous sandstone have locally developed karst forms of different sizes. The studied part of the slope is built by considerably weathered sediments.

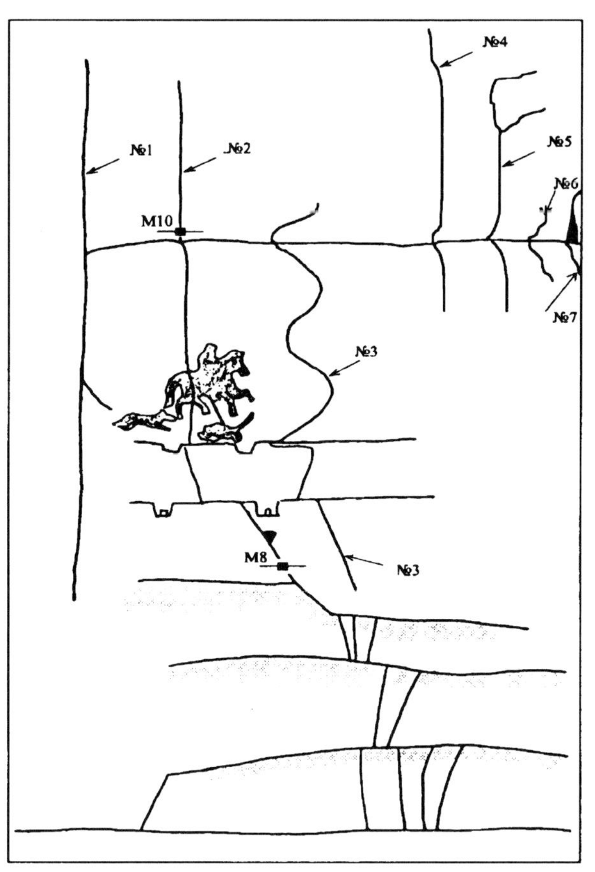

Fig. 33.7. View from the frontal part of the rocky slope in The Horseman of Madara monument (according Kostak et al. 1998). *No. 3:* Fracture; *M8:* point of extensometric measurement

The fractures in the Madara Plateau slope are numerous. Some of them are in close vicinity with The Horseman of Madara monument.

A part of the slope fractures is included in the monitoring measurements (Fig. 33.7). The obtained data propose important information for the block fragmentation of the Madara Plateau and its slope. They propose information also for the events with importance for the deformation processes.

The local monitoring system for the observations of the block disintegration has been built successively since 1990 till 1995. The measurements of the deformation are accomplished regularly by TM-71 gauges made in Czech. The gauges represent 3D extensometers (Kostak 1991). The instruments are recommended by Dr. B. Kostak for the measurement in the contemporary active territories of Bulgaria. The measurements are based on the principle of mechanic-optical interference.

The instrumental's accuracy is of 0.01 mm. The measurements permit to establish the main trends, amplitudes and the directions of the fracture's movements in the investigated rock massif (Fig. 33.7).

The faults and fractures separate the stable central part of the Plateau from the unstable one of its periphery. *The monitoring point M9* (Fig. 33.6) is situated 30 m below the Plateau margin. The axes of the gauge define the following information:

- axis X indicates the extension (+) or the compression (−) of the fracture;
- axis Y indicates the relative displacement along the fracture;
- axis Z indicates the movement in vertical directions down (+) or up (−).

Four measurements are made in the M9 during 2000 (Fig. 33.8) and its results are the subject of the recent interpretation. Any measurements are not realized during the period of December 1995 to September 1997. The displacements during the cited period, when an human-influence could not be excluded, are not interpreted. The measurements (Matova et al. 2001) illustrate very well that the most significant jumps in the measured movements are related to the 17 August 1999 Izmit, Turkey earthquake with $M = 7.4$ (Fig. 33.8).

The measured jumps are of the following values: 8 mm for the axis X, 47 mm for the axis Y and 10 mm for the axis Z. That is a considerable influence of the strong Turkish earthquake of 1999 in the eastern part of Bulgaria, including the slope of the Madara Plateau. The seismogenic movements of 1999 are followed by a relative stabilization of the rock massif during 2000.

The monitoring measurements show an routine amplitude of rock temperature fluctuation of value ≈ 1 mm along the axis X. Some bigger values are not excluded, but winter measurements are not doing. The dilatation of the fracture with approximately value of 0.40 mm yr^{-1} is established on the basis of six measurements since the 1999 Izmit earthquake till now (Fig. 33.8).

Certain seasonal temperature fluctuation of ≈ 0.7 mm is measured along the axis Y. A stable tendency of movement to the SSE with rate of 0.69 mm yr^{-1} is also established.

The amplitude of registered displacement of ≈ 0.3 mm are measured along the axis Z. The trend of relative uplift of several rock slices is influenced probably by the new processes of the Plateau's block disintegration.

Three fractures are distinguished in the block with the investigated monument (Venkov and Kossev 1974). They are included in the monitoring measurements (Figs. 33.5, 33.6):

- *The fracture no. 1* is with vertical position in the Plateau margin, where the measurement point M9 (Fig. 33.5) is installed. The fracture is known as fracture X.
- *The fracture no. 2* affects the frontal part of the rock slope. It separates the block of the monument The Horseman of Madara in two parts. The measurements of the displacements show the compression of approximately 1 mm in the lower part of the fracture and of 0.37 mm in its upper part.
- *The fracture no. 3* is situated in the front of the monument. The fracture was opened and after that closed due to the temperature fluctuation during this year. The amplitude of the displacements is 0.8–1.0 mm.

33.7 Conclusions

The Horseman of Madara monument and the Madara Reserve area are situated in the region of the Madara Plateau. They had preserve very important traces of the long-term evolution of the civilization in the eastern part of the Balkan Peninsula and of the foundation of the Bulgarian state.

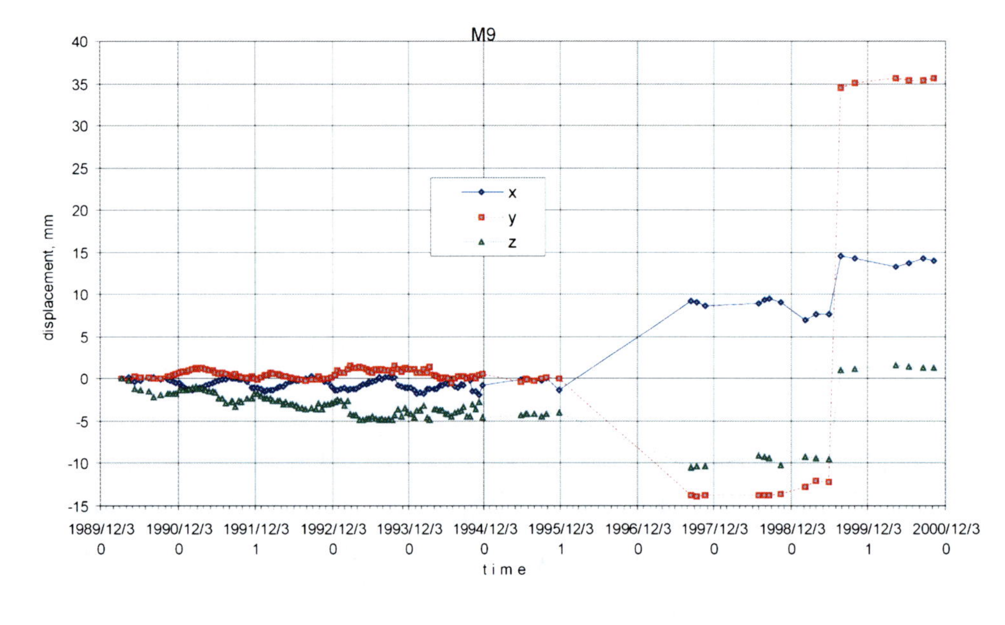

Fig. 33.8. Measured movements in the point of the monitoring measurement M9 (its position is indicated on the Fig. 33.6)

The NW slope of the Madara Plateau with The Horseman of Madara is primary destabilized generally by the weathering, the erosion, the karst, the fracturing, the faulting and the block disintegration. Different types of slope phenomena (toppling and lateral spreading) are developed. They could be interpreted as a result of the above mentioned permanent geological processes. They are accidentally intensified by the earthquakes of the region, as well as of the surrounding countries. The earthquake-induced phenomena accelerate the slope processes and the destruction of the Plateau slope.

The contact of the hydraulically permeable calcareous sandstone of the Madara Formation with the hydraulically impermeable marl of the Gorna Oryahovitsa Formation predestines the main evolution of the slow lateral spreading. The considerable local and regional seismic mobility cause more rapid movements in the top and the bottom of the slope – above and below The Horseman of Madara monument.

The extensometric monitoring of the deformation in the marginal zone of the Madara Plateau indicates the significant influence of the 1999 Izmit earthquake ($M = 7.4$) in the point M9, situated in the upper part of the Madara Plateau slope.

The above mentioned geological situation creates the permanent existing geological hazards for the NW slope of the Madara Plateau and for The Horseman of Madara monument. The study could be used for the mitigation of the local destructive phenomena in the reserve area and for the protection of the world and the national cultural heritage.

References

Frangov G, Ivanov P, Dobrev N, Iliev I (1992) Stability problems of the rock monument "Madara Horseman". Proceedings of the 7th International Congress On Deterioration and Conservation of Stone, Lisbon, Portugal, pp 1425–1435

Frangov G, Broutchev I, Ivanov P, Dobrev N (1998) Seismogravitational effects in Northern Bulgaria. Proceedings of the 8th International IAEG Congress, Balkema, Rotterdam, Netherlands, pp 703–708

Gocev P, Matova M (1977) The present fault mosaic in Bulgaria and the seismic activity. Geotectonics, Tectonophysics and Geodynamics 6:32–47 (in Bulgarian, with English abstract)

Iliev-Bruchev I, Avramova-Tacheva E (1988) Engineering geology and problems of preservation of historical sites and monuments in Bulgaria. In: Marinos PG, Koukis GC (eds) Engineering geology of ancient works, monuments and historical sites. Balkema, Rotterdam, pp 1397–1406

Kostak K (1991) Combined indicator using moiré technique. Proceedings of the 3rd International Symposium on Field Measurements in Geomechanics, Oslo, Norway, pp 53–60

Kostak B, Dobrev N, Zika P, Ivanov P (1998) Joint monitoring on a rock face bearing an historical bas-relief. Q J Eng Geol 30(1):37–45

Matova M (1999) Seismogenic landslides and rockfalls in the vicinity of the Horseman of Madara (NE Bulgaria). Reports and Sub-Project Proposals of IUGS-UNESCO Project no. 425 "Landslide Hazard Assessment and Mitigation for Cultural Heritage Sites and Other Locations of High Societal Value", Paris, France, pp 111–115

Matova M (2002a) Dangerous influence of Large Vrancea earthquakes on cultural heritage in NE Bulgaria. International Conference "Earthquake Loss Estimation and Risk Reduction", 24–26 October 2002, Bucharest, Romania, pp 114

Matova M (2002b) Strong 1901 Shabla earthquake with destructive effects in NE Bulgaria and SE Romania. International Conference "Earthquake Loss Estimation and Risk Reduction", 24–26 October 2002, Bucharest, Romania, p 87

Matova M (2003) Geoenvironment of karst caves of Madara (NE Bulgaria). Proceedings of 4th Symposium on karst protection, Beograd, Serbia-Cherna Gora, pp 37–39

Matova M, Dobrev N, Kostak B (2001) Certain extensometric data for the Influence of the 1999–2000 Turkish earthquakes in Bulgaria. Fourth International Symposium "Turkish-German Joint Geodetic Days", Berlin, Germany, vol. 2, pp 769–776

Ovcharov D (1982) Byzantine and Bulgarian castles during 5th–10th c. The BAS Publ. House, Sofia, Bulgaria, pp 1–171 (in Bulgarian)

Ranguelov B, Gospodinov D (1994) The seismic activity after the 31.03.1901 earthquake in the region of Shabla-Kaliakra. Bulg Geophys J 29(2):49–55 (in Bulgarian, with English abstract)

Venkov V, Kossev N (1974) Investigations of the rocks in the limits of the Horseman of Madara related to the monument's conservation. Investigation and conservation of the cultural monuments in Bulgaria, Sofia, Bulgaria, pp 83–97 (in Bulgarian)

Part IV Sustainable Disaster Management

Chapter 34

Landslide Hazard Mapping and Evaluation of the Comayagua Region, Honduras

Raúl Carreño · **Susana Kalafatovich**

Abstract. The central region of Honduras was strongly affected by the hurricane Mitch in October 1998. This disaster put in evidence the abundance of large sub-active landslides whose acceleration or reactivation caused important human losses and damages to infrastructure. In the same way, most of the thousands of debris flow and rock falls induced by the intense rainfalls were related to this type of landslides. In this region we also discovered the one that up to now consider the largest instability phenomenon in Latin America: the Ajuterique-Playón landslide, which area surpasses 25 km^2. The most of the large scale landslides have been developed mainly in Cretaceous calcareous formations and in the Tertiary volcanic formations. In some areas it the substratum of red layers or sedimentary rocks of continental origin (sandstones and lutites) is also involved, as well as the Paleozoic metamorphic rocks. The results of a preliminary landslide hazard mapping and evaluation carried out between 2000 and 2003 are presented here, including a general appreciation of the danger levels they represent.

Keywords. Sub-active landslide, landslide, Comayagua, Honduras, Mitch hurricane, natural disaster

34.1 Introduction

The hurricane Mitch affected several countries of Central America and the Caribbean Region in the last week of October 1998, putting in evidence the existence of wide areas covered by giant landslides, until then unknown. The intense precipitations induced the reactivation and/or acceleration of many landslides. Thousands of debris flows and mudflows occurred on the surface of those large instable masses, especially on deforested areas. Until now, several of those landslides have continued slow activity, generating wide active fronts of erosion, which consequences affects the agriculture, the biomass of the rivers and the coastal reefs of Fonseca Gulf; those ecosystems have been seriously affected by the extraordinary solid charge generated by the reactivation or acceleration of large landslides.

The region of Comayagua had some antecedents of catastrophic occurrences; related to the strong seismic activity of this area, due to the border geodynamics and to the volcanic eruptions in the neighbors countries. The hurricanes have been always the other source of slope failure, although never with the magnitude reached by the hurricane Mitch in October 1998.

In the seventies and eighties of twentieth century, Honduras and the neighboring countries were affected by other extremely powerful hurricanes (even more powerful than the Mitch), like the Camille, Gilbert, Allen and Davis (intensity = 5, the maximum in the International scale of hurricanes) (INETER 1998), but their destructive consequences were not as important as Mitch's geodynamic sequel. That is one of the reasons because the permanent and slow instability phenomena were unknown, impeaching their prevention.

An interesting historical antecedent of a catastrophic acceleration of the Ajuterique-Playón landslide exists, probably induced by the 14 October 1774 strong earthquake that affected the most of the current Central American countries. This episode destroyed several villages and even the town of Quelepa, former capital of the municipality of Ajuterique (Reina Valenzuela 1968). During the Mitch, many of the secondary blocks of this giant landslide suffered a strong acceleration, mainly in its central part, which triggered dozens of flow-slides, debris flows, mudflows and rock falls, creating large fronts of erosion

34.2 Geological Setting

34.2.1 Geomorphology

The quadrangle of Comayagua is located in the central part of Honduras (Fig. 34.1); belonging to the medium basin of the river Humuya. Three main geomorphological units can be distinguished: the mountainous range of Montecillos in the southern part, the plain of Humuya, in the middle part, and the mountain of Comayagua, at north, which is a natural protected area (National park of Montaña de Comayagua).

34.2.2 Structural Geology

From a structural point of view, the large plain of Humuya constitutes a graben defined by regional parallel faults which direction is N-S. The mountain of Comayagua is a horst that separates the grabens of Comayagua and Yojoa,

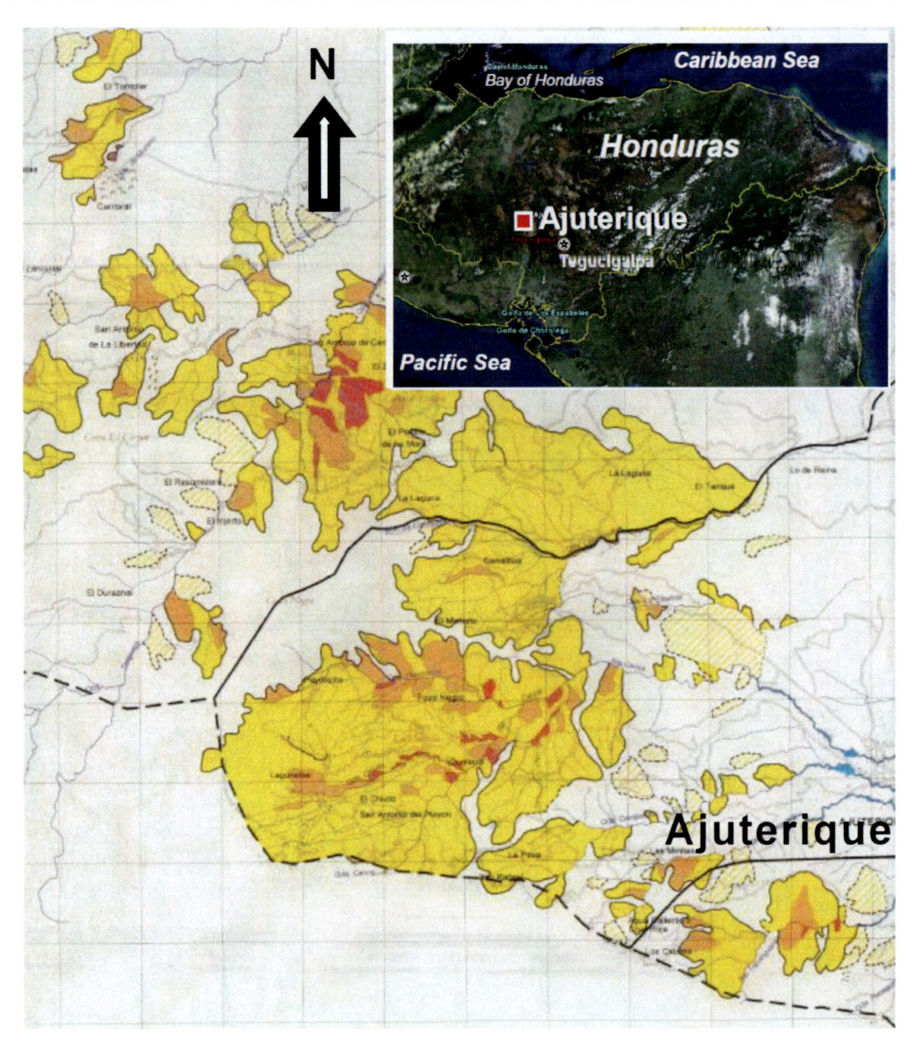

Fig. 34.1.
Extract of the Comayagua quadrangle landslide hazard map. The colors indicate the different relative levels of danger: *red* (high level), *orange* (medium level), and *yellow* (low level)

located toward the east, where the lake of the same name is located. The geological units were affected by important faulting and folding processes during the Laramide Orogeny (Upper Tertiary).

34.2.3 Lithostratigraphic Units

The rocks involved in the geodynamic processes correspond fundamentally to metamorphic rocks of the Middle Paleozoic and marine and continental sedimentary rocks of Mesozoic. Following the description proposed by Elvir (1974) the basement is constituted by metamorphic rocks, especially the Cacaguapa schists (chlorite and muscovite schists), related with the so-called Palacagüina metamorphics whose probable age is Silurian, according to a radiometric datation Rb/Sr made by Puskard (in Everett 1970) that gave 412 Ma. Those schist outcrops in the northern part of the quadrangle, intercalated with phyllites, marble and quarzites of the Las Marías member.

In the oriental region, near the city of Comayagua, a batholith of gabbros and granodiorite outcrops (whose age is unknown) related to the auriferous mineralization and the silicification of limestones. In the base of the mountainous range of Montecillos the formation Todos Santos (probable age Jurassic) outcrops; it's constituted by sandstones, lutites and conglomerates, with a characteristic and predominantly brown-reddish coloration.

The more important lithostratigraphic unit (from the geodynamic perspective) corresponds to the Cretaceous Yojoa Group, divided in four formations: Cantarranas (Neocomian-Lower Albian), Atima (Aptian-Albian), Ilama and Guare (both of the upper Cretaceous) (Fig. 34.2). It's composed by marine sedimentary rocks basically, where the limestones and dolomites prevail, intercalated with levels of sandstones, siltstones and lutites. The landslides have been developed mostly in the limestones and sedimentary rocks of Atima and Ilama Formations.

In some places of the northern area limestones and other red beds of the Valle de Angeles Group (probably

Upper Cretaceous–Lower Tertiary) also outcrop, especially the Esquias Formation.

In many places, the Matagalpa Formation (probably of the Oligocene-Miocene, and mainly constituted by andesitic and basaltic lavas intercalated with pyroclastics levels,), and the Padre Miguel Group (ignimbrites and tuffs, with lacustrine and fluvial layers) is covered thanks to a general unconformity. Fakundiny (1970) and Williams and McBirney (1969) attributed the Padre Miguel Group to Eocene-Pliocene. Some remains of basaltic Quaternary volcanic cones also exist (Elvir 1974) (Fig. 34.3).

34.3 Methods

Until 1998, only 1 : 50 000 topographical maps existed in Honduras. The restitution made after the hurricane Mitch induced a new topographical restitution to obtain 1 : 10 000 scale maps, more useful for the field work. Due to the existent conditions in Honduras before or just after the Mitch hurricane, only the traditional mapping methods could be used, combining the satellite images and air-photos analysis with the field trips.

Fig. 34.2. A view of Ajuterique-Playón landslide that involving rocks of Yojoa Group. In second plan, the mass mobilized by one of the last acceleration episodes

34.4 Results

According to a preliminary analysis, around 5–7% of the area of the quadrangle of Comayagua is affected by large sub-active landslides (Fig. 34.1). If we only take into account the Montecillos mountainous area (excluding the Comayagua plain), this percentage is increased until almost 30%. Sixty percent of the landslides have been developed mostly in calcareous terrains of Cretaceous, around 35% in volcanic materials (especially cinerites with strong hydrothermal alteration); the rest is related to the Quaternary materials (Carreño 2003).

The risk analysis shows that the main threat in this quadrangle is related to the reactivation or acceleration of some secondary blocks of the giant landslides; this kind of episodes could affect and destroy villages and infrastructures in Montecillos Range, La Sampedrana and Las Moras. The damming and overflowing risk also exists in some narrow valleys

The largest landslides of the quadrangle (Ajuterique-Playón and Misterio) are threatening important towns growing on the unstable masses. The Ajuterique and Lejamani Towns are seriously threatened by eventual overflowing, as consequence of the formation of natural dams in the rivers and gullies coming from the Montecillos Range, which is due to the landslides development favored by the deforestation and the erosion.

Fig. 34.3. A view of a secondary block of a landslide developed in the Padre Miguel Group

References

Carreño R (2003) Landslide hazard map of Comayagua quadrangle. Consulting report, BID-CATIE-AMHON, Tegucigalpa (in spanish)

Elvir R (1974) Geología de Honduras. Dirección General de Minas e Hidrocarburos. Tegucigalpa

Everett JR (1970) Geology of the Comayagua quadrangle, Honduras, Central America. PhD dissertation. The University of Texas at Austin

Fakundiny RH (1970) Geology of the EL Rosario quadrangle, Honduras, Central America. PhD dissertation. The University of Texas at Austin

INETER (1998) Meteorological information on the hurricane Mitch. Internal technical report, Managua

Reina Valenzuela J (1968) Comayagua Antañona, 1537–1821. Revista de la Sociedad de Geografía e Historia de Honduras, Colección León Alvarado, Impr. La República

Williams H, Mc Birney AR (1969) Volcanic history of Honduras, vol. 85. University of California publication in Geological Sciences

Swift Action Taken by the Geographical Survey Institute to Analyze and Provide Landslide Information on the Mid Niigata Prefecture Earthquakes of 23 October 2004

Haruo Tsunesumi* · Manabu Hasegawa · Hiroshi P. Sato

Abstract. At around 5:56 in the afternoon of 23 October 2004, a magnitude 6.8 earthquake struck the Chuetsu district of Niigata Prefecture. That was followed by a M 6.0 quake at 6:11 P.M. and a M 6.5 quake at 6:34 P.M. As a result, 46 people were killed, about 4 700 people were injured, and more than 15 000 homes were partially or completely damaged by structural collapse, landslide, or other cause (18 March 2005 survey of the Fire and Disaster Management Agency, Japan). The Geographical Survey Institute (GSI) quickly publicized models of crustal movement and faults by using continuous observation data at the GPS-based control stations. It has also been taking various responses to this seismic damage such as conducting emergency on-site surveys, taking emergency aerial photographs, compiling orthophoto, providing information about damaged areas through the Internet, compiling seismic damage maps and maps for earthquake countermeasures and distributing these products to related organizations (GSI 2005a). This paper provides an introduction to our efforts in these GSI's disaster responses. First, we were able to quickly identify areas and types of damage arising from landslides, compile the data into geographic information and disaster maps, and provide them to the public and relevant organizations. Second, we compiled two types of 1 : 25 000 earthquake damage maps. One is detailed version maps which assist in restoration and recovery efforts, and provide a record of the damage that can be of use in landslide research. The other is color-shaded version maps which show with great precision the patterns and locations of damage using high-precision elevation data obtained from airborne laser scanning.

Keywords. GSI, landslide, 1 : 25 000 earthquake damage maps, airborne laser scanning, Digital Japan, "Denshikokudo Web System"

35.1 Topographic Features of Damaged Areas

The area that experienced many landslides is the hilly area on the eastern side of the Niigata Plain, which is separated from the Sea of Japan by another hilly area. Folding of the hills in this area has created an anticline along a roughly NNE-SSW axis (Economic Planning Agency 1968; Niigata Prefecture 1976). The ground is composed of hardened and semi-hardened sandstone and mudstone from the Neogene and later.

The area is also one of the snowiest regions in Japan, and landslide topography is clearly seen throughout the hills, making it one of the least stable areas in landslide-prone Japan. Within the places where landslide occur are secondary and tertiary talus slopes, gentle slopes or sunken places typical of landslide areas, terraced fields, and even terraced ponds which are used for raising more golden carp than anywhere else in Japan (GSI 1991). Such topographical characteristics and land uses may have helped to expand the earthquake damage.

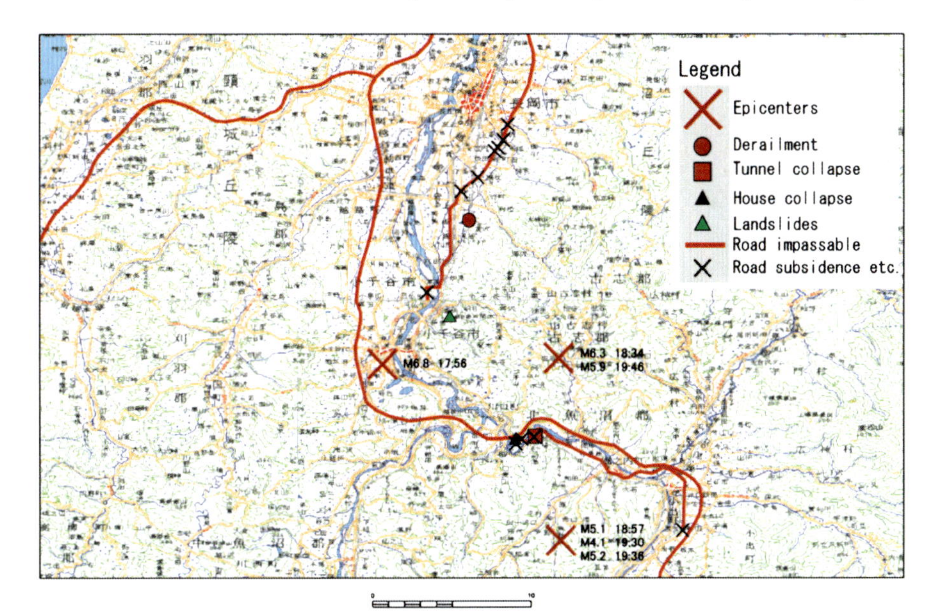

Fig. 35.1. A 1 : 200 000 damage map (reproduced here with a scale of ca. 1: 475 000)

35.2 Addressing the Issues

35.2.1 Providing Damage-Related Information over the Internet Immediately after the Quake

On the evening of 23 October, shortly after the occurrence of an earthquake of Japanese seismic intensity of 6+ (later determined to be 7 at Kawaguchi Town), local TV stations interrupted their regular programs to bring up-to-the-minute news about earthquake damage, which included the derailment of a Shinkansen train (the first since the launch of the Shinkansen in 1964), massive subsidence and upheaving on the Kan'etsu Expressway, subsidence of National Route 17 over a wide area, collapse of people's homes, and the occurrence of debris flow throughout the Niigata Prefecture, and power outages affecting 280 000 households. As reports came in rapid succession, it became apparent that the damage was devastating.

The GSI established a disaster headquarters immediately after the earthquakes and began to collect information on the areas where damage occurred. Although this was an "evening earthquake" and information was being only sporadically provided by TV news, the Internet, and other media, when damage was confirmed in an area, it was plotted on a 1 : 25 000 topographic map.

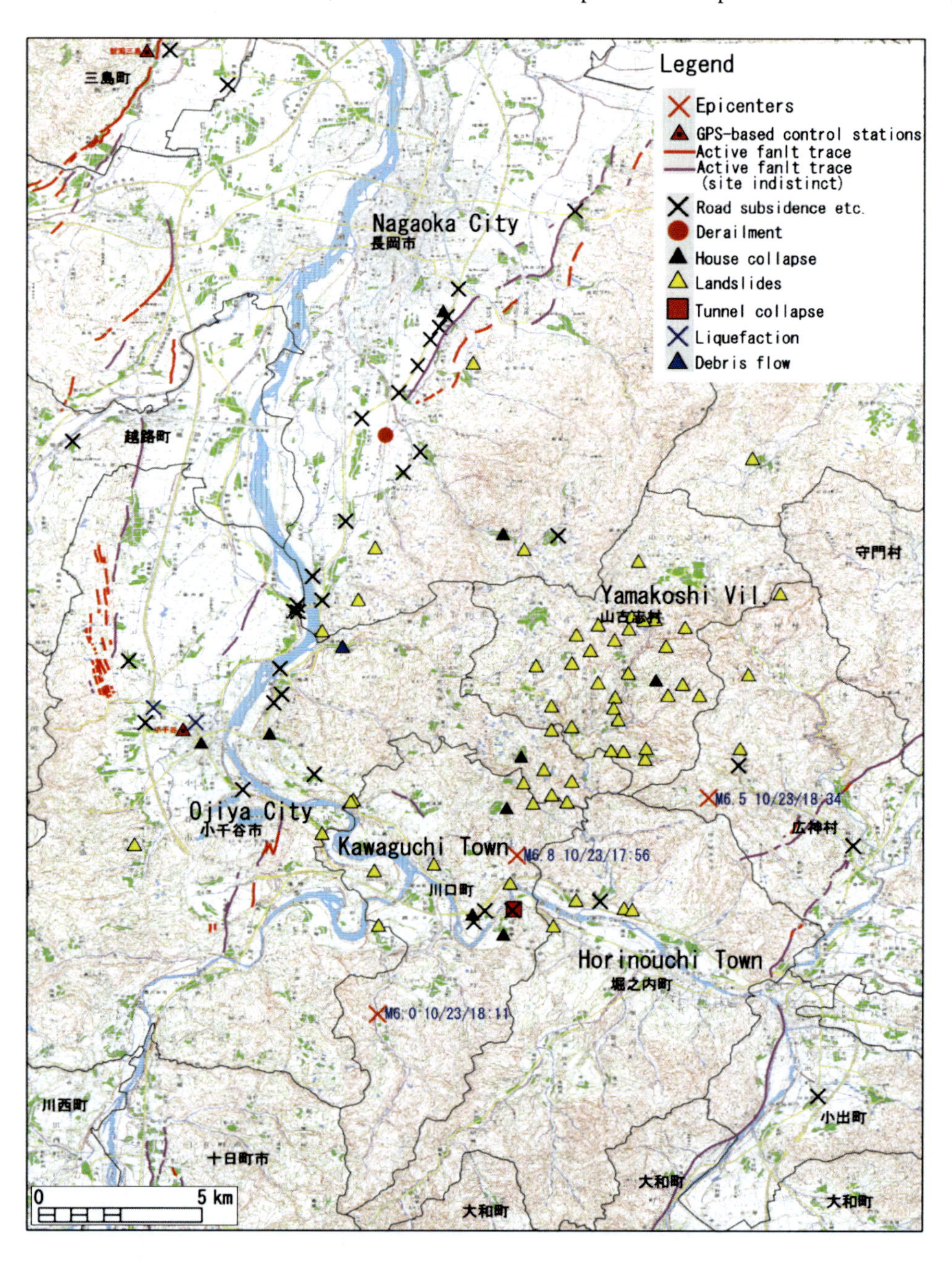

Fig. 35.2. A 1 : 50 000 damage map (reproduced here with a scale of ca. 1: 220 000)

At 1:00 A.M. on the morning of 24 October, the information, which was not always accurate, was hastily compiled into a 1 : 200 000 regional map that showed the epicenters, the location of the derailment, tunnel collapse, road subsidence, and other damage that had been reported by 9:00 P.M. on the 23 October. This damage map was made available to the public via the GSI's Internet home page (Fig. 35.1). At daybreak on the 24 October, the intensive damage in the former Yamakoshi Village (municipalities that merged after the earthquake will be referred to by their former names) became quite apparent. The information on the damage map was updated based on information obtained by TV news teams, newspaper reporters, and other media sources.

At 2:00 P.M. on 26 October, a more detailed 1 : 50 000 topographic map showing locations of road subsidence, debris/landslides, collapsed houses, etc., was added to the Internet home page as damage map. Disaster information on the map was updated until 29 October (Fig. 35.2) (GSI 2005a).

Table 35.1. Aerial photos used for preparing the damage map

Date taken	Resolution	Taken by
October 24, 2004	1 : 10 000	GSI
October 24, 2004	1 : 12 500	Pasco Corp.
October 28, 2004	1 : 10 000	GSI
November 8, 2004	1 : 10 000	ORIS, Inc. and Nakano AI System, Inc.

35.2.2 Interpretation of Disaster Conditions from Aerial Photographs and Provision of Damage Information by the Digital Japan Web System, or "Denshikokudo Web System"

At the same time, an effort was made from 26 October to interpret aerial photographs taken on 24 October (Table 35.1) to determine areas of slope failure (landslides, surface destruction, base rock destruction, etc.), locations of road and railroad damage, and so on. The interpretation work was completed by 28 October. Then, using the Denshikokudo Web System (a GSI Web Mapping System of 1 : 25 000 scale topographic vector data, on which anyone can add geographic information and make it open in his/her Web site), the patterns and locations of landslides and the location of road and railroad damage were symbolized and turned into digital data.

On 29 October, the locations of these damaged areas, etc., were publicly released on the Denshikokudo Web System (Fig. 35.3) as a damage map which covers roughly 640 km^2. At the same time, these data were used to plot the locations of damaged areas on a 1 : 30 000 topographic map titled "State of Damage from the Mid Niigata Prefecture Earthquakes" and distributed to relevant organizations. (Any part of GSI's 1 : 25 000 topographic map data can be extracted and offset printed on demand as a map. This GSI system enables us to print 3 000 sheets in about

Fig. 35.3. A damage map on the Denshikokudo web system (URL: http://zgate.gsi.go.jp/niigatajishin/index2.htm)

2 hours.). The following information was revealed by these maps:

1. Landslides were concentrated in the Uonuma Hills on the right banks of the Shinano and Uono Rivers, and were particularly severe around the former Yamakoshi Village.
2. The regions of concentrated landslides lay along a NNE-SSW axis.
3. There were 1 353 confirmed locations of landslides.
4. Debris originating from landslides in the Imo River Basin stopped up streams, creating landslide dams in scattered locations.
5. The largest scale of landslides occurred within a roughly 600 × 600 m area in the Shiodani area of Ojiya City.

On 1 November, the natural damming of river channels by landslides led to the formation of dammed lakes and flooding of rural communities. There was concern that the landslide dams would break and carry debris that would damage houses, roads, etc., there. Thus, aerial photographs taken on 28 October (Table 35.1) were used to determine areas of landslide dams and dammed lakes, then the results were put on the Denshikokudo Web System as additional disaster information for public release on. At the same time, updated damage maps were printed on demand and quickly distributed to relevant organizations.

Dammed lakes continued to expand after that, so the damage map and the Denshikokudo Web system were updated on 12 November based on aerial photographs taken on 8 November (Table 35.1) and released in the manner described previously. It should be noted that 45 locations of landslide dams and 29 locations of dammed lakes were determined from aerial photo interpretation. The largest of these covered an area of 26 ha (0.26 km^2) in the Higashi-takezawa area of the former Yamakoshi Village (Fig. 35.4) (GSI 2005a).

35.2.3 Creation and Distribution of Base Map on Avalanche Hazard

35.2.3.1 Creation of Snow Depth Data Using Airborne Laser Scanning

A heavy snowfall was recorded in the disaster area for the first time in 19 years. Snow avalanches became prone to

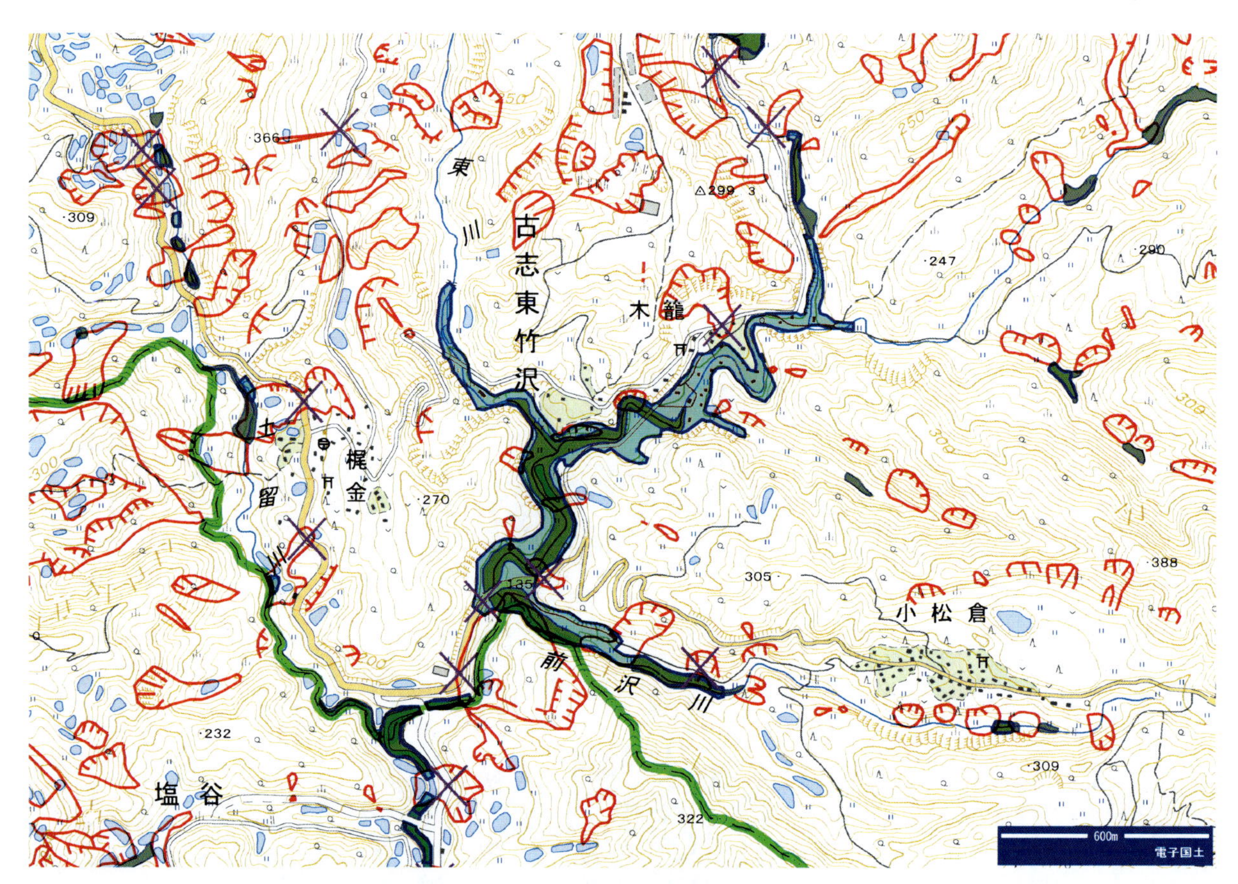

Fig. 35.4. Updated map by adding dammed lakes

take place due to the increase of bare land caused by slope failure and earthquake-induced landslides, etc., making it necessary to take anti-avalanche measures for the snowmelt season. Therefore, the GSI, in an effort to understand the condition of snow accumulations, used airborne laser scanning to create a snow-depth database for a roughly 42 km^2 area in the Imo River watershed centered around the former Yamakoshi Village (Fig. 35.5).

The snow-depth database was created as follows: the GSI compiled accurate digital elevation data of accumulated snow that were derived from airborne laser scanning taken on 18 February 2005, the date of expected maximum snowfall accumulation. These data were then compared with the elevation data obtained from airborne laser scanning conducted by Kokusai Kogyo Co., Ltd. during a snowless period (3–10 December 2004). The difference between the two data sets was used as the snow depth.

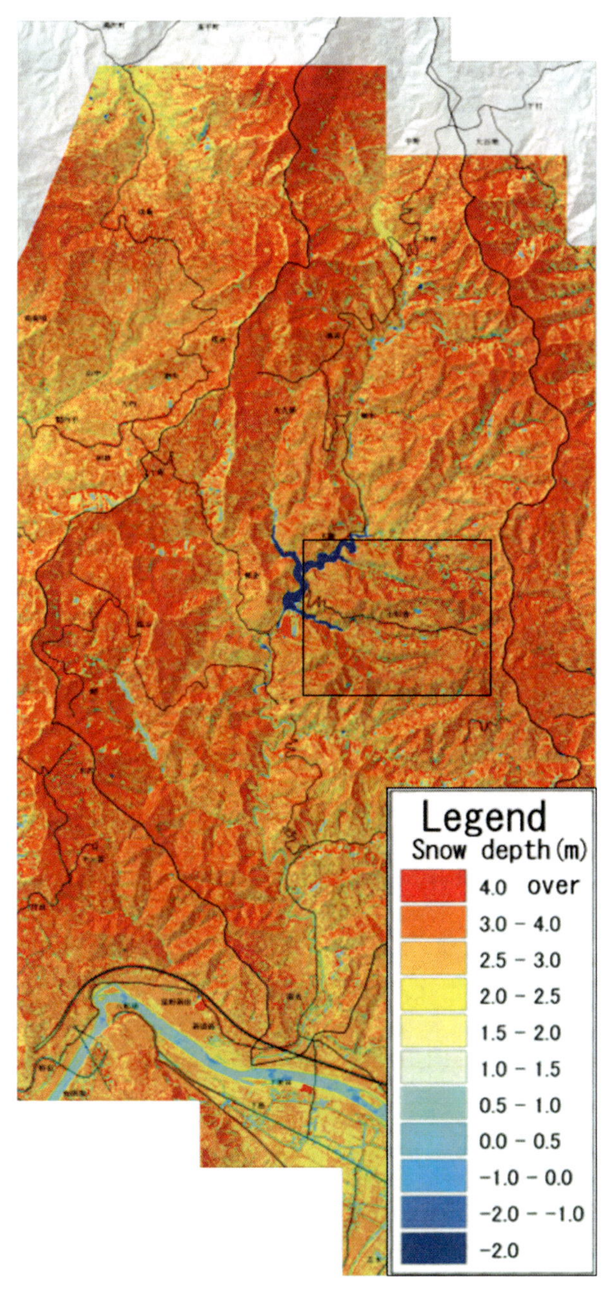

Fig. 35.5. A snow-depth map used airborne laser scanning

The results indicated that the snow depth on 18 February in the mountainous areas of the Imo River watershed was over 3.0 m (shown in orange) in almost all places, and averaged 2.9 m. The total volume of snow cover in this watershed was approximately 104.2 million m^3 (GSI 2005b). Snow-depth color maps based on these data were publicly released on the Internet and, to help prevent disaster during the snowmelt season, these materials were distributed to relevant organizations such as the authorities of the former Yamakoshi Village.

35.2.3.2 Base Map on Avalanche Hazard

The snow-depth data were compiled from airborne laser scanning. Moreover, there was an increased risk that loose sediment and other debris would become mixed in with moving snow during the snowmelt season which would trigger further soil movement (landslides). Thus, the Base Map on Avalanche Hazard (classification map of accumulated points on conditions triggering avalanches (Fig. 35.6) and landform classification map on conditions triggering avalanches) were compiled as a means of evaluating the possibility (risk) of the occurrence of snow avalanches (Sato et al. 2005).

These maps were created as follows: Each 5 m grid was assigned evaluation scores on four factors that were considered to be closely related to the occurrence of snow avalanches: (1) accumulated snow depth, (2) slope, (3) land cover, and (4) avalanche furrow. These data were then added grid by grid using GIS to create a classification map of accumulated points on conditions triggering avalanches. High-risk spots were shown in shades from pink to red, in increasing order; low-risk spots were shown in shades of blue to light blue, in decreasing order. As a result, it was determined that some slopes along roads linking various communities had a high risk of avalanche (A in Fig. 35.6). (The legend is based on the accumulated scores according to the conditions triggering avalanches. The scores are assigned in each condition such as inclination, snow depth, avalanche furrow, and land cover. The assigned scores are shown in Table 35.2. The higher the score is, the higher the possibility that a snow avalanche may occur (Sato et al. 2005).) This map was publicly released on the Internet and distributed to the Ministry of Land, Infrastructure and Transport, relevant local governments, universities, to serve as materials for preventing further disasters in stricken areas.

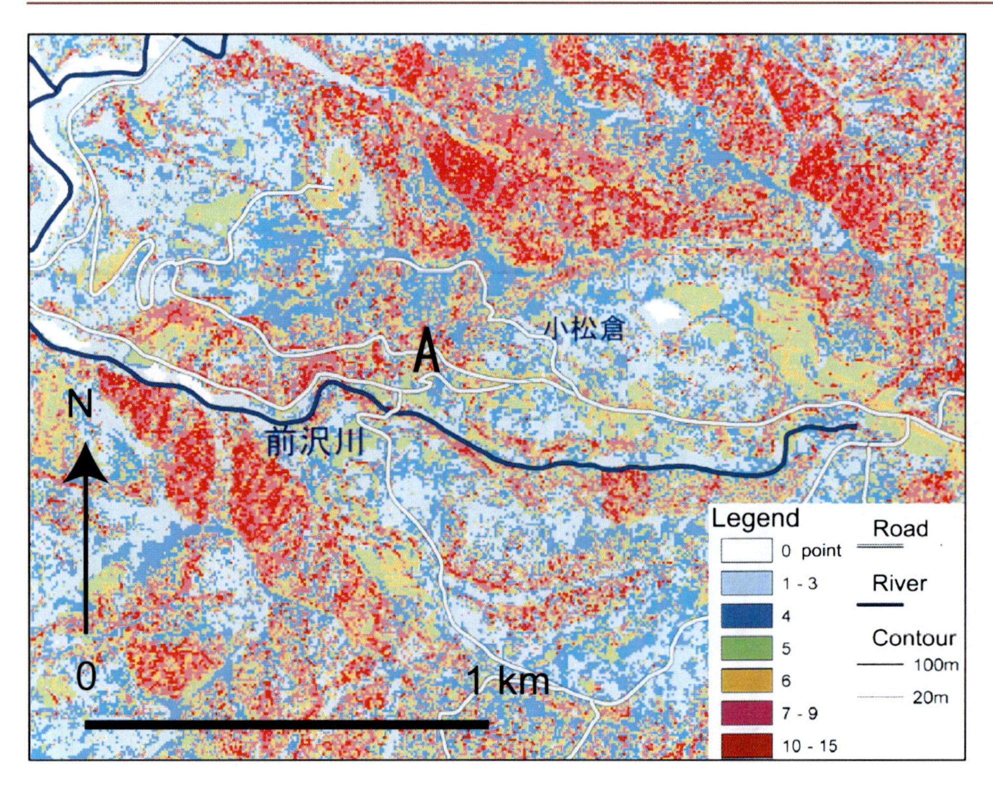

Fig. 35.6. A classification map of accumulated points on conditions triggering avalanche

Table 35.2. The scores according to avalanche-triggered conditions

Conditions		Score
Inclination (°)	<32.5	0
	32.5 – 37.5	1
	37.5 – 42.5	2
	42.5 – 47.5	3
	47.5 – 52.5	4
	52.5 – 57.5	5
	>57.5	6
Snow depth (m)	<1	0
	1 – 2	1
	2 – 3	2
	3 – 4	3
	>4	4
Avalanche furrow	(No interpretation)	0
	A little clearly interpreted	1
	Clearly interpreted	2
Land cover	Water body and so on	0
	Forest	1
	Shrub and grassland	2
	Bare, including paddy, fish-breeding pond, and building	3

35.2.4 Published (Detailed) Version of the "State of Damage from the Mid Niigata Prefecture Earthquakes"

Some of the goals of the Geographical Survey Institute regarding the earthquake disaster were to support the restoration of stricken areas, record the state of damage, and contribute to slope disaster research. To achieve these goals, it published two types of highly accurate maps: 1:25 000 damage maps (detailed versions, see Fig. 35.7) and 1:25 000 seismic damage maps (color-shaded) that accurately depicted the locations and topographic conditions of based on high-precision elevation data obtained from airborne laser scanning (Fig. 35.8) (GSI 2005c). The following are brief descriptions of these maps, which are distributed to relevant organizations when the need arises.

1. Area covered by the maps
   In order to show most of the damaged areas, the maps covered a 758 km^2 area which included the most severely damaged municipalities (part of Nagaoka City, Ojiya City, Tokamachi City, Kawaguchi Town, the former Koshiji Town, the former Oguni Town, the former Yamakoshi Village, the former Horinouchi Town, and part of the former Hirokami Village). Each map was 768 mm wide × 1 081 mm long and covered 1/3 of the 758 km^2 disaster area.
2. Compilation of disaster information, etc., and types of aerial photographs used

The damage map includes various types of disaster information such as the locations and scope of slope failure (landslides, surface destruction, base rock destruction), road and railroad damage, landslide dams, which are described in more detail in the next section. Aerial photographs were interpreted (uncertain areas were checked on the ground) to identify disaster information which was depicted by symbols in the exact locations on topographic maps and prepared as digital data. The aerial photographs used for this purpose are listed below (Table 35.3).

3. Items and information displayed in the maps
   - [Map A: 1 : 25 000 damage map (detailed version)]
     This 1 : 25 000 topographic map was used as the base in map A. This map depicts the disaster information shown with the ◆ below with symbols and shapes, and colored landform classification.
   - [Map B: 1 : 25 000 damage map (color shading version based on laser scanning)]
     This map uses a color-shaded map compiled using 1 : 25 000 airborne laser scanning (ground elevation) data as background. Map B depicts the disaster information shown with the ◆ below with symbols and shapes.

   It should be noted that the backs of both maps contain the following information: descriptions of the topography in the target region, explanation about the information presented in the map, field survey results, examples of disaster-related damage, and other relevant information.

   a Disaster-related information
     - Locations of epicenters: The locations of epicenters of quakes of M 6.0 and above, and from M 5.0 up to but not including M 6.0 are shown.
     - Seismic faults: Sections of faults that appeared on the surface as a result of these quakes are shown with dotted lines.
     - Locations of road and railroad damage: Locations and sections where roads and railroad tracks were made impassable due to landslides, subsidence, etc., are depicted. These are shown by symbols and broken lines.
     - Locations of tunnel collapse: Symbols are used to show locations where tunnels collapsed, seismic heave occurred in roadways, and so on.
     - Major damage to private homes: The ratio of building damage in each residential zone is depicted by varying sizes of symbols.
     - Areas of landslide: Information depicted includes the parts of mountains/hills that tumbled downward in clumps, areas acted upon by hill slope failure or debris sediments, without regard to speed of movement.
     - Dammed lakes created by landslides: Information depicted includes areas of flooding and sections where debris sediments derived from landslides, etc., dammed up stream channels.
     - Areas of liquefaction: Information depicted includes areas of ground fissure, sand volcano, etc., were caused by liquefaction that could be identified from aerial photographs.

   b Information about landform classification, etc.
     Using the Geographical Survey Institute's Urban Active Fault Maps and other materials, as well as reinterpretation of aerial photographs, the results of landform classification and other topographic features, such as active faults, landslide topography, slopes of rugged terrain, uplands and terraces, piedmont accumulations, alluvial fans, floodplains, and so on, were depicted by colors.

   c Information about disaster prevention organizations
     The locations of regional development bureau of MLIT, prefectural public works offices, municipal offices, police stations, fire stations, etc., are shown with symbols.

4. Special features of these maps
   Map A shows the types and areas of damage immediately after the quakes determined from interpretation of aerial photographs. This information was accurately and comprehensively expressed on 1 : 25 000 topographic maps. Map A also contains information about landform classification, past active faults (including presumed ones), and landslide topography to enable the user to make visual note of the relationship between damage and topographic conditions.
   In regard to map B, data from airborne laser scanning conducted from May 2005 after the snow had melted were used to create an accurate color-shaded map for disaster information which allowed the user to gain a more detailed understanding of the relation between damage and topography.

5. Digital data preparation
   In the compilation of damage maps, symbols and patterns for damaged areas were also created as digital data. In addition, high precision digital elevation data (1 m grid ground elevation data) based on airborne laser scanning were prepared. More than merely

Table 35.3. Aerial photographs used

Date taken	Resolution	Taken by
1989, 1991, 1994	1 : 20 000	GSI
October 24, 28, 2004	1 : 10 000	GSI
October 24, 2004	1 : 12 000	Asia Air Survey Co., Ltd.
October 28, 2004	1 : 10 000	Aero Asahi Corp.
October 28/29, 2004	1 : 5 000	ORIS, Inc. and Nakano AI System, Inc.
October 29, 2004	1 : 10 000	Pasco Corp.

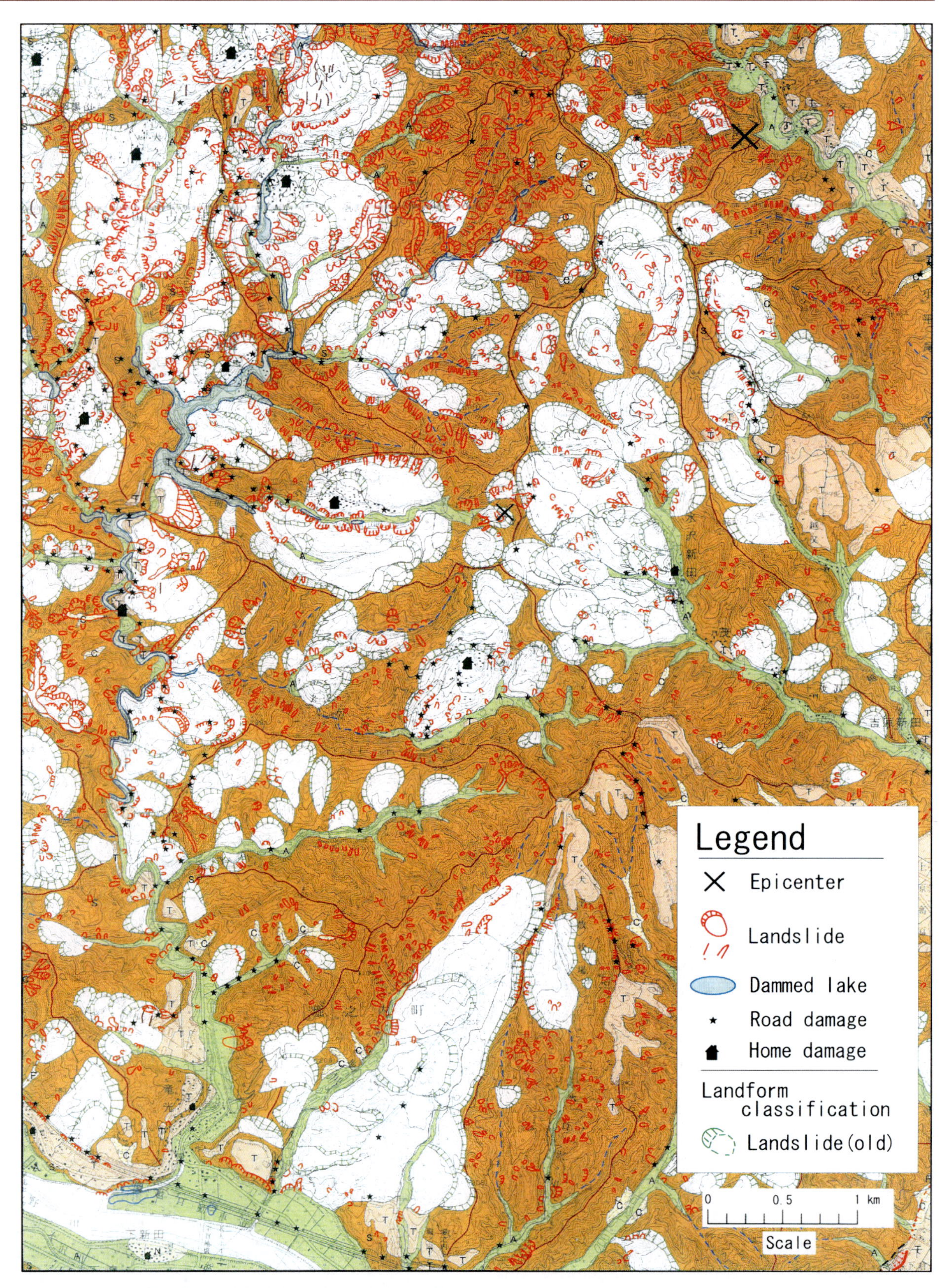

Fig. 35.7. Map A: a damage map (detailed version)

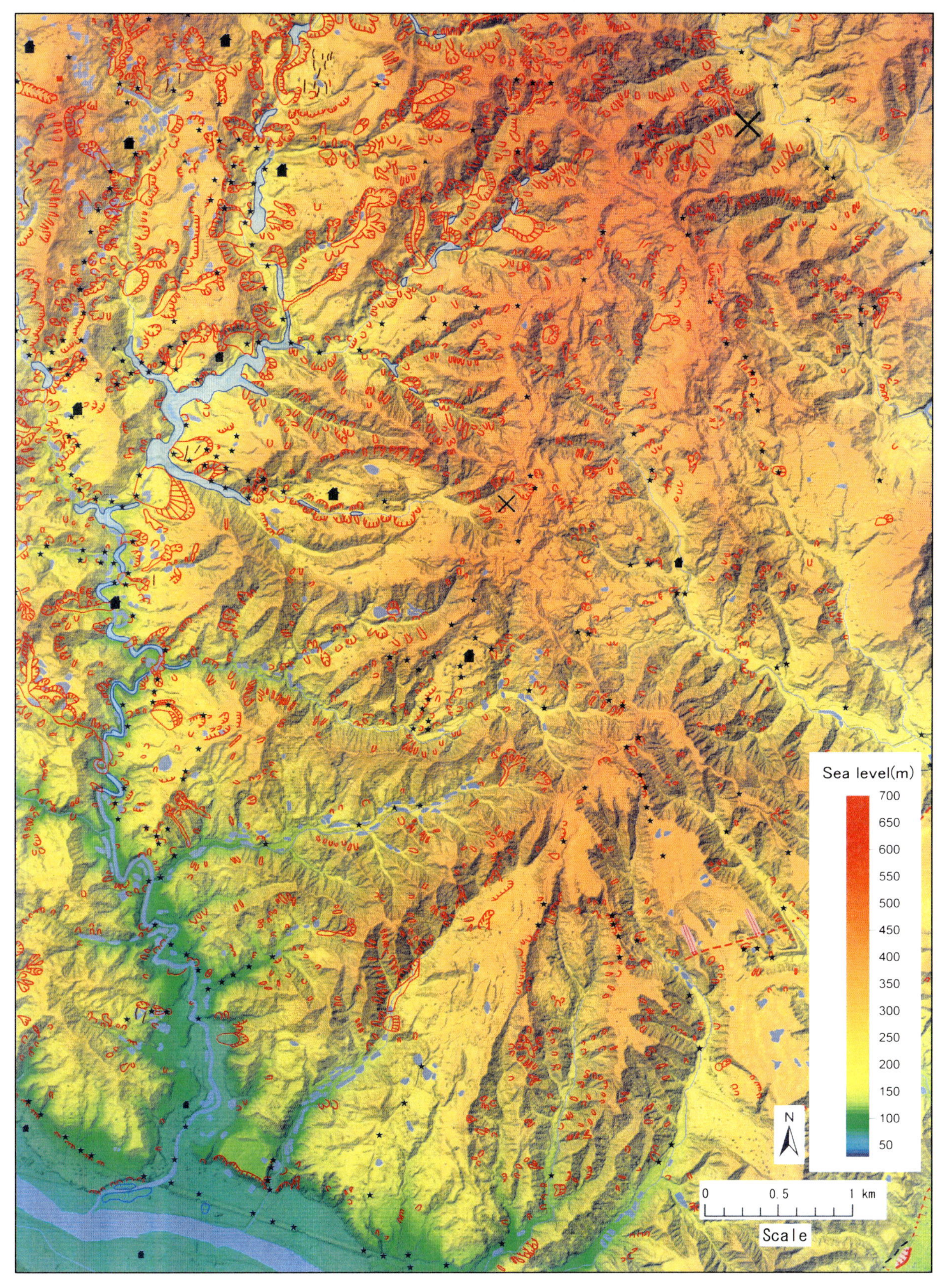

Fig. 35.8. Map B: a damage map (color-shaded version based on laser scanning)

leaving a record of the damage, analyses of these data in conjunction with topographic condition data (such as slope zones, landform classes, geological structure, etc.) can be applied to research that aims to identify hazard areas and clarify the conditions under which damage is likely to occur. The high precision digital elevation data obtained from airborne laser scanning can be used for such purposes as topographic survey for restoration work and monitoring of hazardous slopes.

It took 6 days after the earthquakes to provide comprehensive damage information. Thus, it is necessary to create a system that can provide such information in a more timely manner and a system which can quickly reproduce and distribute aerial photographs that are essential for understanding the state of damage. In addition, we would like to investigate methods for quickly grasping the state of damage using airborne laser scanning data.

35.3 Concluding Remarks

The Mid Niigata Prefecture earthquakes were the most disastrous seismic event in Japan since the Great Hanshin-Awaji earthquake of 1995. However, taking full advantage of experience and knowledge obtained in the aftermath of the Hanshin-Awaji earthquake, the GSI was able to react quickly to the Niigata quakes, making emergency printings of damage maps and distributing them to relevant organizations in record time. The GSI's information was widely acclaimed for its speedy distribution and high reliability.

The maps showing the state of damage from the Mid Niigata Prefecture earthquakes (Figs. 35.7 and 35.8) are scheduled for distribution to all relevant organizations. They are expected to be widely used to support restoration and recovery efforts and to provide information for slope disaster research.

References

Economic Planning Agency (1968) Land classification survey of 1:50000 "Nagaoka" (in Japanese)

Geographical Survey Institute (GSI) (1991) Land condition map of 1:25000 "Nagaoka" (in Japanese)

Geographical Survey Institute (GSI) (2005a) Actions of GSI taken for the disaster "The Mid Niigata Prefecture Earthquakes in 2004." Technical Report of the GSI, A/1 no. 299 (in Japanese)

Geographical Survey Institute (GSI) (2005b) Materials from press release made at 14:00 on 4 March 2005

Geographical Survey Institute (GSI) (2005c) Compiled report entitled "State of Damage from the Mid Niigata Prefecture Earthquakes" (detailed version)

Niigata Prefecture (1976) Land classification survey of 1:50000 "Ojiya" (in Japanese)

Sato HP, Sekiguchi T, Kojiroi R, Kamiya I, Hasegawa H (2005) Overlaying landslides distribution on the data of geology, snow depth, and others: the Mid Niigata Prefecture earthquakes in 2004, Japan. Journal of the Geographical Survey Institute 107 (in press) (in Japanese)

Chapter 36

Early Warning and Prevention of Geo-Hazards in China

Lijun Zhang* · Wei Shan

Abstract. China is experiencing fast social and economic development risked by serious geological hazards. This paper reviewed the ongoing actions of mitigating geological hazards in China from some aspects. At first, the paper presented 290 counties' statistic results of phase I of national geological hazards survey program on the basis of each county. The results indicated that slope hazards including landslide, rock avalanche and unstable slope and debris flow were major geo-hazard types, accounting for 92% of total geo-hazards; that the factor of triggering landslide, rock avalanche and debris flow was mainly rainstorm; that the dead was 8 486 and the direct economic loss was 3.27 billion Yuan resulting from slope hazards; and that the population and property values were respectively 3.46 million and 15.09 billion Yuan threatened by potential slope hazards. Then, the paper introduced the method of geo-hazard risk assessment and zoning which was applied in a research project in China. Subsequently, the principle and successful effects of meteorological early warning of landslide in China were expounded. The Three Gorges Reservoir area was threatened most seriously by wide various geo-hazards such as landslides, rock avalanches and debris flows. Therefore, prevention and control of geo-hazards were especially introduced in the paper. At last, the paper proposed some recommendations for the national geo-hazard mitigation strategy in China.

Keywords. Geo-hazard survey, meteorological early warning, risk zoning, China

36.1 Introduction

China is one of the countries with serious geo-hazards in the world. All geo-hazards have caused substantial human and financial losses (Figs. 36.1 and 36.2), estimated at the annual dead of about 1 000 people and the annual cost of approximately 10 billion Yuan (~U.S.$1.2 billion). The most serious geo-hazard happened in 1998, which was incurred by a heavy flood. During that event, landslides, rock avalanches and debris flows occurred in 180 000 sites; 447 persons missed; 1 157 persons died; more than 10 000 persons were injured; more than 500 000 houses were destroyed; and the economic loss of 27 billion Yuan was caused.

During IDNDR (1990–2000) period, China took substantial actions on investigating, researching, monitoring and preventing and controlling of geological disaster, and effectively reduced the loss which geological hazards brought about. Synthesized ability against geo-hazards in China has been enhanced distinctly. Mitigating effect of geo-hazard not only is the government's heavy responsibility but also has aroused the public's concern.

In next 20 years, the Chinese urbanization and economy will enter the high-speed development phase. Development strategy implementation in the west of China, industrial base promotion in the northeast of China as well as Middle China's rising with high speed development in the east of China all make China be confronted with the high risk of geo-hazards.

Facing stern geo-hazard situation, the Ministry of Land and Resources of P.R. China promulgated "Year 2001–2015 Prevention and Control Plan Guideline of National Geology Disaster" in 2000. The examined and approved sys-

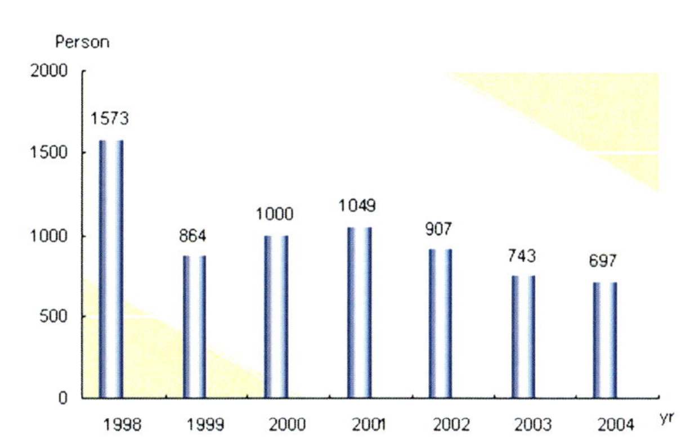

Fig. 36.1. The dead caused by geological hazards in 1998–2004 (data from Ministry of Land and Resources of China 2004)

Fig. 36.2. Economic loss resulting from geological hazards in 1998–2004 (data from Ministry of Land and Resources of China 2004)

tem of geological hazards for the construction project has been put into operation since 2002. The State Council promulgated "The Regulations on the Prevention and Control of Geological Hazards" in 2004. The formulation of the "National Plan for Prevention and Control of Geological Hazards" and "National Master Plan for Unexpected Geological Hazards Emergencies" was completed (Geological Environment Bureau of the Ministry of Land and Resources of China 2004). Comprehensive prevention and control of geological hazards in China are being executed in scientific style and management.

36.2 National Geo-Hazards Survey Program on the Basis of Counties in China

In order to investigate the distribution pattern of geological hazards in China thoroughly, especially the distribution pattern of potential dangerous geological hazards, MLR of P.R. China has implemented the national program of geo-hazards survey in 700 counties that are prone to geological hazards since 1999 (Fig. 36.3). The survey emphasized 6 kinds of geo-hazards, namely landslide, rock avalanche, debris flow, no-stable slope, surface subsidence and surface fissure.

Based on 290 counties' statistic results of phase I of the above survey program, the geo-hazard characteristics of type, distribution pattern, inducing factor as well as casualties and economic loss were analyzed. In the geo-hazard survey completed in 290 counties, altogether 56 112 geo-hazard sites as well as 47 832 potential geo-hazard sites were found out. The survey indicated that slope hazards including landslide, rock avalanche and unstable slope and debris flow were major geo-hazard types, accounting for 92% of total geo-hazards (Table 36.1).

Based on the investigative data, small and middle scale landslides were major types, among which the small scale landslides $<10 \times 10^4$ m^3 accounted for 72% and mainly distributed in the hilly and mountainous areas in the southeast of China. According to the material ingredient, the major landslides were made from soil texture, but in the Qinghai-Tibet Plain area, landslides made from crushed stones ac-

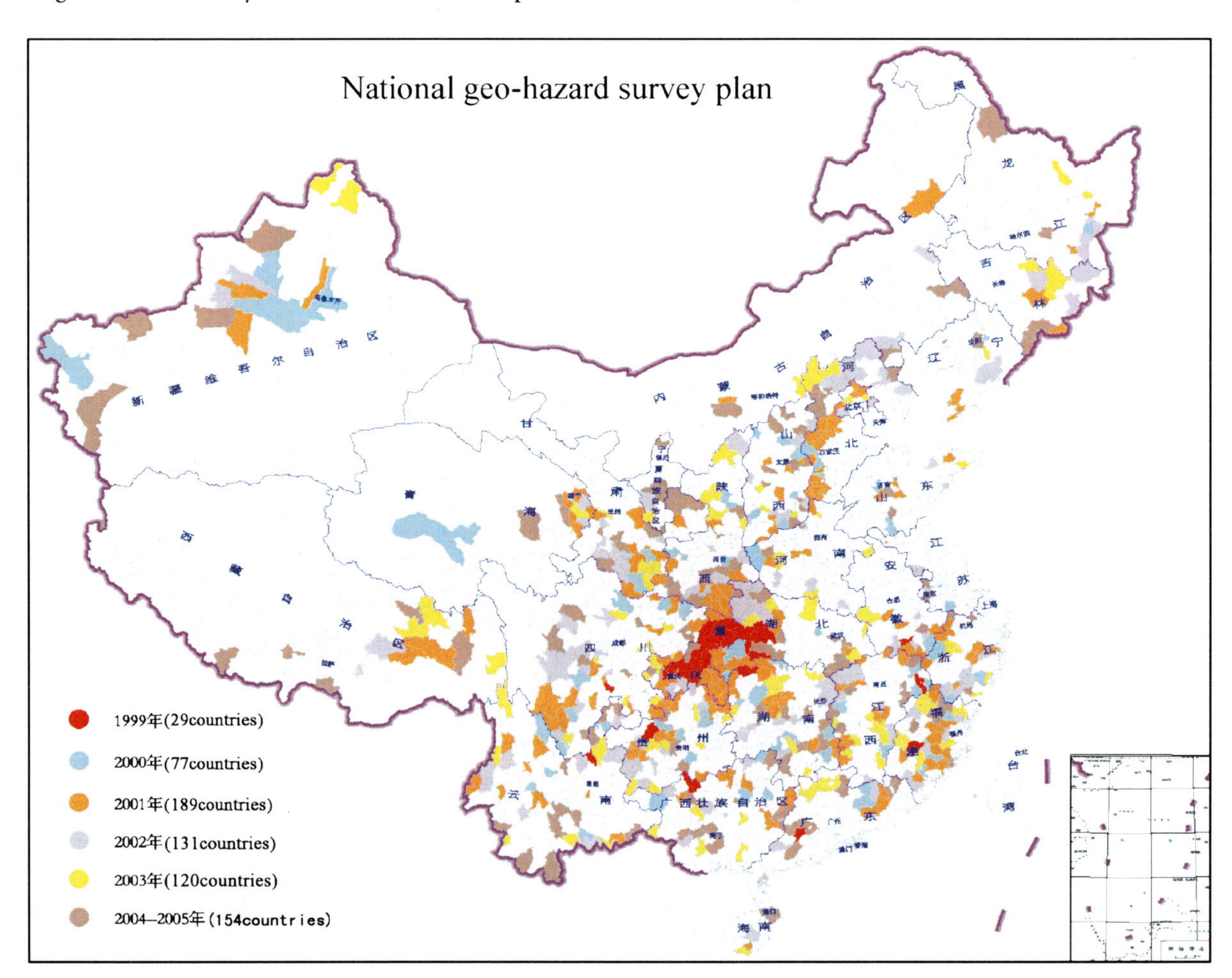

Fig. 36.3. National geo-hazard survey plan arrangement

counted for 31.5% of total landslides. Rock avalanches and debris flows were mainly characterized by small size ones.

The main factor of triggering landslide, rock avalanche and debris flow was rainstorm. The dead totaled to 8486 resulting from landslides (6462), rock avalanches (587) and debris flows (1437). The direct economic loss was total 3.27 billion Yuan resulting from landslides (1.07 billion Yuan), rock avalanches (0.2 billion Yuan), debris flows (2.0 billion Yuan) and unstable slope (0.52 billion Yuan). The population was all together 3.46 million threatened by potential landslides (1.51 million), rock avalanches (0.50 million), debris flows (0.98 million) and unstable slope (0.47 million). Property values totaled to 15.09 billion Yuan threatened by potential landslides (6.27 billion Yuan), rock avalanches (0.54 billion Yuan), debris flows (4.34 billion Yuan) and unstable slope (3.94 billion Yuan).

36.3 Geo-Hazard Risk Assessment and Zoning

The geo-hazard risk degree is determined mainly by two aspects: one is the geological hazard, namely the environmental condition and their intensity of geo-hazards; the other is the vulnerability of elements at risk. These two aspects together determine the geo-hazard risk level (Fig. 36.4). Therefore, geo-hazard risk is expressed as:

$$R = H \times V$$

Where, R represents risk, which means elements at risk possibly receive the direct and indirect economic loss, personnel casualty, the environment destruction.

Table 36.1.
Statistics of geo-hazard types based on 290 counties in China

Geo-hazard types	Number of occurrence sites	Proportion (%)	Number of potential sites	Proportion (%)
Landslide	28738	51	24898	52
Rock avalanche	9421	17	8595	18
Debris	4788	8	3406	7
Unstable slope	8891	16	7599	16
Surface fissure	1576	3	1034	2
Surface subsidence	2698	5	2300	5
Total	56112	100	47832	100

Data from the database of national geo-hazards survey program in China and LI Yuan et al. (2004).

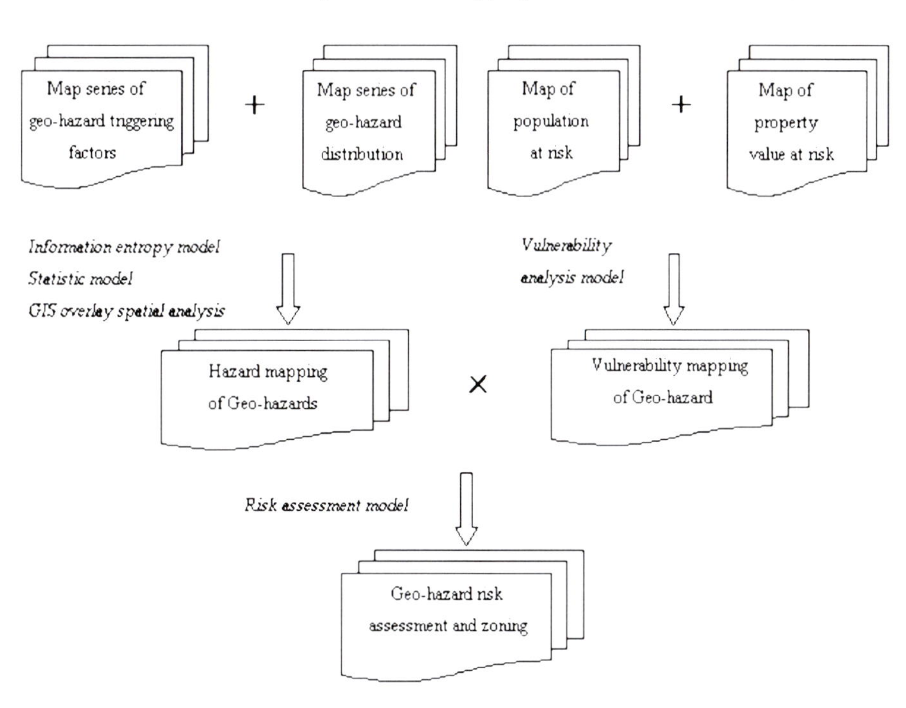

Fig. 36.4.
The process of geo-hazard risk assessment and zoning (modified on Chinese Academy of Land and Resources Economics 2003)

H represents the hazard, namely probability of a potential geo-hazard occurring with the specific extent in the given time. The property of rocks, the landform type, the terrain slope, the earthquake intensity, the precipitation condition and the humanity activities are considered as major factors during hazard assessment.

V represents vulnerability, namely loss degree of elements at risk such as population and property, which is expressed by percentage of maximum loss value and appraisal value of elements at risk before a disaster occurs. Therefore, property vulnerability of geological hazard is expressed by percentage of maximum loss value of assets and appraisal value of assets before a disaster occurs. Population vulnerability of geological hazard is expressed by percentage of maximum death and population numbers before a disaster occurs.

36.4 Meteorological Early Warning of Landslide

According to the statistic data of landslides occurred in China, 2/3 unexpected landslides resulted from the rain induced by or/and correlated directly to climatic factors. Therefore, meteorological early warning of landslide based on raining forecast is crucial for preventing and reducing casualties and loss caused by landslides.

Based on the geo-environmental conditions and climate factors causing geo-hazards, the mainland of China is divided into seven regions and twenty-eight early-warning sub-regions. Many researches indicated, the occurrence of landslides and debris flows is not only related closely to the rainfall amount in the same day, but also to the rainfall amount in the preliminary process. Therefore, according to the statistic analysis of the actual rainfall and the raining process fifteen days before geo-hazards occur at occurrence sites of geo-hazards in history, criterion curves (Fig. 36.5) for geo-hazard early warning are established for every warning sub-region as well as criterion correction diagrams

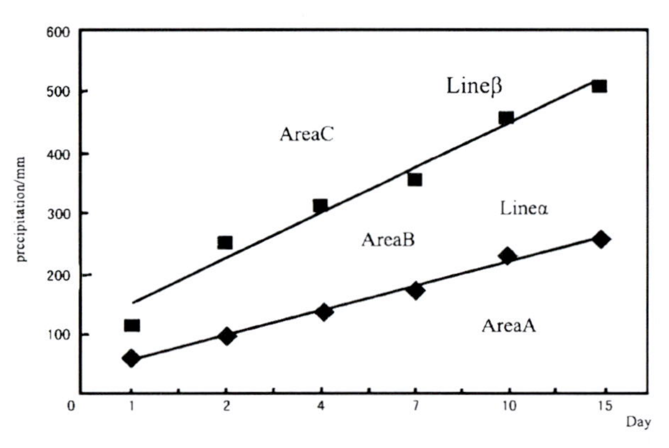

Fig. 36.5. Criterion curves for landslide early warning (from Liu et al. 2004). *A:* No issuing area; *B:* forecasting area; *C:* warning area; α: line-forecasting critical line (dividing line between 2 and 3 grade); β: line-warning critical line (dividing line between 4 and 5 grade)

based on the existing data. On this basis, the MLR and the National Meteorological Bureau of China jointly established an early-warning system of geological hazards based on the precipitation forecasting in 2003. During the flood season of 2003, 56 pieces of real-time information on early-warning of geological hazards forecasting were broadcasted by CCTV (Fig. 36.6) and 109 pieces of information on early-warning of geological hazards forecasting were released at the website of the MLR and the China Geological Environment Information Website. The early-warning system of geological hazards has achieved remarkable effects (Figs. 36.7 and 36.8).

Fig. 36.6. Geological hazards forecasting broadcasted by CCTV

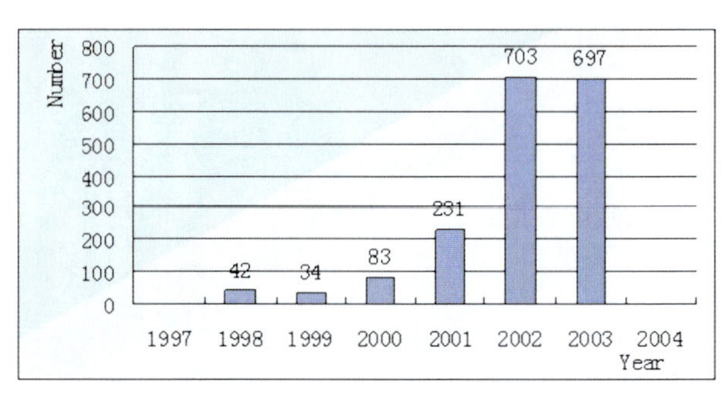

Fig. 36.7. Successful geo-hazard prediction numbers in recent years in China (data from Ministry of Land and Resources of China 2004)

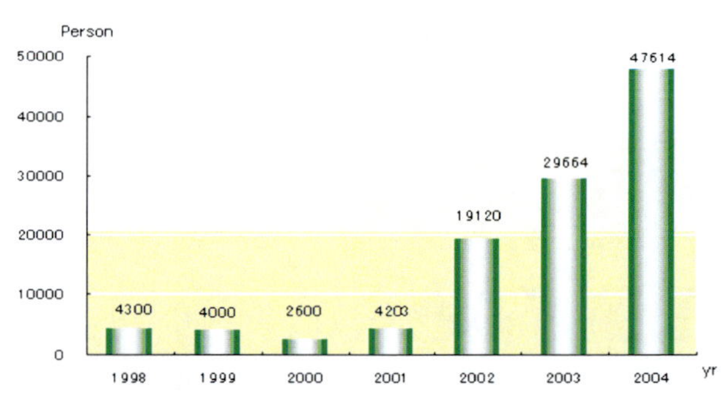

Fig. 36.8. Possible casualties reduced by successful prevention of geological hazards in 1998–2004 (data from Ministry of Land and Resources of China 2004)

36.5 Prevention and Control of Geological Hazards in Three Gorges Reservoir Area

Until 2004, the second stage of the project of prevention and control of geological hazards in Three Gorges Reservoir area was completed and passed the primary check and acceptance, while the planning for the third stage of the project of prevention and control of geological hazards was launched. The geological hazard monitoring and early-warning system in Three Gorges Reservoir area took shape and achieved primary effects. 37 landslides were warned successfully after the water level of the reservoir in front of Three Gorges Dam reached 135 m, resulting in effective protection of lives and properties of nearly 10 000 people. 129 large landslides were put under professional monitoring. 3-order GPS monitoring networks of geological hazards in the reservoir area were established preliminarily (Fig. 36.9). 1 216 spots for public monitoring and prevention of collapse and landslide groups were set up, and public monitoring and prevention teams of 2 300 persons were organized.

Fig. 36.9. GPS monitoring for the landslide control engineering in Three Gorges Reservoir area of China

36.6 Recommendation for National Geo-Hazard Mitigation Strategy

Although China has made every effort to combat geo-hazards, much work is still left for improvement, such as capacity of identifying potential geo-hazards, fundamental theory research of geo-hazard triggering mechanism, efficient and economic technology application in investigating, monitoring, warning, preventing and controlling geo-hazards as well as in public education and awareness. Base on international experiences and lessons (USNRC 2003), the MLR of China proposed national geo-hazard mitigation goal: geo-hazard occurring rate and the loss reduce by 50% and personnel casualty reduces by 70%. In order to achieve the proposed goal, following work should be strengthened:

1. To further implement National Geo-hazards Survey Program and to carry out geo-hazard risk mapping based on the above program.
2. To improve public monitoring and preventing network system of geo-hazards. To promote advanced technology application in geo-hazard investigation, monitoring, preventing and warning, such as high accuracy GPS and RS as well as automatic and real-time monitoring instruments.
3. To build up comprehensive capacity of emergency preparedness, response and rescue to geo-hazards.
4. To construct the information sharing service system of the geo-hazard mitigation. To provide an effective system of information collection, interpretation and dissemination for decision making and public awareness and education.

References

Chinese Academy of Land and Resources Economics (2003) Project report of geological hazard risk zoning of China

Geological Environment Bureau of the Ministry of Land and Resources of China (2004) National master plan for unexpected geological hazards emergencies

Geological Environment Bureau of the Ministry of Land and Resources of China (2004) National plan for prevention and control of geological hazards

LI Y, Meng H, Dong Y, Hu CE (2004) Main types and characteristics of geo-hazard in China – based on the results of geo-hazard survey in 290 counties. Chin J Geol Hazard Contr 15(2)

Liu CZ, Wen MS, Tang C (2004) Meteorological early warning of geo-hazards in China based on raining forecast. Geol Bull China 23:303–309

Ministry of Land and Resources of China (2004) Communiqué on geological environment of China

Ministry of Land and Resources of China (2004) Communiqué on land and resources of China

U.S.A. National Research Council (2003) Partnerships for reducing landslide risk-assessment of the national landslide hazards mitigation strategy. http://www.nap.edu/catalog/10946.html

Chapter 37

Landslide Hazard Zonation in Greece

G. Koukis · N. Sabatakakis · N. Nikolaou* · C. Loupasakis

Abstract. Landslide occurrences in Greece, covering a long time period (1950–2004) were recorded and digitally stored using a relational database management system. The first evaluation of engineering geological data and the geographical distribution of the recorded cases led to the determination of most critical landslide prone geological formations, regarding lithology and structure and to a landslide hazard map compilation at a national scale.

Keywords. Hazard zonation, database management, Greece

37.1 Introduction

A systematic inventory of data concerning landslides and their quantitative expression in Greece, through a relational database management system, was firstly attempted at the end of 1980s by Koukis and Ziourkas (1991), following the general trends of similar studies in other countries. Furthermore, statistical analyses of the main factors affecting landslide phenomena and their interrelations were established by Koukis et al. (1994, 1996, 1997), based upon data recordings through the database of landslide movements covering the as above period of more than 45 years.

In the last decade, the landslide activity is increasingly high as a result of increased urbanization and development (transportation routes, dams and reservoirs, industrial and urban activities) in landslide-prone areas, continued deforestation and extreme meteorological events. An increment of about 20% of the landslide occurrences has been recorded during the last five years all having serious socio-economic consequences resulted in a significant increase of total economic losses, fortunately without casualties.

After those events, the authorities have focused their attention on mitigating the problem by avoiding the hazards or reducing the damage potential. To aid this planning the Department of Geology of Patras University and the Institute of Geology and Mineral Exploration (IGME) were commissioned to undertake an extensive study regarding landslide hazard zonation. The techniques used were in accordance with the required national and regional scale of zonation of extensive areas according to their susceptibility to slope instability phenomena and they mainly included: *(a)* detailed inventory of slope instability processes and their study in relation to engineering geological setting, *(b)* analyses of causing and triggering factors and *(c)* spatial distribution of these factors.

The aim of this paper is to evaluate some first considerations regarding the influence of the geological conditions on landslide frequency distribution in the Greek territory using simple statistics and direct mapping methodology for landslide hazard zonation.

37.2 Geological Setting and Landslide Phenomena

In Greece, which has been experienced a complicated history of geophysical development during the Alpine orogenic cycle, several geotectonic zones or regions can be distinguished by fundamental differences both in lithology and structure (Figs. 37.1 and 37.2). The elongation of those zones more or less coincides with the trend of the main mountain ranges of the country and represents distinct geological structural units. In most cases they are separated by major thrust boundaries or by transition belts.

The eastern part of the country is occupied by the "internal" zones which are composed of geological formations extended in age from Palaeozoic to recent including metamorphic rocks. A variety of older and younger igneous rocks such as granites and extrusives is also included.

The "external" geotectonic zones occupy the western and central part of the country and are characterized by the presence of strong E-W tangential tectonic movements, having as a result a westward-directed ductile thrusting and intense folding and fracturing of the alpine formations.

By the end of Alpine orogenesis, post-alpine neogene sediments were deposited into trenches that were created by faulting tectonism. The "inherent weakness" of the geological formations of the "external" zones constitutes an important landslide causal factor. The landsliding events that affected the regional area are closely related to the fault tectonics and to the existence of still geodynamically active grabens. The presence of the tectonically highly sheared and weathered geological formations of the alpine basement as well as of the neogene sediments, contributed to the periodically induced instability phenomena triggered by heavy rainfalls and earthquakes (Figs. 37.3–37.6).

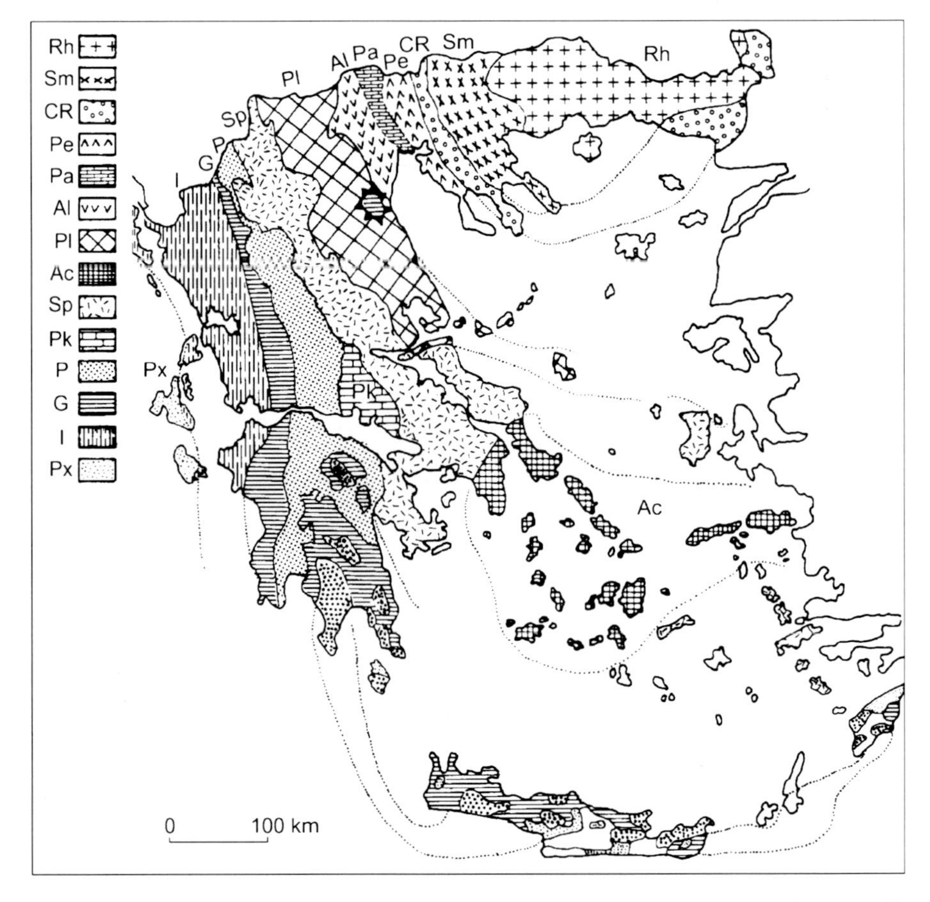

Fig. 37.1.
Geotectonic map of Greece. *Rh:* Rhodope Massif; *Sm:* Serbomacedonian Massif; *CR:* Circum Rhodopic Zone (*Pe:* Peonias Zone; *Pa:* Paikou Zone; *Al:* Almopias Zone) = Axios Zone; *Pl:* Pelagonian Zone; *Ac:* Attiko - Cycladic Zone; *Sp:* Subpelagonian Zone; *Pk:* Parnassos Zone; *P:* Pindos Zone; *G:* Gavrovo Zone; *I:* Ionian Zone; *Px:* Paxos Zone (Mountrakis et al. 1983)

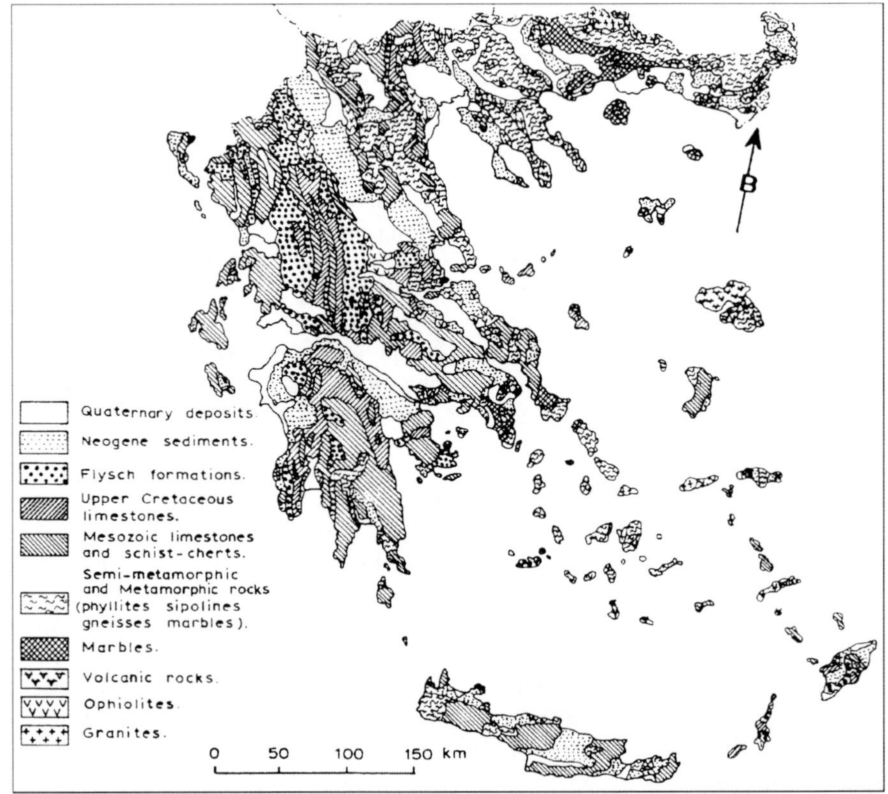

Fig. 37.2.
Engineering geological map of Greece (Koukis 1988)

Fig. 37.3.
Landslide of Panagopoula instability zone in western Greece (April 1971). The landslide took place on the lightly sheared transition zone to flysch. The national road of Athens to Patras failed into a zone of about 500 m

Fig. 37.4.
Composite landslide of Karya Village instability zone, western Greece. The landslide took place on the weathered and fractured flysch formations covered by thick limestone debris

37.3 Landslide Data Recording and Storage

A large number of technical reports and studies including landslide occurrences, obtained from IGME and Ministry of Environment, were analyzed. After modifications, mainly to standardize the terminology, 1 300 landslide cases covering a long time period (1950–2004) were recorded and digitally stored. A Statistical Inventory Form (S.I.F) modified from that initially proposed by Koukis and Ziourkas (1991) was used for data codification mainly based on Landslide Report (WP/WLI 1990) including the former suggestions regarding landslide causes (WP/WLI 1994), rate of movements (WP/WLI 1995) and remedial measures (WP/WLI 2001).

A relational database management system was designed in MS Access to allow rapid retrieval and evaluation of the data in selected unit areas. The interplay between the database system and Geographical Information System (GIS) was established with the defined coordinates of the locations of existing landslide occurrences.

Fig. 37.5. Rock falls occurring in the highly fractured (mylonitized) and weathered zone of limestones, triggered by strong earthquake, in Lefkada Island (August 2003)

Fig. 37.6. Malakasa landslide, February 1995. Damages on the new highway from Athens to Thessaloniki. The landslide manifested in the tectonized and weathered zone of schists covered by debris

Table 37.1. Frequency distribution of landslides in relation to the geotectonic zones

Geotectonic zone	Frequency of landslides (%)
Rhodope	3.00
Serbomacedonian	0.50
Circum Rhodopic	0.00
Axios	2.00
Pelagonian	39.00
Parnassos	4.00
Pindos	41.00
Gavrovo	4.00
Ionian	4.50
Paxos	2.00

Fig. 37.7. Translational slide along the contact of sandstone and argillaceous phases of flysch

Fig. 37.8. Rotational slide in the fine neo-gene sentiments in Skiros Island, triggered by heavy rainfalls

37.4 Statistical Approach

Simple statistical methods were used for first evaluation and quantitative expression of the engineering geological data regarding lithology and structure of the affected materials. The Olonos-Pindos geotectonic zone exhibits the highest frequency of landslide occurrences in Greece, exceeding 40% of the total cases recorded (Table 37.1). The most critical landslide prone geological formations regarding lithology and structure are flysch (Fig. 37.7) and neogene sediments (Fig. 37.8) with 30% and 28% frequency of recorded occurrences through the whole country respectively (Table 37.2). The estimated relative frequency of landslide occurrences (frequency normalized to the real area covered by each lithological type) shows that although flysch predominates over the other geological formations, shists and cherts significantly contribute to the recorded landslide phenomena.

Table 37.2. Frequency and relative frequency distribution of landslides in relation to the lithological type (Koukis et. al. 1994)

Lithological type	Frequency of landslides (%)	Area (%)	Relative frequency of landslides (%)
Loose Quaternary deposits	20.65	15.87	12.99
Neogene	28.20	24.00	11.74
Flysch	30.35	8.48	35.75
Schist-cherts	3.62	1.22	29.64
Limestones, marbles	4.85	19.50	2.48
Metamorphic	9.32	18.35	5.07
Volcanic	3.00	12.58	2.37

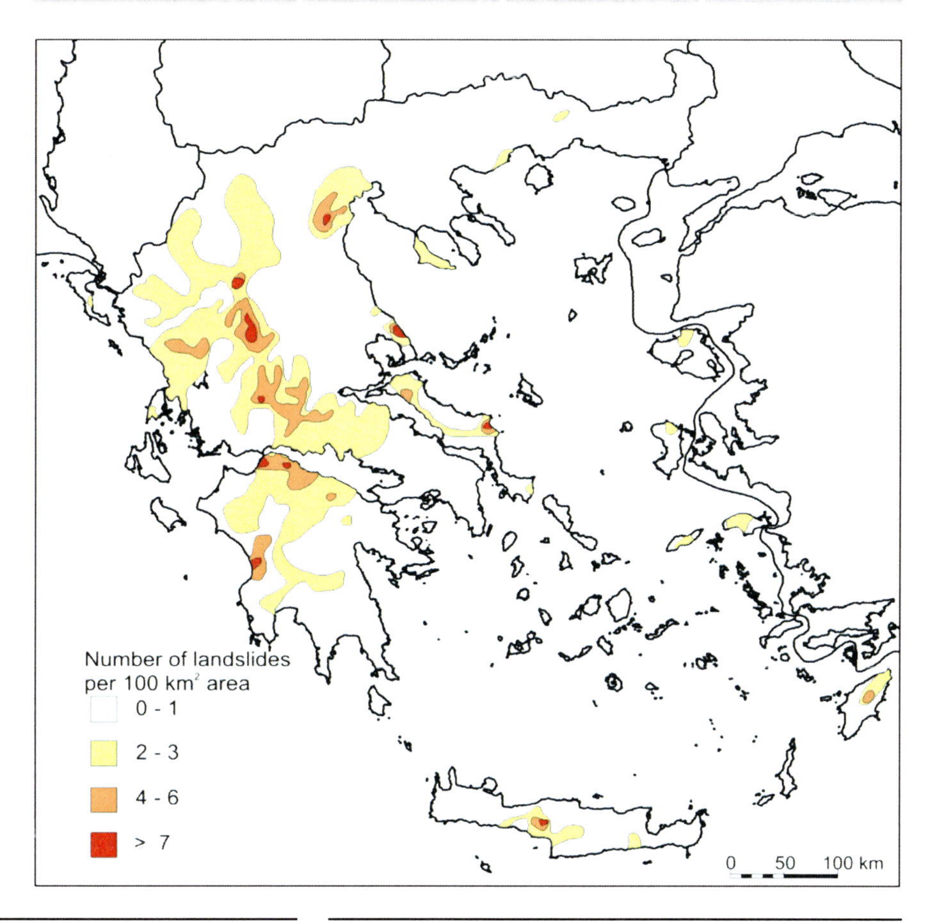

Fig. 37.9. Landslide hazard zonation map of Greece

37.5 Landslide Hazard Zonation

The geographical distribution of the recorded occurrences and the evaluation of the processed data led to the compilation of a landslide hazard zonation map at an original scale of 1 : 500 000 (Fig. 37.9). The national scale of this map (IAEG 1976) is intended to give a general inventory of problem areas for the country that can be generally used to inform the general public. The mapping methodology was based on the experience-driven approach (direct mapping methodology) by employing data from existing landslide sites.

37.6 Conclusions

The first data elaboration regarding lithology and structure of the affected geological formations showed that flysch is the most critical prone geological formation while the most landslide occurrences are mainly located in the western part of the country, where serious causing factors and triggering effects favor the manifestation of such phenomena on a broad scale. During the next stages of the study and after the completion of data base, indirect mapping approaches and specific GIS-based landslide hazard zonation techniques will be used for regional scale mapping preparation.

References

International Association of Engineering Geology (IAEG) (1976) Engineering geological map: a guide to their preparation. UNESCO Press, Paris, 79 pp

Koukis G (1988) Slope deformation phenomena related to the engineering geological conditions in Greece. In: Bonnard C (ed) Proceedings of 7th International Symposium on Landslides. Balkema, Rotterdam

Koukis G, Ziourkas C (1991) Slope instability phenomena in Greece: a statistical analysis. Bull IAEG 43:47–60

Koukis G, Tsiambaos G, Sabatakakis N (1994) Slope movements in the Greek territory: a statistical approach. In: Proceedings of 7th International IAEG Congress, Balkema, Rotterdam, pp 4621–4628

Koukis G, Tsiambaos G, Sabatakakis N (1996) Landslides in Greece: research evolution and quantitative analysis. In: Senneset K (ed) Proceedings of 7th International Symposium on Landslides, Balkema, Rotterdam, pp 1935–1940

Koukis G, Tsiambaos G, Sabatakakis N (1997) Landslide movements in Greece: engineering geological characteristics and environmental consequences. In: Proceedings of International Symposium of Eng. Geol. and the Envar, IAEG, Balkema, Rotterdam, pp 789–792

Mountrakis D, Sapountzis E, Kilias A, Elefteriadis G, Christofides G (1983) Paleogeographic conditions in the western Pelagonian margin in Greece during the initial rifting of the continental area. Can J Earth Sci, 20:1673–1681

WP/WLI (1990) International Geotechnical Societies' UNESCO working party on world landslide inventory (Chairman: Cruden DM) A suggested method for reporting a landslide. Bull IAEG 41:5–12

WP/WLI (1994) International Geotechnical Societies' UNESCO working party on world landslide inventory. Working group on landslide causes (Chairman: Popescu ME) A suggested method for reporting landslide causes. Bull IAEG 50:71–74

WP/WLI (1995) International Geotechnical Societies' UNESCO working party on world landslide inventory. Working group on rate of movement (Chairman: Bonnard C) A suggested method for describing the rate of movement of a landslide. Bull IAEG 52:75–78

WP/WLI (2001) International Union of Geological Sciences working group on landslides, commission on landslide remediation (Chairman: Popescu M) A suggested method for reporting landslide remedial measures. Bull Eng Geol Env 60:69–74

Chapter 38

Landslides Risk Reduction and Monitoring for Urban Territories in Russia

G. P. Postoyev · V. B. Svalova*

Abstract. Landslides process is one of the most widespread and dangerous processes in the urbanized territories. In Moscow the landslips occupy about 3% of the most valuable territory of city. In Russia many towns are located near rivers on high coastal sides. There are many churches and historical buildings on high costs of Volga River. The organization of monitoring is necessary for maintenance of normal functioning of city infrastructure in a coastal zone and duly realization of effective protective actions. Structure of monitoring system for urban territories is elaborated.

Keywords. Landslide, monitoring, cultural heritage, Moscow, Russia

38.1 Introduction

The geological environment is complex dynamic system which change occurs in result of slowly and deep developing processes. The mechanism of landslide process is quite often predetermined by structural-tectonic conditions of territory and lithological-petrographic characteristics of geological composition of rocks. Features of the mechanism of various genetic types of landslips in many respects cause the deepness and volumes of deformation of massifs of rocks, character and speeds of their moving at various stages of development of landslides process. Formation of an initial landslip of the certain type is preceded with usually long period of its preparation during which there is a rearrangement of a site of a slope, decomposition of rocks in slope massif, change of key parameters of a field of pressure, structure, a condition and properties of the breeds composing a slope. Zones of active development of explosive and shift infringements, processes of superficial and deep creep are formed. After overcoming peak resistance of breeds to shift progressing destruction in a zone of landslide displacement with transition of process of deformation in a stage of the basic displacement, quite often catastrophic character with big destructive energy begins. System of environmental monitoring is needed to forecast and prevent such events.

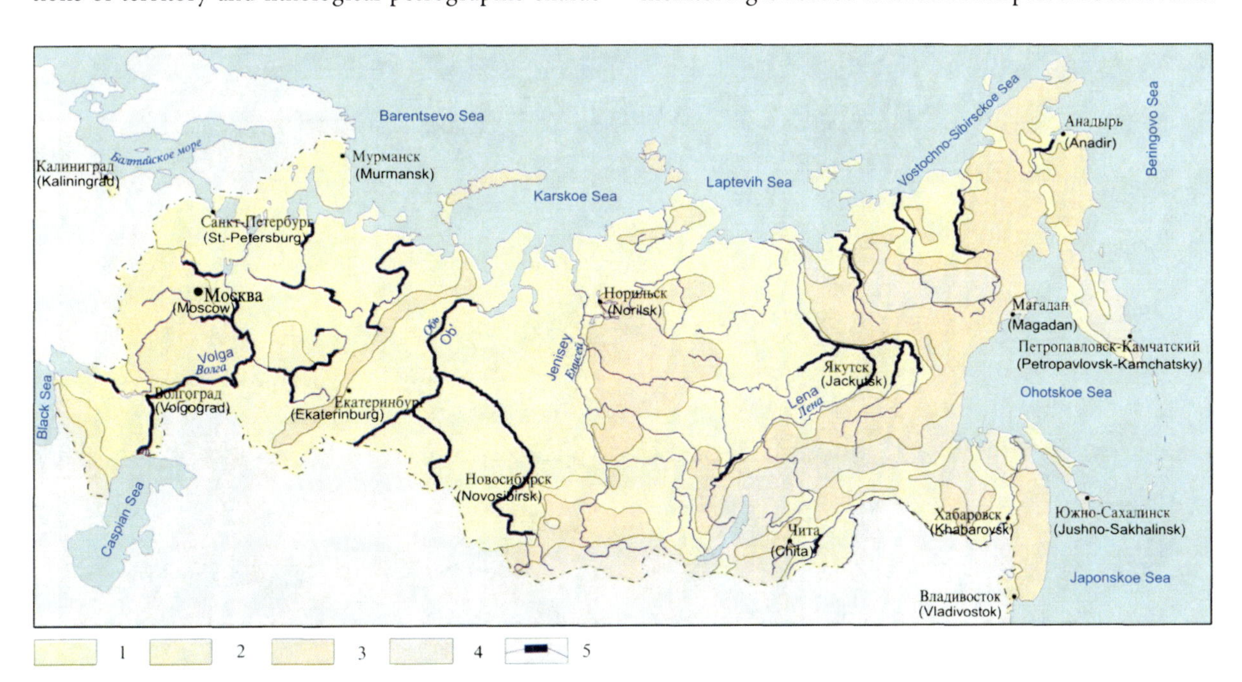

Fig. 38.1. Landslides in Russia (Mjagkov SM in Osipov et al. 2002). Plain areas: *1:* few and seldom landslides; *2:* many and often landslides. Mountain areas: *3:* often landslides; *4:* possible very big seismic landslides. *5:* landslides of river valleys

38.2 Monitoring Organization for Urban Territories

Landslides process is one of the most widespread and dangerous processes in the urbanized territories. In Russia many towns are located near rivers on high coastal sides. There are many churches and historical buildings on high costs of Volga River (Figs. 38.1–38.8).

In Moscow with landslips it is struck about 3% of the most valuable territory of city (Fig. 38.9). First, this territory has value from the historical point of view as the majority of ancient settlements concentrated on coast of the river Moscow and its inflows. In particular, in a south-east part of Moscow in landslides zone there was a museum-reserve "Kolomenskoye" with church and manor constructions of sixteenth to seventeenth centuries. In the central part of Moscow, opposite to sports complex "Luzhniky" on landslide slope of Vorob'evy Mountains it is located Andreevsky Monastery (seventeenth century). Second, coast of the river Moscow and its inflows are built up in many places by modern multi-store buildings of various purpose, occupied with the industrial, scientific and sports-cultural complexes supposing a mass congestion of people on or close to landslides slopes. In zones of landslides influences there is a number of bridge transitions, numerous communications of city (Kutepov et al. 2002, 2004).

The organization of monitoring is necessary for maintenance of normal functioning of city infrastructure in a coastal zone and duly realization of effective protective actions at a modern level of system of the automated monitoring of landslides process in Moscow according to the general concept of ecological monitoring of underground waters and dangerous geological processes of territory of Moscow.

Fig. 38.2. Church and settlement on the shore of Volga River at Mishkin Town

Fig. 38.3. Town Uglich: Kremlin

Fig. 38.4.
Town Ples: view of the Kholodnaya Hill. In the *forefront*, houses of merchants Groshev and Podgorny where I. I. Levitan's studio was located in 1889. To the *right*: St. Barbara Church

Fig. 38.5.
Town Ples: the summer church of the Trinity

Fig. 38.6.
Town Ples: view of Zarechye and the Shokhonka River

Stationary supervision over landslips in territory of Moscow are conducted since 1954. For expired time the extensive material on conditions of development and dynamics of landslips is saved up. Value represents not only experience of realization of regime supervision for landslides process and the factors determining its mode, but also the revealed laws of behavior of such complex landslips, as deep block landslips of compression and landslips of expression, similarly to a site of Vorob'evy Mountain.

Last years the regime supervision spent on landslides sites, are considerably reduced, besides the used complex of supervision does not correspond to modern requirements on structure and a level of automation.

On the basis of features of development and a mode of landslips of various types the system of monitoring most full adequate to problems of monitoring of landslides process and requirements to an observant network is offered.

The purpose of monitoring of landslides processes of territory of Moscow are a supply with information of authorities, managements of an ecological condition of the geological environment of city about a condition and dangerous development of landslips for good safety the population and normal functioning of urban services.

Primary goals of monitoring are the following.

1. Revealing sites of active development of landslides process.

   Activity is estimated on: to sizes of landslides deformations determined by tool methods of supervision; to results of inspection with mapping of fresh landslides cracks, water-displays; to geophysical parameters of development of landslides motions.
2. The control, including automated, of conditions of landslides slopes and landslides-prone sites.

Fig. 38.7.
Town Tutayev: resurrection cathedral built in seventeenth century

Fig. 38.8.
Town Tutayev: Kazan Mother of God church on the left high landslide prone shore of Volga River

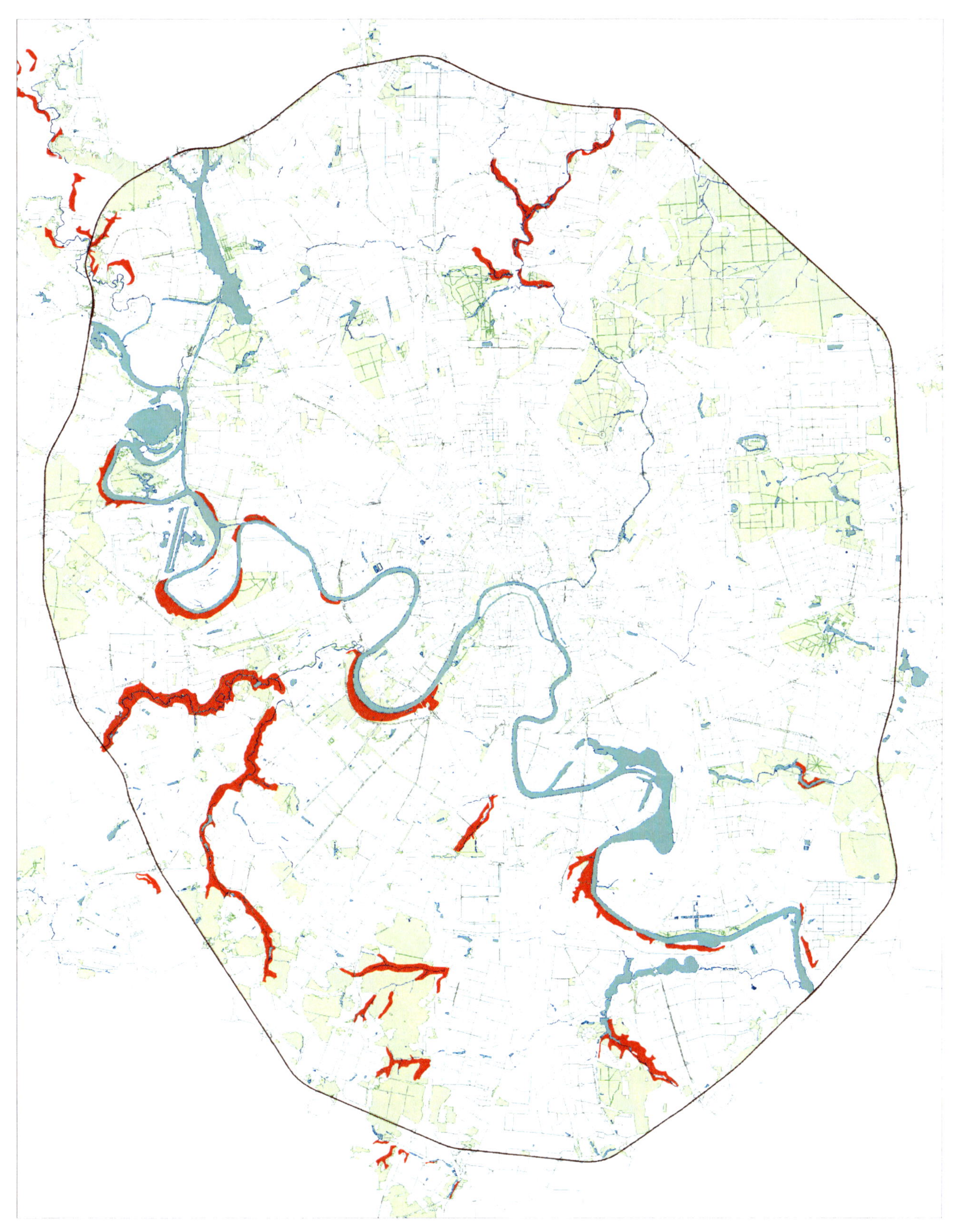

Fig. 38.9. Landslides prone zones in Moscow (*red color*). From map of IEG RAS. "Landslides hazards" (Osipov et al. 2002)

In Moscow it is revealed 18 sites with display deep block landslips representing the greatest danger to a city infrastructure. Means of monitoring should allow to identify a condition of local sites and a slope as a whole (changes of controllable parameters at a stage of preparation of motions, local displays of integument landslips, deep motions etc.).

3. Revealing laws and the mechanism of development of landslides process and, in particular, deep block landslips.

   On the basis of the analysis of the given regime supervision capacity of a landslip, time and conditions of occurrence deep landslides motions, deformation behavior of various parts of landslides slope, interrelation between landslides-forming factors and landslides sites, influence of man-caused factors is established and specified.

4. The prevention of possible influence of landslides deformations on constructions and the communications of city located in landslides zones.

   By results of the automated monitoring and regime supervision operative estimations and forecasts of a condition of landslides slopes on sites of an arrangement of various constructions are made. The appropriate preventions are dispatched under the established rules.

5. Gathering, processing, a data storage on controllable parameters and sites, use given others monitoring systems (private, departmental etc.).

6. An estimation and forecasting of a dangerous condition of landslides slope. Calculations of stability and a condition of landslides slope in view of conditions of development of landslides process by the current moment and the required period of anticipation.

   On the basis of the given supervision the site of a surface of sliding, a level of underground waters is specified, engineering – geological cross-sections on observably ranges are under construction. Calculations of stability of landslides slope and landslide pressure are made in view of conditions of development of landslide process by the current moment and possible changes of values of influencing factors. By results of change of the controllable parameters determining a condition of landslides slope, the forecast of activity of landslips in Moscow is done.

7. Development of recommendations for acceptance of managing decisions, to increase of stability and stabilization of landslides slopes, to optimum development of slope territories.

   During conducting monitoring, accumulation of the information on activity of landslides process, the criteria determining a degree of activity and activation of landslips, dangers for existing in landslides zone of constructions are developed. Recommendations, technical decisions on stabilization of landslides slopes are developed in view of the information received by results of monitoring on features of the mechanism and dynamics of development of landslides deformations, change of risk of construction of various objects in landslides zone.

8. An estimation and the control of efficiency of against-landslides actions and a condition of armed landslides slope.

Means of monitoring of landslides deformations, stress-deformed state of landslides slopes, landslides-forming factors, supervise change of a condition of landslides slope and its parts, characterizing a degree of activity of landslides process and efficiency of carried out stabilizing works.

Scheme of structure of monitoring post for landslide slope is performed in Fig. 38.10. At present the project of monitoring network for observation over the landslide slopes in the Moscow River Valley is being developed. This network includes the arrangement of the automatic observation system at two slopes (Vorob'evy Hills and Kolomenskoye settlement) and the performance of recurrent geodetic survey at other slopes. The following observations will be performed at the landslide slope in Kolomenskoye Region and in Vorob'evy Hills:

- the control over the rock massif displacement by the extensometer data and geodetic reference marks;
- the control of the groundwater level according to special observations;
- the visual description of the slope conditions.

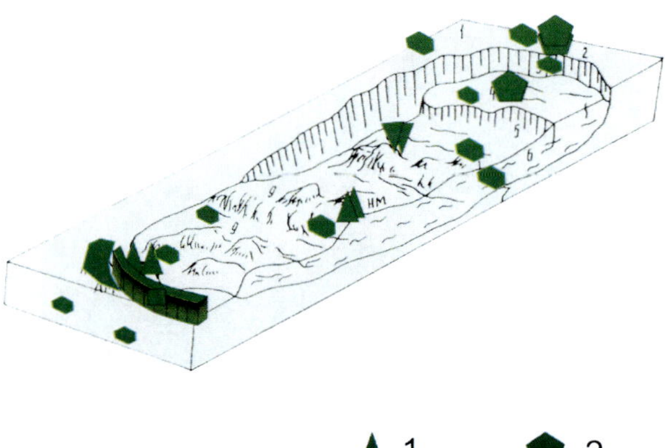

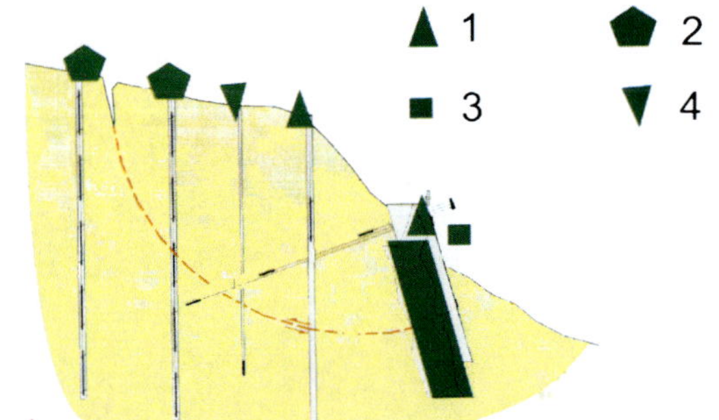

Fig. 38.10. Monitoring post structure for landslide process. *1:* Inclinometer; *2:* acoustic seismometer; *3:* deformometer; *4:* water-level meter (Osipov et al. 2002)

Monitoring system of Moscow can be recommended for other Russian towns especially along rivers.

38.3 Results

Monitoring system for Moscow is elaborated. The most dangerous places are connected with landslide prone slops of Moscow River. The experience of monitoring organization can be used in other towns and urban territories in Russia and abroad.

References

Kutepov VM, Sheko AI, Anisimova NG, Burova VN, Victorov AS et al. (2002) Natural hazards in Russia. Exogenous geological hazards. Moscow, "KRUK", 345 pp

Kutepov VM, Postoev GP, Svalova VB (2004) Landslide hazards estimation on sites of modern and historical constructions in Moscow. Proceedings of 32 IGC, Italy, Florence

Osipov VI, Shojgu SK, Vladimirov VA, Vorobjev YuL, Avdod'in VP, et al. (2002) Natural hazards in Russia. Natural hazards and society. Moscow, "KRUK", 245 pp

Chapter 39

Numerical Analysis on Slope Stability under Variations of Reservoir Water Level

Hongjian Liao* · Jie Ying · Shihang Gao · Qian Sheng

Abstract. The water level of the Yangtze River Three Gorges Reservoir is adjusted periodically, and the stability of the slide masses on both banks is affected seriously. Against the background of the Three Gorges projects, the influence of drawdown speed of water level on the stability of slide mass was analyzed. The variation of the slide mass stability during the drawdown period was calculated and analyzed considering the seepage fields. Through numerical calculation, the relationships between the slide mass stability, hydraulic conductivity and water level drawdown speed were obtained. It might have some reference value for the study on slope stability in reservoir districts.

Keywords. Reservoir water level, slide mass, stability, hydraulic conductivity

39.1 Introduction

It is significant to study the slide mass stability in the reservoir areas to ensure that hydraulic and hydroelectric engineering can be built without danger and work normally. Specially, after the landside of Vaiont Reservoir in Italy (1963), geologists and geological engineers began paying more attention to the interaction between human engineering activities and surrounding geological environment.

Compared to general landslides in mountainous region, the landslides in reservoir district are special, because their activities are related to the variation of reservoir water level, the erosion of reservoir water wave, and the immersion by the water. With academic, economic, environmental, and even social significance, the research on landslides on banks of the reservoir must be paid enough attention (Cui and Li 1999; Pan 1980; Duncan 1996; Lam and Fredlund 1993; Zhu et al. 2002).

In Japan, about 60% of reservoir landslides occurred in the period of sudden drawdown of water level, and other 40% occurred in the period of water rising, including the storing of initial water. After testing and analyzing reservoir landslides, Japanese scholar thought that rainfall, water soaking and sudden drawdown of water level are major factors to induce reservoir landslides (Nakamura 1990). The water level of a river, a reservoir or a canal will draw down fast after floods, and that will happen to the level of sea water when a tide is down. Investigating results indicated that sudden drawdown of water level could cause landslides in the slope facing water. During the process of water level drawdown, if the pore water pressure could not disappear fast with the outside water drawing down, large shearing force would appear in the slope and landslide would occur (Bishop 1955).

In China, two large landslides of 259 m and 210 m occurred at the middle and southern parts of Yuecheng Reservoir in 1968 and 1974 due to sudden drawdown of water level (Mao et al. 1981). The Yangtze River Three Gorges Reservoir, which is still under building, is a long and narrow watercourse reservoir. The reservoir district will submerge 27 counties in Hubei Province and Chongqing City, and the water level of that district will rise about 100 m. The reservoir will apply a periodical adjusting mode, storing water in winter and discharging water in summer, and the variation of water level will be 30 m. The reservoir will store water from the low water level to the normal level, and discharge water from the normal level to the low level. This change must disturb the seepage field in both banks. If the water level of the reservoir suddenly drew down, while the water level in both banks could not draw down simultaneously, the seepage force caused by the water between the high and low water level would become a seriously adverse factor to the stability of the slope, and it might cause disasters to the hydraulic and hydroelectric engineering.

Against the background of Three Gorges Project and combined with the geological investigation report (Integrating Reconnaissance Bureau of Yangtze River Hydraulic Committee 1999; Yangtze River Reconnaissance, Programming and Designing Academe 2000), the stability of soil slope with different hydraulic conductivities and under different velocities of water level drawdown was numerically calculated and analyzed during the period when the water level draws down 30 m, adopting the slope stability analyzing software *Geo-Slope*. The possible effect on the stability of bank slope caused by periodical adjusting of water level in the reservoir district was also analyzed. In addition, the relationship between drawdown velocity, hydraulic conductivity and slope stability was obtained. It might have some reference value for the research on slope stability in reservoir districts.

39.2 Calculating Theory

39.2.1 Seepage Calculating Basic Theory and the Hydraulic Conductivity Ascertaining

The continuity equation of groundwater movement can be thought about at the aspect of mass conservation. The difference between the flow (flux) entering and leaving an elemental volume at a point in time is equal to the change in the volumetric water content. More fundamentally, the sum of the rates of change of flows in the x- and y-directions plus the external applied flux is equal to the rate of change of the volumetric water content with respect to time. Combined with Darcy law, the following equation can be obtained:

$$\frac{\partial}{\partial x}\left(k_x \frac{\partial H}{\partial x}\right)+\frac{\partial}{\partial y}\left(k_y \frac{\partial H}{\partial y}\right)+Q=\frac{\partial \Theta}{\partial t} \tag{39.1}$$

where H = total head, k_x = hydraulic conductivity in the x-direction, k_y = hydraulic conductivity in the y-direction, Q = applied boundary flux, Θ = volumetric water content, and t = time.

As water flows through soil, certain amounts of water are stored or retained within the soil structure. The amount of water stored or retained is a function of the pore-water pressure and the characteristics of the soil structure. For a seepage analysis, it is convenient to specify the stored portion of the water flow as a ratio of the total volume. This ratio is known as the volumetric water content. In equation form:

$$\Theta = V_w / V \tag{39.2}$$

where Θ = volumetric water content, V_w = volume of water, and V = total volume. When the degree of saturation is 100%, the volumetric water content is the porosity of soil.

Variations of volumetric water content are dependent on the properties of the soil and changes of the stress state. In seepage calculation, the total stress and the pore-air pressure are assumed to be constant. Consequently, the change of volumetric water content is only dependant on variations of pore-water pressure.

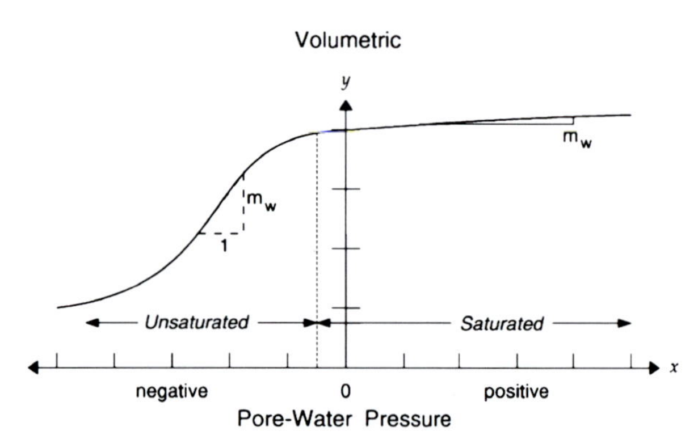

Fig. 39.1. Function of volume water content

Table 39.1. Hydraulic conductivities for calculation

Soil	Hydraulic conductivity adopted in calculation (m s^{-1})			
Standard sand	7.5e–4	5e–4	2.5e–4	1e–4
Sand	7.5e–5	5.4e–5	2.5e–5	1e–5
Fine sand	7.5e–6	4.3e–6	2.5e–6	1e–6
Silt	7.5e–7	5e–7	2.5e–7	1e–7

Figure 39.1 shows the relationship between the volumetric water content and the pore-water pressure, which is also known as the soil-water characteristic function. A change in volumetric water content can be related to a change in pore-water pressure by the equation (Fredlund and Rahardjo 1993):

$$\partial\Theta = m_w \partial u_w \tag{39.3}$$

where m_w = slope of the storage curve, and u_w = pore-water pressure.

Hydraulic conductivity is a parameter indicating the capability of soil conducting water and dependant on the volumetric water content, so it is a function of volumetric water content. And volumetric water content is a function of pore-water pressure, so hydraulic conductivity is the indirect function of pore-water pressure. Because it is more complicated to obtain the hydraulic conductivity of unsaturated soil than the soil-water characteristic curve, many methods were developed to predict the hydraulic conductivities from the soil-water characteristic curve (Ho 1979).

The process to ascertain the hydraulic conductivities adopted in this paper are as follows. First of all, according to the conducting grade sorts of rock and soil mass in Applied Standard Criterion in Civil Engineering (Wang 2001), the hydraulic conductivities were selected from 7.5e–4 to 1e–7 m s^{-1} in four magnitudes. Then, adopting the standard experimental parameters provided by the software and selecting four basic corresponding soil-water characteristic curves, 16 hydraulic conductivities (in Table 39.1) were ascertained and the slope stability in different hydraulic conductivities was numerically analyzed separately to reflect the influence of variation of hydraulic conductivities.

The total hydraulic head can be defined as:

$$H = \frac{u_w}{\gamma_w} + y \tag{39.4}$$

where u_w = pore water pressure, γ_w = unit weight of water, and y = elevation head.

Substituting Eq. 39.4 into 39.3, the following equation can be obtained:

$$\partial\Theta = m_w\gamma_w(H - y)\partial \quad (39.5)$$

Substituting Eq. 39.5 into 39.3, since the elevation is a constant, the derivative of y with respect to time disappears, leaving the following governing differential equation:

$$\frac{\partial}{\partial x}\left(k_x\frac{\partial H}{\partial x}\right) + \frac{\partial}{\partial y}\left(k_y\frac{\partial H}{\partial y}\right) + Q = m_w\gamma_w\frac{\partial H}{\partial t} \quad (39.6)$$

Using the finite element method, the seepage governing equation of finite element can be obtained (Segerlind 1984):

$$(\Delta t[K] + [M])\{H_1\} = \Delta t\{Q_1\} + [M]\{H_0\} \quad (39.7)$$

where Δt = time increment, $[K]$ = element characteristic matrix, $[M]$ = element mass matrix, $\{H_1\}$ = head at end of time increment, $\{Q_1\}$ = nodal flux at end of time increment, and $\{H_0\}$ = head at start of time increment.

39.2.2 Stability Calculating Theory and the Seepage Calculating Model

The calculation of slope stability adopted Morgenstern-Price method, one of general limit equilibrium methods. In Morgenstern-Price method (Morgenstern and Price 1965), it is assumed that the relationship between the interslice normal and shear force is functional with respect to the x-coordinate of slide mass. A Half-Sine function is applied to satisfy the condition that the interslice force is zero at the left and right boundary of slide mass.

Employing the software *Geo-Slope*, the dynamic seepage field is considered with the limit equilibrium analysis. The transient seepage problem is firstly analyzed in the software *Seep* and the hydraulic head distribution of seepage field at different time is obtained, then the hydraulic head values are inputted to the slope stability calculating software *Slope*, and the stability coefficients of slide mass are calculated with the method of slices in the software *Slope*.

In the calculating process, for the sake of transforming the finite element head values in *Seep* into *Slope*, firstly the finite element mesh from *Seep* is brought into *Slope*, and the stability analysis model is rebuilt on this mesh, including ascertaining the slide surface and the number and the method of slices in soil mass. Then the center of the slice base and the element that exists at the center of the slice base are found on every potential slide surface. Using the nodal head information which *Seep* transforms,

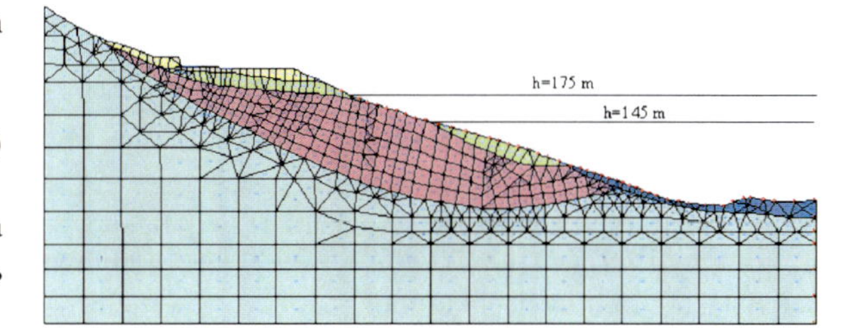

Fig. 39.2. Calculating model of seepage

Slope can find the corresponding local coordinates of the center of slice base and the matrix of interpolation functions. The hydraulic head distribution within an element in terms of the heads at the element nodes is expressed by the equation:

$$h = \langle N\rangle\{H\} \quad (39.8)$$

where h = head anywhere in the element, $\langle N\rangle$ = matrix of interpolation functions, and $\{H\}$ = matrix of heads at the element nodes.

In this way, the dynamic seepage field is combined with the limit equilibrium analysis.

Combined with part of the geological investigation report in Fengjie County in Three Gorges Project Reservoir district, the calculating profile and model illustrated in Fig. 39.2 were adopted. The small vacant circles on the boundary indicate the outside hydraulic conditions under three different reservoir water level drawdown velocities. It is thought that the flux on the point is zero if the elevation of the point is higher than the water level. Other boundaries which are not explained are non-permeable boundaries.

39.3 Numerical Calculation and Analysis

To analyze the effects of drawdown velocity of water level and hydraulic conductivity on safety factor, the height of water level (h) was assumed to be drawn down 90 m, at velocities of 1, 0.5, and 0.33 m d^{-1} (the corresponding drawdown time is 3, 6, and 9 months).

39.3.1 Relationship between Reservoir Water Level Descending Rate and Safety Factor Descending Rate

It is supposed that the hydraulic conductivities of slide mass are the same in every calculation. 16 hydraulic conductivities (in Table 39.1) were selected, and seepage fields in three different water level drawdown velocities were calculated. The calculating step of seepage fields is 12 h (43 200 s).

The influence of different drawdown velocities of water level on slope masses made of the same material with different hydraulic conductivities was compared. Figure 39.3 illustrate the calculating results. The x-coordinate is the percentage of the height of water level drawdown in the total descending height, and y-coordinate is the percentage of variations of safety factor in relation to the initial safety factor. In the calculation for standard sand, four hydraulic conductivities were selected: $k = 7.5e{-}4$, 5e–4, 2.5e–4 and 1e–4 m s^{-1}. Because curves of $k = 7.5e{-}4$ and $k = 5e{-}4$ m s^{-1} (or $k = 2.5e{-}4$ and $k = 1e{-}4$ m s^{-1}) are approaching under different drawdown velocities, the results of $k = 7.5e{-}4$ and $k = 1e{-}4$ m s^{-1} were only showed in Fig. 39.3a for clarity. Similarly, for other materials, Fig. 39.3b–d only showed the results when hydraulic conductivity is the maximum and minimum.

For slope masses with larger hydraulic conductivities, the safety factor increased fast after reaching the bottom, even the last safety factor could be larger than the initial. As the hydraulic conductivity became smaller, trends of curves became just descending. The stability variations of slope masses with the same permeability magnitudes are similar, and stability curves of slope masses with the same hydraulic conductivity are also similar under different drawdown velocities.

In addition, the minimum safety factor (the most dangerous hydraulic condition) occurred during the descending period of the water level, but not at the highest or the lowest height, which is accordant with the calculating result of some scholars. Moreover, slope masses with smaller hydraulic conductivities need longer time to reach the minimum safety factor.

39.3.2 Effects of Different Water Level Drawdown Velocities on Slope Stability

The stability of four sorts of slope masses was calculated and compared under different velocities of water level: 1, 0.5 and 0.33 m d^{-1}. It is found that the results of slope masses made of standard sand and silt were in the middle of results of slope masses made of sand and fine sand. To be clear, Fig. 39.4 only showed the results of slope masses made of standard sand and silt. The x-coordinate is the drawdown time of reservoir water level, and y-coordinate is the safety factor.

From the curves in Fig. 39.4, the safety factors of slope masses made of standard sand decrease with the drawdown of water level at first, increase after reaching the bottom, and then become smooth and approach certain constant value. The larger the hydraulic conductivities the faster the safety factors of slope masses increase after decreasing. And the larger the drawdown velocities the shorter the time reaches the bottom of curves. Moreover,

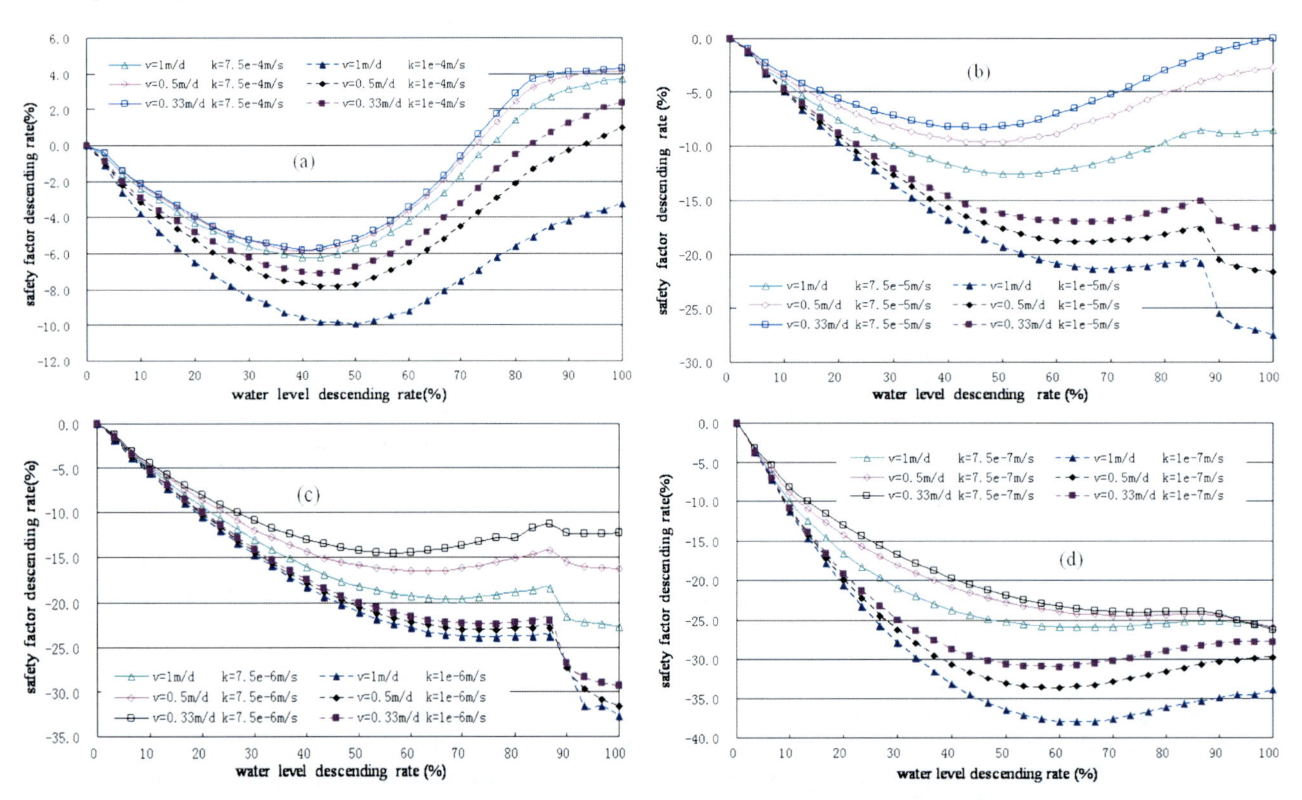

Fig. 39.3. Relationship between reservoir water level descending rate and safety factor descending rate when $h = 90$ m: **a** standard sand; **b** sand; **c** fine sand; **d** silt

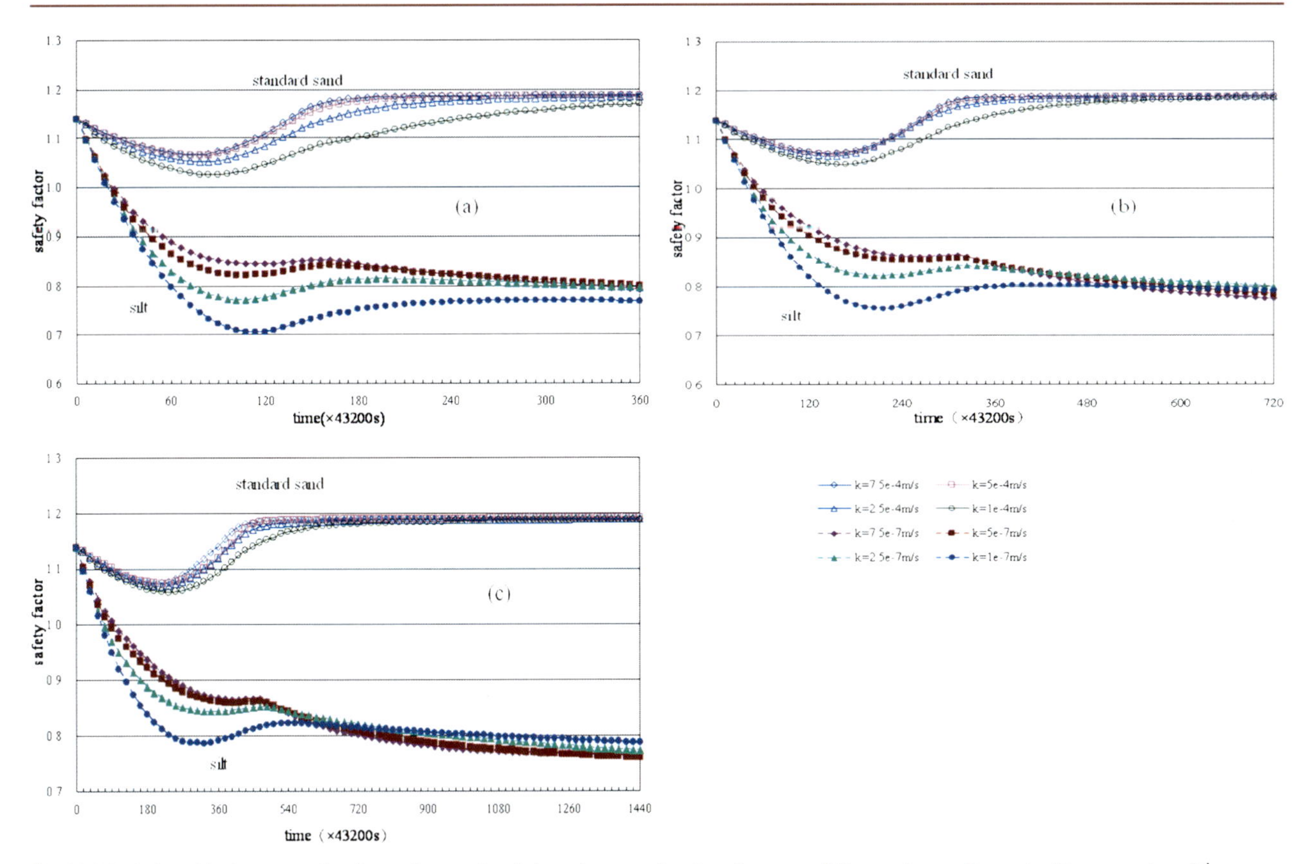

Fig. 39.4. Relationship between the time of water level drawdown and safety factor at different descending velocities: **a** $v = 1$ m d^{-1}; **b** $v = 0.5$ m d^{-1}; **c** $v = 0.33$ m d^{-1}

for slope masses made of standard sand, safety factors are close and larger than 1.0. It indicates that the stability of slope masses made of standard sand is favorable and could not be easily affected by variations of water level.

For slope masses made of silt, safety factors decrease quickly with the drawdown of water level, increase slightly after reaching the bottom of the curves, but decrease at once mostly and become smooth and approaching. If drawdown velocity is larger, the time to reach the minimum safety factor is shorter. Curves of safety factor have a descending trend with the time. Curves descend fast at first, but the range of change become slighter after reaching the bottom. In all, the descending range of safety factors of slope masses made of silt is large, and landslides may occur.

From the properties of curves obtained, the following interpretation can be concluded. Water in slope masses with larger hydraulic conductivities could discharge fast when water level draws down. Pore-water pressure could be eliminated, so safety factors do not descend much. On the contrary, if the hydraulic conductivities of slope masses are small, water in slope masses could not discharge in short time when water level descends. Large shear force caused by pore-water pressure could induce landslides. Therefore, attention should be paid to stability of slope masses made of materials with small hydraulic conductivities.

39.4 Case Study

In fact, the water level of Three Gorges Reservoir is designed to be adjusted from 175 to 145 m. Therefore, in the following calculation, the water level was assumed to descend 30 m in 1, 2 and 3 months, in other words, the drawdown velocities are still 1, 0.5 and 0.33 m d^{-1} separately.

Because the drawdown height of water level was much changed in calculation, the descending rate of safety factor changed a lot. Figure 39.5 showed the relationship between reservoir water level descending rate and safety factor descending rate when $h = 30$ m.

From the results shown in Fig. 39.5, it can be found that the change trend of slope stability is nearly related to the hydraulic conductivities of slide masses. No matter what kind of slide mass, the safety factor descending rates of slope masses with the same hydraulic conductivity apparently increase while the water level drawdown velocities increase. The change curves of stability under different water level drawdown velocities are similar. From standard sand to silt, the safety factor descending rates increase as the hydraulic conductivities of slide masses decrease. For slide mass of every material, the safety factor descending rates apparently become larger when the water

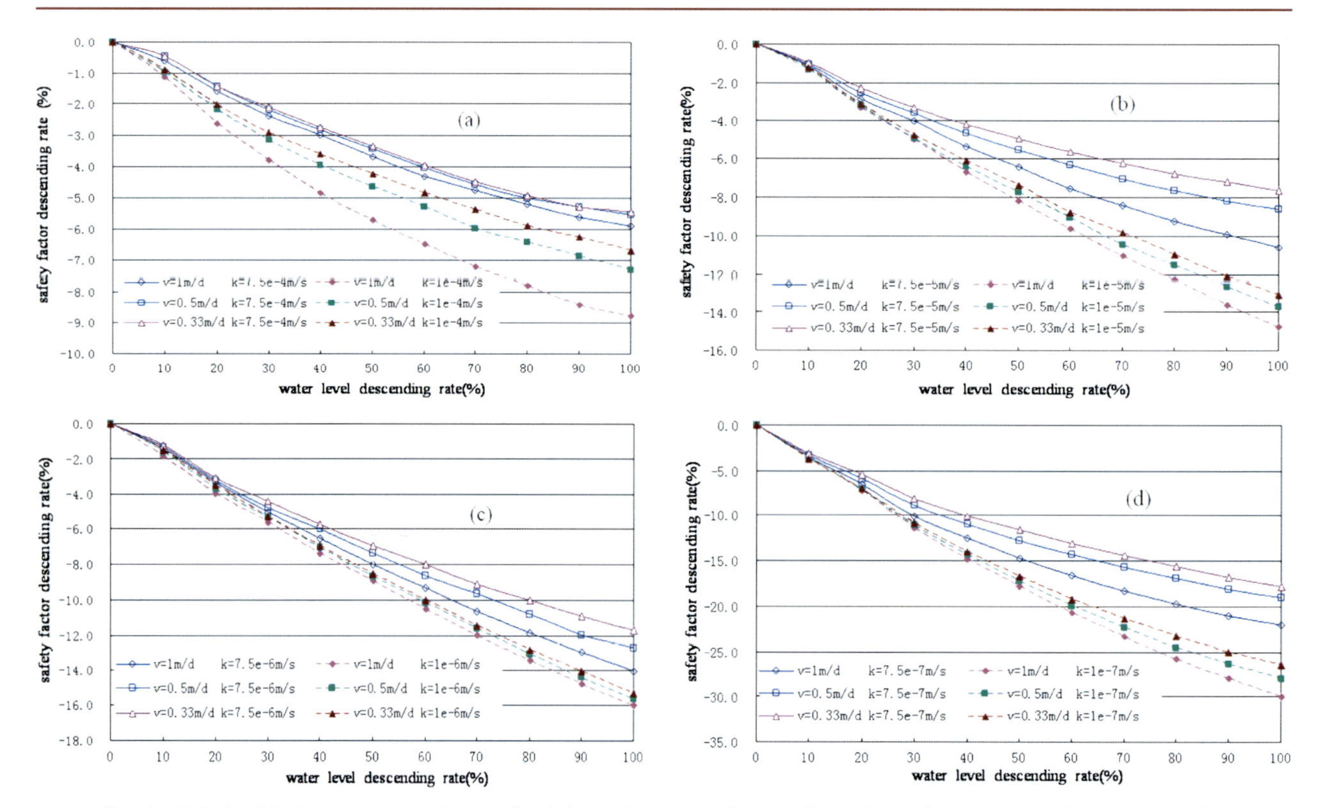

Fig. 39.5. Relationship between reservoir water level descending rate and safety factor descending rate when $h = 30$ m: **a** standard sand; **b** sand; **c** fine sand; **d** silt

level drawdown velocities increase, but the safety factors of slide masses with larger hydraulic conductivities are approaching under different water level drawdown velocities, in other words, the effect of drawdown velocity on stability is little. The safety factor descending rates become larger when the hydraulic conductivities of slide masses decrease with the same water level descending rate. According to the calculating results, the safety factor descending rate of silt is about 3 times as large as that of standard sand. In summary, due to the diversity of geological environment where practical engineering locates, corresponding preventing and curing measures should be adopted for slope masses with different hydraulic conductivities under different variation velocities of water level.

39.5 Conclusions

The following conclusions can be drawn by analyzing the calculating results:

1. The change trend of slope stability is nearly related to the hydraulic conductivities of slide masses during the period of reservoir water level drawing down. The safety factor descending rates increase as the hydraulic conductivities of slide masses decrease. The trend of slope stability change is nearly related to hydraulic conductivities. Stability behaviors of slope masses with different permeability magnitudes are different. Trends of stability change of slope masses with the same permeability magnitudes are similar. Stability curves of slope masses with the same hydraulic conductivity are similar under different drawdown velocities. From standard sand to silt, when the reservoir water level draws down to the same height, the descending rate of safety factor increases as hydraulic conductivity of slope masses decreases.
2. When $h = 30$ m, for slide mass of every material, the safety factor descended as the water level drawdown velocity increased. When $h = 90$ m, the safety factor of slope mass with larger permeability increased fast after reaching the bottom, even the last safety factor could be larger than the initial. As the hydraulic conductivity became smaller, the safety factor became just descending.
3. The reservoir water level drawdown velocity has certain influence on the change of safety factor. The safety factor descending rates apparently increase while the water level drawdown velocities increase for slide mass with the same hydraulic conductivity. The change curves of stability under different water level drawdown velocities are similar. When the drawdown height of water level is 90 m, the minimum safety factor (the most dangerous hydraulic condition) occurred during

the descending period of the water level, but not at the highest or the lowest height. For slope masses of the same materials, if water level draws down faster, the time to reach the minimum safety factor is shorter. Moreover, at the same drawdown height of water level, if water level draws down faster, the safety factor of reservoir bank will be smaller, and landslides might occur more possibly. It indicates that the time to draw down reservoir water level should be extended as long as possible, and the sudden drawdown of water level should be avoided.

Acknowledgment

This paper is supported by the national nature science fund of China (50379043) and open fund of key laboratory of Wu Han rock and soil mechanic of Chinese academy of science (z110302).

References

Bishop AW (1955) The use of the slip circle in the stability analysis of slope. Geotechnique 5:7–17

Cui Z, Li N (1999) Slope engineering: recent advances in theory and practice. China WaterPower Press, Beijing

Duncan JM (1996) State of the art: limit equilibrium and finite element analysis of slopes. J Geotech Eng-ASCE 22:577–596

Fredlund DG, Rahardjo H (1993) Soil mechanics for unsaturated soils. John Wiley & Sons, Inc.

Ho PG (1979) The prediction of hydraulic conductivity from soil moisture suction relationship. B.Sc. Thesis, University of Saskatchewan, Saskatoon, Canada

Integrating reconnaissance bureau of Yangtze River hydraulic committee (1999) Geological engineering reconnaissance report of BaiYi'An slide in managing and programming phase in Fengjie County, Three Gorges area

Lam L, Fredlund DG (1993) A general limit equilibrium model for three-dimensional slope stability analysis. Can Geotech J 30: 905–919

Mao C, Duan XB, Li ZY (1981) Numerical calculation and analysis in seepage. HoHai University Press, Nanjing, China

Morgenstern NR, Price VE (1965) The analysis of the stability of general slip surface. Geotechnique 15(1):79–93

Nakamura K (1990) On reservoir slide. Bull Soil Water Conserv 10(1):53–64

Pan JZ (1980) Stability against sliding of building and landslide analysis. Hydraulic Press, Beijing

Segerlind LJ (1984) Applied finite element analysis. John Wiley & Sons, Inc.

Wang YB (2001) Applied standard criterion in civil engineering. Xi'an Map Press, Xi'an

Yangtze Rive Reconnaissance, Programming and Designing Academe (2000) Feasibility research report of retaining wall engineering at new Fengjie City Zone in Three Gorges area

Zhu DL, Ren GM, Nie DX, Ge XR (2002) Effecting and forecasting of landslide stability with the change of reservoir water level. Hydrogeology and Engineering Geology 3:6–9

Chapter 40

Displacement Monitoring and Physical Exploration on the Shuping Landslide Reactivated by Impoundment of the Three Gorges Reservoir, China

Fawu Wang* · Gonghui Wang · Kyoji Sassa · Atsuo Takeuchi · Kiminori Araiba · Yeming Zhang · Xuanming Peng

Abstract. The Three Gorges Dam construction on the Yangtze River in China is the largest hydro-electricity project in the world. After the first impoundment in June 2003, many landslides occurred or reactivated. Shuping landslide is one of the most active landslides among them. In this paper, the deformation of the Shuping landslide monitored by GPS, extensometers, and crack measurements are summarized. Also, for the investigation of the groundwater situation, 1 m depth ground temperature measurement was conducted, and the groundwater veins were estimated. Based on the monitoring data and exploration results, a deformation model of the landslide caused by impoundment of reservoir was proposed.

Keywords. Landslide, Three Gorges, impoundment, monitoring, groundwater

40.1 Introduction

The Three Gorges Dam construction on the Yangtze River in China is the largest hydro-electricity project in the world. The dam site is located at Sandouping Village near Maoping Town, the capital of Zigui County, Hubei Province. The designed final dam height is 185 m, the final length 2 309.5 m, and the designed final highest water level 175 m. When dam construction is finished, the Three-Gorge Reservoir will reach Chongqing City, about 660 km upstream from the dam. The first impoundment started from 95 m on 1 June 2003, and reached 135 m on 15 June 2003. As soon as the water level reached 135 m, many slopes began to deform and some landslides occurred (Wang et al. 2004). For example, in the early morning at 00:20 14 July 2003, Qianjiangping landslide occurred in Shazhenxi Town (Fig. 40.1) at the bank of Qinggan-he River, a tributary of the Yangtze River (Zhang et al. 2004a).

Aiming to study the influence of impoundment on landslide deformation, we selected Shuping landslide, which is located at the main stream of the Yangtze River (Fig. 40.1) in Shazhenxi Town, just about 3.5 km from the Qianjiangping landslide, as our research and monitoring field. Figure 40.2 is an oblique photograph of the Shuping landslide, and Fig. 40.3 is the plane of the landslide. The landslide ranged its elevation from 65 m to 500 m. Its width was about 650 m, the estimated thickness of the sliding mass was 40 m to 70 m according to the bore hole data, and the total volume was estimated as 2.0×10^7 m^3. The toe part of the landslide was under the water level of the Yangtze River. The slope is gentle at the upper part and steep at the lower part with a slope angle of 22 degrees and 35 degrees respectively.

40.2 Features of the Shuping Landslide

The Shuping landslide is an old landslide which composed of two blocks. This can be confirmed in the photograph (Fig. 40.2). After the first impoundment of the Three Gorges Reservoir ended on 15 June 2003, obvious deformation phenomenon appeared at the slope, and it became intense from 8 February 2004. Also, the two blocks shows different deformation rate at slope surface. The serious de-

Fig. 40.1. Location map of the Shuping and Qianjiangping landslide in Three Gorge Water Reservoir area, Hubei Province

Fig. 40.2. Shuping landslide consisting of two blocks at the main stream of the Three Gorges Water Reservoir

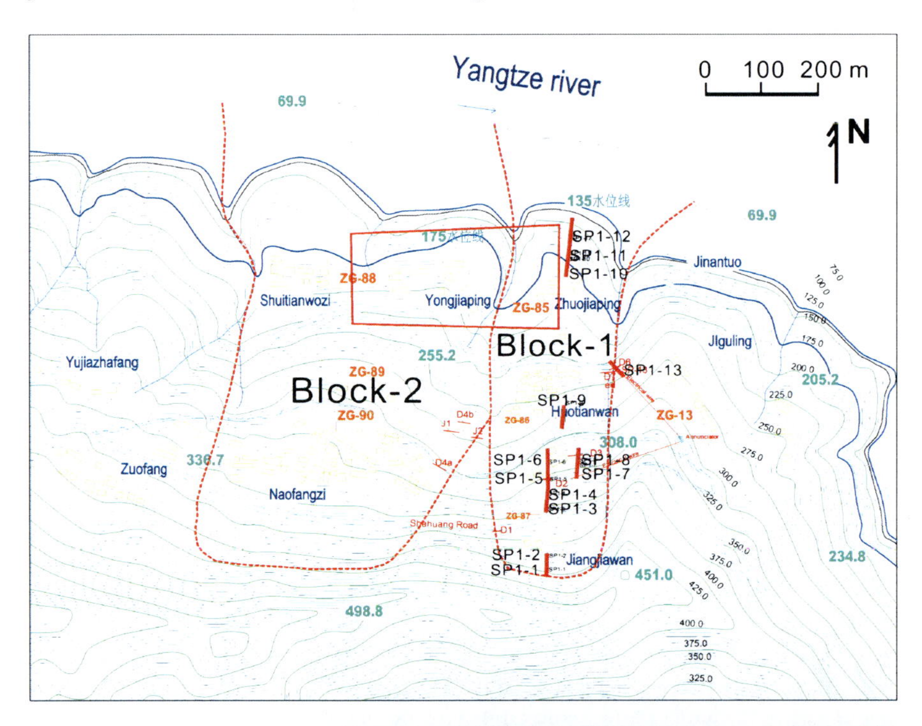

Fig. 40.3. Plan of the Shuping landslide and locations of monitoring and measurement works (locations of extensometers and measurement area of 1 m depth ground temperature are shown)

formation situation made 580 inhabitants and 163 houses in danger directly, and all of the inhabitants were asked to live in the disaster prevention tents which were provided by the central government. Until May 2004, most inhabitants moved their houses out of the landslide area.

Figure 40.4 shows a crack at the right boundary of block 1 outcropped at a local road. The right-hand side is the sliding mass consisting of red muddy debris of old landslide, and the left-hand side is bed-rock of sandy mudstone, muddy siltstone of Badong Formation of Triassic period (T3b).

Figure 40.5 shows muddy water coming from block 1. It appears at the river even at continual sunny days showing no relationship with surface water erosion, but the underground water erosion from the inner part of the landslide.

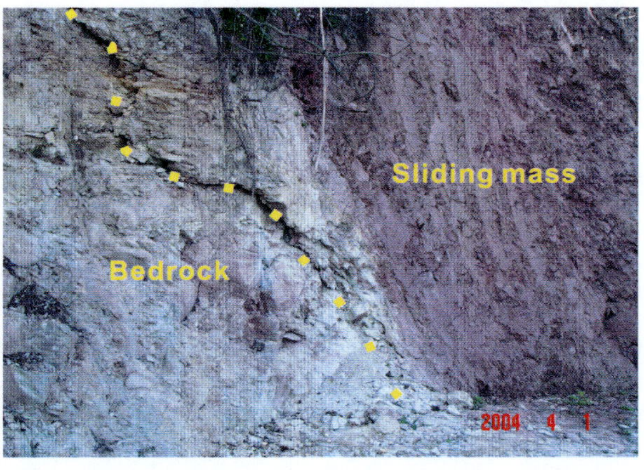

Fig. 40.4. Crack at the right boundary of block 1 outcropped at a roadside

Fig. 40.5. Muddy water seeped out of the toe part of the Shuping landslide

40.3 Slope Deformation Characters of the Shuping Landslide

Because the landslide area is a densely populated area and its active deformation occurred just after the occurrence of the Qianjiangping landslide, the deformation situation was observed by the inhabitants and reported to the local government promptly. According to the urgent investigation report (Gan et al. 2004), the typical deformation behaviors were recorded as follows.

From the end of October to the beginning of November 2003: Cracks became obvious at the slope surface, especially at the upper part. These cracks were enlarged from January to February 2004.

On 5 January and 8 February 2004: The water at the toe part of the landslide became very muddy. From March, the muddy water appeared almost everyday. Figure 40.5 shows the muddy water situation in April 2004. This phenomenon is a very dangerous sign for slope deformation, because it may mean that the newly sheared soil at the sliding surface was eroded by underground water gradually.

On 25 January and 8 February 2004: Sharp noises coming from underground were heard by the inhabitants at night for two times. The noise is possibly caused by shearing at the sliding zone.

Because of the serious deformation situation, local government decided to monitor the cracks distributing in the slope from 12 February 2004. The inhabitants were asked to measure the width change of the cracks near their houses (see Fig. 40.3). For measurement, two small piles were set across a crack, and the distance between the two piles was measured three times one day. Figure 40.6 shows a part of the measured results of the cracks. However, because the inhabitants are moving out of this area, the measurement points are decreasing gradually.

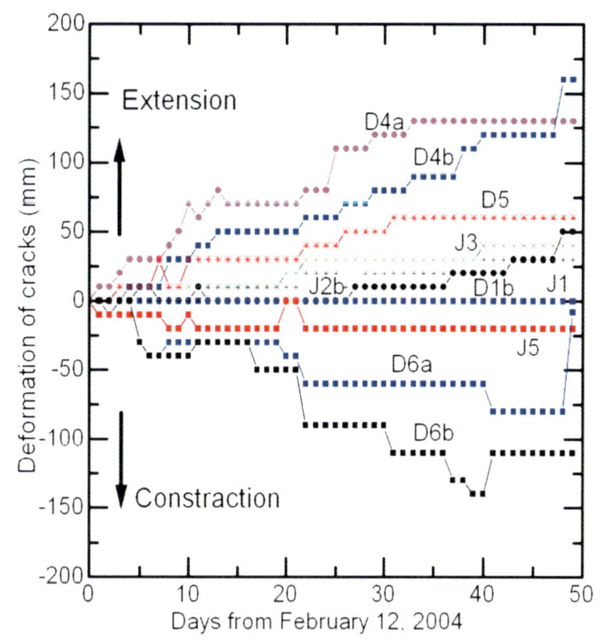

Fig. 40.6. Measured results of crack deformation in block 1

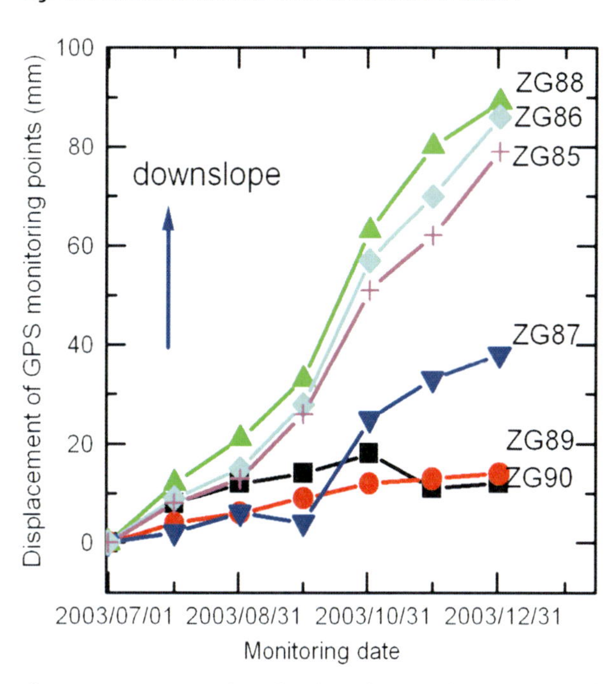

Fig. 40.7. GPS monitored results along the central section lines of block 1 and block 2

Roughly, the measurement results of the crack deformation shows that the extension behavior occurred at the boundary and inside landslide, and contraction deformation occurred at the other parts in the landslide block. For the 50 days period from 24 February 2004, the maximum displacement including extension and contraction respectively reached about 140 mm, showing an active deformation.

40.4 GPS Monitoring Results

Two GPS monitoring lines were arranged at the central longitudinal section of the two blocks by China Geological Survey. Each monitoring line has three GPS monitoring points, i.e., ZG85, ZG86 and ZG87 from toe to upper part in block I, and ZG88, ZG89 and ZG90 from toe to upper part in block II (see Fig. 40.3). The monitoring started in July 2003, just one month after the first impoundment. The measurements were conducted one time each month by Rockfall and Landslide Research Institute of Hubei Province.

Figure 40.7 shows the monitored results of the GPS monitoring at the first six months after the impoundment. The displacement rate of block I increased rapidly after October 2003, and other two tendencies are also very clear. (1) The displacement at the lower part is larger than that at the upper part, this may be caused by water buoyant of the impoundment of the reservoir; (2) The displacement of block I is more active than block II, showing that the two blocks are independent from each other.

Fig. 40.8. Extensometer installed in Shuping landslide. The deformation is recorded on the rolled paper and also saved in a memory (visible in the right side)

40.5 Installation of Extensometer and the Monitoring Results

Until April 2004, the displacement monitoring of the Shuping landslides included crack measures conducted three times each day, and the GPS monitoring conducted one time each month. However, because of the evacuation of the inhabitants, the crack monitoring was interrupted gradually. Located at the main stream of the Yangtze River,

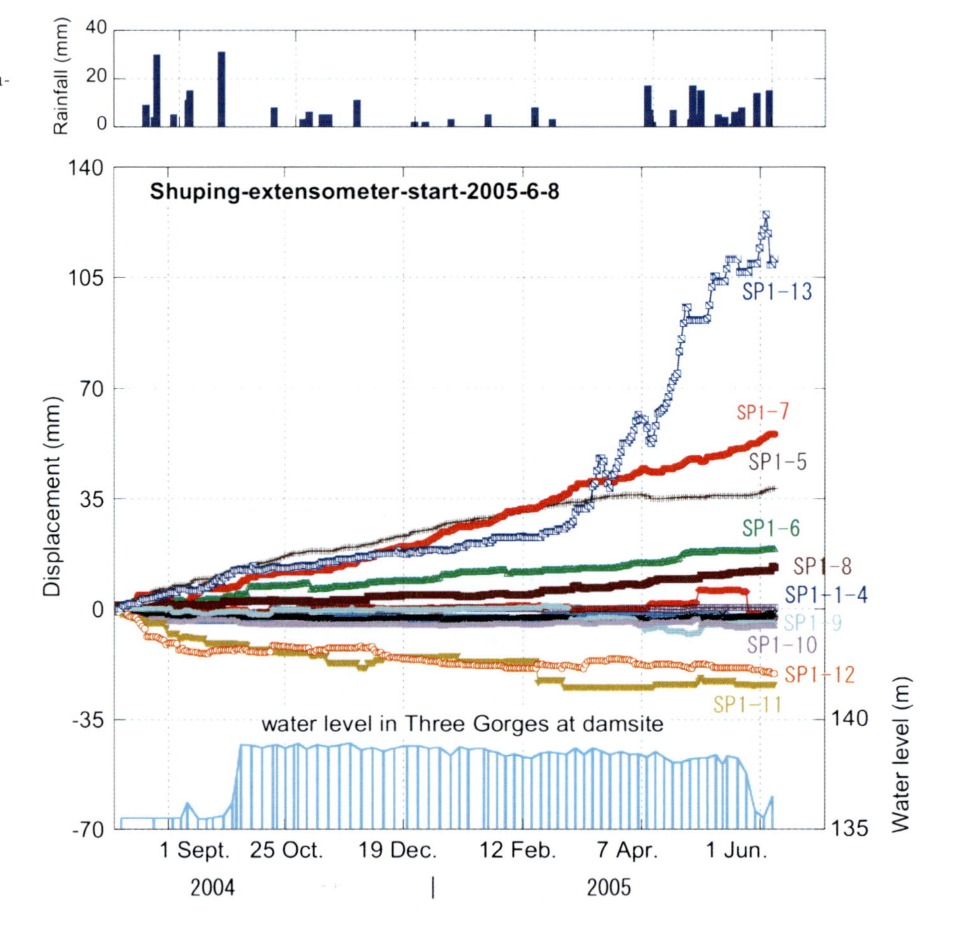

Fig. 40.9. Monitored results of the extensometers (*middle*), the precipitation data in Yichang City (*top*) and water level in the Three Gorges Dam site (*bottom*)

it is not enough for the Shuping landslide to be monitored only with GPS when considering the safety of shipping along the Yangtze River. Although GPS monitoring has high precision, the time interval of measurement is too large. Facing this situation, two extensometers donated by Kowa Co. LTD., a Japanese company, were installed in the block I of the Shuping landslide in April 2004. The extensometer is Sakata Denki style. The monitoring can continue for one week or one month automatically. A warning system is also connected with the extensometer. When the displacement rate exceeds 2 mm h^{-1}, a warning will be announced.

The automatic monitoring with the extensometer was confirmed to be working well (Zhang et al. 2004b). However, two extensometers are not enough for such a large landslide. In August 2004, another 11 extensometers were installed along the central line of the longitudinal section of block 1, with emphasizing on the serious deformation parts. Also because of the limit of funds, the extensometers cannot form a continual longitudinal section line.

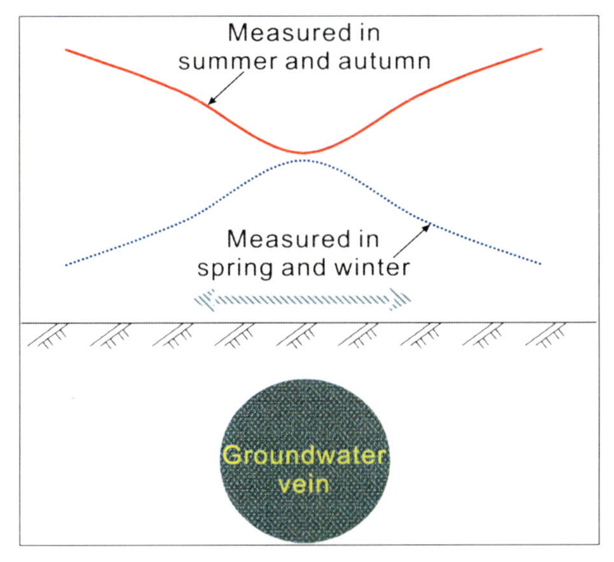

Fig. 40.10. Principle of the 1 m depth ground temperature measurement

Figure 40.8 shows the extensometer installed in Shuping landslide. The positions of all of the thirteen extensometers were shown in Fig. 40.3 as "SP1-x". Among them, SP1-1 and SP1-2 were set across the main scarp; SP1-3 to SP1-6 were set below the Shahuang road which has a high traffic. SP1-7 and SP1-8 were set almost parallel with SP1-5 and SP1-6. SP1-9 was set at the low part. Then, SP1-10, SP1-11, and SP1-12 were set near the Yangtze River at the toe part of the landslide. SP1-13 was set at the right boundary of block 1 shown in Fig. 40.4, because the crack extension is obvious.

Figure 40.9 is the monitoring results of all the thirteen extensometers from August 2004 to June 2005, companying with the water level in the Three Gorges at dam site showing at the bottom, and the rainfall records of this area showing at the top.

The monitoring results show some tendencies of the landslide displacement. (1) The SP1-1 and SP1-2 at the main scarp did not record obvious displacement. One possible reason is that the setting positions of the two extensometers did not cross the main scarp. (2) The deformations at SP1-5, SP1-6, SP1-7 and SP1-8 were the most active ones showing extension. Because of local failure, the SP1-13 showed extremely great extension. (3) The toe part of the landslide showed compression behavior. Comparing with the largest displacement at the lower part by GPS monitoring in the first six months after the first impoundment, it may estimate that the toe part moved down faster at the first period and became silent now; the upper part moved slowly at the first stage, and now followed the movement of the lower part and compressed the lower part. An exact examination will be conducted with the

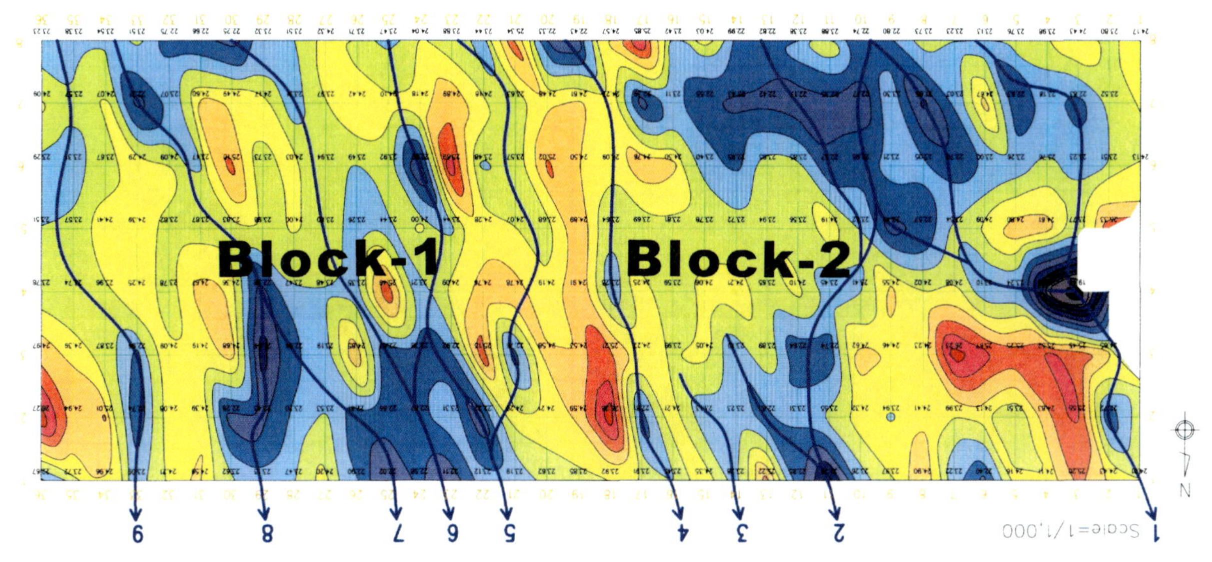

Fig. 40.11. Groundwater veins distribution estimated from 1 m ground temperature measurement

comparison of the recent GPS monitoring data along the central longitudinal section, which is not open currently.

Another important tendency was recorded in the monitoring results. In mid-September, the water level of the Three Gorges was raised for about 3 m, corresponding to this water level raising, the displacement velocity of SP1-5, SP1-6 and SP1-7 in-creased obviously, reflecting the influence of the impoundment on the slope deformation. Also, during the rainy season after April 2005, displacement acceleration tendency at some point (SP1-6, 7, 8) was monitored, showing the displacement of the landslide is also influenced by groundwater conditions.

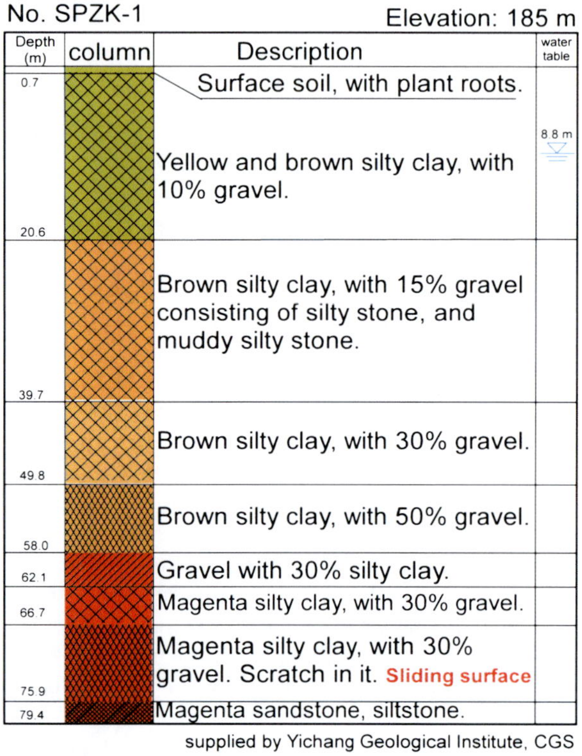
No. SPZK-1
Elevation: 185 m
Depth (m)
column
Description
water table
0.7
Surface soil, with plant roots.
8.8 m
Yellow and brown silty clay, with 10% gravel.
20.6
Brown silty clay, with 15% gravel consisting of silty stone, and muddy silty stone.
39.7
Brown silty clay, with 30% gravel.
49.8
Brown silty clay, with 50% gravel.
58.0
Gravel with 30% silty clay.
62.1
Magenta silty clay, with 30% gravel.
66.7
Magenta silty clay, with 30% gravel. Scratch in it. Sliding surface
75.9
79.4
Magenta sandstone, siltstone.
supplied by Yichang Geological Institute, CGS

Fig. 40.12. Column diagram of borehole SPZK-1 in block 1 of Shuping landslide

40.6 One Meter Depth Ground Temperature Measurement for Groundwater Veins

Takeuchi (1972) developed a method for the investigation of groundwater veins through 1 m depth ground temperature measurement. This method is widely applied in the practice of groundwater exploration in landslide area, especially in Japan (Takeuchi 1996).

Figure 40.10 shows the principle of the 1 m depth ground temperature measurement for groundwater vein. Comparing with the ground without water, the ground with water always has a temperature similar to that of the groundwater. Generally, groundwater temperature does not change so much around a year. While, the temperature in the ground without water is controlled by the atmosphere temperature, which will be higher than groundwater in summer and autumn, and lower in spring and winter. Through measuring the temperature distribution in an area, the distribution situation of groundwater vein can be estimated.

One meter depth ground temperature measurement was conducted to detect the groundwater veins in the lower part of the Shuping landslide. The measured area was shown as a square in Fig. 40.3. Figure 40.11 is the measured results. From the ground temperature distribution, two independent groundwater vein groups were estimated existing in block 1 and block 2. The exit of the groundwater vein in block 1 is lower that that in block 2. It is estimated that the groundwater veins no. 8 and no. 9 correspond to the muddy water seeped from block 1.

To confirm the above estimation, and to explore the sliding surface, a borehole drilling was conducted at SPZK-1, as shown in Fig. 40.11. Figure 40.12 is the column diagram of

Fig. 40.13. Longitudinal section of the Shuping landslide

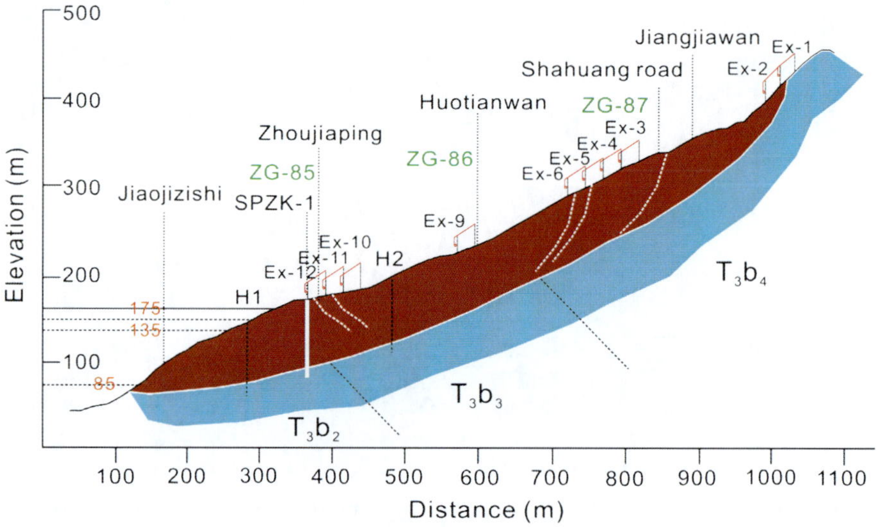
500
400
300
200
100
Elevation (m)
Jiangjiawan
Shahuang road
Huotianwan
Zhoujiaping
Jiaojizishi
SPZK-1
ZG-85
ZG-86
ZG-87
Ex-1
Ex-2
Ex-3
Ex-4
Ex-5
Ex-6
Ex-9
Ex-10
Ex-11
Ex-12
H1
H2
175
135
85
T_3b_2
T_3b_3
T_3b_4
100 200 300 400 500 600 700 800 900 1000 1100
Distance (m)

this borehole. The groundwater table was found at 8.8 m depth, and the sliding zone formed at the depth between 66.7 and 75.9 m. The sliding zone consisted of silty clay with 30% gravel. Scratches caused by sliding are rich in the zone.

40.7 Summaries and Conclusive Remarks

Through the GPS monitoring, crack displacement measurements, extensometer monitoring along the longitudinal section of block 1, the deformation style of the block 1 of Shuping landslide can be sketched as Fig. 40.13. Sooner after the impoundment of the water reservoir, the toe part displaced downward faster than the upper part. In the current stage, two years after the first impoundment, the slope deformation style changed. The displacement of the lower part almost terminated while the upper part displaced downward gradually, and compressed the toe part.

For the Shuping landslide, reactivating from an old landslide and with rich groundwater in it, the influencing factors on the slope deformation is complicated. From about one year monitoring with the extensometers, it is very clear that the slope displaced soon after the impoundment of the water reservoir. It is very important to keep the monitoring continued especially during the next stage of impoundment which will be conducted in June 2006 (water level will be raised from 139 m to 156 m).

Acknowledgments

Deep thanks are given to Mr. Masahiro NAGUMO in Kowa Co. LTD., Japan for donation of two extensometers. The local government of Shazhenxi supplied measurement results of the crack displacement. The research fund from Sabo Technical Center, Japan is highly appreciated.

References

Gan YB, Sun RX, Zhong YQ, Liao SY (2004) Urgent investigation report on the Shuping landslide in Shazhenxi Town, Zigui County, Hubei Province. (in Chinese)

Takeuchi A (1972) Ground temperature measurement in landslide area. Journal of Japan Landslide Society 8(4):29–37 (in Japanese)

Takeuchi A (1996) Flowing groundwater investigation by temperature measurement. Kokon Syoin Press (in Japanese)

Wang FW, Zhang YM, Huo ZT, Matsumoto T, Huang BL (2004) The July 14, 2003 Qianjiangping landslide, Three Gorges Reservoir, China. Landslide 1(2):157–162

Zhang YM, Liu GR, Chang H, Huang BL, Pan W (2004a) Tectonic analysis on the Qianjiangping landslide in Three Gorges Reservoir area and a revelation. Yangtze River 35(9):24–26 (in Chinese)

Zhang YM, Peng XM, Wang FW, Huo ZT, Huang BL (2004b) Current status and challenge of landslide monitoring in Three Gorge Reservoir area, China. Proc. of Symp. on Application of realtime information in disaster management, JSCE, pp 165–170

Chapter 41

Capacity Enhancement for Landslide Impact Mitigation in Central America

Oddvar Kjekstad* · Farrokh Nadim

Abstract. Central America is a region plagued with different kinds of natural disasters, such as earthquakes, floods, landslides, volcanic eruptions and droughts. Following the Hurricane Mitch Disaster in 1998, which set many of the Central America countries back 10 years in economic development, the Government of Norway has supported a number of project to reduce the catastrophic consequences of future events.

The paper summarizes the experience with two Norwegian-supported institutional cooperation programs for strengthening the capacity in Nicaragua and El Salvador to deal with the landslide hazard. A regional training program with participation from 6 Central America countries is also highlighted. This program is executed by CEPREDENAC with funding from Norway.

Keywords. Landslide mitigation, capacity building, Central America, institutional strengthening, natural hazards, training programs

41.1 Introduction

Geographical location and geological conditions make Central America highly prone to natural hazards. These include earthquakes, floods, landslides, as well as volcanic eruptions. Quite often these hazards turn into major disasters as high level of poverty, uncontrolled urban growth, lack of proper land use planning and disaster preparedness plans make the population extra vulnerable. A recent study carried out by the ProVention Consortium of the World Bank (Dilley et al. 2005) concluded, for instance, that Guatemala is among the countries that are most exposed to multiple natural hazards. Altogether 40% of the population is exposed to these hazards and the area affected amounts to 21% of the county's land area.

Major disasters in Central America in recent years are the Hurricane Mitch in 1998 that hit Nicaragua, Honduras and El Salvador strongly; and the earthquakes in El Salvador in January and February of 2001. Following these disasters the Norwegian Government supported several projects in Central America with the aim of reducing the likelihood that new events should have such catastrophic consequences. The experience with the capacity building programs on landslide risk mitigation are summarized below.

41.2 Landslides and Regional Setting

Geology and climate in Central America contribute to the prevalence of landslides. Steep slopes, rapid weathering, large deposits of unconsolidated soils, cultivated land with minor forest cover combined with intense storms in the wet season, serve as major factors for triggering landslides in the region. The Casita landslide (Fig. 41.1), represents the largest landslide that has happened in the region in recent years.

Large earthquakes are also a major trigger of landslides. The magnitude 7.8 Guatemala City earthquake in 1976 triggered more than 2 000 registered landslides and the El Salvador January ($M = 7.7$) and February ($M = 6.7$) earthquakes in 2001 triggered more than 500 registered landslides. The most dramatic of these earthquake-induced landslides was the one that occurred at Las Colinas, near the capital city of San Salvador (Fig. 41.2). In the recent "Natural Disaster Hotspots – A Global Risk Analysis" study carried out by the Prevention Consortium of the World Bank (Nadim et al. 2005), it was found that parts of Guatemala, Honduras, El Salvador, Costa Rica and Panama rank among the highest in the world in terms of the annual risk of people getting killed by landslides.

Fig. 41.1. Casita landslide, Nicaragua 1998, triggered by heavy rain during the hurricane Mitch. The landside killed more than 2 500 people

Fig. 41.2. Las Colinas landslide in El Salvador, January 2001, triggered by an earthquake (magnitude = 7.7). The landslide caused more than 600 fatalities

Fig. 41.3. The Pan America Highway in El Salvador was badly damaged by a landslide during the January 2001 earthquake

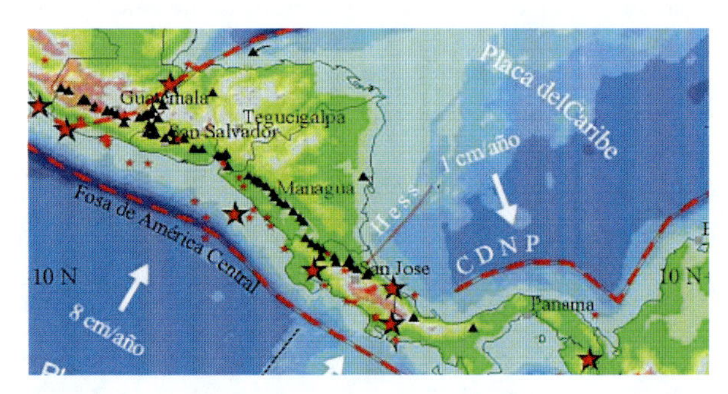

Fig. 41.4. Central America with the chain of volcanoes shown by *black triangles*

An important feature of the region is the chain of volcanoes, approximately parallel to the Pacific coast, extending all the way from Guatemala-Mexico border down to Costa Rica. (Fig. 41.4). The volcano slopes are of major importance for cultivation of coffee plants, and a significant part of the population in the region live in towns and cities built in the shadow of the volcanoes. Many of these settlements are highly exposed to landslides. Landslides triggered by heavy rain have their peak intensity in May at the start of the wet season, and in September and October when the effect of antecedent rain becomes an important factor. The average annual rainfall decreases from about 4 000 mm in the south to 2 000 mm in the North, but it shows great local variation. At some mountain locations in Costa Rica the annual rainfall averages 6 000 mm per year.

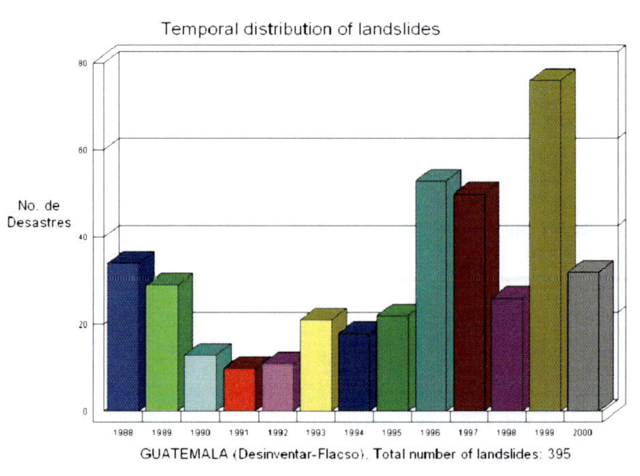

Fig. 41.5. Number of registered landslide events in Guatemala, derived from DesInventar-Flasco

Typical economic consequences of landslides in the region are damaged houses, blocked roads (Fig. 41.3) and loss of valuable agricultural land. During the El Salvador earthquake in 2001, it is reported that 20% of the country's coffee plants were damaged (World Bank 2001).

Usually, it is the poor group of the population that suffer most from the landslides. For instance in Tegucigalpa, the capital city of Honduras, it is reported that about 20% of the population have suffered damage from landslides during the last 5 years (Fay et al. 2003).

Unfortunately inventories for historic landslide events and their consequences in the Central American countries are incomplete and partly missing (King 1989; Stuart 1989; Rymer and Randall 1989; Mora 1989). The international natural disaster databases (EMDAT-CRED 2005; DesInventar/LA RED 2005) are also somewhat incomplete. The landslide profile for Guatemala derived from the database DesInventar-Flasco is shown on Fig. 41.5 and Fig. 41.6 in terms of reported number of events and causalities per year in the period 1988–2000. There are reasons to believe that the profile is somewhat similar for Costa Rica, El Salvador and Honduras. The number of events and their consequences are likely to be less for Nicaragua and Panama.

It is, however, clear that the recorded landslide data in the databases are grossly underestimated. The reason is that losses from many landslide events are recorded as losses from earthquake and floods when they have been the primary triggering mechanism for the landslides.

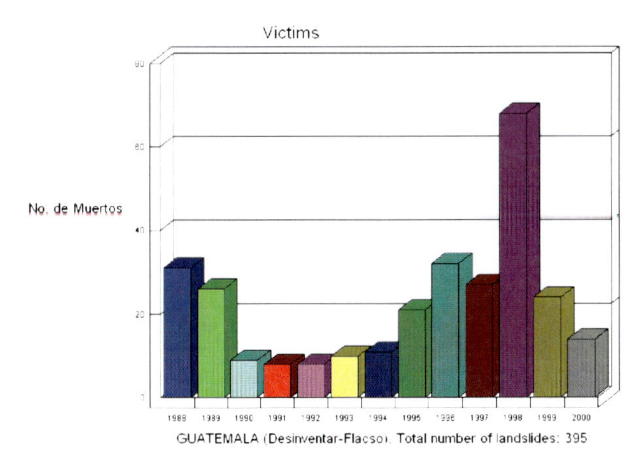

Fig. 41.6. Number of people reported killed by landslides in Guatemala, derived from DesInventar-Flasco

41.3 Institutional Structure for Natural Disaster Prevention, Preparedness and Response

In the period after the Hurricane Mitch Disaster, many governments in the region took initiatives to strengthen their institutional structure to deal with the management of risk caused by natural hazard, both on technical and operational matters. In Nicaragua, the Nicaraguan Institute for Territorial Studies, INETER, which is the national center for all technical matters related to natural hazards, decided to build up a new landslide division. Almost at the same time the national secretariat for operational management of natural disasters, SINAPRED (Secretaria Ejecutiva del Sistema Nacional para la Prevencion, Mitigacion y Atencion de Desastres) was established.

In El Salvador, the earthquake disasters in 2001 led the government to establish a new national center, SNET (Servicio Nacional de Estudios Territorales) to deal with natural hazards, including the landslides. The function of this center is somewhat similar to INETER in Nicaragua. The operational unit responsible for emergency in El Salvador is the National Emergency Committee (COEN).

Guatemala, Honduras and Panama have somewhat similar institutional structures dealing with natural disaster prevention and response. The entities on national level are typically:

a A national center, dealing with the technical aspects of natural disasters, often supported by the geological and civil engineering departments of the national universities.
b A governmental umbrella organization in charge of the emergency situations.
c A Civil Defence unit which often coordinates the practical aspects of evacuation of people when necessary in critical situations.

The governments of the Central America countries realized already in the '90s that they had much in common in the efforts to reduce the consequences of natural hazards. This was the basis for the establishment of CEPREDENAC (Centro de Coordinacion para la Prevencion de los Desastres Naturales en America Central). This secretariat, now located in Guatemala City, undertakes a series of regional programs in risk prevention and mitigation.

41.4 International Support

The Hurrican Mitch Disaster in 1998 was an eye-opener for the international community. The inflicted losses from this disaster, which totalled more than 10% of the Central American GDP for 1998, led to a number of international support programs. These included support for reconstruction, assistance for better risk identification and disaster mitigation measures, including capacity building.

Within the area of landslide mitigation, the countries that gave support were U.S.A., Japan, Germany, Norway, the Czech Republic, Switzerland, Sweden and Spain. The work supported by these countries have included geological mapping, assistance for landslide hazard and risk mapping both on national and on municipality levels, establishment of GIS models for recording of landslide inventories, supply of instrumentation, as well as capacity building efforts. In addition comes the reconstruction and natural disaster mitigation programs undertaken by the World Bank, the Inter-American Development Bank and the European Commission.

41.5 Norwegian-Supported Institutional Cooperation Programs

With the newly-established landslide division within INETER in Nicaragua, NGI was called upon in 2001 to assist in a 3-year institutional cooperation program under the name of "measures to prevent landslides". The program was funded by the Norwegian Government, as it was realized that the growth of this division would not be fast enough without international support.

Major elements in the INETER-NGI institutional cooperation program included:

- Investigations of the landslide hazard on 4 volcano slopes: Cristobal, Casita, Mombacho and Concepcion.
- Build-up and installation of a pilot early warning system for landslides based on critical values for rainfall that can trigger landslides.
- Investigations of mechanical models that can explain triggering and sliding mechanisms for typical landslides in the country.
- Training of the INETER staff in Norway and Nicaragua.

All activities were carried out as joint work between INETER and NGI personnel both in terms of field work and desk studies in the office. The basic philosophy behind the program was that capacity building is best achieved using application-oriented approaches where the challenge is to find solutions to national problems.

Fig. 41.7. Meteorological station installed in the Concepcion Volcano in Nicaragua, forming a part of a pilot landslide early warning system

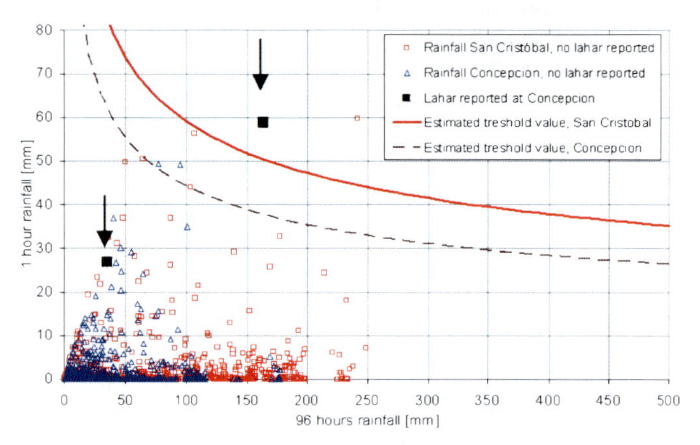

Fig. 41.8. Rainfall in Nicaragua measured from October 2001 until January 2003. The lahar events at Concepcion Volcano May 2002 are indicated together with estimated threshold rainfall values for triggering of lahars in the region (Heyerdal et al. 2003)

The project also included a pilot project of a landslide early warning system. Automatic meteorological stations were installed on slopes of 4 different volcanoes, with satellite transmission of data to the INETER office in Managua, Fig. 41.7. In May 2002 the Concepcion Volcano slope experienced a severe rainfall situation that triggered a major lahar. As can be seen from the plot on Fig. 41.8, this event fitted fairly well with the estimated threshold values for warning that were initially set when the system was installed.

A part of the work program for NGI was also to make recommendations for how INETER could set priorities for their work in the years to come.

The institutional cooperation program between SNET in El Salvador and NGI, executed in the period 2002–2004, had a somewhat similar design compared to the one in Nicaragua. The landslide hazard evaluation and proposal for mitigation measures were limited to one volcano slope on the Northwestern part of San Vicente. In this area more than 3 000 people were considered to be exposed to high landslide risk. The program in El Salvador also included a pilot project for landslide early warning, with one of the stations located at elevation 850 m on the San Vicente Volcano slope. This station proved to be very important when the Hurricane Adrian struck El Salvador in May 2005. Based on recorded rainfall from this station, SNET issued several warning bulletins as Adrian swept over the country.

Another activity in the institutional cooperation program in El Salvador was the review of national building codes and construction regulations to see if the rather high landslide risk profile for the country was adequately taken care of. The review identified several areas for improvements.

41.6 Norwegian-Supported Regional Training Program

Dialogue between the Norwegian Development Agency (NORAD) and CEPREDENAC led to an initiative to start a regional training program on the management of risks caused by landslides. The program, which started in early 2005, runs over 4 years. It is referred to as RECLAIMM, Capacity Building on the Management of Risks caused by Landslides in the Central America Countries. The rationale for the initiative was that most of the countries in Central America are exposed to the same type of landslide problems and have to deal with the same challenges when implementing mitigation measures, which justified a regional approach.

The objectives of the program are as follows:

- improve the knowledge and skills of professionals from a selection of relevant organizations in the region who are dealing with landslide hazards,

- create a forum and network where representatives from the Central America countries can exchange experience, derive common methodologies and assist each other on practical issues when needed,
- form mechanisms that secure dissemination of knowledge and methodologies generated in the capacity building program to a larger audience.

NGI was selected by CEPREDENAC to serve as the technical executing organization for the program, which had its first session in San Jose, Costa Rica in April 2005. Altogether 20 candidates from Nicaragua, El Salvador, Honduras, Guatemala, Costa Rica and Panama were invited to take part in the program. They represented a variety of organizations such as national technical centers, civil defence, universities, national emergency organizations and urban planners from city authorities. The philosophy behind this mix of participants is to create a multidisciplinary forum where the challenges of landslide mitigation measures can be discussed in a broad setting.

RECLAIMM will have a one-week training session once a year over a 4-year period. Key elements in the one week training programs are: knowledge exchange based on real case histories that the participants bring with them to the session, field work including site visits to landslide areas for discussion of appropriate mitigation measures (Fig. 41.9), guest lectures and discussion of best practice on selected subjects.

In the period between the yearly sessions, the participants are organized in work groups dealing with selected common subjects related to mitigation methodologies. During the first training session in Costa Rica, the following 4 work groups were formed: *(a)* Methodologies for hazard and risk assessment, *(b)* Early warning systems and monitoring, *(c)* Land use planning and regulations, and *(d)* Prevention and mitigation measures. The outcome of the work in the working groups will serve as source material for discussions in the 2006 training session. Another feature to the yearly one-week sessions, is an additional one-day event in form of a "National Dissemination Seminar", where a broad representation of national organizations and representatives from the international natural disaster agencies are invited.

41.7 Lessons Learned

Institutional cooperation and capacity building are challenging activities that are very well received in the region. There is a considerable amount of local knowledge on how to deal with natural hazards, and some of the regional organizations are highly advanced when it comes to use of GIS. Identification, appreciation and utilization of the existing local knowledge are fundamental factors for achieving good progress in capacity building. Having this in place, external international assistance can serve more as facilitators than a top down teaching body.

The experience gained in the above described programs indicates that international assistance is helpful in the following ways:

- Increase the knowledge level for how to apply practical landslide mitigation measures.
- Encourage the responsible organizations for landslide mitigation measures to be more proactive rather than reactive.
- Serve as mediators to bring together the different national stakeholders, which do not necessarily communicate frequently.

Fig. 41.9. The RECLAIMM participants taking part in training program in San Jose, Costa Rica in April 2005, discuss practical methods for protecting the city of Orosi from future landslide disasters

- Contribute to create regional networks for sharing of problems and solutions across country borders in the region.
- Provide a link to international professional entities.

There is a heavy burden placed on the national organizations in charge of the technical aspects for natural hazards. To serve municipalities and city authorities properly, they need for instance to be fully updated on sub-national initiatives on hazard and risk mapping. In the period just after Mitch, many assistance programs were linked directly to municipalities without national coordination. This created some problems, but practice has now been improved. The experience is that national technical organizations in charge of natural hazards must be involved and continuously updated on sub-national initiative on hazard identification and assessment.

Acknowledgment

The authors would like to acknowledge the support from the Norwegian Government by NORAD (the Norwegian development Agency) and by the Ministry of Foreign Affairs for the initiatives which have been taken in Central America regarding natural disaster mitigation. Special thank goes also to CEPREDENAC, INETER and SNET for a fruitful collaboration over many years.

References

Bommer J, Rodriguez C (2002) Earthquake-induced landslides in Central America. Eng Geol 63:189–220

EMDAT-CRED (2005) Centre of Research on Epideminology of Disasters in Brussels, http://www.cred.bc

DesInventar/LA RED (2005) www.desinventar.org

Dilley M, Chen RS, Deichman U, Lerner-Lam AL, Arnold M, Agwe J, Kjek-stad O, Lyon B, Yetman G (2005) Natural disaster hotspots, a global risk analysis. The World Bank, Disaster Risk management Series no. 5

Fay M, Ghesquiere F, Solo T (2003) Natural disasters and the urban poor. En Breve no. 32, Octobre 2003, The World Bank

Heyerdal H, et al. (2003) Rainfall induced lahars in volcanic debris in Nicaragua and El Salvador. Proceedings International Conference on Fast Slope Movements – Prediction and Prevention for Risk Mitigation, Sorrento, May 2003

King AP (1989) Landslides, extent and economic significance in Honduras. In: Brabb EE, Harrods BL (eds) Landslides: extent and economic significance. Balkema

Mora S (1989) Extent and socio-economic significance of slope instabilities Costa Rica. In: Brabb EE, Harrods BL (eds) Landslides: extent and economic significance. Balkema

Nadim F, Kjekstad O, Peduzzi P (2005), Assessment of global landslides hazard and risk hotspots. Proceedings of the sixteenth International Conference on Soil Mechanics and Foundation Engineering, Osaka 2005

Rymer MJ, White RA (1989) Hazards in El Salvador from earthquake induced landslides. In: Brabb EE, Harrods BL (eds) Landslides: extent and economic significance. Balkema

Steward RH, Steward JL (1989) Slides in Panama. In: Brabb EE, Harrods BL (eds) Landslides: extent and economic significance. Balkema

World Bank (2001) Country Assistance Strategy for El Salvador

Chapter 42

Interpretation of the Mechanism of Motion and Suggestion of Remedial Measures Using GPS Continuous Monitoring Data

Zieauddin Shoaei* · Gholamreza Shoaei · Reza Emamjomeh

Abstract. Population growth and human activities as its consequence, development of residential zones, life lines such as power lines, water canals, gas and oil pipelines, and road construction, bring the population close to mountainous areas. Unfortunately, due to lack of detailed study of slope stability the number of landslide events triggered by earthquakes and heavy rainfalls has increased in such regions. In 1990, Fatalak landslide (north-western Iran) triggered by Roudbar earthquake caused 173 casualties, and in 1998, Abikar landslide (western Iran) in Chahar-Mahal Province caused 53 casualties after a heavy rainfall period (Shoaei and Ghayoumian 1997). Predicting the time to failure in susceptible slopes and evaluating the risk of their occurrence including evaluating the casualties and damages to people properties are the main responsibilities of landslide researchers. In addition, description, analyzing, and interpretation of landslide mechanism are some of the major steps for approaching the best method to control such disaster. Unconformity between the results of landslide hazard zonation maps and the occurred natural events in the field is one of the problems for an accurate landslide investigation by applying common methods. A major reason in such failure would be the complexity of mechanism and the high cost of detail investigations on each effective factor. It is often that such complexity prevents applying the result of an investigation in a landslide to another neighbor slide. Thus, monitoring of slopes behavior includes all effective controlling factors, seems necessary for a detail research. To investigate the results reliability in such monitoring that suggests the most effective remedial measures and prevention method, an active landslide at the north of Tehran City (Iran) was selected. Akha Village landslide is located along one of the major roads connecting Tehran to Caspian coast at the north of Iran. Based on GPS monitoring and controlling the accuracy of results by field reconnaissance, in addition to basic data and information, some remedial measures and prevention works were suggested. This paper is a brief description out of the final report of this investigation, which was carried out by Soil Conservation and Watershed Management Research Institute (Tehran, Iran) and as a part of IGCP-425 project in the country.

Keywords. Landslide, GPS monitoring, mechanism, Central Alborz, Akha

42.1 Introduction

Due to the complexity of mechanism, effective factors, movement initiation, and travel distance of the material, and its relation to different factors such as essential characteristics of the material, dip of the slopes, and other external effective factors, which is impossible to model in laboratories, landslide monitoring in natural and manmade slopes are considered by researchers. One of the major reasons of different behaviors of slides is the mechanism circumstance of stresses disturbance in sliding prone masses. Soil structures and mineralogy of the soil are other significant factors. Recognition of such effective factors is necessary to predict the possible behavior of the mass and choosing the best remedial measures and prevention works.

For example, one of the most effective factors in landslides behavior is the inherent characteristics of the mass materials and soil structure. Soils with high porosity and collapsible structure form a stable structure and high shear strength before shearing. In this type of soils, due to its capacity of collapsibility, in saturation and semi-saturation condition, when effective stress exceeds to shearing strength, the structure of soil collapses and almost the whole stress changes to pore water pressure. The behavior of such soil in saturated and semi-saturated condition after the initial displacement will be a debris flow (Shoaei and Sassa 1994) as Ontake landslide (occurred in western Japan in 1984). Therefore, in such situation in both saturated and semi-saturated condition, landslides occur suddenly without any pre-caution of the movement. Soils with granular structures and good sorting and roundness, has the potential of liquefaction when sliding.

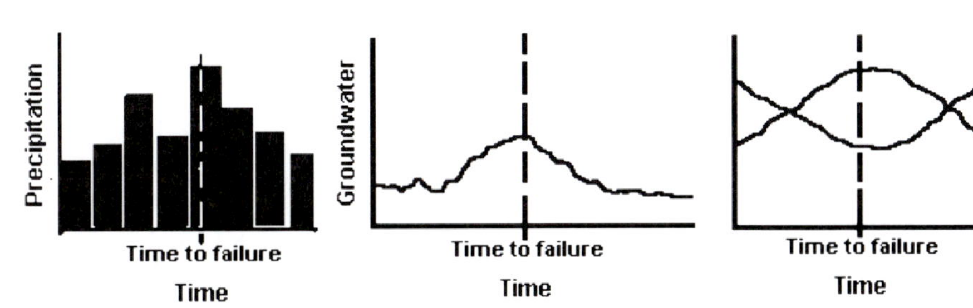

Fig. 42.1. Variable factors in landslide occurrence

Nevertheless, in a soil with a poor sorting and roundness, due to good condition of draining, soil mass stops after a short travel distance and after a relatively small movement.

Besides the significant effects of soil structure, mineralogy of soil grains is another effective factor in controlling the behavior of soil masses. By summarizing the results of researches, Skempton (1964) showed that increasing in clay percentage from 2% to 5% has a low affect on soil characteristics. Even though, after 5% of clay, soil strength drops rapidly. Shoaei (1991) showed that not only the percentage of clay, but also the type of clay has a major effect of internal friction angle of the soil. He showed that in existence of non-active clay minerals such as kaolinite, the behavior of soil is as silty soils, whereas, in existence of active clay such as smectite group, the behavior of soil is as clay soils and sliding initiates with more pre-cautions. Sufficient data about the composition of soil and its effect assists for an accurate analysis and selection of the best preventing method (Shoaei et al. 2001).

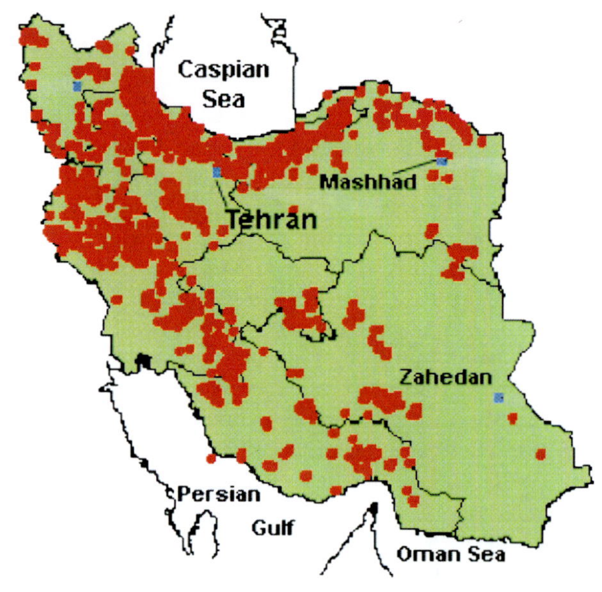

Fig. 42.2. Landslide distribution map within Alborz and Zagross Ranges

Besides all inherent characteristics of the soil, some other external factors are effective to initiate a landslide. Rainfall has a great role by different affects: soil mass saturation increases the weight of upper soil layer and consequently increases shear stress. In the other hand, rainfall changes the level of groundwater as one of the major effective factors, and also water changes some geotechnical characteristics of the soil.

Earthquakes, traffic, development of infrastructures, road construction and cutting slopes, are other external factors affecting mass movements and the post failure behavior of the soil masses. Though, human activities have no direct effect on the occurrence of landslides, because they affect stability and instability of slopes, they must be considered precisely. For example overloading by development of infrastructures, un-loading the toe of sloes for roads and life lines construction, deforestation and changing of land use, change the balance condition of stresses in slopes which affects slope instability.

Thus, essential characteristics of the soil mass and external factors are the causes that have a great role on the time of occurrence and post-failure behavior of landslides. Regarding all soil properties and external factors, mass movement in slopes shows different level of precaution indicators. Monitoring the behavior of soil mass, besides analyzing the variable effective factors (Fig. 42.1) assists the prediction of time to failure and post behavior of landslide.

Fig. 42.3. Location of study area

42.2 Surface Monitoring Systems

There are some different systems to record the landslide movements. These systems are divided into two groups of surface and sub-surface systems. Due to the reason of monitoring, surface monitoring systems are designed as 1D recording and 3D recording systems. One of the newest 3D systems is GPS (Global Positioning System) monitoring system. Application of GPS and the reliability of its data for monitoring landslide displacement were investigated in this work.

42.3 Study Area

Most of the landslides take place along two main mountain ranges in Iran, Alborz Range with NE-NW trend and Zagross Range with NW-SE trend (Fig. 42.2).

Concerning climate condition, economy, and tourist attractions, the landslide risk along Alborz Range specifically in Central Alborz has a higher risk than other regions. For this reason and regarding the aims of the investigation and the necessity of choosing an active land-

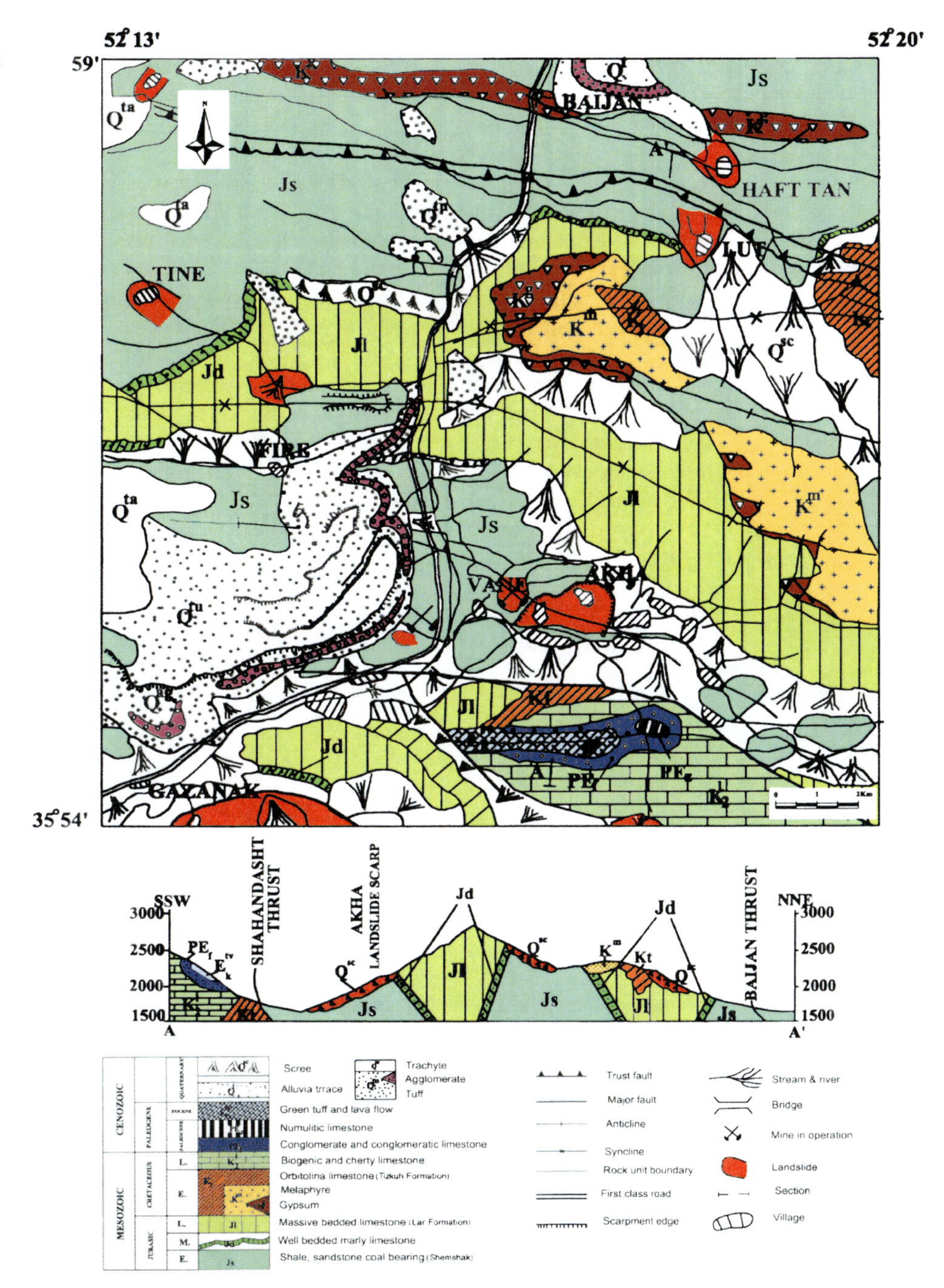

Fig. 42.4. Geological map of study area and a profile across the landslide area

slide for monitoring, a relatively wide region in the Central Alborz had been investigated. Then, a landslide prone zone had been chosen through this study results. At final stage, one of the most active landslides was selected for GPS monitoring as the representative of such common slides in the region.

The study area was selected along Haraz Highway, one of the most crowded highways from Tehran to Caspian Seaside. This highway is located in Central Alborz (Figs. 42.3 and 42.4) between Gazanak and Baigjan (52°13' E to 52°20' E and 35°54' N to 35°59' N).

42.4 Regional Study

42.4.1 Regional Geology in Study Area

Large scale geological map of the region was prepared by detail field survey. The geological map of the region is shown in Fig. 42.4. The geology of the area as a part of southern flank of Central Alborz is very complex. Due to the aim of this paper, among all formations, four most prevailed formations that covered the study area are described:

1. *Shemshak Formation* (lower Jurassic): this formation in study area composed of shale, dark sandstone, with organic component and coal layers. Because of the lithology, most of the landslides and slope instability have occurred in this formation. There are some small mining activities in this formation as the west of Akha Village.
2. *Lar Formation* (upper Jurassic): this formation in study area comprises of thick layer massive limestone with 150 to 200 m diameter.
3. *Tiz-Kuh Formation* (lower Cretaceous): this formation in study area shows light fine grain limestone. Morphology of Tiz-Kuh Formation forms steep slopes in study area similar to Lar Formation.
4. *Quaternary Formations*: Alluvial terraces with slight tilting to the valley are the most common Quaternary Formation around study area. Scree and colluviums are also widely distributed in the area and form soft deposited material at the toe of steep mountains.

42.4.2 Structural Geology in Study Area

Structural elements such as faults, joints, and bedding layers, except folding forms, appear as weakness surfaces and discontinuities which have some effective role of slope instabilities. Anticlines and synclines with W-E trend are sequentially repeated in study area (Fig. 42.4). Most of the old or young landslides are located close to major faults. It seems that activity of these faults has some effect on instabilities in the region.

42.4.3 Seismology

Earthquakes around study area are mostly large and shallow. According to the historical seismological studies of the region, the depth of earthquakes ranges widely from 10 to 20 km. Most of the quakes in study area are between 0.0 to 5.9 in magnitude (Ms) (Amberseys and Melville 1982). It can be concluded that 79.46% of earthquakes in the region are less than 5 in magnitude (Ms), and approximately 20% of them are larger than 5 (Ms > 5). Historical studies showed that quakes larger than 5 had been destructive in this area. The acceleration of 17 active faults within 100 km from study area ranges from 0.03 g to 0.45 g for average earthquake magnitude of 6.

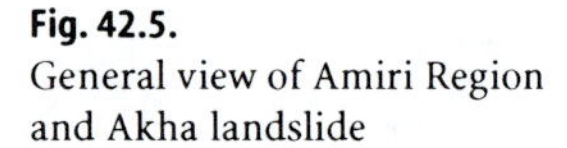

Fig. 42.5.
General view of Amiri Region and Akha landslide

42.4.4 Climate of the Study Area

Based on the analysis of 11 years data of meteorological station, the hottest month of the year is July with average temperature as 25.26 °C, and the coldest month is January with average temperature as –7.7 °C. Based on this period of data, maximum and minimum annual temperatures are 12.2 °C and 3.8 °C.

During the same 11 years data, the maximum precipitation was 720.9 mm in 1996, and the minimum precipitation was 347.1 mm in 1990. Maximum precipitation is 119.5 mm in March and the minimum is 2 mm in September. At the same period of time, the maximum precipitation is belonged to winter with 85.5 mm and the minimum is belonged to summer with 6.9 mm. Regarding the average precipitation, the total precipitation in winter is more than spring. The average monthly precipitation in the area is 45.55 mm.

42.4.5 Landslides in the Region

The study area is known as landslides prone area. Some of large landslides are shown in Fig. 42.4. In this study, recent active landslides, old inactive and reactivated landslides were studied. For the purpose of monitoring, an active landslide should be selected. Therefore, Amiri Region was selected for detail studies. This area is located 65 km south of Amol City along a minor branch of Haraz River (Fig. 42.3). A general view of the area is shown in Fig. 42.4. Some villages and residential sites are located on this area. The most active block, which is named as Akha landslide (Fig. 42.4), was selected for monitoring. Coordinates of this area is as 52°16'38" E to 52°17'37" E and 35°55'15" N to 35°55'38" N. The population of Akha Village is 1 100 peoples and increase to 2 000 during farming season. This area has been suffered mostly from damages of the landslides activities. Akha landslide has the characteristics of creeps.

Landslide activity evidence is clear as tensional cracks on ground surface and rural houses, bending of power line and trees, and disturbance of morphology. Other evidences prove that the slide is slow as a creep. Akha landslide was selected for more and detail investigation and GPS monitoring. Length of the slide is about 800 m long and the width is approximately 950 m.

Location of this landslide and a profile along the region is shown in Fig. 42.4. Generally, the economical losses of the landslide consists of house damages up to 100%, power lines, public infrastructures, cracks in gardens guard walls, disturbance of farms irrigation canals, and damages in bridges. In some blocks of landslide, severe damages to the houses forced people to evacuate. Activity of landslide then was proved based on villagers experience and other landslide evidences during last decade. By interviewing the villagers some conclusion had been discovered:

1. Generally the activity of slide is very slow and could be a creep.
2. Some times, the sliding appears as sudden subsidence of blocks in small scales.
3. Activity of different part of the slide is quite different.
4. Sometimes ground roaring at midnight, when villagers sleeping.

42.5 Complementary Study and Monitoring

First a topographical map of landslide area with the scale of 1 : 500 was prepared (Fig. 42.6). By field surveying and plotting all tensional joints and cracks and other instability evidences on the map, and comparison the relative surface level changing, the boundaries of main and sub-blocks of the landslide had been discovered.

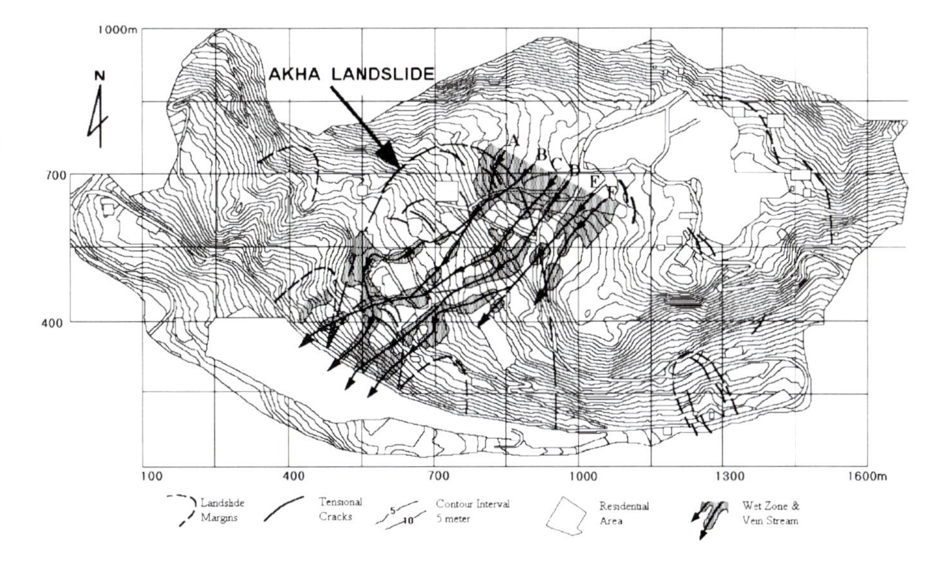

Fig. 42.6. Topographical map of Amiri Region and location of Akha landslides, groundwater condition of Akha landslide is shown

This field survey showed that not only the slide is divided into many sub-blocks, but also each sub-block has a specific displacement direction, velocity, and amount of displacement. Study on groundwater of the landslide shows that the groundwater in the active landslide mass is composed of a series of veins (Takeuchi et al. 1997).

42.5.1 Global Positioning System

Global positioning systems are powerful tools to detect the displacement of points on the ground precisely by utilizing satellites data. They have different precision due to their models and selected method of data recording.

Different accurate and applicable methods have been developed during last years due to the development of technology and higher precision. In the study area of this research, due to rough topographical characteristics, applying common field instrumentation was unaffordable. Thus, a GPS monitoring system was set up to survey the displacement. One of the advantages of using this method is that it is applicable in sites like Akha landslide, where due to the topographical characteristics, points are not visible from each other locations. Another advantage of applying this method is collecting 3D data from each point. Besides, there are some disadvantages such as the high cost of the instruments, impossibility of permanent installing, requirements of sending surveying groups to the field periodically, and complexity of data interpretation. Though, some of these problems are solved in new models of GPS instruments.

Not only the accuracy of instrument, which ranges from 100 m to 1 mm, but also the method of installation networks, reading data, and the number of instruments applying simultaneously affects the accuracy of data. Real time, rapid static (constant short term) and static (constant long term) are some common methods for surveying. Static method (fixed long term) and reading each point for about 1 hour offers more accuracy. Some of new systems, which were used in this research, offer more accuracy by applying Dual Frequency. Three instruments from Leica Co. (known as 300 series) were used simultaneously for this surveying.

42.5.2 Installation of Bench Marks

Concerning the accuracy of methods and systems and local characteristics of the area, the best networks of data reading bench marks were planned. To avoid any mistakes in data recording, a concrete bench mark with dimensions of 50 × 50 × 80 cm were planned to make the collecting possible repeatedly in each data recording period (Fig. 42.7). To collect reliable accurate data, reading took place, when (1) sufficient numerous satellites signals were received, (2) each point was read during a proper period, (3) static method was applied constantly for all stations, and (4) a reference bench mark out of sliding zone was read simultaneously with other point within the landslide zone.

Due to the difficulty of filed surveying repeatedly in all seasons, the team tried to read the data on the points with an interval not longer than 3 months. Data collecting was done from summer 2000 to spring 2004 shown in Table 42.1.

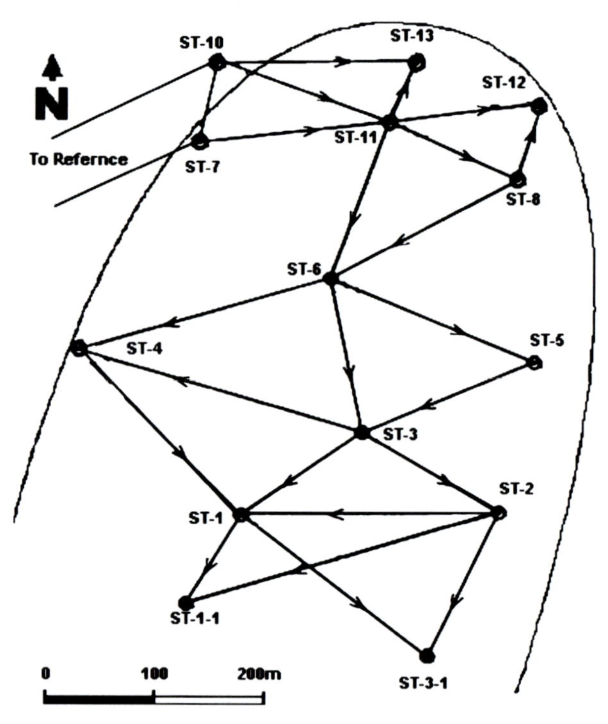

Fig. 42.7. Location of selected benchmarks and recording plan

Table 42.1. Distribution of data recording in 14 stages

No.	Date	Period (month)
1	07/06/2000	3.0
2	22/10/2000	2.5
3	20/12/2000	1.0
4	13/03/2001	2.0
5	26/04/2001	2.5
6	10/07/2001	2.5
7	06/10/2001	2.5
8	18/03/2002	4.5
9	25/05/2002	3.0
10	10/08/2002	2.5
11	24/12/2002	4.5
12	21/02/2003	2.0
13	22/10/2003	8.0
14	03/03/2004	4.5

42.5.3 Monitoring Results

Data of the positioning had been read in three dimension of *X*, *Y* and *Z*. Then the horizontal displacement was measured on main directions of N-S and W-E. At final step, the resultant vector of displacements was measured using displacement along N-S, W-E and *Z*-direction. This result was plotted on *XY*-plain. According to the results, *XY*-plain has the biggest displacement to down hill and all analysis and interpretations are based on this assumption.

The amount of displacement in each data reading period was calculated and some maps of displacement – showing the vectors of movement – were created for 14 stages. A combination of these maps is shown in Fig. 42.8. Figure 42.9 shows the amount of total displacement for each station.

42.6 Data Analysis and Interpretation of the Mechanism of Motion

It was described that the accuracy of GPS instruments and data is reliable. The results of measured movement of 14 steps for 14 stations are shown in Figs. 42.8 and 42.9. Figure 42.8 is the vector map of monitored displacement. The cumulative value of displacements is shown in Fig. 42.9. Concerning the monitoring data and results, basic data, and other surface evidence, interpretation of the mechanism of motion was done as follows:

1. The amounts of displacement during 14 stages of data recordings show no relationship between the length of interval time and displacement. High diversity of precipitation during the period of data collecting would be the main cause of such scattered results. Scattered amount of motion in various stations confirmed that minor small blocks are active inside the selected large block.
2. Detected pattern of groundwater in landslide mass by surface thermometry method (Fig. 42.6) confirmed that underground condition in the landslide mass formed the aquifer as water veins instead of a unified aquifer.
3. Even though the water movement direction is usually from top of the slope to down, groundwater direction in this slope is from east to west, approximately perpendicular to the direction of sliding.
4. Not only the precipitation of Amiri Basin affects the volume of water in water veins through the landslide mass, but also because of limestone surrounded the area, groundwater condition in the slope is affected by neighbor basins.

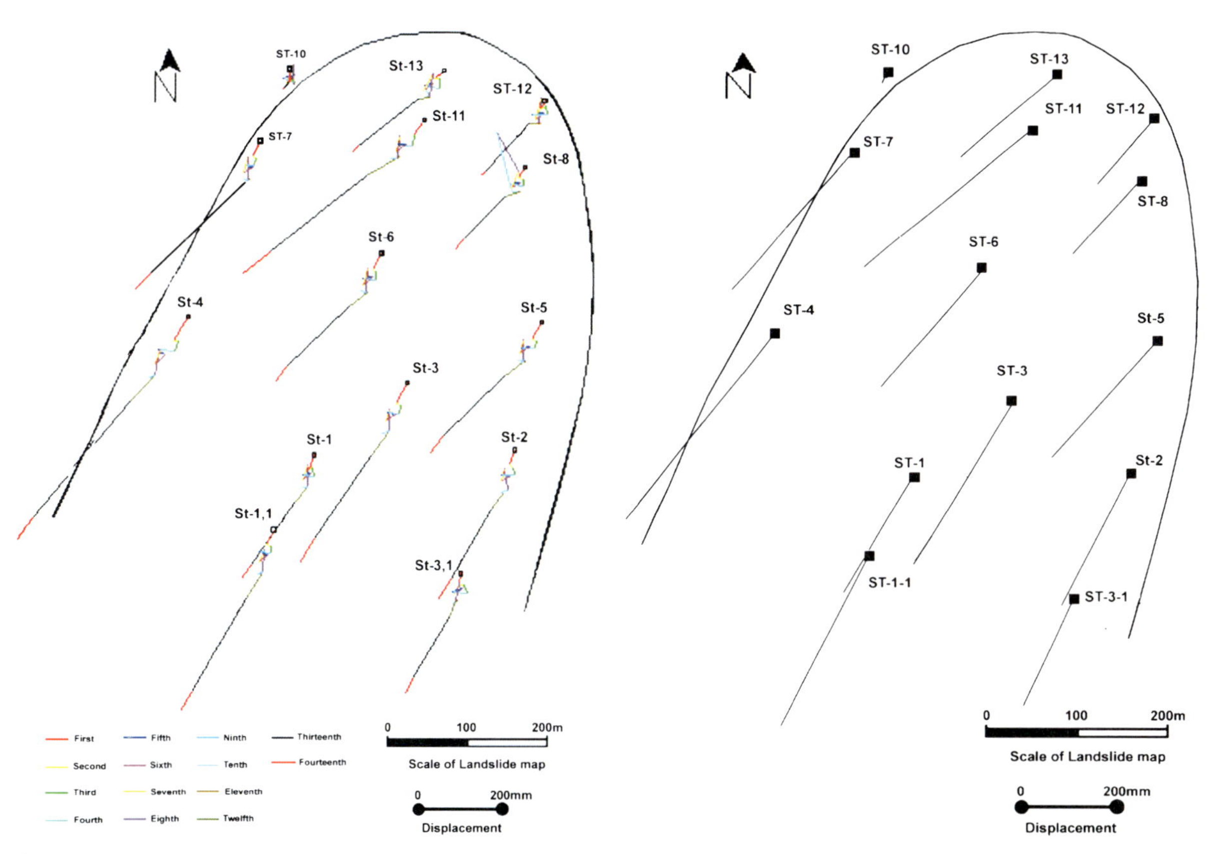

Fig. 42.8. Monitoring results and vector map of displacement of *XY*-plain

Fig. 42.9. The amount of total displacement of 14 stages in *XP*-plain

5. Regarding the above condition, effective stress that changes due to the groundwater table fluctuation is not effective in slope instability. Therefore, the effect of precipitation on slope activity is due to water saturation of the landslide mass by the water charging the slope from higher levels and neighboring basins.
6. Another major effective factor in activating some landslides in the region is the acceleration of numerous earthquakes. Shoaei and Sassa (1993) show that the effect of earthquake acceleration is not obvious in the material with a grain size less than silt. It is already proved in clay and clay rupture surfaces. Concerning the eroded material of this landslide mass from Shemshak clay-rich formation, it doesn't seem that earthquakes would be effective regarding a rapid slide. It is also proved by interviewing resident people on this slope, where there was no rapid motion after large shocks. Even tough, due to lack of a large earthquake during the period of monitoring it could not be concluded by monitoring data.

Summarizing the results of data analyses and interpretations, soil saturation is in charge of sliding in Akha landslide. To describe this effect, data collected from monitoring was analyzed.

As it is shown in Figs. 42.8 and 42.9, displacement during first data collection period differs from 25 mm to 68 mm. Maximum displacement was 68 mm in St-4 station, where the minimum displacement was 25 mm in St-12 station. It shows that from up to down except St-11, St-3, and St-4, the displacement was not much different. This data collection was done at the end of winter and spring as wet seasons in the region, when the precipitation exceeds to 200 mm in the meteorological station of the area. Results showed that the displacement in all stations was low (about 10 mm). Moreover, precipitation was low during summer. The characteristic of displacement during this time was that the amount of movement exceeds to about 10 mm to east. Due to reliable collected data, and the confirmation of eastward movement, the whole landslide mass must be displaced a few millimeters. This matter must be related to the regional tectonic activity of central Alborz.

Occurrence of a long drought during the wet seasons of 2000–2001 and the first half of 2001–2002, caused a less amount of displacement of the landslide. Data collected from displacement mostly show the movement of tectonic activity in Central Alborz. The results of monitoring from winter 2000 to the beginning of autumn 2001, which was an unusual drought, show that there was no noticeable displacement. Average precipitation was reported from the meteorological station of the area as 150 mm.

Relatively good precipitation by the second half of wet seasons of 2001–2002, and after drought period in autumn 2002, caused the activity of Akha landslide accelerated again. As shown in Figs. 42.8 and 42.9, the largest displacement in Akha landslide is belonged to this period and was during 12, 13, and 14 stages of data collecting. Precipitation in this period is 856 mm, recorded in the meteorological station of the area. Unconformity between the precipitation recorded in the meteorological station and the displacement of the landslide would be due to the difference of rainfall between the location of meteorological station and the landslide, and also the effect of neighbor basins to the landslide basin.

Amount of displacement collected from stage 12 on 23 February 2002 after a rainy autumn was evident from surface morphology. As it is shown in Fig. 42.9, the amount of displacement in stations st-7 and st-11 located on upper portion of the main landslide, and stations st-4 and st-3 at the middle of the main slide, form more active minor blocks displaced at the southwest of the main slide. Data collection of stages 13 and 14, which took place at 22 October 2003 and 3 March 2004, with 8 and 4.5 months interval after rainy winter and spring seasons, shows the same evidence. Precipitation during this period in the area was 450 mm.

42.7 Conclusions

1. Landslides in Amiri Region and Akha landslide is categorized as very slow slide (creep), which is due to the amount of clay in eroded colluviums at the toe of slops (Shoaei et al. 2001). Another possible reason of such mechanism would be the location of rupture surface in bed rock through structural and tectonical disconformities.
2. Unconformity between the precipitation on the site of slide and the activity of landslide shows the effect of precipitation in neighbor basins.
3. Due to the specific subsurface geological characteristics and the lack of unified aquifer through the landslide mass, the effect of hydrostatic pressure and reduction of effective stress is negligible. The most effective factor activating this landslide is the saturation of the sliding mass by precipitation of the region and other neighbor basins.
4. In spite of the occurrence of numerous and large earthquakes in the region, there is no proved evidence showing the relation between them and the activity of the landslide.
5. Regarding the composition of the material and subsurface geology, occurrence potential of rapid slides without pre-cautions is not expected.

42.8 Recommendations

1. Concerning the width of landslide with different behavior, it is recommended to develop monitoring stations all over the area of sliding.

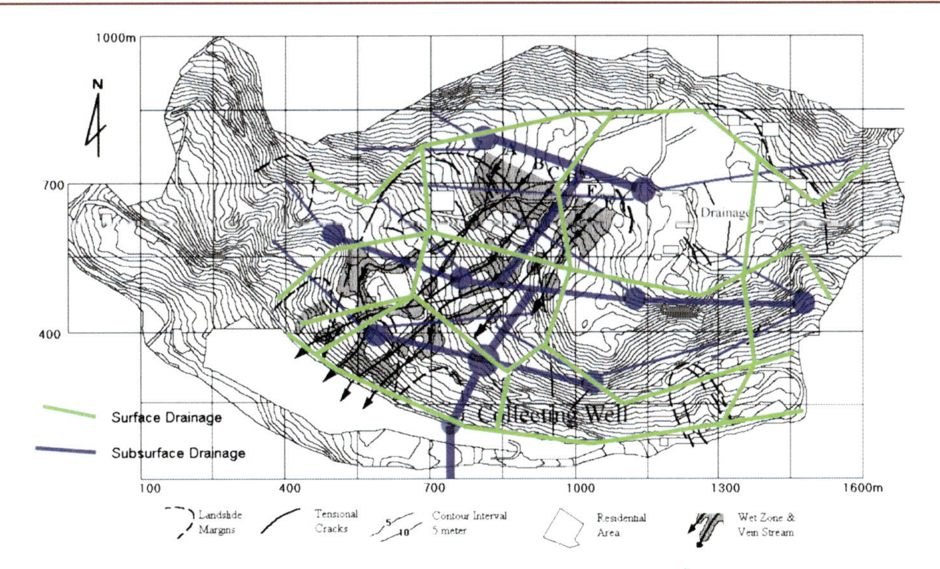

Fig. 42.10. Recommended surface and subsurface drainage systems

2. As mentioned above, the main factor activating the slide is saturation and increasing in groundwater inlet (in water veins). Therefore, as the first priority it is recommended to build up a proper subsurface drainage system. Figure 42.10 shows a primary subsurface drainage pattern applying the minimum number of wells and subsurface tunnels.
3. Even though, the level of groundwater is affected by the precipitation in other basins, but as the second priority, surface drainage would be recommended to drain run-off and floods by heavy rainfalls. Regarding the topographical characteristics of the area, a surface drainage pattern is suggested as shown in Fig. 42.10 concerning the minimum requirements of the area.
4. Irrigation system must be changed from traditional submerging system to trickle system for less water consumption on slope.
5. Appropriate designing and construction of houses and infrastructures utilizing lighter material and unified foundation is recommended too.
6. Correction of geometry of the slope for decreasing the steepness, decreasing the driving forces and for easier surface drainage would be the last recommendation. In spite of simplicity and effectiveness of this method, it is costly in the slopes, where gardens, farms, and residential houses are developed. In the study area, a geometry correction is recommended for stations st-11 and st-4 areas by avoiding any negative effect on neighbor blocks.

References

Amberseys NN, Melville CP (1982) A history of Persian earthquakes. Cambridge University Press London

Shoaei Z (1991) Mechanism of landslide in the Yokote Region, Akita-Japan. M.Sc. Thesis, Institute of Mining Geology, Mining College, Akita University-Japan

Shoaei Z, Ghayoumian J (1997) Landslide in North and Western Iran. International Symposium on Natural Disaster Prediction and Mitigation, Kyoto-Japan, pp 293–297

Shoaei Z, Sassa K (1993) Mechanism of landslide triggered by the 1990 Iran earthquake. Bull Disaster Prev Res I, Kyoto University 43:1–29

Shoaei Z, Sassa K (1994) Basic study on the shear behavior of landslides during earthquake (excess pore pressure generation in the undrained cyclic loading ring shear tests). Bull Disaster Prev Res I, Kyoto University 44:1–43

Shoaei Z, Ghayoumian J, Shariat Jafari M (2001) Landslide along weathered layers of Jurassic-Shemshak Formation, Northern Iran. 4th Asian IAEG, Yugjakarta, Indonesia, pp 139–142

Skempton AW (1964) Long term stability of clayey slopes. Geotechnique 14:77–101

Takeuchi A, Tanaka H, Shoaei Z (1997) Underground temperature survey for detecting the ground vein-stream. 8th International Conference on Rainwater Catchments System, Tehran, Iran, pp 1180–1183

Chapter 43

On the Use of Ground-Based SAR Interferometry for Slope Failure Early Warning: the Cortenova Rock Slide (Italy)

Dario Tarchi · Giuseppe Antonello · Nicola Casagli · Paolo Farina* · Joaquim Fortuny-Guasch · Letizia Guerri · Davide Leva

Abstract. This contribution illustrates the capabilities of ground-based SAR interferometry (GBInSAR) to be used as an early warning for the detection of precursory ground displacements that can suggest the imminent occurrence of a slope failure. SAR data were acquired by a ground-based SAR system, belonging to the LISA interferometer series designed by the Joint Research Centre of the European Commission, over the Cortenova rock slide (Regione Lombardia, Italy) and interferometrically analyzed in near-real time. The system was used to provide, during the 2002–2003 emergency caused by the landslide reactivation, the civil protection authorities with an operational tool for the assessment of the mass movement temporal evolution. After the main rupture occurred at the beginning of December 2002, which caused severe damage to the Bindo Village, destroying several houses and factories, interrupting one key connection road and partially damming a river, concern over the occurrence of further collapses of the still unstable slope led to the evacuation of 900 people living close to the run-out area. Such a situation induced the civil protection authorities to arrange a real-time monitoring system. Measurements of ground displacements continuously collected by the radar system up to May 2003, besides detecting the portions of the slope affected by movement, revealed the gradual deceleration of the residual movements passing from 5 cm d^{-1} to 0.3 cm d^{-1}.

Keywords. Rock slide, ground-based SAR, interferometry, remote sensing, emergency management

43.1 Introduction

Civil protection and local authorities are frequently faced to the problem of managing emergencies deriving from slope instability that threats built-up areas. Decision makers involved in the forecasting of risk scenarios require effective landslide monitoring systems to predict the phenomenon evolution and draw up the emergency plans. To this aim the availability of monitoring systems characterized by real-time capabilities and flexibility is mandatory. Among all the measurable parameters connected to slope movements the measurement over time of superficial ground displacements represents one of the most effective mean for slope failure prediction. Conventional geodetic and geotechnical instrumentations, such as total stations, GPS receivers, extensometers, etc., do not properly match the emergency requirements. In fact, these instruments need to access the monitored area for being installed and are able to provide measurements only over a few points and not over the whole unstable area. SAR interferometry implemented using ground-based systems (GBInSAR) has demonstrated to be a powerful monitoring instrument for the retrieval of accurate displacement measurements over unstable slopes with a high image acquisition rate (Tarchi et al. 2003a,b; Antonello et al. 2004). Here we illustrate how the same technique, through a near real-time processing, can be successfully employed during an emergency as an operational tool for the early detection of ground movements potentially leading to catastrophic slope failures.

The paper relies on the results obtained during a 5-months radar permanent monitoring of the Cortenova rock slide (northern Italy). At the end of 2002 the occurrence of a large and destructive rock slide forced the local civil protection authorities to evacuate the portion of the village close to the potential run-out area and to set up a monitoring system able to provide warnings of sudden accelerations of the still unstable slope with sufficient time to take prevention actions. The radar system, installed few days after the parossystic event, revealed the presence of residual movements affecting a large portion of the slope with velocities of about 2–3 cm d^{-1}, that along the monitoring period gradually decreased reaching at the end of March values of few mm d^{-1}. To respond to the civil protection needs for the emergency management, the monitoring data were also used to implement an empirical method for estimating the possible slope time failure.

43.2 Study Site

The Cortenova rock slide is located in the Lombardia Prealps, to the east of the Lecco Lake (northern Italy) along the Valsassina Valley (Figs. 43.1). After two weeks of prolonged and intense rainfall, with values up to 128.4 mm d^{-1}, as recorded at the Introbio pluviometric station, during the night of 1 December 2002, between 3:00 and 5:00 A.M., a large rock slide, which involved about 8×10^5 m^3 of rock, came down on the Bindo Village, wiping out 15 houses and 3 factories, and causing severe damage to infrastructures and services. Further, the mass movement created a temporary landslide dam along the Pioverna stream flow-

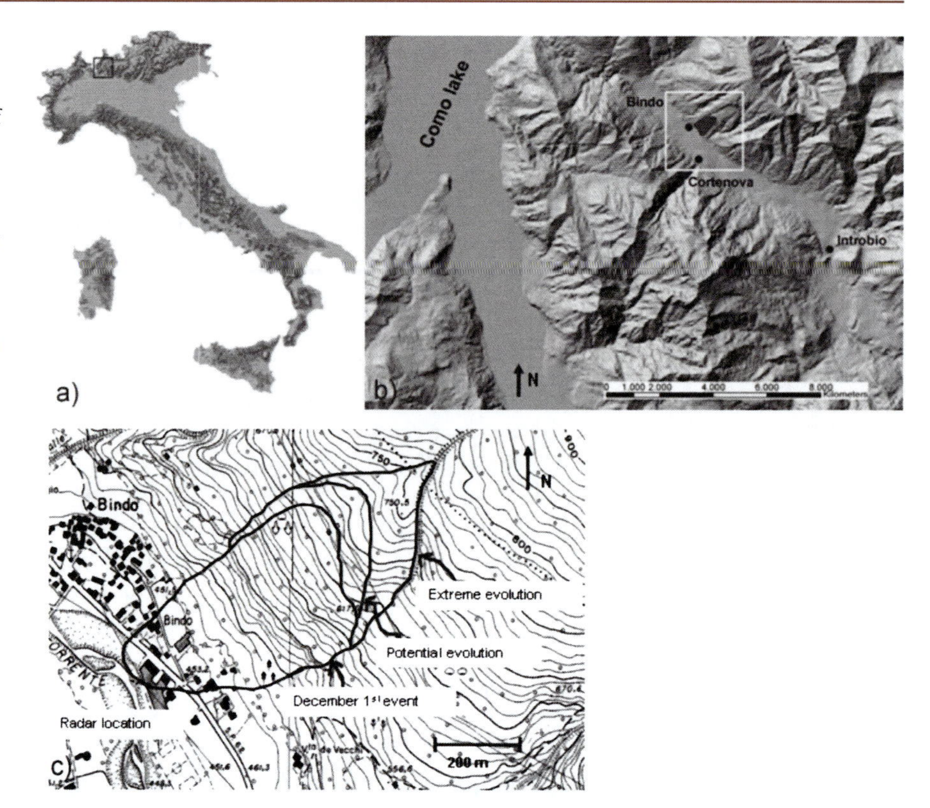

Fig. 43.1.
Locations of Cortenova rockslide in the Valsassina Valley, northern Italy. **a** Location map of the Cortenova rock slide, in the Italian Alps. **b** Shaded relief of the Como Lake area. *White box* places the Cortenova rock slide in context. **c** Topographic map of the Bindo area. *Black lines* indicate respectively from west to east the boundaries of the 1 December event, the area interested by movements during the monitoring period and the area potentially interested by the retrogressive evolution of the slide

Fig. 43.2.
Picture of the unstable area, as viewed from the Cortenova Village

ing close to the village. The potential landslide evolution required the evacuation of about 900 people living close to the run-out area.

The rock slide represents the re-activation of an ancient landslide accumulation zone. The geological setting of the area, which consists of conglomeratic deposits belonging to the Verrucano Lombardo, with a thickness ranging from several meters up to 100 m, settled on the metamorphic bedrock, together with the high conductivity of the debris deposits respect to the substratum, have been addressed as the main causes of the landslide occurrence (Ambrosi and Crosta 2003). The phenomenon is located on the lower part of the west facing slope of the valley, on the hydrographic left of the Pioverna stream, extending from the valley bottom at 450 m a.s.l. to an altitude of 700 m a.s.l. The unstable area is delimited on the northern part by the main scarp, while on the southeast border by a deep-incised stream. Figure 43.2 shows the unstable slope after the collapse occurred on 1 December. A further failure of the slope, connected to a potential retrogressive evolution of the rock slide, was threatening the eastern part of the Bindo Village and local authorities evacuated about 900 residents. Before carrying out mitigation strategies (Crosta et al. 2005), during the emergency a monitoring of ground movements was necessary. Difficulties in accessing the unstable area because of the electric cables under strain due to piles tilting, placed few tens of meters above the slide scarp, and the atmospheric conditions, frequently affected by a dense fog, suggested the use of a radar system, able to remotely collect data in every atmospheric condition, for the monitoring of ground displacements.

43.3 Methods

Synthetic Aperture Radar (SAR) images were acquired by a ground-based system, belonging to the LISA interferometer series designed by the Joint Research Centre of the European Commission (Rudolf et al. 1999) (Fig. 43.3). The system consists of a continuous-wave step-frequency (CW-SF) radar working in the Ku frequency band. A set of antennas is displaced along a 3 m long rail at steps of 6.5 mm, synthesizing a linear aperture. At each position, the radar collects the backscatter return, at 1 601 discrete frequencies ranging from 16.70 to 16.76 GHz. The synthesis of the image is achieved by coherently summing signal contributions relative to different antenna positions and different microwave frequencies. The frequency and radar displacement steps have to be fine enough in order to avoid ambiguities, respectively in range and cross-range. The LiSA system, installed at an average distance of 600 m from the target area and pointed up towards the unstable slope, collected data over the Cortenova landslide since 14 December 2002 up to May 2003. The spatial resolution of the resulting SAR images was of about 3×3 m. The time needed for the acquisition of a single SAR image under the described operational parameters was about 20 minutes. Standard telephone lines were used to transfer these data to JRC's venue, in order to be processed in near-real time. This allowed for the production of a sequence of displacement maps useful to follow the temporal evolution of the landslide movements.

Fig. 43.3. Picture of the LiSA system installed in front of the Cortenova rock slide. The radar rail was placed on a concrete wall specifically built for the monitoring

43.4 Results

Along the monitoring period more than 8 000 SAR images have been acquired. The near-real-time processing of data, consisting of the SAR focusing of raw data and the combination of consecutive images to create an interferogram, allowed us to observe every 25 minutes the displacement field of the observed scenario as occurred during the spanned time interval between the two acquisitions. Through this approach radar measurements revealed the presence of residual movements affecting the unstable slope. In particular, the detachment zone of the slide, located between 680 m and 550 m a.s.l., recorded high deformation rates, up to 30 mm d^{-1} along the system line-of-sight (L.O.S.) during the first days of the monitoring, while the rock slide accumulation zone resulted stable along the monitored time interval. Movements gradually decreased from December 2002 to March 2003, as evident from the sequence of interferograms displayed in Fig. 43.4. Interferograms spanning a time interval from 1 day to 3 days in different months (January, February and March) are shown. Each color cycle represents 9 mm of ground displacement along the radar L.O.S. direction. Negative values indicate a decreasing range or movement towards the radar location, while positive values indicate an increasing range or movements away from the radar location. The phase is still wrapped and deformation values are, therefore, affected by the intrinsic ambiguity of phase measurements: if ground displacement towards the radar exceeds the end of the scale, i.e. $-0.25\lambda = -4.5$ mm, the successive values will restart from the opposite scale end, i.e. $+0.25\lambda = 4.5$ mm. After a ground displacement of $0.5\lambda = 9$ mm the values are in phase (value = 0) again. From January to March is clearly visible how the number of interferometric fringes decreases, indicating a global deceleration.

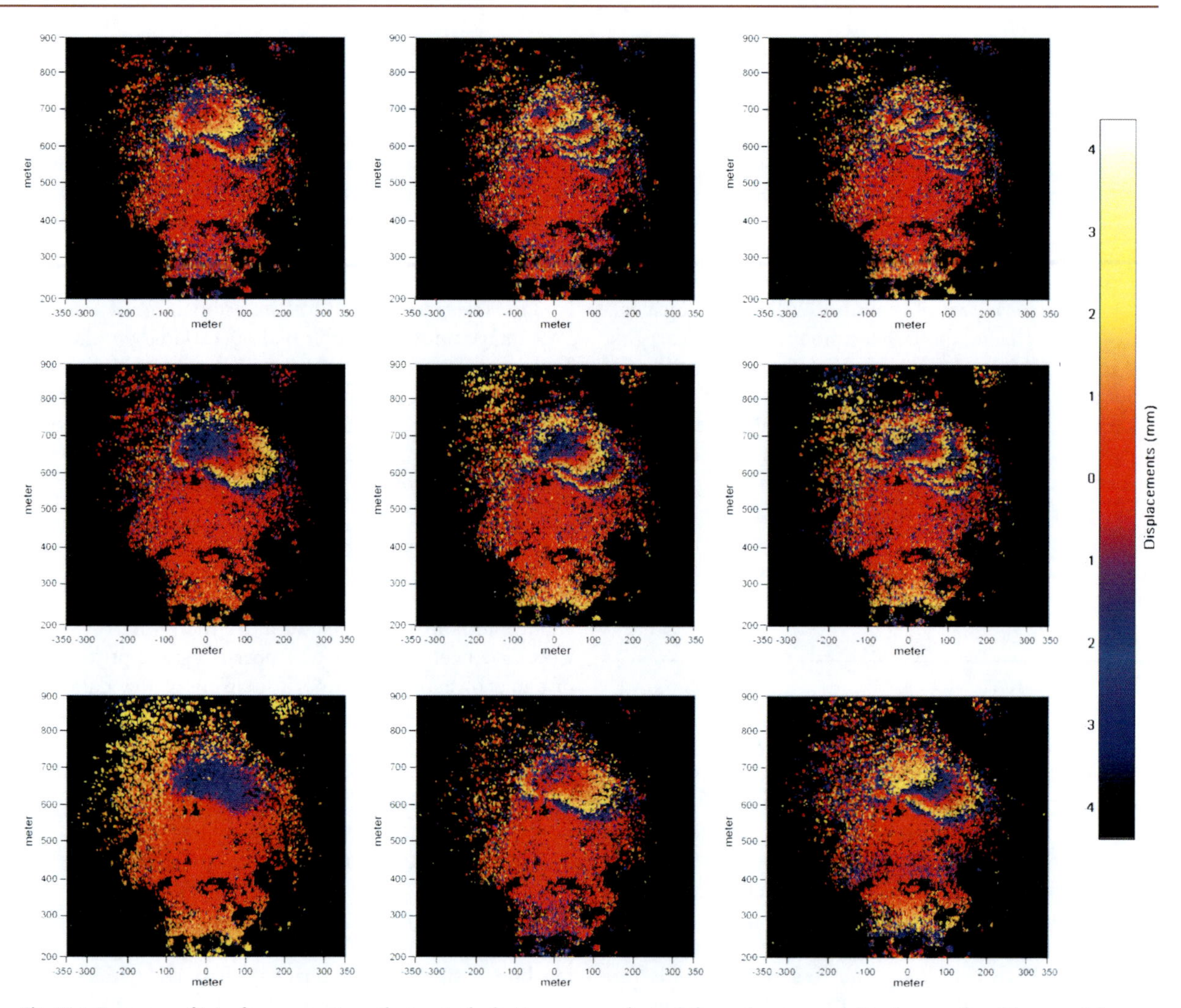

Fig. 43.4. Sequence of interferograms. From the top to the bottom maps refer to deformations occurred in the months of January, February and March 2003. For each month, the three displacement maps (from the left to the right) have been produced by cross-correlating a reference SAR image with three SAR images acquired after 1, 2, and 3 days, respectively

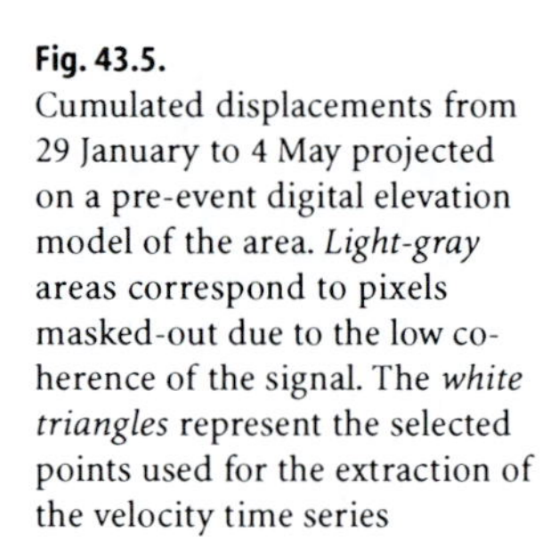

Fig. 43.5.
Cumulated displacements from 29 January to 4 May projected on a pre-event digital elevation model of the area. *Light-gray* areas correspond to pixels masked-out due to the low coherence of the signal. The *white triangles* represent the selected points used for the extraction of the velocity time series

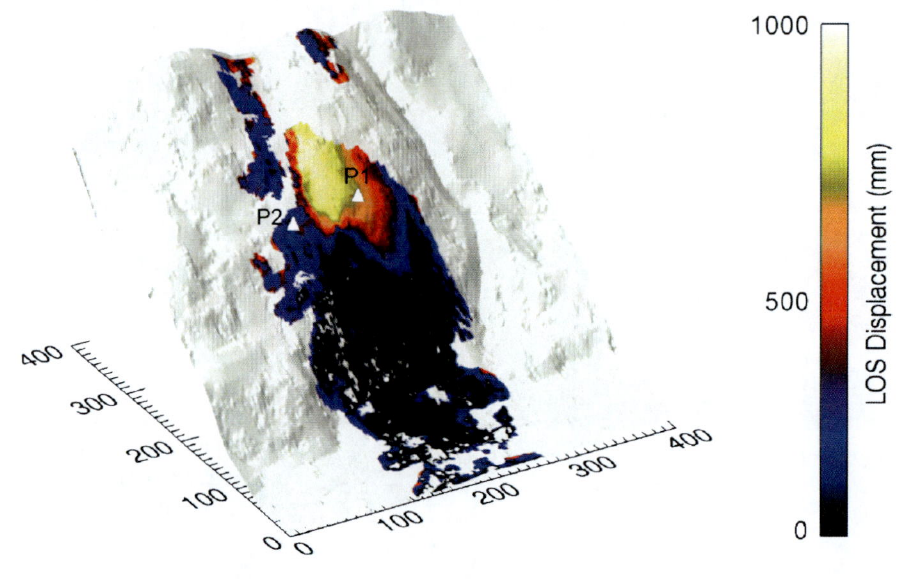

Moreover, to increase the signal to noise ratio, quite low over the upper portion of the slope due to the dense vegetation coverage, the InSAR processing has been implemented combining images resulting from the average of 10 acquisitions, each one spanning a temporal interval of 3 h and 20 m. Thanks to such an analysis deformations affecting the upper portion of the slope, above the slide main scarp, characterized by slope movements up to 5–6 mm d^{-1} and related to a retrogressive evolution of the rock slide, have been detected. The averaged SAR images have been also used for the generation of an unwrapped sequence of cumulated displacements (Fig. 43.5). To locate unstable areas this sequence has been georeferenced and projected on a Digital Elevation Model of the slope. Field surveys, carried out after the removal of the electrical cables, revealed the presence of trenches and tilted trees in the upper part of the slope, suggesting an involvement of this area related to a retrogressive evolution of the movement, as confirmed by the radar data (Fig. 43.6).

Fig. 43.6. Picture of a trench located in the upper part of the slope, on the slide right hand

From the cumulated displacement sequence we extracted the velocity history of a few selected points located in different sectors of the slope (Fig. 43.7a). This procedure, updated every day, allowed us to observe significant changes of the displacement rate, indicating possible variations in the stability conditions. Indeed, the implementation of an early warning system, that relies on a power law between deformation rate and its variations (Fukuzono 1985, 1990; Cornelius et al. 1995) was used to estimate slope time failure. Using the velocity measurements of few selected points, the method has been implemented providing the civil protection authorities with an operational tool for the early warning of slope failure. The plots of the inverse velocity vs. time (Fig. 43.7b), indicate along the monitored period a trend towards stable conditions.

43.5 Conclusions

This paper describes the application in near-real-time of GBInSAR for the monitoring of a rock slide during an emergency. Conventional geodetic instrumentation provides only point-wise information and requires the access to the unstable area for the installation of benchmarks, limiting the possibilities of detecting ground de-

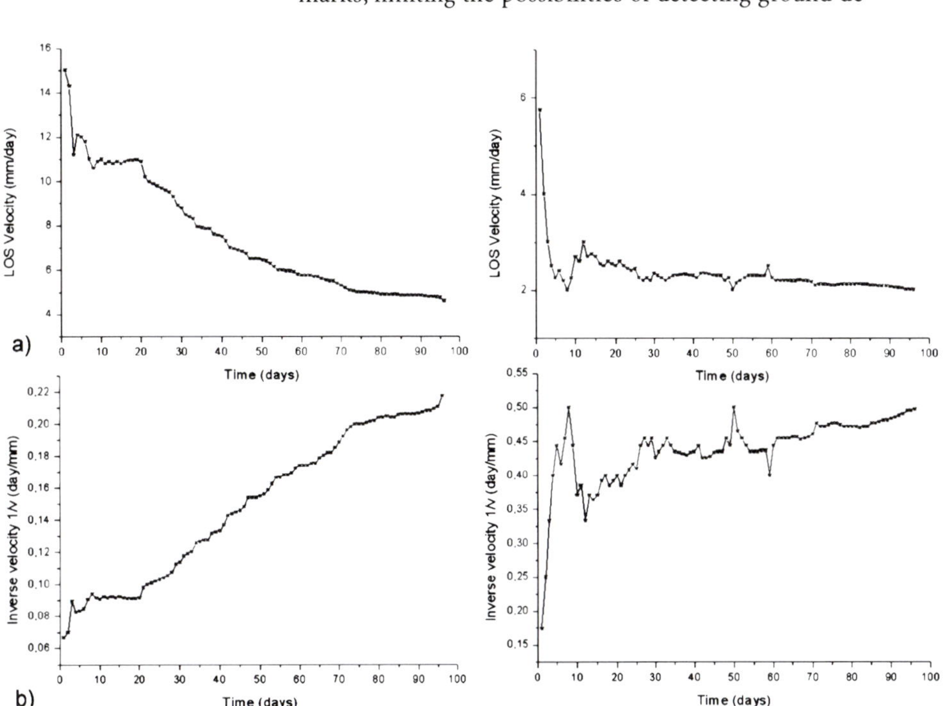

Fig. 43.7. Graphs of the LOS velocity and inverse velocity of the selected points over the slope. **a** Velocity versus time plot respectively, from left to right, of the P1 and P2 points; **b** inverse velocity versus time plot respectively, from left to right, of the P1 and P2 points

formation potentially leading to slope failures. In contrast, the proposed ground-based radar instrument has demonstrated its capability to continuously acquire accurate displacement measurements over wide areas. Its high image acquisition rate and the ability to provide displacement maps with sub-millimeter accuracy are specifically suited for assessing slope instability problems. Moreover, the near-real-time processing of the data has provided civil protection authorities with an operational tool for the detection of ground movements to be used as an early warning system.

Acknowledgments

The results described in the present paper have been obtained within the framework of a contract funded by the Comunità Montana della Valsassina. The authors wish to thank Mr. Gregorio Mannucci (ARPA, Regione Lombardia), Mr. Massimo Ceriani (Struttura Rischi Idrogeologici, Regione Lombardia) for the strong support they gave to the project along with Prof. Giovanni Crosta (Dipartimento Scienze Geologiche, Università di Milano) for having provided the Digital Elevation Model of the monitored area.

References

Ambrosi C, Crosta GB (2003) Rilevamento, modellazione e valutazione della pericolosità di grandi frane del bacino del Pioverna. Proc. 1° Convegno Nazionale AIGA (Chieti, Italy), pp 31–45

Antonello G, Casagli N, Farina P, Leva D, Nico G, Sieber AJ, Tarchi D (2004) Ground-based SAR interferometry for monitoring mass movements. Landslides 1(1):21–28

Cornelius RR, Voight B (1995) Graphical and PC software analysis of volcano eruption precursors according to the materials failure forecast method (FFM). J Volcanol Geoth Res 139:295–320

Crosta GB, Frattini P, Cheng J, Fugazza F (2005) Cost-benefit analysis for debris avalanche risk management. EGU General Assembly 2005, Geophys Res Abstr, vol. 7, EGU05-A-08732

Fukuzono T (1985) A new method for predicting the failure time of a slope failure. Proc. 4th Int. Conf. and Field Workshop on Landslides, Tokyo (Japan), pp 145–150

Fukuzono T (1990) Recent studies on time prediction of slope failure. Landslide News 4:9–12

Rudolf H, Leva D, Tarchi D, Sieber AJ (1999) A mobile and versatile SAR system. Proc. IGARSS 99, Hamburg, pp 595–594

Tarchi D, Casagli N, Leva D, Moretti S, Sieber AJ (2003a) Monitoring landslide displacements by using ground-based SAR interferometry: application to the Ruinon landslide in the Italian Alps. J Geophys Res 108:2387–2401

Tarchi D, Casagli N, Fanti R, Leva D, Luzi G, Pasuto A, Pieraccini M, Silvano S (2003b) Landslide monitoring by using ground-based SAR interferometry: an example of application to the Tessina landslide in Italy. Eng Geol 68:15–30

Chapter 44

Preservation from Rockfall of the Engraved Wall in the Fugoppe Cave, Hokkaido, Japan

Tadashi Yasuda* · Hiromitsu Yamagishi · Hideji Kobayashi

Abstract. The Fugoppe Cave, Hokkaido, Japan is known as a unique engraved walls designated as National Cultural Heritage formed in ca. 1 600 yr B.P. The Fugoppe Cave is composed of Neogene tuffaceous sandstones and has been weathered to easily fall and be spalled off along many cracks. Hence, according to the advices of the Committee for the Preservation of the Fugoppe Cave (Chairman: Masami Fukuda), we were making the photographic images of the three walls, one of which were divided into 116 grids of 50 cm long, and doing geological and geotechnical researching for each wall, such as evaluation of rockfall susceptibility of the walls.

This paper is describing the process and the results of the researching on the three walls in the Fugoppe Cave.

Keywords. Engravings, rockfall, separation, grid map

44.1 Introduction

The Fugoppe Cave is located at Maruyama Hill, Yoichi Beach, Hokkaido, Japan (Fig. 44.1), and is designated as a National Cultural Heritage. The cave (Fig. 44.2) has very unique engraved walls, which are known to have been sculptured ca. 1 600 yr B.P. The engravings of the wall were discovered about 50 years ago by high school students who were visiting the beach.

The engraved walls of Fugoppe Cave have been weathered and overgrown with blue-green algae because of lights used for exhibition in the cave. As a result, the engravings (Fig. 44.3), such as human bodies and boats, have become obscure. Furthermore, the 1996 Toyohama rockfall (Yamagishi 1997) have called attention to be afraid of rockfall from the engraved walls, because many cracks have been appearing on the wall (Fig. 44.4).

These factors have led to establishment of the Committee for the Preservation of the Fugoppe Cave (Professor Masami Fukuda of Hokkaido University, chairman). For the three years, the committee has been discussing the problem and possible solutions for preservation of the engraved walls in Fugoppe Cave. As results, many problems have arisen related to preservation of the walls. Ac-

Fig. 44.2. Aerial overview of the Maruyama Hill including Fugoppe Cave

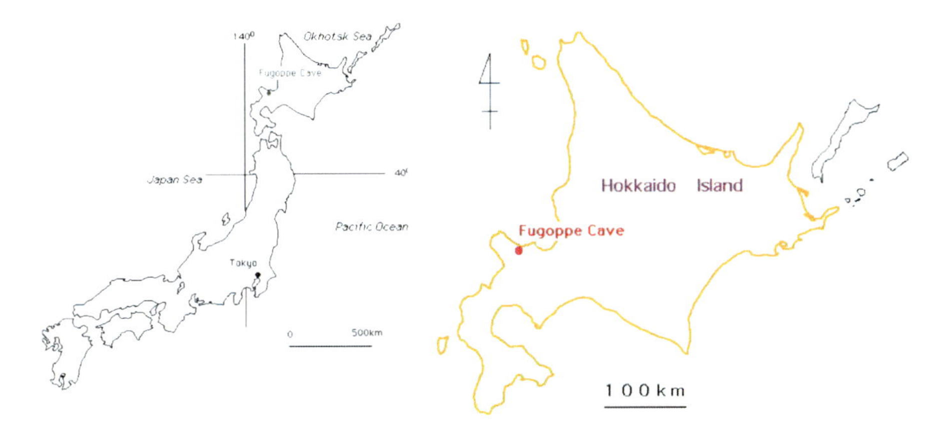

Fig. 44.1. Map showing location of the Fugoope Cave

Fig. 44.3. Photo showing the engravings

Fig. 44.4. Photo showing the cracks

cording to the results by the committee, since 1998, we have been researching the problems of rockfall and separation (i.e., spalling) of the walls. This is a preliminary report of the results of the research.

44.2 Photographs for Image Processing

The committee planned, first of all, to obtain fundamental data, such as detailed topographic photo images of quadrangle grids. Based on the photographic images, we evaluated the possibility of rock fall and wall separation, and installed the equipment to monitor the crack opening induced by earthquakes and other factors. Hence, we took photographs (Fig. 44.5) of the northern walls, which have many engravings, and are in places subject to rockfall and separation.

The northern wall is made up of the B wall associated with small A wall which is the flank of the B wall. The field for the photograph of the B wall covered by the photograph (Fig. 44.5) is 3.5 m high, 8.5 m wide, and its area is 24 m^2. The A wall is 3.5 m high at the maximum, 1.5 m wide, and the area is 5 m^2. Before taking the photographs, we set 116 grids 50 cm long by threads, followed by placing targets on each intersection of the grid. Each grid was photographed using simple negative films and stereo-paired positive films, the latter of which were taken by 60% overlapping. The cameras were appropriately moved vertically and parallel to a standard line set on the wall.

44.3 Geological Features

44.3.1 Geology of Maruyama Hill

Maruyama Hill, including Fugoppe Cave, is composed of Miocene tuffaceous sandstones deposited several million years ago. These sandstones are interbedded with pumice layers and volcanic breccias, and dip gently to the west (Fig. 44.6).

Fig. 44.5. Photograph of the *A* and *B* walls of the northern walls of the Fugoppe Cave

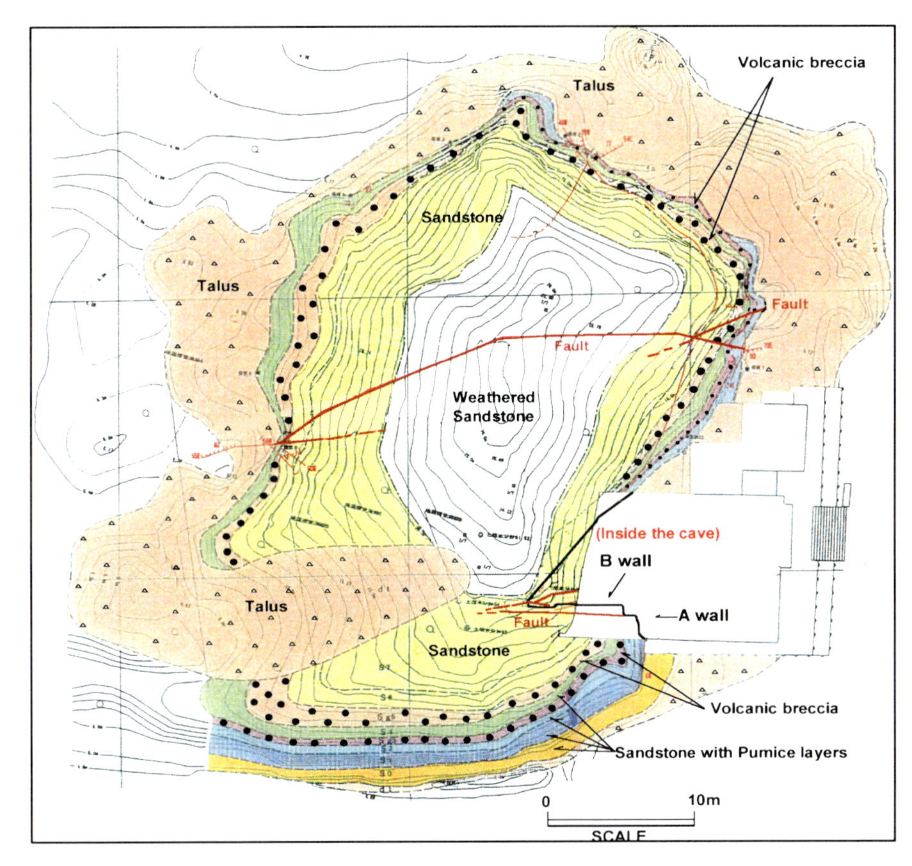

Fig. 44.6. Geologic map of Maruyama Hill

As shown in Fig. 44.7, the average dip of bedding planes is around 5~20°NW~SW (Fig. 44.7A), and fault system is characterized almost east-west direction (Fig. 44.7B), In particular, minor faults indicate N70E~EW~N70W and some of them show conjugate sets, while they are distributed almost parallel to the faults (Fig. 44.7C). Such geometric are discontinuities as bedding planes, faults and minor faults, caused large rockfalls (the blocks are up to 5 m in diameter) especially at northern slope of the Maruyama Hill.

44.3.2 Division of Sandstone Facies of the Walls

Inside the cave, on the A and B walls, sandstone facies can be observed in detail. They are tuffaceous sandstones consisting of massive parts in the lower and bedded parts in the upper, the latter of which are overlain by convoluted lamination layers. Pumice layers are interbedded and are up to several centimeters thick; they are composed of scat-

N
W
E
S
A:Beddig planes
N
W
E
S
B:Faults
N
W
E
S
C:Minor faults

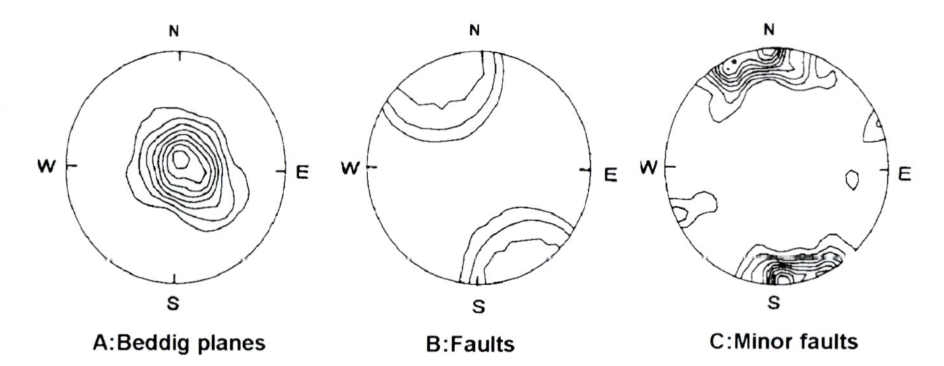

Fig. 44.7. Stereo net diagram of bedding planes, faults and minor faults (projected to lower hemisphere)

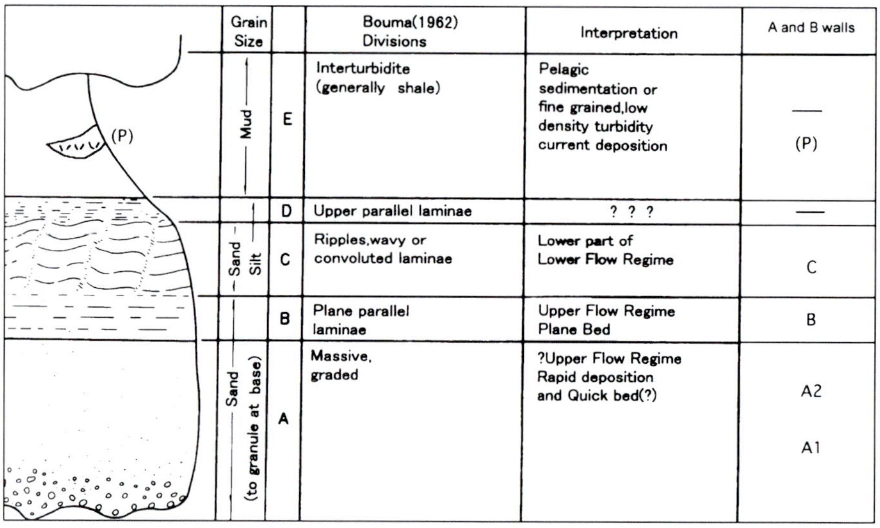

(P)
Grain Size
Bouma(1962) Divisions
Interpretation
A and B walls
Mud
E
Interturbidite (generally shale)
Pelagic sedimentation or fine grained,low density turbidity current deposition
(P)
D
Upper parallel laminae
? ? ?
Sand – Silt
C
Ripples,wavy or convoluted laminae
Lower part of Lower Flow Regime
C
B
Plane parallel laminae
Upper Flow Regime Plane Bed
B
Sand (to granule at base)
A
Massive, graded
?Upper Flow Regime Rapid deposition and Quick bed(?)
A2
A1

Fig. 44.8. Schematic figure of the Bouma Sequence model (modified from Middleton and Hampton 1976)

A wall (The eastern wall)
B wall (The northern wall)
a
b
c
d
8
9
10
11
12
13
14
15
16
17
18
19
20
21
22
23
24
E
F
G
H
I
J
K
L
50cm
50cm
A2
P
B
A1
C
Legend
+ The direction of the northern wall is N70E.
Discontinuities
F1; Fault
J2; Joint
Cr2; Crack
Be; Bedding plane
Facies
A2; Masssive coarse grained sandstone with granules
P; Pumice layer
C; Convolute laminated sandstone
B; Parallel laminated sandstone
A1; Massive coarse grained sandstone with scattered pumice
Result of beating test
Rank 1; Dead-wooden sound (Very high potential for separation)
Rank 2 ;Alive-wooden sound (High potential for separation)
Rank 3 ;Metallic sound (Stable)
Failure Scarp ; (Already separated)

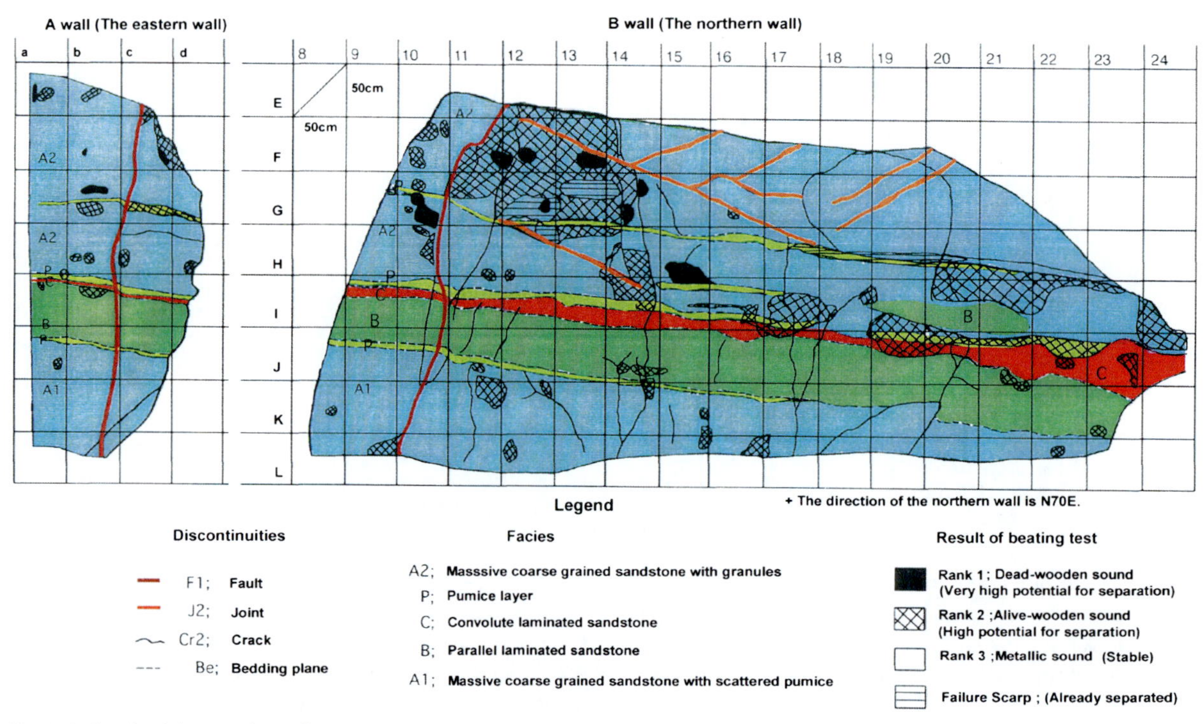

Fig. 44.9. Sketch of the *A* and *B* walls

tered pumice fragments in a fine-grained pumiceous matrix, showing faint reverse grading. Such scattering of the pumiceous materials indicates that they originated by falling through water and were erupted by subaerial or submarine shallow-water eruptions. Volcanic breccias are underlain by the massive parts of the sandstones, and show remarkable reverse grading, suggesting a debris flow mechanism. Rock properties of the sandstones are as considerably soft to engrave and are highly permeable, porous and low in density.

Based on the Bouma Sequence model (Fig. 44.8; Middleton and Hampton 1976), we divided rocks of the A and B walls into the five facies as shown in Fig. 44.9.

44.4 Identification of Rockfall-Prone Areas of the Walls

44.4.1 Rockfall

We studied geometric distribution of the discontinuities on the walls for evaluation of the possibility of rockfall. Based on the distribution patterns, we classified the discontinuities into such four categories, as faults, joints, cracks, and bedding planes (Fig. 44.9). The faults were regarded as normal faults distributed throughout the Maruyama Hill. The joints were recognized as short (1–10 m) and straight open discontinuities. They show regular distribution. The cracks are curved and are irregularly distributed. Areas showing high possibility of rockfall are closely encircled by joints, cracks and bedding planes. As the result, several areas were evaluated to have high potential of rockfall, as shown in Fig. 44.10.

In detail, block no. 1, 2, 3 and 4 are completely separated by discontinuities, while block no. 5 and 6 are stuck incompletely. Block no. 1 and 2 are prone to falling down because they were dissected clearly by open cracks. As shown in Fig. 44.11, block no. 3 is prone to sliding inside the wall, while no. 4 is prone to moving outwards. Block no. 5 and 6 will topple down if the crack (Cr2) connects the fault. The potential for rockfall may be no. 1 and 2, no. 4, no. 3, no. 5 and 6, in order.

As shown in Fig. 44.12, the areas prone to rock falling (Fig. 44.10) and the grids indicating high density of discontinuities are well correlated with each other. The area including namely no. I~K and no. 17~20 in grid map has no separated block, but there is possibility to be separated blocks in the future if the crack (Cr3) grows. Potential for rockfall depends on the crack density.

44.4.2 Separation

In order to evaluate the separation possibility of the engraved wall rock, we investigated the wall surfaces by beating with a small pen. Thus, we classified the surfaces into three categories based on the sound, in response to beating as follows: metallic sounding areas, alive-wooden sounding areas, and dead-wooden sounding areas. The area of metallic sounding is stable. The alive-wooden sounding area shows a higher possibility of separation than the stable

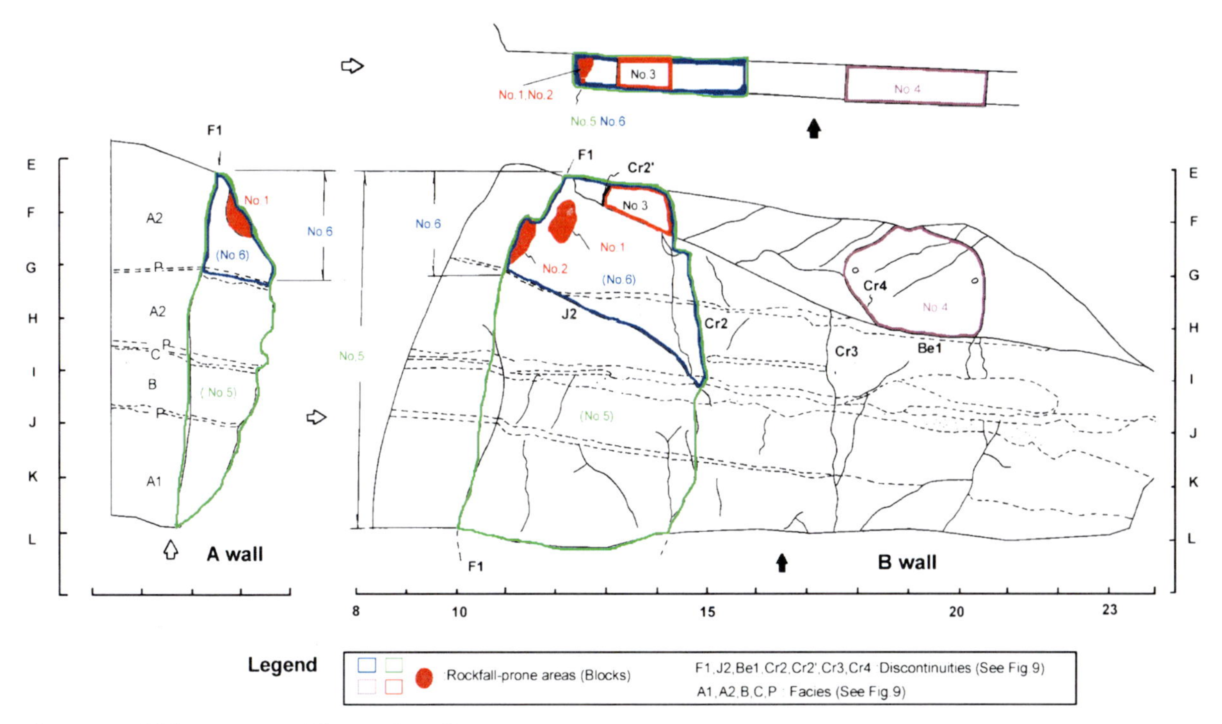

Fig. 44.10. Rockfall-prone areas of the *A* and *B* walls

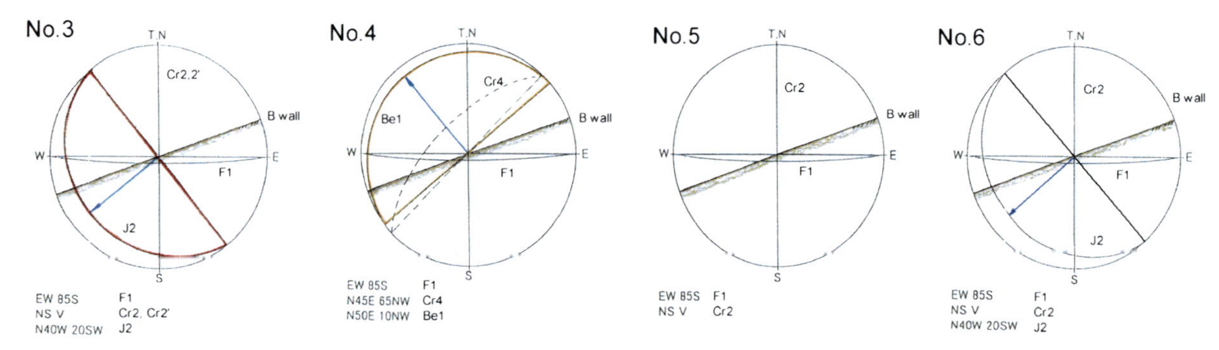

Fig. 44.11. Stereo net diagrams of geometry of discontinuities (projected to lower hemisphere)

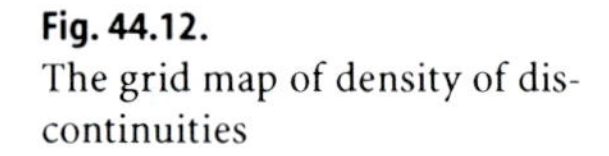

Fig. 44.12. The grid map of density of discontinuities

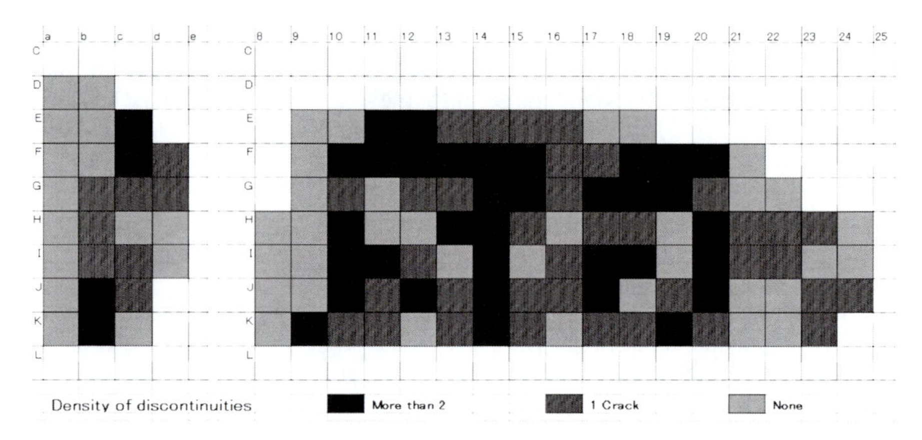

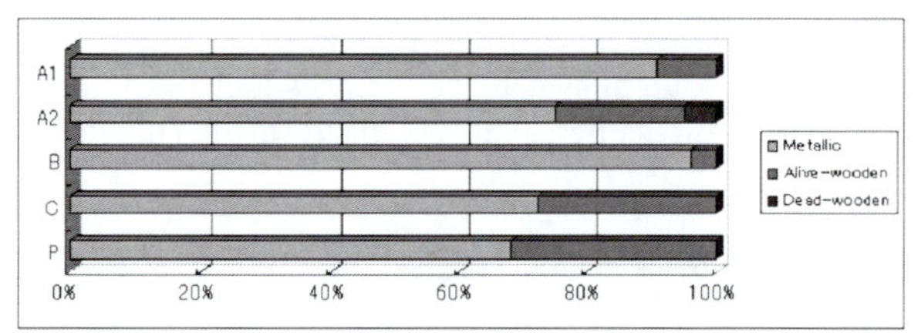

Fig. 44.13. Graph showing proportion of the results by beating in each facies

areas, although cracks were seldom recognized. The dead-wooden sounding area shows signs of already separation from the bedrock, because of existence of cracks behind the surface. Therefore, the dead-wooden sounding areas were identified as very high potential for separation; in some areas the surface have already separated and shown a failure scarp (see Fig. 44.9). Figure 44.13 shows the proportion of the results of beating in each facies.

As shown in Fig. 44.13, areas with higher potential for separation, are limited to certain rock facies. The areas with high and very high potential corresponded to the massive sandstone with granules (A2 facies), the convolute laminated sandstone (C facies) and the pumice (P facies). The massive sandstone with granules (A2 facies) showed better sorting and contained a smaller amount of cementing materials than the massive sandstone with scattered pumice (A1 facies.) The convolute laminated sandstone (C facies), which was characterized by repeated wavy laminations, has many cracks. The pumice (P facies) includes many pebble-sized pumice fragments that are easily extracted from the wall to form holes that resulted in hairline cracks. On the contrary, the parallel laminated sandstone (B facies) is recognized as almost stable, probably because laminated sandstone layer is resistant against separation because of a few cracks.

Besides Hitting, we investigated also water content and temperature of the wall surfaces in order to check external factors of separation. Water content is measured by surface TDR (Time Domain Reflectometry technology) probe. The air temperature in the cave, which is almost constant throughout one year, is 10 to 20 °C. Both are expressed on the grid maps (Fig. 44.14).

In the grid map of facies (Fig. 44.14A), the convolute laminated sandstone (C facies) and the pumice (P facies) are neglected because of small distribution on the wall. As shown in Fig. 44.14, correlating maps with each other show such tendency as follows;

1. Within the area of the massive sandstone with granules (A2 facies), dead-wooden sounding areas coincide with the areas of lower water content. On the contrary, metallic or alive-wooden (less than 50%) areas coincide with the areas of higher water content.
2. The temperature is lowest (17.8 °C) at grid no. E-24, where water seepage is observed, and continues to the grids below, and it becomes higher from right hand to

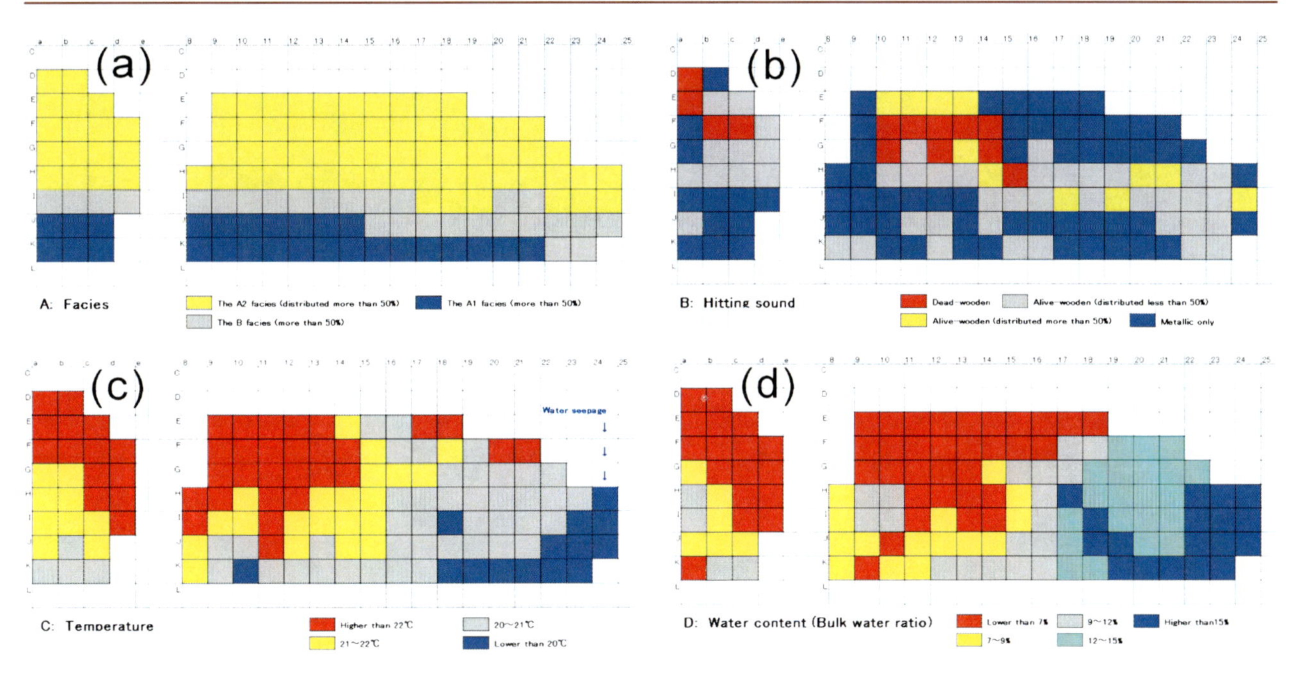

Fig. 44.14. The grid maps showing water content, temperature, beating sound, and facies

left hand. It indicates that water flow is existing behind the wall, where temperature is low.

3. The pattern of the grid maps of water content and temperature are very similar to each other, this similarity is suggesting that water content is supplied by water flow inside the wall.

As mentioned above, the massive sandstone with granules (A2 facies) shows high possibility of separation, which is caused by decrease in water content. It depends on water flow inside the wall.

44.5 Concluding Remarks

We concluded that the rockfall-prone areas are the upper left hand and right hand of the B wall, where the density of discontinuities is high, and if much distribution of discontinuities is prone to sliding or toppling, the possibility of falling becomes higher. We also concluded that the separation-prone areas are the upper left hand of the A and B walls, where the massive sandstone with granules (A2 facies) is predominant and water content is low. Separation is caused by cracks opening behind the walls. If the properties of the separation-prone areas become dry, it will cause not only growing hair lined cracks which induce separation of the surface of the walls, but also producing cracks inside the wall, to trigger rockfall. To preserve the engraved walls from rockfalls, it is important to keep appropriate water content in the rock of the walls. For determination of appropriate water content, it is necessary to examine actually the strength against cracking in certain water content of rock specimen, and to measure temperature of the rock in order to estimate water-flowing behind the walls and their surroundings. Engineering methods such as rock anchor or crack cementing for stabilizing blocks are not recommended now, because of "removable" or "non-destructive" policy which is a principal of preserving cultural heritage.

Acknowledgment

We would like to express our thanks to those who co-operated this study, especially Mr. Mori and Mr. Asano of the Museum of Yoichi Town.

References

Middleton GV, Hampton MA (1976) Subaqueous sediment transport and deposition by sediment gravity flows. In: Stanley DJ, Swift DJP (eds) Marine sediment transport and environment management. Wiley-Intersci. Publ., New York, pp 197–218

Yamagishi H (1997) Giant rockfall at Toyohama Tunnel along the coast of Hokkaido, Japan; 2: Geological background. Landslide News 10:10–11

Landslide Hazard and Mitigation Measures in the Area of Medieval Citadel of Sighisoara, Romania

Cristian Marunteanu* · Mihai Coman

Abstract. Sighisoara Fortress, named in the past "the pearl of Transylvania", was considered the most beautiful and well-preserved town fortress from the Central and southeastern Europe. This medieval complex, with military, ecclesiastic and civil architecture, was well preserved during many centuries. Nowadays Sighisoara Citadel, situated on the hill, is affected by some landslide instability phenomena, endangering the medieval walls and towers and other constructions. The main objectives of the paper are the study of the geological engineering conditions in the area of the medieval fortress, the human impact and its influence on the instability phenomena and on the causes of walls and towers cracking and collapse. Landslide hazard assessment and mitigation measures in the affected areas are also presented.

Keywords. Landslide hazard, mitigation measures

45.1 Introduction

Sighisoara is located in the southeastern part of the Tarnava Plateau, belonging to the Transylvanian Basin. The settlement of Sighisoara, the work of German colonists, has been set up in the twelfth century. The medieval complex has been built on the Citadel Hill in more stages between the twelfth and the seventeenth century.

In the year 1367 Sighisoara is named "civitas" (town) for the first time. Between 1431–1436 Vlad Dracul (Vlad Tepes's father) lived in the "watching nest" of Sighisoara. In this period the Romanian name of the town appears in documents for the first time.

During the seventeenth century a series of natural catastrophes, fires and epidemics befell the town. The most dramatic event remains the great fire from 1676, which spread into the citadel, Low Town and outskirts. After the fire, the greatest part of the wooden works and towers were destroyed, being afterwards rebuilt with stone and bricks.

Nevertheless, the fortified area and ecclesiastic buildings from the thirteenth to fifteenth century and build-

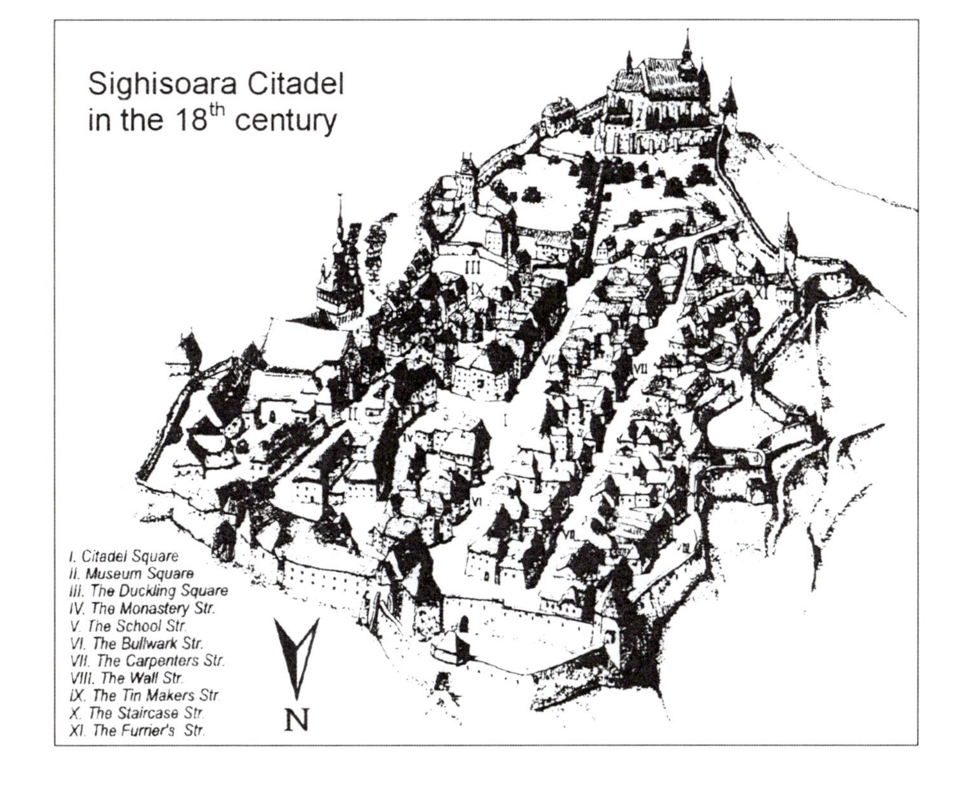

Fig. 45.1. Sighisoara Citadel in the eighteenth century

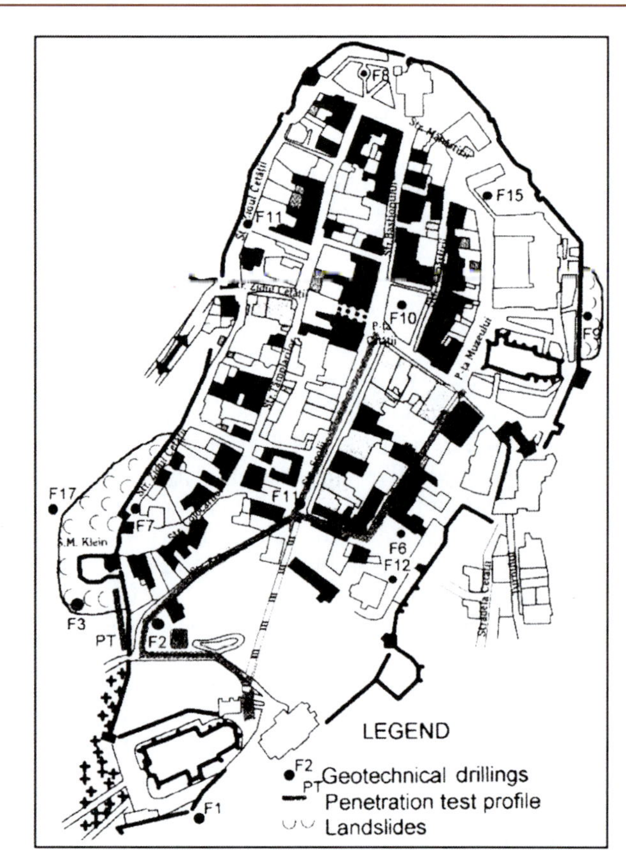

Fig. 45.2. The fortified area with some of the buildings from the twelfth to seventeenth century (in *black*), including landslides and some of the geotechnical works

ings from the fifteenth to seventeenth century are still well preserved: the Clock Tower, the Furriers' Tower, the Butchers' Tower and Bastion, the Ropemakers' Tower, the Tinsmiths' Tower and Bastion, the Leather Dresser Tower, the church from the Hill, the Monastery Church, the Vlad Dracul House, etc.

A sketch of the Sighisoara Citadel from the eighteenth century is shown in the Fig. 45.1. The fortified area and some of the buildings from the twelfth to seventeenth century are presented in the Fig. 45.2.

45.2 Morphological and Geological Conditions

The geological background is constituted of Pannonian sedimentary deposits, represented by quasi-horizontal clays, marly clays and clayey sands alternating with fine to coarse sands. These deposits are covered with Quaternary deluvial formations and anthropogenic deposits, up to 9 m in thickness.

Morphologically, the Fortress Hill is a witness of Tarnava Mare River erosion. The maximum level of the hill is of 432 m, with a relative height of 80–85 m.

45.3 Meteorological and Hydrological Data

The characteristic of the region is the moderate temperate-continental climate. The mean multiannual temperature varies between –3.9 °C in January and 18.9 °C in July, with the mean number of frost days of 135, from October to April. The data referring to precipitations show a mean multiannual value of 614.4 mm, with a weight of 424.5 mm in the warm seasons (April–September) and of 189.9 mm in the cold seasons (October–March). In the period 1963 and 2000 the annual rate of precipitations varied between 383.9 mm in 1986 and 884.6 in 1980. The mean monthly values are different from month to month (23.4 mm in February and 93.0 mm in June).

The surface runoff was estimated as 150 $l\ s^{-1}\ ha^{-1}$, with a probability of 5%. The velocity of the runoff can exceed 3 $m\ s^{-1}$, especially on the northeastern slope of the hill, with inclination of 0.5–0.6.

45.4 Hydrogeological Conditions

The hydrogeological configuration in the structure of the hill was obtained by hydrogeological drillings. The drillings, with the depth of investigations of 10.0–35.0 m, intercepted silty sands and clayey sands alternating with silty and marly clays. The underground water was found at different depths proving a discontinuous and suspended aquifer, drained to the foothill. The chemical analysis of the waters evidences a high content of nitrates and ammonium. These results provide the conclusion that the rainfall storages, the leakage from the supply and sewerage pipes or from cesspool exfiltration (leakages) and the irrigation of the gardens are the main sources of the underground water.

45.5 Instability Phenomena

The physical-geological phenomena that affect the medieval fortress hill occur as shallow slides, and erosion gullies. One landslide affects the vicinity of the Monastery Church and the Smith's Tower (drilling F9, Fig. 45.2). The main scarp is 1.5–2.0 m high and the width about 85 m. This seems to be the result of the leakage of a sewerage pipe situated above.

Another landslide is described in the eastern part of the hill (drillings F3 and F7), with the scarp of 4.0 m high and the width of 120 m. The cause of this landslide is the exfiltration from the water storage.

A landslide produced in 1974 because of the deforestation is now semi-stabilized by benches and plantations but the retaining walls is fissured. The landslide was investigated by the drilling F15.

Fig. 45.3. a, b Fortified walls affected by landslides

Fig. 45.4. Earth pressure on a retaining wall

The landslide from the western part of the hill (drilling F3) destroyed a part of the medieval fortified wall in the zone of the cemetery and of the Butchers' Tower (Fig. 45.3). Because of local ground movements and earth pressure, the retaining wall of the Leather Dresser Tower is affected (Fig. 45.4) and the Tinsmiths' Tower is fissured.

Two erosion gullies are still active on the eastern slope of the hill.

45.6 Geotechnical Works

To determine the geotechnical conditions in the area of interest, 17 manual and mechanical drillings with the depth between 10.0 and 35.0 m have been performed by ISPIF Bucuresti 1991, 1999 (Fig. 45.2). Some of them are equipped with inclinometers. Geotechnical parameters of the main soils in the area were determined. From the investigations of the drillings the following main results have been obtained:

- clayey topsoil with 0.20–0.30 m thickness covers sporadically the hill;
- a great part of the investigated area is filled with fragments of bricks, sandstones and gravels in a loamy matrix, with the thickness of 0.50–9.00 m that were deposited during the centuries;
- deluvial soils of 1.00–4.70 m lays at the surface or under the fills, constituted predominantly of clay and silty clay. The mean physical-mechanical characteristics of the soils are presented in the Table 45.1.

Table 45.1. Mean physical-mechanical characteristics of the deluvial soils

Physical-mechanical index	Mean value
Plasticity index, I_p (%)	28.20
Consistence index, I_c	0.89
Wetness, W (%)	19.80
Unit weight, γ (kN m^{-3})	20.20
Porosity, n (%)	38.60
Porosity index, e	0.67
Saturation degree, S_r	0.70
Friction angle, ϕ_u (degrees)	21
Cohesion, c_u (kN m^{-2})	30

Table 45.2. Recommended values for the design of the stabilization works

Recommended value	Type of soil					
	Filling	Deluvium (clay, silty clay)	Clayey silt, clayey sand	Sand and silty sand	Sand and sand-stone, sandy gravel	Marly clay
Friction angle, ϕ_{uu} (degrees)	7	12	15	25	32	18
Cohesion, c_{uu} (kN m^{-2})	5.0	20.0	15.0	0	0	30.0
Unit weight, γ (kN m^{-2})	19.0	20.0	19.5	20.0	20.0	20.5

- the bedrock (Pannonian deposits) is formed by marly clay, sandy clay, fine-grained sands and sometimes sandstones. The Pannonian deposits are constituted sometimes by gravels and silty sands.

The design of the retaining works to stabilize the unstable areas could be made using the parameters values from the Table 45.2.

45.7 Landslide Risk Assessment

From the geological point of view, the Fortress Hill is constituted of quasi-horizontal Pannonian deposits, represented by clays, marly clays with intercalations of clayey sands, silty sands and sands weak cemented. They are covered by deluvial deposits (clays and silty clays), 1–5 m thick and, here and there, by anthropogenic fills composed by fragments of bricks and ceramics in a clayey matrix, 0.9–9.0 m thick.

In the covering deposits, the accumulations of the water provided by the rainfall storages, the leakage from the supply and sewerage pipes or from cesspool exfiltration and by the irrigation of the gardens produce an important reduction of the shear resistance of the soils and of the filling determining the slope instability.

Taking into consideration the geological, geomorphological, hydrogeological and climatic background and the identifiable human induced factors as main conditions that cause landsliding, some elements of the landslide risk can be assessed (Marunteanu and Coman 2000):

- the Fortress Hill is not affected by large and deep landslides that could produce the general stability of the citadel;
- the landslides (active or semi-stabilized) have the depth of 2–4 m and the width of 85–120 m and affect only the covering deposits (deluvium or filling);
- the induced increasing of the moisture is considered the main cause of the instability phenomena;
- the seismicity of the region is estimated to 7 intensity degrees (MSK scale), but does not affect the slope stability.

45.8 Mitigation Measures

The results of the investigations allow of some mitigation measures in the view to decrease the risk of degradation or destroy of the historical site of the Sighisoara Citadel:

- Rehabilitation of the medieval walls in the zone affected by landslides and stabilization by retaining walls and counterforts. The recommended values for the design of the stabilization works are given in the Table 45.2;
- Consolidation of the towers and sealing works of the pavements around;
- Rehabilitation or repair of the supply and sewerage pipes and cesspools.

Field tests have been carried out in 1999 and mitigation measures have been proposed in the most affected zone of the citadel wall, between the Ropemakers' Tower and the Butchers' Tower by AGISFOR Bucuresti, 1999 (Fig. 45.2).

This zone suffered in the last years developing processes of degradation like outside sloping of the wall with 0.3–0.7 m and breaking down of the wall on a length of about 20 m (Fig. 45.3a).

Generally, because of the important differences between the ground level inside the citadel and the outside level, the citadel wall used to get also the function of retaining wall.

The phenomenon of progressive sloping and rotation of the wall is considered an effect of the earth pressure,

Fig. 45.5. The damage produced by a landslide on the citadel wall

amplified by the steep slope of the hill and the accumulation of the underground infiltration water.

Dynamic penetration tests (6 profiles) and exploration ditches (2 manual excavations) inside the precincts have been carried out to evaluate the nature and the properties of the foundation ground and the depth and type of the wall foundation. The damage produced by a landslide on the citadel wall (Fig. 45.5) and some of the penetration profiles obtained in the landslide area are shown (Fig. 45.6). A resulting characteristic geotechnical profile provided by the penetration tests shows at the surface 2–3 m of loose sands and soft silty-clays, then compact sands alternating with consistent clays to the depth of 8–9 m. A lens of plastic clay develops at 3–4 m depth.

The wall of the citadel is built of row stone and lime mortar. The uncovering excavation at the basement of the wall evidenced that the foundation of the wall consists also of row stone and lime mortar having a width of 55–60 cm. The foundation depth is 1.50–1.70 m and the foundation ground is composed of sands and silty clays with low consistence. The pressure of the wall was estimated to 270–300 kN m^{-2} adding also the earth pressure component of 50 kN m^{-2}, these values being very close to the ultimate resistance of the founding soils. The outside sloping of the wall is a result of the situation presented above.

Mitigation and consolidation measures have been proposed and designed (AGISFOR Bucuresti 1999) in this affected area (Fig. 45.7):

- stabilization of the ground by 2–3 rows of piles on the both parts of the wall, with the heads consolidated by concrete slabs in the foundation of the wall;
- vertical recovery of the inclined zones of the wall with the help of hydraulic jacks and consolidation of the wall by reinforcing with vertical concrete pillars inserted and hidden in the masonry;

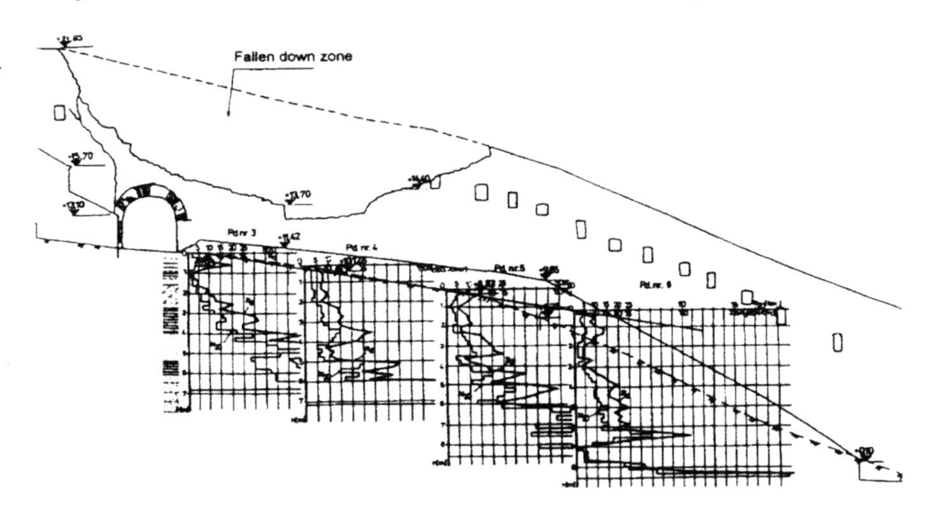

Fig. 45.6. Geotechnical profile provided by penetration tests in the unstable zone from the Fig. 45.5

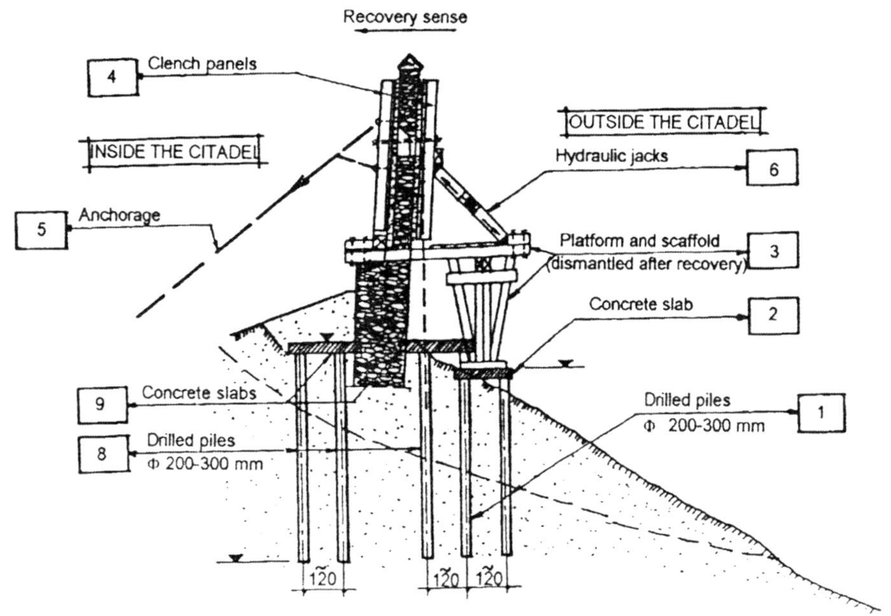

Fig. 45.7. Recovery and consolidation measures

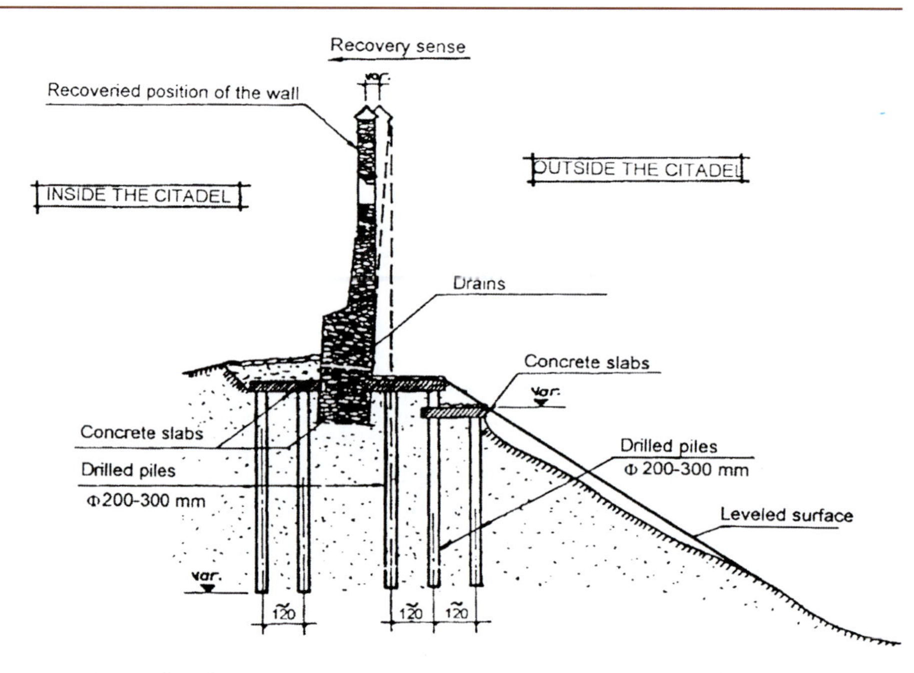

Fig. 45.8.
Final situation of the mitigation works

- draining of the infiltration waters by surface facilities and horizontal drains through the wall.

The final situation of the proposed mitigation measures and works that stabilized the walls is presented in the Fig. 45.8.

References

AGISFOR Bucuresti (1999) Mitigation and consolidation of the wall between Rope makers' Tower and Butchers' Tower (in Romanian). Unpublished geotechnical report

ISPIF Bucuresti (1991) Informative geotechnical study for Sighisoara Town especially on the instability phenomena affecting the medieval fortress hillsides, Mures County. Unpublished geotechnical report (in Romanian)

ISPIF Bucuresti (1999) Geotechnical and hydrogeological study concerning the mitigation of the landslides on the territory of the town Sighisoara, Mures County. Unpublished geotechnical report (in Romanian)

Marunteanu C, Coman M (2000) Natural and human induced instability phenomena in the area of medieval citadel of Sighisoara. Proc. GeoBen 2000 Congress "Geological and Geotechnical Influences in the Preservation of Historical and Cultural Heritage", 8–9 June 2000 CNR-IRPI Torino, CNR-GNDCI & UNESCO IGCP-425, Torino, pp 6

A Hazard Assessment of Settlements and Historical Places in the Upper Volga River Region, Russia

Yuri A. Mamaev · Ivan B. Gratchev* · Dmitri A. Vankov

Abstract. The problem of natural hazard to historical places has received a great deal of attention from the world community in recent years. This issue has been discussed at international meetings such as the International Congress on Engineering Geology in Canada (1998) and the UNESCO Conference in Paris (1999). In response to the worldwide effort to protect historical places from natural disasters such as landslides, a field investigation was conducted by the Institute of Geoscience of the Russian Academy of Science under the International Geological Correlation Program, IGCP-425, "Landslide hazard assessment and cultural heritage". The purpose of that work was to perform a hazard assessment of the Volga River Region, including a landslide inventory and detection of places where landslides posed a threat to historical landmarks. The results indicated that a number of landslides occurred on the high banks of the Volga River due to river erosion. The characteristics of those landslides as well as their mechanisms were studied. It was revealed that several historical buildings were in a potential danger due to slope processes, and it was suggested that further slope deformation be monitored. Other exogenous geological processes such as river erosion, abrasion and suffusion were also studied. It was pointed out that the creation of a cascade of water reservoirs has caused great damage to the environment of the studied area. In addition, many examples of bank destruction were observed at the Rybinskoe and Uglich Reservoirs.

Keywords. Landslide, historical buildings, Volga River

46.1 Introduction

Russia's long history has yielded numerous historical landmarks. A number of them are scattered in the central part of the country, particularly in the Volga River valley. In the last few decades, this area has suffered significant hydrological changes due to intense urban development. Such interference has disturbed the river slope stability and caused the activation of river erosion, posing a potential danger to settlements as well as to the historical buildings. The local authorities have acknowledged the problem of bank destruction in the upper Volga River Region, which has been aggravated by the creation of a cascade of great water reservoirs (Ivankovskoe, Uglich and Rybinskoe Reservoirs). In order to perform a hazard assessment of the area along the Volga River (Fig. 46.1), a field investigation was conducted by a scientific group from the Institute of Geoscience of the Russian Academy of Science under the International Geological Correlation

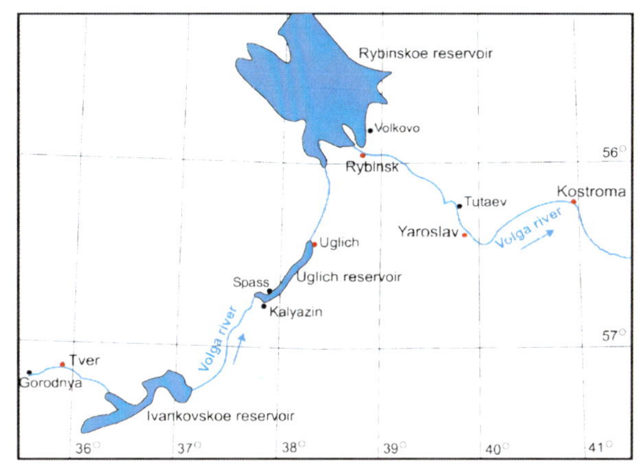

Fig. 46.1. Study area. The investigation was conducted along the Volga River from Gorodnya to Kostroma

Program (IGCP), Project no. 425, "Landslide hazard assessment and cultural heritage" (Mamaev and Gratchev 2000; Mamaev et al. 2001). Special attention was given to slope processes such as landslides which were found to be common phenomena in the studied area. The results of field observation revealed two distinct types of landslides to be dominant on the banks of the Volga River. Also, a few places where landslides posed a potential danger to certain historical buildings were detected. The scientific group studied not only landslides, but also other exogenous geological processes actively operating on river slopes, including erosion, abrasion and suffusion.

46.2 Landslide Activity in the Studied Area

During the field investigation, several places affected by landslides were discovered. The analysis of the landslides characteristics revealed two major types of slides in the studied area. The first type was a shallow slide in colluvium mostly caused by river erosion. The colluvium, which resulted from weathered moraine, was transported by rain or melted snow down to the base area of relatively high (up to 20 m) and steep (up to 60°) slopes (Fig. 46.2). In these areas, individual, usually translational, slides with volumes up to 50 m^3 occurred due to river erosion.

The second type of landslide formed in the glacial moraine deposits, which were comprised of layers of sandy and clayey soils. It was generally a rotational individual slide (up to 10 m depth) with an average volume of landmass of about 2 000 m^3 (Fig. 46.3), however, a small number of retrogressive multiple slides were also found in the moraine deposits (Fig. 46.4). Many shallow landslides in the moraine deposits occurred on the high (40–45 m) and steep (30–35°) slopes of the Volga River in the vicinity of the towns Tutaev and Kostroma. In the Spas Village, which is located on left bank of the Uglich water reservoir, landslide activity has caused the bank to retreat at a rate of 1–3 m per year. Currently, the bank has nearly approached (3–5 m) several houses in the village, threatening the local people's safety. A few other examples of landslides posing a threat to the local people as well as to the safety of historical buildings will be discussed below.

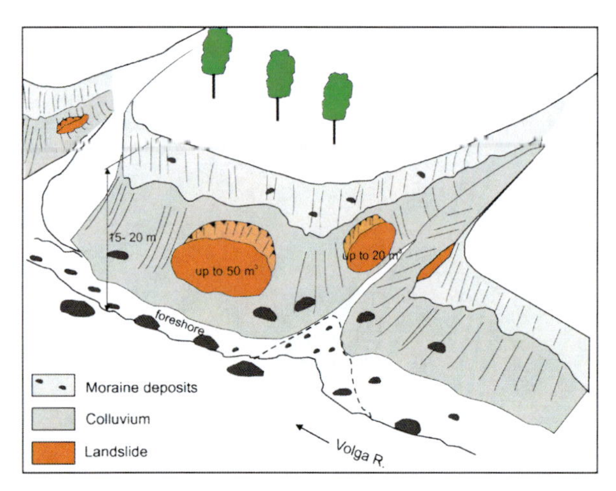

Fig. 46.2. Formation of shallow landslides in colluvium on the high slopes of the Volga River

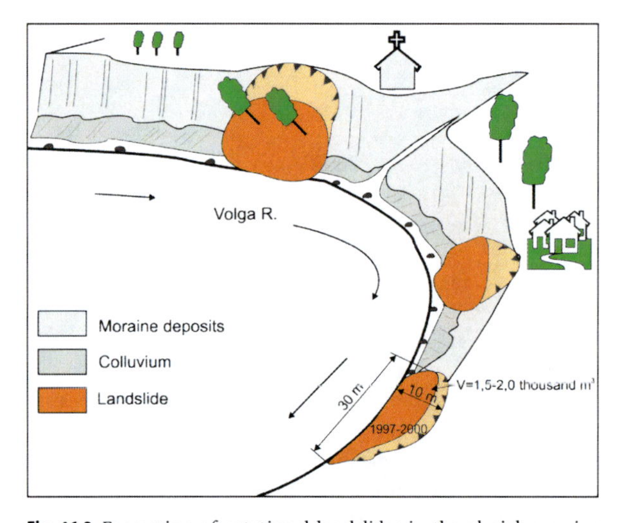

Fig. 46.3. Formation of rotational landslides in the glacial moraine deposits on the high slopes of the Volga River

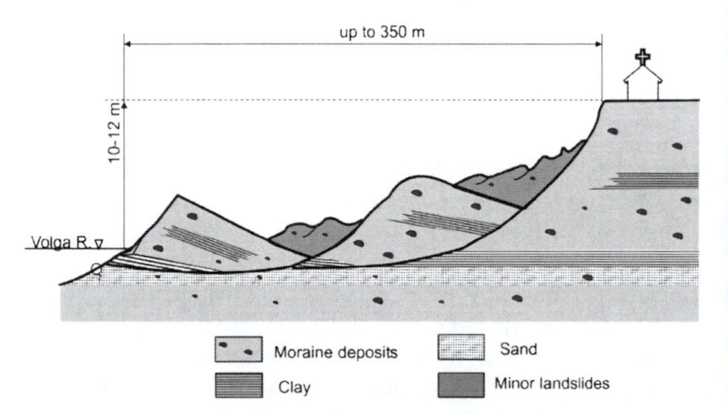

Fig. 46.4. A schematic profile of a retrogressive, multiple rotational landslide in the moraine deposits in the village of Gorodnya

46.2.1 Gorodnya Village

In the village of Gorodnya, a large retrogressive landslide was observed near the fifteenth century Blessed Virgin Church. The total volume of the landslide was estimated to be about 2.5 million m^3, and the distance between the landslide crown and its tip to be as long as 350 m, as schematically shown in Fig. 46.4. The retrogressive landslide occurred in the moraine deposits, likely due to river erosion. Two rotational landslide bodies from 50 to 70 m wide, as well as several minor landslides between and above the main displaced blocks were clearly observed. The sliding surface was assumed to be located below the river water line. Groundwater discharge was detected on the surface in different places. The landslide appeared to be active because of a few fresh cracks observed in the landslide crown, potentially affecting the stability of the Blessed Virgin Church, which is erected on the high slope remnant near the Volga River (Fig. 46.5). Although it is protected by the retaining wall and embankment and separated from the native slope by the deep (8 m) man-made ditch, some deformations of the church walls were observed.

Fig. 46.5. The Blessed Virgin Church (fifteenth century, the village of Gorodnya) located on the remnant of a high slope affected by landslide process

46.2.2 Tutaev Town

In the town of Tutaev, which is located on both banks of the Volga River approximately 40 km downstream of Rybinsk (Fig. 46.1), several landslides threatening the sta bility of two churches were observed. The Resurrection Orthodox Cathedral (seventeenth century), which is regarded as a masterpiece of the Yaroslavl architectural school, suffered significant deformation of one of the Cathedral's walls, and tilting of the bell tower (Fig. 46.6).

The church of Kazanskoy God Mother is situated on the opposite bank of the Volga River, on one of the landslide steps near the river water line. The church was tilted and turned towards the river (Fig. 46.7).

Fig. 46.7. The church of Kazanskoy God Mother (the town of Tutaev) located on the landslide bank of the Volga River

46.3 Other Geological Processes Developed in the Studied Area

The creation of a series of water reservoirs (Ivankovskoe, Uglich and Rybinskoe) has set off a number of geological processes in the area. A vast area with settlements and the architectural monuments was flooded by water from the reservoirs provoking a negative change in the environment. One particularly striking example of the dramatic changes that occurred is the monastery bell tower surrounded by water in the center of the town of Kalyazin (Fig. 46.8). The construction of the largest Rybinskoe Reservoir (Fig. 46.1) has caused a number of geological processes such as erosion, abrasion, and landslides. The maximum stretch of the water area of this reservoir is equal to 130 and 57 km from the NW to SE and from SW to NE, respectively; and more than half of it consists of shallow water up to 3–6 m in depth. In the village of Volkovo on left bank of the Rybinskoe water reservoir (Fig. 46.1), strong waves with the height of up to 2 m has been continuously destroying the shores. Figure 46.9 shows the result of only two summer storms, when the coast was reported to retreat by 8–10 m.

Fig. 46.6. The Resurrection Cathedral (seventeenth century, the town of Tutaev). Wall deformation and tilting of the bell tower

Fig. 46.8. The monastery bell tower surrounded by water in the center of the town of Kalyazin (the Uglich water reservoir)

Fig. 46.9. The eastern bank of the Rybinskoe water reservoir (the village of Volkovo). The coastal destruction caused by storms

In the Uglich Town a suffusive sink-hole of 4 m in diameter and 1.5 m in depth was observed in the territory of the Uglich Kremlin near the Saviour and Transfiguration Cathedral, located 20 m away of the Uglich Reservoir. In Kostroma (Fig. 46.1), the Ipatevskiy Monastery (fifteenth century) standing on the leveled surface of the low terrace of the Kostroma River was subjected to flooding and consecutive deformation of the subsoil.

46.4 Summary

The field investigation revealed two distinct types of landslides: (1) a shallow, translational slide in the colluvium, and (2) a rotational slide in the glacial moraine deposits. The main cause of both types of landslides was river erosion. It was also pointed out that a few historical buildings were in potential danger due to landslide activity. Further detailed investigation, including monitoring, is required to secure the safety of historical landmarks from natural hazards.

The exploitation of historical monuments for the sake of tourism business should be limited, and the local government finances should be redistributed in support of the maintenance of architectural heritage.

References

Mamaev YA, Gratchev IB (2000) Estimation of landslide hazard to historical places and settlements in the area of Upper Volga water reservoirs. In: Proceedings of the 1st International Scientific and Practical Symposium "Environmental Conditions of Construction and Preservation of the Russian Orthodox Cathedrals", Sergiev Posad, pp 132–134 (in Russian)

Mamaev YA, Gratchev IB, Vankov DA (2001) The landslide hazard to the Russian Orthodox Church Monuments in the upper Volga Region of Russia. In: Sassa K (ed) Proceedings of UNESCO/IGCP Symposium "Landslide Risk Mitigation and Protection of Cultural and Natural Heritage", Tokyo, pp 117–127

Appendix 1

ICL Statutes and Structure

A1.1 International Consortium on Landslides – Statutes

I Denomination

1. ICL: The International Consortium on Landslides, hereinafter named "ICL" is an international non-governmental and non-profit making scientific organization.

II Objectives

2. The principal objectives are to:
   a) promote landslide research for the benefit of society and the environment, and capacity building, including education, notably in developing countries;
   b) integrate geosciences and technology within the appropriate cultural and social contexts in order to evaluate landslide risk in urban, rural and developing areas including cultural and natural heritage sites, as well as to contribute to the protection of the natural environment and sites of high societal value;
   c) combine and coordinate international expertise in landslide risk assessment and mitigation studies, thereby resulting in an effective international organization which will act as a partner in various international and national projects; and
   d) promote a global, multidisciplinary program on landslides.

III Background and Domicile

3. The International Consortium on Landslides (ICL) was created in 2002 as a result of several international initiatives by specialists in the field of landslides:
   - the international newsletter "Landside News", published since 1986 by the Japan Landslide Society in cooperation with UNESCO and other international organizations and experts;
   - the 1999 Memorandum of Understanding between UNESCO and the Disaster Prevention Research Institute, Kyoto University, Japan (DPRI/KU) concerning cooperation in research for landslide risk mitigation and protection of the cultural and natural heritage as a key contribution to environmental protection and sustainable development in the first quarter of the twenty-first century; and
   - the 2001 Tokyo Declaration "Geoscientists tame landslides" in the UNESCO/IGCP Symposium on Landslide Risk Mitigation and Protection of Cultural and Natural Heritage.

   The present domicile of ICL is Kyoto, Japan, where the Secretariat is located. The official languages of the Consortium are English and Japanese.

IV Members

4. **Members** are those organizations that support the objectives of ICL intellectually, practically and financially. Membership is for a minimum period of two years. Members will come from one of four categories:
   a) Intergovernmental organizations;
   b) Non-governmental organizations;
   c) Governmental organizations and public organizations;
   d) Other organizations and entities.

V Observers

5. Relevant UN entities and Governmental entities may wish to delegate **Observers** as Special Supporting Organizations to the Steering Committee and the Board of Representatives.

VI Associates

6. **Associates** are organizations and individuals who support the objectives of ICL and meet the appropriate financial obligations but do not qualify for Member status.

VII Supporters

7. **Supporters** are other organizations and individuals who support the objectives of ICL and provide funds for its activities.

VIII Board of Representatives

8. Full power for the management of the affairs of the Consortium is vested in the Board of Representatives, which will meet at least annually. The quorum and internal regulations are defined by the bylaws.
9. The Board of Representatives shall be composed of representatives of the Member organizations. Each Member organization shall designate one Representative and one Alternative Representative.
10. In the absence of a Member's Representative from any meeting of the Board of Representatives, the Alternative Representative may attend the meeting and exercise all the rights, powers and privileges of the absent Representative. Alternatively, the Representative may delegate his rights, powers and privileges to another Member of ICL for that particular meeting, or authorize him/her to act and vote on his behalf.
11. The Board of Representatives shall:
    a) determine general policy;
    b) initiate scientific programs and decide on future priorities for the activities of ICL;
    c) approve or change, if necessary, the budget and accounts;
    d) examine and decide on each application for Member, Associate or Supporter status;
    e) elect the Officers of ICL in accordance with the Bylaws;
    f) terminate the status of any Member, Associate or Supporter of ICL which has failed to fulfill any of its obligations or when the association is no longer considered appropriate, in accordance with the Bylaws;
    g) change the Statutes and Bylaws;
    h) deal with other items which may be referred to it.
12. Voting will be decided on a simple majority. Each Member shall have one vote. Normally the President of ICL will not vote but in the event of a tie, the President may have the casting vote.

IX Steering Committee of the Board

13. The Steering Committee of the Board shall consist of:
    a) the President, between 2 and 4 Vice Presidents, the Executive Director, the Treasurer and the Immediate Past President;
    b) Co-opted Members from the board recommended by the President to act as Assistants to the President during his term of office; and
    c) a limited number of "ex officio" observers from the United Nations Educational, Scientific, Cultural Organization (UNESCO), the World Meteorological Organization (WMO), the Food and Agriculture Organization of the United Nations (FAO), the United Nations International Strategy for Disaster Reduction (ISDR) Secretariat, the Ministry of Education, Culture, Sports, Science and Technology (MEXT) Japan, and other organizations as may be considered appropriate.
14. The Steering Committee, which duly reflects the international character of the Consortium, reports to the Board of Representatives. It oversees the operations of ICL and recommends the direction and priorities of ICL to the Board of Representatives.
15. The Steering Committee shall meet as often as necessary and at least annually. Its duties are to:
    a) prepare the agenda for the meetings of the Board of Representatives;
    b) present to each annual meeting of the Board of Representatives a report of the scientific and administrative activities of ICL since the previous ordinary meeting of the Board of Representatives;
    c) propose a draft budgetary outline for consideration by the meeting of the Board of Representatives and recommend the scale of annual membership fees to be paid by Members and Associates for the ensuing 3-year period; and
    d) review the scientific activities of ICL and make appropriate recommendations to the meeting of the Board of Representatives.

X Secretariat

16. The Secretariat is responsible for the day-to-day operations of ICL. It consists of the Executive Director, the Treasurer and other secretarial members. The number of secretarial members may vary and will depend on the extent of the activities of ICL.
17. The Secretariat prepares for and reports to the meetings of the Steering Committee including:
    a) preparation of an annual work plan for the scientific and administrative activities of ICL;
    b) preparation of an annual budget and financial report for ICL;
    c) preparation of an annual report of the scientific and administrative activities of ICL; and
    d) dissemination of the results of landslide studies undertaken by/through ICL.

XI General Assembly

18. In order to report and disseminate the activities and achievements of ICL, a General Assembly shall be convened once every three years by inviting Mem-

bers, Associates and Supporters of ICL, individual members within those organizations and all levels of co-operating organizations and individual researchers, engineers and administrators. The General Assembly will receive reports on ICL's activities and provide a forum for open discussion and new initiatives from all participants.

19. Symposia and Field Workshops shall be organized annually or as appropriate at/or between the General Assemblies, in order to present the scientific and technological progress achieved through ICL activities and to disseminate new proposals and new scientific initiatives.

XII Officers

20. The Officers of ICL shall consist of the President and Vice Presidents, the Executive Director, the Treasurer and the Immediate Past President. They shall meet and communicate as often as is deemed necessary.
    a) The **President** of ICL shall preside at all meetings of the General Assembly, the Board of Representatives and the Steering Committee and shall perform such other duties and exercise such other powers as shall be assigned by the Board of Representatives.
    b) The **Vice Presidents** shall assist the President and in his absence preside at meetings and exercise the powers of the President in his place.
    c) The **Executive Director**, except as otherwise provided by the Board of Representatives, shall be the chief executive officer of the Consortium and execute contracts and agreements with external parties on behalf of ICL. The Executive Director, upon the approval of the Steering Committee, may appoint secretaries, working groups or committees to assist in carrying out the business of the Consortium.
    d) The **Treasurer**, in accordance with the financial regulations to be developed, approved by the Board of Representatives and set out in the Bylaws, shall collect and receive and have charge and custody of the funds and securities of the Consortium. The accounts of the Consortium shall be prepared at the end of each calendar year and submitted by the Treasurer to the Board of Representatives after having been audited by two authorized auditors appointed by the Board of Representatives.
    e) Election and Terms of Office
    The President and Vice Presidents shall be elected by the Board of Representatives, in accordance with the Bylaws and hold office for a term of three years, beginning from the end of the ordinary meeting of the Board of Representatives at which he or she has been elected. The President and Vice Presidents may be re-elected but may not hold the same office for more than two consecutive terms.
    The Executive Director and Treasurer shall each be recommended by the President and the Vice Presidents and approved by the Board of Representatives. They shall hold office for three calendar years beginning from 1 January of the year following the meeting of the Board of Representatives at which they have been approved. The Executive Director and the Treasurer may each be re-elected for no more than three consecutive terms.

XIII Finance

21. The funds of ICL are obtained from:
    a) membership fees from Members;
    b) contributions from Special Supporting Organizations;
    c) membership fees from Associates;
    d) funds from Supporters;
    e) other subventions, donations and financial support; and
    f) funds for research and investigation projects on landslide risk mitigation which are requested by third parties and accepted by ICL.

XIV Membership Fees

22. Annual membership fees for Members, Associates and Supporters are decided by the Board of Representatives.

XV Termination of Membership

23. Notification of termination of Membership must be given to the Treasurer at least one year in advance.

XVI Modification of the Statues

24. Changes to the Statutes require approval by a quorum of the Board of Representatives with a minimum two-thirds majority of the votes cast.
25. Changes to the Bylaws require approval by a quorum of the Board of Representatives with a simple majority votes cast.

A1.2 International Consortium on Landslides – Bylaws

1. This Bylaw will define the internal regulation for the management of the International Consortium on Landslides.

2. Quorum and Internal Regulations of the Board of Representatives defined by Chapter VIII: Board of Representatives, Article 8 are:
   1) Quorum is defined as half of the members; and
   2) Decisions of the Board of Representatives are made by majority vote. When the votes are equally divided, the chairperson decides.
3. Alternative representation defined by the Board of Representatives described in Chapter VIII: Board of Representatives, Article 10 is:
   1) If a representative of ICL will be absent and asks his alternative or another member to act and vote on his behalf, the president must be informed in writing before the meeting of the Board of Representatives.
4. Election of Officers defined by XII: Officers, Article 20 are:
   1) President shall be elected by a quorum of the Board of Representatives with a simple majority of votes cast.
   2) Vice presidents shall be recommended by the nominating committee, the latter consists of five individuals who shall be approved by the Board of Representatives.
5. Membership fees by Chapter XIV: Article 22 are:
   1) Membership fees for Members are U.S.$5 000 for full membership;
   2) Membership fees for developing countries are U.S.$500, U.S.$1 000 or U.S.$2 000;
   3) Membership fees for Associates to be examined;
   4) Membership fees for Supporters are U.S.$500 or more;
   5) The financial year of ICL starts on 1 January and ends on 31 December; and
   6) Membership fees for the current year must be paid one month before the ordinary BOR meeting.

Note:

Conditions for Members and Supporters

Members have the right to participate and vote in the Board of Representatives, which has full power for the management of the affairs of the Consortium. Members will receive a minimum of two hard copies of the journal and access to the web version of the journal once it is published.

Supporters will receive information, news and reports on ICL and IPL. After the publication of ICL Journal, the list of supporters shall appear in each issue of the journal and they will receive one hard copy of the journal and access to the web version of the journal. They may attend the Board of Representatives as observers, if invited.

A1.3 Officers of ICL for the First Term (2002–2005)

President

Kyoji Sassa (Kyoto University, Japan)

Vice Presidents

Peter Bobrowsky (Geological Survey of Canada)
Paolo Canuti (University of Firenze, Italy)
Peter Lyttle (U.S. Geological Survey)
Romulo Mucho (Instituto Geologico Minero y Metalurgico, Peru)

Executive Director

Kaoru Takara (Kyoto University, Japan)

Treasurer

Claudio Margottini (Italian Agency for New Technologies, Energy and Environment = ENEA, Italy)

Assistants to the President

Rafi Ahmad (University of the West Indies, Jamaica)
Nicola Casagli (University of Firenze, Italy)
Yasser Elshayeb (Cairo University, Egypt)
Hiroshi Fukuoka (Kyoto University, Japan)
Oddvar Kjekstad (International Centre for Geohazards, Norway)
Zieaoddin Shoaei (Soil Conservation and Watershed Management Research Institute, Iran)
Alexander Strom (Shear-holding Company "Institute Hydroproject", Russia)
Fawu Wang (Kyoto University, Japan/China)

ICL Committees and Coordinators

IPL Review Committee
Yasser Elshayeb (Cairo University, Egypt), Coordinator
Jan Vlčko (Comenius University, Slovakia), Assistant Coordinator

Committee for Capacity Building
Zieaoddin Shoaei (Soil Conservation and Watershed Management Institute, Iran), Coordinator
Alexander Strom (Institute of Geospheres Dynamics, Russian Academy of Sciences, Russia), Assistant Coordinator
Rafi Ahmad (University of the West Indies, Jamaica), Assistant Coordinator

Task Force for Resources Mobilization
Oddvar Kjekstad (International Centre for Geohazards, Norway), Coordinator
Peter Lyttle (U.S. Geological Survey), Assistant Coordinator
Paula Gori (U.S. Geological Survey), Assistant Coordinator
Claudio Margottini (ENEA, Italy), Assistant Coordinator
Kaoru Takara (Kyoto University, Japan), Assistant Coordinator

Committee for ICL Library and Information Dissemination
Hiroshi Fukuoka (Kyoto University, Japan), Coordinator
Nicola Casagli (University of Firenze, Italy), Assistant Coordinator

ICL Award Committee (Research, Practice and Service)
Peter Bobrowsky (Geological Survey of Canada), Coordinator
Paolo Canuti (University of Firenze, Italy), Assistant Coordinator

A1.4 Members and Supporting Organizations

The ICL Member Organizations

1. Geological Survey of Canada, Peter Bobrowsky/Baolin Wang
2. Chengdu Institute of Mountain Hazards and Environment, Chinese Academy of Sciences, China, Tianchi Li/Peng Cui
3. Chongqing Seismological Bureau, China, Tieliu Chen/Qiang Wang
4. Jilin University, Environmental Geological Disaster Research Institute, China, Binglan Cao/Fawu Wang
5. Lishan Landslide Prevention and Control Office, Xian Municipal Government, China, Yong Wang/Yongjin Tian
6. Northeast Forestry University, China, Wei Shan
7. Shanghai Jiotong University, School of Civil Engineering and Mechanics, Geotechnical Engineering Group, China, Xingchun Huang/Dexuan Zhang/Souji Du
8. Xian Jiaotong University, Department of Civil Engineering, Geotechnical Engineering Group, China, Hong-Jian Liao
9. The University of Hong Kong, Department of Civil Engineering/Jockey Club Research and Information Center for Landslip Prevention and Land Development, China, C. F. Lee/J. Yang
10. National Taiwan University of Science and Technology, Ecological and Hazard Mitigation Engineering Research Center, China: Taipei, H. J. Liao
11. Universidad Nacional de Colombia, Colombia, Carlos Eduardo Rodriguez/Gonzalez Garcia
12. Charles University, Research Center of Earth Dynamic, Czech Republic, Vit Vilimek/Jiri Zvelebil
13. Cairo University, Faculty of Engineering, Rock Engineering Laboratory, Egypt, Yasser Elshayeb/Hany Helal
14. Mekelle University, Ethiopia, Kurkura Kabeto/Trufat Hailemariam
15. Institute of Geology and Mineral Exploration (IGME), Greece, Nikos Nikolaou/Eleftheria Poyiadji
16. Indian Institute of Technology, Roorkie, India, A. K. Pachauri
17. Building & Housing Research Center, Iran, S. H. Tabatabaei/M. H. Tofigh Rayhani
18. International Institute of Earthquake Engineering and Seismology (IIEES), Iran, Mohammadreza Mahdavifar/Ebrahim Haghshenas
19. Soil Conservation and Watershed Management Research Institute (SCWMRI), Iran, Zieaoddin Shoaei
20. ENEA (Italian Agency for New Technologies Energy and Environment), Italy, Claudio Margottini/Guiseppe Delmonaco
21. European Commission's Joint Research Centre, IPSC/HSU, Italy, Alois Sieber/Dario Tarchi
22. University of Firenze, Earth Sciences Department, Italy, Paolo Canuti/Nicola Casagli
23. Instituto Nationale di Oceanografia e di Geofisica Sperimentale, OGS, Italy, Daniel Nieto Yabar/Emanuele Lodolo
24. International Association of Geomorphologists (IAG), Italy, Mario Panizza/Andrew Goudie
25. University of the West Indies, Jamaica, Rafi Ahmad
26. Ehime University, Faculty of Engineering, Japan, Ryuichi Yatabe/Netra P. Bhandary
27. Forestry and Forest Product Research Institute, Japan, Kiyoshi Tanaka/Hirotaka Ochiai
28. Geographical Survey Institute, Japan, Haruo Tsunesumi/Manabu Hasegawa
29. Japan Landslide Society, Japan, Hiromitsu Yamagishi/Toyohiko Miyagi
30. Kyoto University, Disaster Prevention Research Institute, Research Centre on Landslides, Japan, Kyoji Sassa/Hiroshi Fukuoka
31. Kyoto University, Disaster Prevention Research Institute, Flood Section, Japan, Kaoru Takara/Roy C. Sidle
32. Niigata University, Research Institute for Hazards in Snowy Areas, Japan, Hideaki Marui/Naoki Watanabe
33. University of Tokyo, Department of Civil Engineering, Geotechnical Engineering Group, Japan, Ikuo Towhata
34. University of Tokyo, Institute of Industrial Science, Japan, Kazuo Konagai
35. Korea Institute of Geoscience and Mineral Resources (KIGAM), Won-Young Kim/Byung-Gon Chae
36. Mara University of Technology, Malaysia, Roslan Zainal Abidin/Yusof Abd. Rahman

37. International Centre for Integrated Mountain Development (ICIMOD), Nepal, Binayak Bhadra
38. International Centre for Geohazards (ICG) in Oslo, Norway, Oddvar Kjekstad/Farrokh Nadim
39. Grudec Ayar, Peru, Raul Carreño
40. Instituto Geologico Minero y Metalurgico (INGEMMET), Peru, Romulo Mucho/Antonio Guzman
41. Proexrom S.R.L. Technical University, Civil Engineering Faculty, Romania, Nicolae Botu/Dan Carastoian
42. Federal State Unitary Geological Enterprise Scientific Centre "HydGeo", Russia, Oleg Zerkal/Julia V. Frolova
43. Institute of Geospheres Dynamics, Russian Academy of Sciences, Russia, Alexander Strom/Nikolai Syrnikov
44. Institute of Environmental Geoscience (IEG RAS), Russian Academy of Sciences, Russia, Victor Osipov/Valentina Svalova
45. Open Joint-Stock Company Engineering Centre, Unified Energy System of Russia, Alexander Piotrovskiy
46. West-Siberian Regional Center (RC "TomskGeoMonitoring"), SC HydGeo, Russia, Viktor A. Lgotin
47. Comenius University, Faculty of Natural Sciences, Department of Engineering Geology, Slovakia, Rudolf Holzer/Ján Vlčko
48. Swedish Geotechnical Institute, Sweden, Karin Rankka/Bo Berggren
49. Swiss Federal Institute for Snow and Avalanche Research SLF, Switzerland, Walter Ammann/Oliver Korup
50. Ministry of Agriculture and Cooperatives, Land Development Department, Thailand, Chumpol Lilittham/Aniruth Potichan
51. U.S. Geological Survey, U.S.A., Peter T. Lyttle/Randall G. Updike

ICL Supporting Organizations

- The United Nations Educational, Scientific, Cultural Organization UNESCO: Andras Szollosi-Nagy (Deputy Assistant Director-General and Director of Division of Water Sciences), Badaoui Rouhban (Head of the Section for Disaster Reduction), Laurent Levi-Strauss, Christian Manhart (Division of Cultural Heritage)
- The World Meteorological Organization (WMO): Michel Jarraud (Secretary-General), Hong Yan, Maryam Golnaraghi
- The Food and Agriculture Organization of the United Nations (FAO): M. Hosny El-Lakany (Assistant Director-General), Jose Antonio Prado (Director of Forest Resources Division, Forestry Department), Thomas Hofer
- The United Nations International Strategy for Disaster Risk Reduction (UN/ISDR): Salvano Briceno (Director), Pedro Basabe (Technical Advisor)
- The United Nations University (UNU): Hans van Ginkel (Rector, Under-Secretary General of the United Nations), Srikantha Herath, Libor Jansky
- The Ministry of Education, Culture, Sports, Science and Technology of the Government of Japan (MEXT): Takashi Fujii (Director of Office of Disaster Prevention Research), Yasutaka Takeuchi
- The U.S. Geological Survey of the United States Department of the Interior: P. Patrick Leahy (Acting Director of U.S. Geological Survey)
- The International Union of Geological Sciences (IUGS): Peter Bobrowsky (Secretary General)
- Governments of Italy, Canada and Norway

IPL Projects

A2.1 Coordinating Projects

C100

"Landslides": Journal of International Consortium on Landslides (2002–)

Four issues per year, full color, both printed version and web version

Coordinator: Kyoji Sassa

C101

Landslide risk evaluation and mitigation in cultural and natural heritage sites (2002–)

Coordinators: Kyoji Sassa and Paolo Canuti

C101-1

Landslide investigation in Machu Picchu (2002–)

Coordinator: Kyoji Sassa

C101-1-1

Low environmental impact technologies for slope monitoring by radar interferometry: application to Machu Picchu site (2002–)

Office: ENEA (Italian Agency for New Technology Energy and Environment)

Proposer: Claudio Margottini

C101-1-2

Expressions of risky geomorphologic processes in deformations of rock structures at Machu Picchu (2002–)

Office: Research Center of Earth Dynamic, Charles University, Czech Republic

Proposers: Vit Vilimek and Jiri Zvelebil

C101-1-3

Shallow geophysics and terrain stability mapping techniques applied to the Urubamba Valley, Peru: Landslide hazard evaluation (2004–)

Office: Instituto Geologico Minero y Metalurgico, Peru (INGEMMET)

Proposers: Romulo Mucho and Peter Bobrowsky

C101-1-4

A proposal for an integrated geophysical study of the Cuzco Region (2004–)

Office: Istituto Nazaionale di Oceanografia e di Geofisica Sperimentale (OGS), Italy

Proposer: Daniel Nieto Yabar

C101-1-5

Satellite monitoring of Machu Picchu (2005–)

Office: University of Firenze, ENEA, and Politecnico di Milano, Italy

Proposer: Paolo Canuti, Claudio Margottini, and Fabio Rocca

C101-2

Landslides monitoring at selected historic sites in Slovakia (2002–)

Office: Faculty of Natural Science, Comenius University in Bratislava, Slovakia

Proposer: Jan Vlčko

C101-3

The geomorphological instability of the Buddha niches and surrounding cliff in Bamiyan Valley (Central Afghanistan) (2002–)

Office: ENEA (Italian Agency for New Technology Energy and Environment)

Proposer: Claudio Margottini

C101-4

Stability assessment and prevention measurement of Lishan Landslide, Xian, China (2002–)

Office: Lishan Landslide Prevention and Control Office, Xian, China

Proposer: Qing-Jin Yang

C101-5

Environment protection and disaster mitigation of rock avalanches and landslides in Tianchi Lake Region and natural preservation area of Changbai Mountains, Northeast China (2002–)

Office: Environmental Geological Disaster Research Institute, Jilin University, China

Proposer: Binglan Cao

C101-6

Conservation of Masouleh Town (2002–)
Office: Building and Housing Research Center, Iran
Proposer: S. H. Tabatabaei

C102

Assessment of global high-risk landslide disaster hotspots (2002–2004)
Office: International Centre for Geohazards at Norwegian Geotechnical Institute (NGI), Oslo
Coordinator: Farrokh Nadim

C103

Global landslide observation strategy (2004–)
Coordinators: Kauro Takara and Nicola Casagli

A2.2 Member Projects

M101

Areal prediction of earthquake and rain induced rapid and long-traveling flow phenomena (APERITIF) (2002–)
Office: Disaster Prevention Research Institute, Kyoto University
Proposer: Kyoji Sassa

M102

Disaster evaluation and mitigation of the giant Jinnosuke-dani Landslide in the Tedori water reservoir area, Japan (2002–2004)
Office: Geotechnical Engineering Group, Kanazawa University, Japan
Proposer: Tatsunori Matsumoto

M103

Capacity building on management of risks caused by landslides in Central America countries (2002–2004)
Office: International Centre for Geohazards at Norwegian Geotechnical Institute (NGI), Oslo
Proposer: Farrokh Nadim

M104

A global literature study on the use of critical rainfall intensity for warning against landslide disasters (2002–2004)
Office: International Centre for Geohazards at Norwegian Geotechnical Institute (NGI), Oslo
Proposer: Haakon Heyerdal

M105

Hurricane-flood-landslide continuum: A forecast system (2002–)
Office: U.S. Geological Survey
Proposer: Randall Updike

M106

A best practices handbook for landslide hazard mitigation (2002–)
Office: U.S. Geological Survey and Geological Survey of Canada
Proposers: Lynn Highland and Peter Bobrowsky

M107

Landslide risk assessment in landslide prone regions of Slovakia – modelling of climatic changes impact (2002–)
Office: Faculty of Natural Science, Comenius University in Bratislava, Slovakia
Proposer: Rudolf Holzer

M108

Disaster evaluation and mitigation of landslides in the Three-Gorge water reservoir area, China (2002–)
Office: Chongqing Seismological Bureau, China
Proposer: Renjie Ding

M109

Recognition, mitigation and control of landslides of flow type in Greater Kingston and adjoining parishes in Eastern Jamaica, including public education on landslide hazard (2002–)
Office: Department of Geography and Geology, University of the West Indies, Jamaica
Proposer: Rafi Ahmad

M110

Capacity building in landslide hazard management and control for mountainous developing countries in Asia (2002–)
Office: International Centre for Integrated Mountain Development (ICIMOD), Nepal and Research Institute for Hazards in Snowy Areas, Niigata University, Japan
Proposer: Tianchi Li and Hideaki Marui

M111

Detail study of the internal structure of large rockslide dams in the Tien Shan and international field mission: "Internal structure of dissected rockslide dams in Kyrgyzstan" (2002–)
Office: Hydroproject Institute, Russia
Proposer: Alexander Strom

M112

Landslide mapping and risk mitigation planning in Thailand
Office: Land Development Department, Govt. of Thailand and the Japan Landslide Society (2002–)
Proposers: Chaiyasit Aneksamparm and Toyohiko Miyagi

M113

Zone risk map: Towards harmonized, intercomparable landslide risk assessment and risk maps (2002–)
Office: Faculty of Engineering, Cairo University, Egypt
Proposer: Yasser Elshayeb

M114

Landslide hazard assessment along Tehran-Caspian seaside corridors (2002–)
Office: Soil Conservation and Watershed Management Research Institute, Iran
Proposer: Zieaoddin Shoaei

M115

Establishment of a regional network for disaster mitigation, disaster education, and disaster database system in Asia (2003–)
Office: Faculty of Engineering, Ehime University, Japan
Proposer: Ryuichi Yatabe

M116

Standardization of terminology, integration of information and the development of decision support software in the area of landslide hazards (2003–)
Office: Geological Survey of Canada
Proposer: Catherine Hickson

M117

Geomorphic hazards from landslide dams (2003–)
Office: Swiss Federal Institute of Snow and Avalanche Research, Switzerland
Proposer: Oliver Korup

M118

Development of an expert DSS for assessing landscape impact mitigation works for cultural heritage at risk (VIP project) (2003–)
Office: ENEA Agency, Italy
Proposer: Giuseppe Delmonaco

M119

Slope instability phenomena in Korinthos county (2003–)
Office: Institute of Geology and Mineral Exploration (IGME)
Proposer: Nikos Nikolaou

M120

Landslide hazard zonation in Garhwal using GIS and geological attributes (2003–)
Office: Center of Disaster Mitigation – Earth Sciences, Indian Institute of Technology Roorkee, India
Proposer: Ashok Kumar Pachauri

M121

Integrated system for a new generation of monitoring of dynamics of unstable rock slopes and rock fall early warning (2003–)
Office: Research Center of Earth Dynamic, Charles University, Czech Republic
Proposer: Jiri Zvelebil and Vit Vilimek

M122

Inka cultural heritage and landslides: detailed studies in Cusco and Sacred Valleys (2004–)
Office: Grudec Ayar, Cusco, Peru
Proposer: Raul Carreño

M123

Cusco regional landslide hazard mapping and preliminary assessment (2004–)
Office: Grudec Ayar, Cusco, Peru
Proposer: Raul Carreño

M124

The influence of clay mineralogy and ground water chemistry on the mechanism of landslides (2004–)
Office: Institute of Environmental Geosciences, RAS, Russia
Proposer: Victor Ivanovich Osipov

M125

Landslide mechanisms on volcanic soils (2004–)
Office: Universidad Nacional de Colombia
Proposer: Carlos Edusrdo Rodriguez

M126

Compilation of landslide/rockslide inventory of the Tien Shan Mountain System (2004–)
Office: Institute of Environmental Geospheres Dynamics, RAS, Russia
Proposer: Alexander Strom

M127

Development of low-cost detector of slope instability for individual use (2004–)
Office: University of Tokyo, Japan
Proposer: Ikuo Towhata

M128

Development of sounding methodology for a root-reinforced landslide mass (2004–)
Office: University of Tokyo, Japan
Proposer: Kazuo Konagai

Appendix 3

ICL Documents

A3.1 The 1997 Xian Appeal

'97 XIAN APPEAL
for Protection of the IPL Projectsin Xian and Promotion of Worldwide Landslide Hazard Assessment and Risk Mitigation

1. Introduction

The development of knowledge and technology for landslide hazard assessment is one of the most important targets in the International Decade for Natural Disaster Reduction (IDNDR 1990–2000). The Committee for Prediction of Rapid Landslide Motion of the IUGS Working Group on Landslides (IUGS/WGL/RLM) has made efforts to develop an integrated technique to predict initiation and motion of rapid landslides which cause disasters in many parts of the world. For example, the joint Chinese and Japanese landslide research group has investigated the Lishan slope above the historic Huaqing Palace national monument in Xian, China. This Lishan project was initiated during the China-Japan Joint Field Workshop on Landslides held in Xian and Lanzhou in 1987. Landslide Hazard Assessment in Lishan was selected as one of IDNDR's Joint Research Projects between China and Japan in 1991. This site has been the most important location of the IUGS/WGL/RLM research program since the formation of the Committee in 1992. This is because of the outstanding value attached to the site and the historic city of Xian by the Chinese Government and the international community, and because a landslide may not only ruin the site, but also affect tourism. This investigation in Lishan has now attracted worldwide attention, and the pilot project for the prediction of rapid landslides and the protection of the cultural heritage has enjoyed substantial international collaboration. An international symposium on landslide hazard assessment was organized in Xian from 13 to 16 July 1997. This was convened jointly by the IUGS/WGL/RLM, the People's Government of Shaanxi Province, China, the People's Government of Xian Municipality and the Xian University of Engineering Science, with the strong support of the United Nations Educational, Scientific and Cultural Organization (UNESCO), the International Union of Forestry Research Organizations (IUFRO), the State Science and Technology Commission, China, the Ministry of Geology and Mineral Resources, China, the Ministry of Education, Science, Culture and Sports, Japan, the Embassy of Japan in China, the Japan International Cooperation Agency, the Japan Landslide Society, the Korean Geotechnical Society and the Gansu Society on Landslides and Debris Flows, China.

2. Protection of the Historic Huaqing Palace, Xian, China

The Imperial Resort Palace in Lishan was first constructed in the Zhou Dynasty (770 B.C.), and the location was chosen due to the hot springs and the beautiful scenery of the steep scarp of the Lishan slope. An active fault system in this region caused many earthquakes, including the Huaxian earthquake of magnitude 8 on the Richter scale on 23 January 1556. This palace is particularly well known because the Emperor Xuanzhong of the Tang Dynasty enjoyed these hot spring baths with Lady Yang-Gui-Fei. The baths still exist in the historic Palace which is nominated as a national monument. The palace buildings have continued to be used until the present. The Huaqing Palace currently attracts four million tourists per year, a quarter of this number coming from abroad. The hot springs and the palace itself have attracted people and led to the establishment of many resort facilities. The exceptionally active tourism is almost entirely due to the presence of the historic palace and to the Mausoleum of the First Qin Emperor inscribed on the World Heritage List, both located in Lintong County, close to the historic city of Xian which was the capital city of China during thirteen dynasties.

As reported and discussed during the international symposium, the view was expressed that this slope is in the precursor stage of a large scale landslide. The failure of this slope would cause a damage to this important cultural heritage with risk of human casualties. Consequently, it is essential to take all the possible measures to avoid such damage.

It is considered that the current investigation results provide sufficient impetus to obtain funding to initiate a landslide warning system and to develop a mitigation strategy.

3. Worldwide Landslide Hazard Assessment and Risk Mitigation

According to the Scientific Committee of the IDNDR, landslide risk is a serious problem affecting communities located on or near slopes in many parts of the world. This problem has become more serious especially in areas of rapidly growing populations, notably in the developing countries. During the past several decades, landslide hazards have had a widespread negative impact, both economically and socio-culturally, upon the affected communities. Such damage often occurs close to cultural sites and in urban quarters of historical interest, important to the generation of tourist income. Thus, and especially following the 1994 Yokohama Action Plan for a Safer World in the twenty-first century under the IDNDR, it is important to develop a methodology to detect landslide hazard as the basis for effective preventive and mitigation measures. These are the new aims of landslide researchers in the twenty-first century. Therefore, we call for an international joint venture further to develop research and techniques for landslide hazard assessment and risk mitigation, the protection of historic settlements and the cultural heritage, and also to ensure a safer environment, especially in areas of rapid population growth. All individuals and organizations are requested to cooperate with this scientific effort in the coming century. In addition, all the partners in the international community, notably decision makers, opinion leaders, developers, the general public and the media, are invited to draw attention to such hazards and their negative impact. Current international initiatives along these lines include the following.

1. The joint Canada-Japan Research Program on the Occurrence, Mechanisms and Behavior of Catastrophic Landslide, was adopted on 12 June 1997 as the Canada-Japan Science and Technology Inter-Governmental Agreement for 1997~2006. The agreement was based on the preceding joint research program on prediction of rapid landslide motion supported by the governments of Japan (the International Scientific Grant of the Ministry of Education, Science, Culture and Sports) and Canada (the Pacific 2000 program) during 1993–1996.
2. An application is being prepared for submission to the International Geological Correlation Programme. The project will be entitled "Landslide Hazard Assessment and Mitigation for Cultural Heritage Sites". The aim is to begin in the 1998 fiscal year, with support from the Division of Earth Science and the Division of Cultural Heritage of UNESCO, within the framework of the Medium-term Strategy for 1996–2001. IUFRO (International Union of Forestry Research Organizations) will cooperate with this important initiative.
3. The Disaster Prevention Research Institute (DPRI), of Kyoto University will make every possible effort in this international venture, and raise funds on the basis of the results obtained in the current special project on the Assessment of Landslide Hazard in Lishan, China for April 1991–March 1999. It is foreseen that the existing facilities and expertise at the DPRI will be assigned to this part of the work in order to ensure the effective coordination of the rele-vant international activities in close collaboration with all institutions and experts concerned, including UNESCO.

During this International Symposium, symposium panelists discussed landslide hazard reduction methodology, and agreed to release this Xian Appeal to all interested parties around the world.

Date: 16 July 1997
Place: Grand Castle Hotel, Xian, China

Symposium Panelists

Derbyshire, Edward: Chairman of the Scientific Board of the International Geological Correlation Programme (IGCP), Prof. of Royal Holloway College, University of London,UK.

Hungr, Oldrich: Associate Prof. of the University of British Columbia, Canadian Leader of the Canada-Japan Inter-Governmental Joint Research Project on Catastrophic Landslides, Canada.

Kim, Sang-Kyu: Vice President of the International Society for Soil Mechanics and Foundation Engineering, Prof. of Dongguk University, Korea.

Li, Tianchi: Coordinator of Mountain Risk Engineering of the International Center for Integrated Mountain Development, Nepal.

Li, Tonglu: Associate Prof. of Xian University of Engineering Science, China.

Lin, Zaiguan: Prof. of Shaanxi Engineering Investigation & Design Institute, China.

Noguchi, Hideo: Chief for Asia/Pacific and Europe, Division of Cultural Heritage, UNESCO Sector of Culture, France.

Park, Yong-Won: Professor of Myong Ji University, Korea.

Sassa, Kyoji: Chairman of I.UGS/WGL/RLM, Coordinator of IUFRO Div. 8, Vice President of the Japan Landslide Society, Prof. of Kyoto University, Japan.

Towhata, Ikuo: Prof. of University of Tokyo, Japan.

Wang, Gongxian: Professor, Northwest Institute of Railway Science.

Weinmeister, Wolfgang: Prof. of the University of Bodenkultur, Austria.

Yang, Qingjin: Director of the Lishan Landslide Prevention Observatory, China

Zhao, Binglan: Prof. of Changchun University of Science and Technology, China.

A3.2 The 1999 Tokyo Appeal

Natural Hazards, Society and Cultural Heritage[1]: Approaches for the Next Millenium the 1999 Tokyo Appeal

We, as international experts in the earth sciences, are concerned at the deterioration of many of the world's cultural heritage sites under the threat of both natural and man-enhanced hazards.

In consideration of the richness and global diversity of this heritage, and the integral place it occupies in the cultures of all the nations of the world, **we**

1. recognize that protection of the cultural heritage is vital to any balanced socio-economic development;
2. recognize the fundamental role that understanding of natural and human-induced environmental change is bound to play in the achievement of any degree of sustainable development;
3. consider that both natural and human-induced change should be accommodated and managed by monitoring a wide range of Earth-surface changes because of the important insights such changes provide in the anticipation and prognosis of future changes;
4. consider that today's environmental and developmental concerns demand the application of the latest and best scientific research in order to define problems, identify causes and effects, provide sound solutions, and design cost-effective remedial measures.

In the light of these considerations, the Earth science and cultural heritage communities urge that

5. the contribution of the geosciences to the measurement and understanding of global changes, natural and human-induced hazards, and the impact these hazards have upon the delicate balance between natural environment, cultural heritage and human settlements be recognized and enhanced;
6. all concerned (the general public, policy-makers and national and international funding agencies) should be aware of the importance of the Earth sciences in both underpinning sustainable development and the mitigation of natural and human-induced hazards;
7. deliberate efforts be made to narrow the gap between developing and advanced countries in the understanding, management and monitoring of deteriorating heritage sites by means of international collaboration, as exemplified by the current range of projects and programs under the aegis of organisations such as the International Geological Correlation Programme; and
8. national and international expertise in these fields be focused and mobilized for the protection of the world's threatened heritage sites with the aid of better-targetted technical and financial support and enhanced international communication and collaboration.

[1] The UNESCO-IUGS-IGCP Symposium on Natural Hazards and Cultural Heritage (Tokyo, 30 November to 1 December 1998) was organized by the IGCP National Committee of Japan, the Landslide Research Council of Japan, Kyoto University, the Canada-Japan Science and Technology Agreement Groups, the IGCP-425 Project Team "Landslide Hazard Assessment and Cultural Heritage and Other Sites of High Societal Value" and the IGCP-383 Project Team "Natural Disasters in the West Pacific and Asia", with the co-sponsorships of the Technical Committee on Landslides (TC11) of the International Society for Soil Mechanics and Geotechnical Engineering (ISSMGC), Commission no. 2 "Landslides and Other Mass Movements" and Commission no. 16 "Engineering Geology and Protection of Ancient Monuments and Archaeological Sites" of the International Association of Engineering and the Environment (IAEG), Division 8 "Forest Environment" of the International Union of Forestry Research, the Japan Landslide Society, the Japan Society of Erosion Control Engineering, the Japan Society of Engineering Geology, the Geological Society of Japan, the Japanese Geomorphological Union, the ISSMGE Asian Regional Technical Committee "Protection of Cultural Heritage from Landslides" (ATC-9), the Geological Survey of Japan, and the IUGS Commission COGEOENVIRONMENT.

A3.3 The 1999 Memorandum of Understanding between UNESCO and Disaster Prevention Research Institute, Kyoto University

MEMORANDUM OF UNDERSTANDING

BETWEEN

THE UNITED NATIONS EDUCATIONAL, SCIENTIFIC AND CULTURAL ORGANIZATION

AND

THE DISASTER PREVENTION RESEARCH INSTITUTE KYOTO UNIVERSITY, JAPAN

CONCERNING COOPERATION IN RESEARCH FOR LANDSLIDE RISK MITIGATION AND PROTECTION OF THE CULTURAL AND NATURAL HERITAGE AS A KEY CONTRIBUTION TO ENVIRONMENTAL PROTECTION AND SUSTAINABLE DEVELOPMENT IN THE FIRST QUARTER OF THE TWENTY-FIRST CENTURY

- 2 -

The United Nations Educational, Scientific and Cultural Organization (hereinafter referred to as "UNESCO"), represented by its Director-General, Mr Koïchiro Matsuura,

and

the Disaster Prevention Research Institute, Kyoto University (hereinafter referred to as "DPRI"), represented by its Director, Mr Shuichi Ikebuchi,

Taking into account their common goal of disaster reduction, in particular landslide risk mitigation and protection of the cultural and natural heritage and of other fragile treasures of humanity,

Recognizing that the imperatives of environmental protection and sustainable development will be even more pressing in the twenty-first century, and that population growth, increasing urbanization and mountain development will magnify the risk of various kinds of landslide disasters, including the gravitational movement of soil masses and rocks such as rock falls, debris avalanches and slides, earth flows, volcanic mud flows, pyroclastic flows, and coastal and marine landslides caused by tsunamis;

Considering that, in order to meet the requirements of population growth and economic growth accompanied by extensive regional development, it is vitally necessary to improve our understanding of landslide processes, mechanisms and dynamics;

Noting that financial and other resources mobilized to prevent landslide phenomena are limited in terms of the very large number of sites prone to landslide risk;

Recalling that most landslide phenomena can be prevented or mitigated by engineering works;

Wishing to undertake joint research initiatives for landslide risk mitigation and protection of the cultural and natural heritage;

Have agreed as follows:

ARTICLE I: Areas of cooperation

Areas of cooperation shall include:

1. collection and analysis of landslide information worldwide;
2. study of landslide triggering mechanisms and dynamics during motion;
3. development of landslide monitoring including remote sensing and long-distance data-transfer and warning systems;
4. improvement of hazard mapping techniques for landslides;
5. development of landslide disaster mitigation measures to create a viable, sustainable environment as well as to ensure safety.

- 3 -

ARTICLE II: Priority cooperative activities

Within these areas of cooperation the highest priority shall be given to:

a) cooperation within related international frameworks such as the International Environmental Conventions of the United Nations, Agenda 21, and the follow-up to the International Decade for Natural Disaster Reduction (IDNDR);
b) cooperation within the framework of relevant UNESCO programmes such as the International Geological Correlation Programme (IGCP), the International Hydrological Programme (IHP), the Man and the Biosphere Programme (MAB), the Management of Social Transformations Programme (MOST), the World Heritage Convention, the Coastal Regions and Small Islands Programme (CSI) and initiatives of UNESCO's Culture Sector;

c) cooperation with the Landslide Research Council of Japan, particularly in regard to the international newsletter "Landslide News";
d) detection of potential landslide slopes and identification of precursor phenomena;
e) development of high-precision and durable slope monitoring systems which can be used in developing countries;
f) research on reliable landslide hazard assessment and risk evaluation, to serve as a basis for decision-making by administrative authorities;
g) research on economic and practical slope conservation techniques and disaster mitigation measures socially suited to areas suffering from landslide disasters;
h) development of a network for landslide monitoring and data transfer;
i) education of specialists and leaders for landslide risk mitigation and protection of the cultural and natural heritage;
j) increasing public awareness and understanding of, and preparedness for, landslide disasters;
k) incorporation of new knowledge into future activities and integration of the environmental and developmental aspects of landslide research.

ARTICLE III: Implementation

1. Cooperative activities under this Memorandum shall be subject to and dependent upon the financial support and human resources available to the two Parties, who shall work together to secure the funding needed to meet specific objectives.
2. DPRI activities under this Memorandum shall be managed and monitored by its Executive Board.
3. UNESCO activities shall be managed by its Natural Sciences Sector, its Culture Sector and the World Heritage Centre.

ARTICLE IV: Exchange of information

The information and data exchange between the DPRI and UNESCO shall be non-proprietary. In the unlikely event of any change in the status of information or data, the two Parties shall immediately consult with each other concerning this change.

- 4 -

ARTICLE V: Review of activities

The DPRI and UNESCO shall designate representatives who, at times mutually agreed upon by the two Organizations, will review the objectives and activities established under this Memorandum. The designated representatives shall meet as deemed necessary. Activity reports and the programmes for the following year shall be exchanged annually. With the exception of specific projects otherwise decided upon by written agreement, each Party reserves the right to review the cooperative programme during the year. In case of modification of the programme each Party shall inform the other of its decisions and of its intentions regarding continuation of the operations.

ARTICLE VI: Terms of reference of activities

The provisions of this Memorandum shall in practice relate to the progress of research on the generation and movement of landslides, future landslide disaster events and preparedness activities. A written agreement setting out the terms of reference for each specific cooperative activity shall be established, in consultation with their designated representatives by the two Parties prior to undertaking that activity. The terms of reference shall include a work plan, staffing cost estimates, funding sources, and other undertakings, obligations and conditions related to the specific activity.

ARTICLE VII: Disputes

All disputes arising out of, or in connection with the present Memorandum or the breach thereof, shall be settled primarily by mutual agreement. However, if at the expiration of a six-month period starting from the date the dispute arose, no amicable settlement has been made, this Memorandum shall be deemed terminated.

ARTICLE VIII: Obligations

This Memorandum is not intended to create for either Party any obligation under international or domestic law.

ARTICLE IX: Status of the Parties

Neither Party shall be considered to be an agent or representative of the other Party, nor shall either Party be authorized to use the other Party's emblem or logo, nor shall either Party declare or imply that it has an officially recognized affiliation or status with regard to the other Party.

ARTICLE X: Entry into force and termination

This Memorandum shall enter into force upon signature by the two Parties and remain in force until the year 2005. It may be modified or extended by written agreement and may be terminated at any time by either Party upon

- 5 -

ninety (90) days' advance written notice to the other Party. The termination of this Memorandum shall not affect the validity or duration of projects that have been established under mutually-approved terms of reference and initiated prior to such termination.

Done in two copies in the English language

For the United Nations Education, Scientific and Cultural Organization	For the Disaster Prevention Institute, Kyoto University, Japan
Koïchiro Matsuura Director-General	Shuichi Ikebuchi Director
26 Nov. 1999 Date	3 Dec. 1999 Date

A3.4 The 2001 Tokyo Declaration

Geoscientists Tame Landslides – 2001 Tokyo Declaration

With reference to the 1997 Xian and 1999 Tokyo appeals (Report of IGCP Project 425, UNESCO, Paris) we, as international experts in a broad scientific field and knowledgeable for our understanding of processes of landslides and their impact on society as well as for our expertise to mitigate such impact and design remedial and preventive measures,

1) recognize the very significant safety and economic impact landslides may have for humankind, particularly in densely populated areas,
2) recognize the disastrous impact landslides may have on many historical monuments and UNESCO cultural and natural heritage sites, recommend:
   1. to join all international and national scientific and non-scientific efforts to improve understanding of processes and assessment of landslides and to co-operate on landslide prevention and on mitigation of their effects on society and ecology,
   2. to establish an International Consortium on Landslides (ICL),
   3. ICL consist of all interested organizations related to landslides studies and mitigation,
   4. to seek official approval for ICL by UNESCO and the International Union of Geological Sciences (IUGS) and to consider ICL as a joint initiative of both parties,
   5. the Board of ICL create and provide direction to an International Secretariat with an Executive Director, to be situated in the Disaster Prevention Research Institute of the Kyoto University, Japan,
   6. ICL develop a Research Programme, a Communication Plan, an Educational Plan, and a Publication Plan,
   7. to enhance awareness among worldwide public for the landslide risk mitigation and to demonstrate our abilities to reduce landslide impact,
   8. the Secretariat develop Statutes which would include above-mentioned Program and Plans to be approved by the Board,
   9. to focus its Research Program initially on the threatened Machu Picchu World heritage site.

Tokyo, 19 January 2001

A3.5 The 2002 Kyoto Declaration

Establishment of an International Consortium on Landslides – The 2002 Kyoto Declaration

We, international experts in the fields of landslide research, disaster reduction, in particular landslide risk mitigation and protection of cultural and natural heritage, who are gathering in the ICL Foundation Meeting held in the International Symposium on Landslide Risk Mitigation and Protection of Cultural and Natural Heritage organized in January 2002 in Kyoto, discussed the foundation of an international non-governmental and non-profit making scientific organization named as an International Consortium on Landslides (ICL) to promote and coordinate landslide research for the benefit of society and the environment in the global scale, and agreed on the following principal objectives of ICL:

1. To promote landslide research and capacity building including education for the benefit of society and the environment;
2. To integrate geosciences and technology within the appropriate cultural and social contexts with an aim to evaluate landslide risk in urban, rural and developing areas and cultural and natural heritage sites, as well as to contribute to the protection of the natural environment and sites of high societal value;
3. To combine and coordinate international expertise in landslide risk assessment and mitigation studies, thereby resulting in a renowned international organization, which will act as a partner in various international and national projects; and
4. To promote a global multidisciplinary Program on landslides.

Members of ICL shall include, inter alia, *(a)* Inter-governmental entities, *(b)* Non-governmental Organizations, *(c)* Governmental agencies and departments, universities, research institutes and other public institutions and *(d)* Other organizations that support the objectives of ICL intellectually, practically and financially. The United Nations system Organizations, entities and Programs will be invited to provide special support.

Accordingly, we have unanimously agreed and declared to found the International Consortium on Landslides under the Statutes attached.

Date: 21 January 2002
Place: Kyoto, Japan

A3.6 2003 The Agreement of a UNITWIN Cooperation Programme between UNESCO, Kyoto University and ICL

Please see page 376.

A3.7 The 2005 Letter of Intent proposed by ICL and Approved by UNESCO, WMO, FAO, UN/ISDR, UNU, ICSU, and WFEO

Please see page 377.

AGREEMENT

CONCERNING THE ESTABLISHMENT OF A UNITWIN COOPERATION PROGRAMME

BETWEEN

THE UNITED NATIONS EDUCATIONAL, SCIENTIFIC AND CULTURAL ORGANIZATION (UNESCO)

AND

KYOTO UNIVERSITY, KYOTO, JAPAN

AND

THE INTERNATIONAL CONSORTIUM ON LANDSLIDES

AGREEMENT

concerning the establishment of a UNITWIN Cooperation Programme between the United Nations Educational, Scientific and Cultural Organization and Kyoto University and the International Consortium on Landslides

The United Nations Educational, Scientific and Cultural Organization (hereinafter referred to as "UNESCO"), 7 place de Fontenoy, 75352 Paris, France, represented by its Director-General, Mr Koïchiro Matsuura,

and

Kyoto University (hereinafter referred to as "the University"), Sakyo, Kyoto 606-8501, Japan, represented by its President, Dr Makoto Nagao,

and

the International Consortium on Landslides (hereinafter referred to as "the Consortium"), an international non-governmental and non-profit-making scientific organization, 37 Shichiku Shimokosai-cho, Kita-ku, Kyoto 603-8114, Japan, represented by its President, Dr Kyoji Sassa,

Considering that one of the essential factors favouring development in the fields of competence of UNESCO is the exchange of experience and knowledge between universities;

Convinced that joint work by university teachers, researchers and administrators from different regions across the world will benefit the entire academic community;

Bearing in mind the mission and the objectives of UNESCO set forth in its Constitution and the role of UNESCO in promoting inter-university cooperation on an international scale;

Taking into account the experience of the international UNITWIN/UNESCO Chairs Programme as a stimulus to the rapid transfer of knowledge through twinning, networking and other linking arrangements;

Have agreed as follows:

I. Purpose

UNESCO and the University and the Consortium will create a UNESCO/KU/ICL Landslide Risk Mitigation for Society and the Environment Cooperation Programme (hereinafter referred to as "the Cooperation Programme") in the framework of the UNITWIN/UNESCO Chairs Programme.

II. Main objectives

The principal objectives of the Cooperation Programme are to:

- promote an integrated system of research, training, information and documentation activities in the field of Landslides for the benefit of society and the environment and as a key contribution to sustainable development and the protection of the environment on a global scale;
- provide advice and expertise to all countries, particularly the least developed, with a view to:
  - establishing landslide research and education for landslide risk mitigation;
  - facilitating exchange of scientists and engineers;
  - assisting members of the Consortium in developing methods of global landslide monitoring;
  - enhancing landslide experiments;
  - permitting development of a landslide database and digital library as well as of a world digital inventory;

III. Fields and disciplines concerned

The activities of the Cooperation Programme pertain to the field of geosciences, water sciences, engineering sciences, culture, human and social sciences. The main fields of interest are: landslide risk assessment and mitigation studies; environment and cultural heritage. The disciplines concerned are Geology, Geography, Geophysics, Civil and Mining Engineering, Forest and Agricultural Engineering, Informatics, Policy and Administration.

IV. Phasing of the establishment of the Cooperation Programme

IV.1 The official date of launch will be 18 March 2003.

IV.2 The development of the Cooperation Programme will comprise two phases:

- Phase one: twinning of the Consortium activities and those of the Cooperation Programme
- Phase two: identification, in close collaboration with the members of the Consortium and with UNESCO, of ways of extending the Programme to include other participants and/or institutions. All admission of new participants and/or institutions into the Cooperation Programme must be approved in writing by each of the parties concerned.

VII.4. The present Agreement shall enter into force upon its signature and will stay in force for a period of four (4) years on the date of all signatures having been appended. It may be terminated by any party subject to sixty (60) days' written notice to the other parties.

VII.5 Any renewal of the present Agreement shall be put into effect by an exchange of letters between the parties.

VII.6 In the event of a disagreement, the parties shall make an effort in good faith to settle it amicably. In the event that a settlement cannot be reached, any dispute arising out of, or relating to, this Agreement shall be settled by a sole arbitrator appointed by mutual agreement, or failing this, by the President of the International Court of Justice at the request of any party.

In witness whereof the undersigned, duly authorized to that effect, have signed three copies of the present Agreement in the English language.

For the United Nations Educational, Scientific and Cultural Organization	For Kyoto University	For the International Consortium on Landslides
K. Matsuura (signature)	Makoto Nagao (signature)	Kyoji Sassa (signature)
Koïchiro Matsuura Director-General	Makoto Nagao President	Kyoji Sassa President
Date: 10 MAR 2003	Date: 18 March 2003	Date: 18/03/2003

International Consortium on Landslides

The International Consortium on Landslides (ICL) proposed "Letter of Intent" related to **"Integrated Earth system risk analysis and sustainable disaster management"** in the thematic session 3.8 "**New International Initiatives for Research and Risk Mitigation of Floods (IFI) and Landslides (IPL)**" of the United Nations World Conference on Disaster Reduction held on 19 January 2005 in Kobe, Japan.

This is the Letter of Intent with all signatures, which was electronically combined based on the original Letters of Intent, formally approved and signed by the following parties:

United Nations Educational, Scientific and Cultural Organization;
World Meteorological Organization;
Food and Agriculture Organization of the United Nations;
UN International Strategy for Disaster Risk Reduction;
United Nations University;
International Council for Science; and
World Federation of Engineering Organizations.

All of the original Letters of Intent signed by the above seven parties are deposited in the secretariat of the International Consortium on Landslides which is located in the Research Centre on Landslides of the Disaster Prevention Research Institute, Kyoto University.

Mr. Kyoji Sassa
President of the International Consortium on Landslides (ICL)

30 06. 05
Date

Research Centre on Landslides, Disaster Prevention Research Institute
Kyoto University, Uji, Kyoto 611-0011, Japan
Tel: +81 - 774 - 384110, Fax: +81 - 774 - 325597
sassa@scl.kyoto-u.ac.jp, http://icl.dpri.kyoto-u.ac.jp

LETTER OF INTENT

"United Nations World Conference on Disaster Reduction (WCDR)", Kobe, Japan, 18-22 January 2005

This 'Letter of intent' aims to provide a platform for a holistic approach in research and learning on 'Integrated Earth system risk analysis and sustainable disaster management'.

Rationale

- Understanding that any discussion about global sustainable development without addressing the issue of Disaster Risk Reduction is incomplete;
- Acknowledging that risk-prevention policies including warning systems related to Natural Hazards must be improved or established;
- Underlining that disasters affect poor people and developing countries disproportionately;
- Stressing that after years of under-investment in preventive scientific, technical and communicational infrastructure activities it is time to change course and develop all activities needed to better understand natural hazards and to reduce the vulnerability notably of developing countries to natural hazards, and
- Acknowledging that a harmful deficiency in coordination and communication measurements related to Disaster Risk Reduction exists.

Proposal

Representatives of United Nations Organisations, as well as the Scientific (ICSU) and Engineering (WFEO) Communities propose to promote further joint global activities in disaster reduction and risk prevention through

Strengthening research and learning on 'Earth System Risk Analysis and Sustainable Disaster Management' within the framework of the 'United Nations International Strategy for Disaster Risk Reduction' (ISDR).

More specifically it is proposed,

based on the existing structural framework of the ISDR and plan of action of the UN-WCDR, as well as other relevant networks and institutional and international expertise,

to establish specific, goal-oriented 'Memoranda of Understanding' (MoUs) between international stakeholders targeting Disaster Risk Reduction, for example focusing on landslide risk reduction, and other natural hazards.

Invitation

Global, regional and national competent institutions are invited to support this initiative by joining any of the specific MoUs following this letter through participation in clearly defined projects related to the issues and objectives of any of the MoUs.

Signatories:

Mr. Koïchiro Matsuura
Director-General
United Nations Educational, Scientific and Cultural Organization

4 MAR 2005
Date

Mr. Michel Jarraud
Secretary-General
World Meteorological Organization

22. 3. 2005
Date

Mr. Jacques Diouf
Director-General
Food and Agriculture Organization of the United Nations

Date

Mr. Sálvano Briceño
Director
UN International Strategy for Disaster Risk Reduction

19 01.05
Date

Mr. Hans van Ginkel
Rector
United Nations University

19.01.05
Date

Ms. Jane Lubchenco
President
International Council for Science

21. 04. 05
Date

Ms Françoise Come
Executive Director
World Federation of Engineering Organizations

24/ 2/ 2005
Date

Index

D

E

F

G

Q

R

S

T

U

V

W

X

Y

Z

Printed by Publishers' Graphics LLC